Communications in Computer and Information Science 462

T0185353

Minrui Fei   Chen Peng   Zhou Su
Yang Song   Qinglong Han (Eds.)

Computational Intelligence, Networked Systems and Their Applications

International Conference on Life System Modeling
and Simulation, LSMS 2014
and International Conference on Intelligent Computing
for Sustainable Energy and Environment, ICSEE 2014,
Shanghai, China, September 20-23, 2014
Proceedings, Part II

 Springer

Volume Editors

Minrui Fei
Shanghai University, China
E-mail: mrfei@staff.shu.edu.cn

Chen Peng
Shanghai University, China
E-mail: c.peng@shu.edu.cn

Zhou Su
Waseda University, Tokyo, Japan
E-mail: zhousu@ieee.org

Yang Song
Shanghai University, China
E-mail: y_song@shu.edu.cn

Qinglong Han
Central Queensland University, Rockhampton, QLD, Australia
E-mail: q.han@cqu.edu.au

ISSN 1865-0929   e-ISSN 1865-0937
ISBN 978-3-662-45260-8   e-ISBN 978-3-662-45261-5
DOI 10.1007/978-3-662-45261-5
Springer Heidelberg New York Dordrecht London

Library of Congress Control Number: Applied for

Typesetting: Camera-ready by author, data conversion by Scientific Publishing Services, Chennai, India

Printed on acid-free paper

Springer is part of Springer Science+Business Media (www.springer.com)

Preface

The 2014 International Conference on Life System Modeling and Simulation (LSMS 2014) and 2014 International Conference on Intelligent Computing for Sustainable Energy and Environment (ICSEE 2014), which were held during September 20–23, in Shanghai, China, aimed to bring together international researchers and practitioners in the field of life system modeling and simulation as well as intelligent computing theory and methodology with applications to sustainable energy and environment. These events built on the success of previous LSMS conferences held in Shanghai and Wuxi in 2004, 2007, and 2010, and ICSEE conferences held in Wuxi and Shanghai in 2010 and 2012, and are based on large-scale RCUK/NSFC jointly funded UK–China collaboration projects on energy.

At LSMS 2014 and ICSEE 2014, technical exchanges within the research community take the form of keynote speeches, panel discussions, as well as oral and poster presentations. In particular, two workshops, namely, the Workshop on Integration of Electric Vehicles with Smart Grid and the Workshop on Communication and Control for Distributed Networked Systems, were held in parallel with LSMS 2014 and ICSEE 2014, focusing on the two recent hot topics on smart grid and electric vehicles and distributed networked systems for the Internet of Things.

The LSMS 2014 and ICSEE 2014 conferences received over 572 submissions from 13 countries and regions. All papers went through a rigorous peer review procedure and each paper received at least three review reports. Based on the review reports, the Program Committee finally selected 159 high-quality papers for presentation at LSMS 2014 and ICSEE 2014. These papers cover 24 topics, and are included into three volumes of CCIS proceedings published by Springer. This volume of CCIS includes 60 papers covering 5 relevant topics.

Shanghai is one of the most populous, vibrant, and dynamic cities in the world, and has contributed significantly toward progress in technology, education, finance, commerce, fashion, and culture. Participants were treated to a series of social functions, receptions, and networking sessions, which served to build new connections, foster friendships, and forge collaborations.

The organizers of LSMS 2014 and ICSEE 2014 would like to acknowledge the enormous contributions made by the following: the Advisory Committee for their guidance and advice, the Program Committee and the numerous referees for their efforts in reviewing and soliciting the papers, and the Publication Committee for their editorial work. We would also like to thank the editorial team from Springer for their support and guidance. Particular thanks are of course due to all the authors, as without their high-quality submissions and presentations the conferences would not have been successful.

Finally, we would like to express our gratitude to:

The Chinese Association for System Simulation (CASS)
IEEE Systems, Man and Cybernetics Society (SMCS) Technical Committee on
    Systems Biology
IET China
IEEE CIS Adaptive Dynamic Programming and Reinforcement Learning
    Technical Committee
IEEE CIS Neural Network Technical Committee
IEEE CC Ireland Chapter and IEEE SMC Ireland Chapter
Shanghai Association for System Simulation
Shanghai Instrument and Control Society
Shanghai Association of Automation
Shanghai University
Queen's University Belfast
Life System Modeling and Simulation Technical Committee of CASS
Embedded Instrument and System Technical Committee of China
    Instrument and Control Society
Central Queensland University
Harbin Institute of Technology
China State Grid Electric Power Research Institute
Cranfield University

September 2014

Bo Hu Li
George W. Irwin
Mitsuo Umezu
Minrui Fei
Kang Li
Qinglong Han
Shiwei Ma
Sean McLoone
Luonan Chen

Organization

Sponsors

Chinese Association for System Simulation (CASS)
IEEE Systems, Man and Cybernetics (SMCS) Technical Committee on Systems
Biology
IET China

Organizers

Shanghai University
Queen's University Belfast
Life System Modeling and Simulation Technical Committee of CASS
Embedded Instrument and System Technical Committee of China Instrument
and Control Society

Co-Sponsors

IEEE CIS Adaptive Dynamic Programming and Reinforcement Learning
Technical Committee
IEEE CIS Neural Network Technical Committee
IEEE CC Ireland Chapter and IEEE SMC Ireland Chapter
Shanghai Association for System Simulation
Shanghai Instrument and Control Society
Shanghai Association of Automation

Co-Organizers

Central Queensland University
Harbin Institute of Technology
China State Grid Electric Power Research Institute
Cranfield University

Honorary Chairs

Li, Bo Hu (China)
Irwin, George W. (UK)
Umezu, Mitsuo (Japan)

Consultant Committee Members

Cheng, Shukang (China)
Hu, Huosheng (UK)
Pardalos, Panos M. (USA)
Pedrycz, Witold (Canada)
Scott, Stan (UK)
Wu, Cheng (China)

Xi, Yugeng (China)
Xiao, Tianyuan (China)
Xue, Yusheng (China)
Yeung, Daniel S. (Hong Kong, China)
Zhao, Guangzhou (China)

General Chairs

Fei, Minrui (China)

Li, Kang (UK)

International Program Committee

Chairs

Han, Qinglong (Australia)
Ma, Shiwei (China)

McLoone, Sean (UK)
Chen, Luonan (Japan)

Local Chairs

Chen, Ming (China)
Chiu, Min-Sen (Singapore)
Ding, Yongsheng (China)
Fan, Huimin (China)
Foley, Aoife (UK)
Gu, Xingsheng (China)
He, Haibo (USA)
Li, Tao (China)

Luk, Patrick (UK)
Park, Poogyeon (Korea)
Peng, Chen (China)
Su, Zhou (Japan)
Yang, Taicheng (UK)
Zhang, Huaguang (China)
Yang, Yue (China)

Members

Albertos, Pedro (Spain)
Cai, Zhongyong (China)
Cao, Jianan (China)
Chang, Xiaoming
   (China)
Chen, Guochu (China)
Chen, Jing (China)
Chen, Qigong (China)
Chen, Weidong (China)

Chen, Rongbao (China)
Cheng, Wushan (China)
Deng, Jing (UK)
Deng, Li (China)
Ding, Zhigang (China)
Du, Dajun (China)
Du, Xiangyang (China)
Emmert-Streib (Frank)
   (UK)

Fu, Jingqi (China)
Gao, Shouwei (China)
Gao, Zhinian (China)
Gu, Juping (China)
Gu, Ren (China)
Han, Liqun (China)
Han, Xuezheng (China)
Harkin-Jones, Eileen
   (UK)

He, Jihuan (China)
Hu, Dake (China)
Hu, Guofen (China)
Hu, Liangjian (China)
Hu, Qingxi (China)
Jiang, Ming (China)
Jiang, Ping (China)
Kambhampati,
    Chandrasekhar (UK)
Keane, Andrew (Ireland)
Konagaya, Akihiko
    (Japan)
Lang, Zi-Qiang (UK)
Li, Donghai (China)
Li, Gang (China)
Li, Guozheng (China)
Li, Tongtao (China)
Li, Wanggen (China)
Li, Xin (China)
Li, Xinsheng (China)
Li, Zhicheng (China)
Lin, Haiou (China)
Lin, Jinguo (China)
Lin, Zhihao (China)
Liu, Guoqiang (China)
Liu, Mandan (China)
Liu, Tingzhang (China)
Liu, Wanquan
    (Australia)
Liu, Wenbo (China)
Liu, Xinazhong (China)
Liu, Zhen (China)
Luo, Wei (China)
Luo, Pi (China)

Luo, Qingming (China)
Maione, Guido (Italy)
Man, Zhihong
    (Australia)
Marion, McAfee
    (Ireland)
Naeem, Wasif (UK)
Ng, Wai Pang (UK)
Nie, Shengdong (China)
Ochoa, Luis (UK)
Ouyang, Mingsan
    (China)
Piao, Xiongzhu (China)
Prasad, Girijesh (UK)
Qian, Hua (China)
Qu, Guoqing (China)
Ren, Wei (China)
Rong, Qiguo (China)
Shao, Chenxi,(China)
Shen, Chunshan (China)
Shen, Jingzi (China)
Song, Zhijian (China)
Song, Yang (China)
Sun, Guangming (China)
Sun, Xin (China)
Teng, Huaqiang (China)
Tu, Xiaowei (China)
Verma, Brijesh,
    (Australia)
Wang, Lei (China)
Wang, Ling (China)
Wang, Mingshun (China)
Wei, Hua-Liang (UK)
Wei, Kaixia (China)

Wen, Guihua (China)
Wen, Tieqiao (China)
Wu, Jianguo (China)
Wu, Lingyun (China)
Wu, Xiaofeng (China)
Wu, Zhongcheng (China)
Xi, Zhiqi (China)
Xu, Daqing (China)
Xu, Sheng (China)
Xu, Zhenyuan (China)
Xue, Dong (China)
Yang, Hua (China)
Yang, Wankou (China)
Yang, Yi (China)
Yao, Xiaodong (China)
Yu, Ansheng (China)
Yu, Jilai (China)
Yuan, Jingqi (China)
Yue, Dong (China)
Zhang, Bingyao (China)
Zhang, Peijian (China)
Zhang, Qianfan (China)
Zhang, Quanxing
    (China)
Zhang, Xiangfeng
    (China)
Zhao, Wanqing (UK)
Zheng, Qingfeng (China)
Zhou, Huiyu (UK)
Zhou, Wenju (China)
Zhu, Qiang (China)
Zhu, Xueli (China)
Zhuo, Jiangang (China)
Zuo, Kaizhong (China)

Organizing Committee

Chairs

Jia, Li (China)
Wu,Yunjie (China)

Cui, Shumei (China)
Laverty, David (UK)

Members

Sun, Xin (China)
Liu, Shixuan (China)

Niu, Qun (China)

Special Session Chairs

Wang, Ling (China)
Ng, Wai Pang (UK)

Zhang, Qianfan (China)
Yu, Jilai (China)

Publication Chairs

Zhou, Huiyu (UK)

Li, Xin (China)

Publicity Chairs

Wasif, Naeem (UK)
Deng, Jing (UK)

Deng, Li (China)

Secretary-General

Sun, Xin (China)
Liu, Shixuan (China)

Niu, Qun (China)

Registration Chairs

Song, Yang (China)

Du, Dajun (China)

Table of Contents

The Third Section: Pattern Recognition and Machine Intelligence

The Forth Section: Intelligent Modeling, Monitoring, and Control of Complex Nonlinear Systems

The Fifth Section: Communication and Control for Distributed Networked Systems

Research of Metro Illumination Control Based on BP Neural Network PID Algorithm

Yong Shao, Fengbo Wang, Yuting Zhang, and Peng Zan*

School of Mechatronics Engineering & Automation, Shanghai University
Shanghai, 200072, P.R. China
zanpeng@shu.edu.cn

Abstract. This paper presents the metro constant illumination controller, which is designed to solve the current problem that the interior illumination in a moving metro fluctuates drastically with the changes of exterior environment. By detecting the real-time interior illumination values and adjusting the lights in a metro with PID controller based on BP neural network, the controller can keep the metro interior illumination values around the preset value. Simulations and actual test results show that the PID controller based on BP neural network has strong adaptability and robustness in a nonlinear system. It can both save energy and solve the problem of drastically fluctuating illumination in a moving metro, which cannot be achieved in conventional PID controllers.

Keywords: PID Control, BP Neural Network, constant illumination, back-propagation-algorithm.

1 Introduction

In recent years, with the acceleration of industrialization and the development of urban civilization, the metro has been rapidly developed as an effective means of transportation to ease traffic pressure on many large and medium-sized cities in the world. Traditional metro illumination adopts fluorescent lamp, which is continuously flickering, energy-consuming, not adjustable, and often causes discomfort in passengers. To solve this problem, we designed the metro constant illumination controller that can tune the luminance of LED lights real-timely, so that the metro interior illumination is maintained at a relatively appropriate level.

Currently, LED illumination control [1] mainly use linear dimming method and PWM dimming method. PWM dimming method is a good means, but linear dimming method is simpler, and no interference. Of course linear dimming is less flexible, and has low efficiency. In this paper, we take linear dimming to adjust a LED illumination, and improve its performance by using the BP neural network PID algorithm.

PID control is one of the earliest developed control strategies, and it is widely used in industrial process control for its simple algorithm, good robustness, and high reliability.

---

* Corresponding author.

M. Fei et al. (Eds.): LSMS/ICSEE 2014, Part II, CCIS 462, pp. 1–8, 2014.

Conventional PID control is a linear control algorithm, and does not work well for a nonlinear system [2]. Thus many improved PID controllers are developed [3] [4] [5], including the PID controller based on BP neural network [6].

BP neural network (BP-NN), also known as the error back-propagation-algorithm [7], is one of the most widely and successfully applied neural networks. It is a forward neural network composed of non-linear transformation units [8]. With adequate hidden layers and neurons, the neural network has the ability of arbitrary non-linear expression [9]. By combining BP-NN and conventional PID controller in this paper, the deficiencies that exist in traditional PID controls are offset, making the controller achieve constant illumination control effectively.

2   Fundamentals of PID Controller Based on BP-NN

One of the design goals of the PID controller based on BP-NN is to apply PID controller to nonlinear systems. The controller is composed of two parts, a classic PID controller (shown in Fig. 1) and a BP neural network. The classic PID controller has direct closed-loop control over its objects, and the three optimal control parameters of the controller can be tuned through self-learning of the neural network and adjusting of the weighting coefficient.

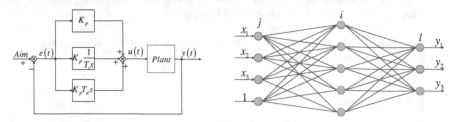

Fig. 1. Block diagram of the classic PID controller   **Fig. 2.** A three-layer BP network structure

In general application, the system employs a sampling-based incremental digital PID to control. Its output expression is as following:

$$\begin{cases} u(k) = u(k-1) + \Delta u(k) \\ \Delta u(k) = k_p \left(e(k) - e(k-1) \right) + k_i e(k) + k_d \left(e(k) - 2e(k-1) + e(k-2) \right) \end{cases} \quad (1)$$

Where k_p, k_i, k_d are respectively the proportional coefficient, the integral coefficient and the differential coefficient. Once they are identified, the control increment can be evaluated according to the deviation of the values in three successive measures. In a nonlinear system, to achieve better control effect, it is necessary for a PID controller to be able to adjust and balance the proportional, integral and derivative control, and find the optimal nonlinear PID parameters from the infinite combinations. Therefore, the non-linear characteristics of BP-NN are used to tune the parameters of PID controller online.

Fig. 2 is the structure of a three-layer (4-5-3) BP neural network [10].The three inputs of the input layer are respectively the sample value y, the desired value Aim and the error $e = Aim - y$. The three outputs of output layer are k_p, k_i, k_d.

The basic principle of online tuning of PID parameters by BP neural network is that the sample signal y, the desired signal Aim, and the error signal e are inputted from the input layer, processed by hidden layer, and then transmitted to the output layer, the three outputs of the output layer are currently taken as the parameters of the PID controller. And then modify the connective weights of neurons in each layer according to BP-NN algorithm. In this process, the state of neurons in each layer only influences that of neurons in the next layer. Thus, the network can tune the new PID parameters based on the next sampling.

3   PID Control Algorithm Based on BP-NN

For the three-layer BP neural network whose structure is 4-5-3, the algorithm is explained as following [11] [12] [13].

The input of the input layer is as following:

$$O_j^{(1)} = x(j)(j=1,2,...,4)$$ (2)

The output of the hidden layer (the middle layer) is as following:

$$O_i^{(2)}(k) = f\left(\sum_{j=1}^{4} w_{ij}^{(2)} O_j^{(1)} \right)(i=1,2,...,5)$$ (3)

Where $w_{ij}^{(2)}$ is the weighting coefficient for the hidden layer; superscripts (1), (2), (3) represent respectively the input layer, the hidden layer and the output layer. The activation function of neurons in the hidden layer is Sigmoid function:

$$f(x) = \frac{e^x - e^{-x}}{e^x + e^{-x}}$$ (4)

The output of the output layer is as following:

$$O_l^{(3)}(k) = g\left(\sum_{j=1}^{5} w_{ij}^{(3)} O_i^{(2)}(k) \right)(i=1,2,3)$$ (5)

$$\left(O_1^{(3)}(k) = k_p, \ O_2^{(3)}(k) = k_i, \ O_3^{(3)}(k) = k_d \right)$$

The output nodes of the output layer (k_p, k_i, k_d) are three adjustable parameters of the PID controller. And the activation function of neurons in the output layer is non-negative Sigmoid function:

$$g(x) = \frac{e^x}{e^x + e^{-x}} \ . \tag{6}$$

The performance index function is as following:

$$E(k) = \frac{1}{2}(Aim(k) - y(k))^2 \ . \tag{7}$$

The weighting coefficients of output layer corrected using gradient descent algorithm are searched and adjusted as per the negative gradient direction of $E(k)$ against the weighting coefficient and an add-on minimal inertia item to make the search quickly converge the situation as a whole:

$$\Delta_{w_{li}}^{(3)}(k) = -\eta \frac{\partial E(k)}{\partial w_{li}^{(3)}} + \alpha \Delta_{w_{ij}}^{(3)}(k-1) \ . \tag{8}$$

Where η is the learning rate and α is the inertia factor.

From the formulas above, the following formulas can be obtained.

$$\begin{cases} \dfrac{\partial \Delta u(k)}{\partial O_1^{(3)}(k)} = e(k) - e(k-1); \dfrac{\partial \Delta u(k)}{\partial O_2^{(3)}(k)} = e(k) \\ \dfrac{\partial \Delta u(k)}{\partial O_3^{(3)}(k)} = e(k) - 2e(k-1) + e(k-2) \end{cases} \tag{9}$$

The learning algorithm of the output layer weighting coefficient is as following:

$$\begin{cases} \Delta_{w_{li}}^{(3)}(k) = \alpha \Delta_{w_{li}}^{(3)}(k-1) + \eta \delta_l^{(3)} O_l^{(2)}(k) \\ \delta_l^{(3)} = e(k) \mathrm{sgn} \left[\dfrac{\partial y(k)}{\partial \Delta u(k)} \right] \cdot \dfrac{\partial \Delta u(k)}{\partial O_l^{(3)} u(k)} g'\left(net_l^{(3)}(k) \right) \end{cases} \ . \tag{10}$$

The learning algorithm of the hidden layer weighting coefficient is as following:

$$\begin{cases} \Delta_{w_{ij}}^{(2)}(k) = \alpha \Delta_{w_{ij}}^{(2)}(k-1) + \eta \delta_i^{(2)} O_j^{(1)}(k) \\ \delta_i^{(2)} = f'\left(net_i^{(2)}(k) \right) \sum\limits_{l=1}^{3} \delta_l^{(3)} w_{li}^{(3)}(k)(i = 1, 2, ..., 5) \end{cases} \ . \tag{11}$$

Where

$$g'(.) = g(x)(1 - g(x)), f'(.) = \left(1 - f^2(x)\right)/2 \ . \tag{12}$$

4   The Algorithm Flow

Above is the algorithm of PID controller based on BP-NN. Its control system structure is shown in Fig. 3, where *Aim* represents the desired value and *y* represents the sampled value [14].

The main design steps of BP neural network PID controller based on this algorithm is as following:

(1) Give the initial weights of each layer, and select the learning rate η and the inertia coefficient α.

(2) Obtain $Aim(k)$ and $y(k)$ by sampling, and calculate the time error $e(k) = Aim(k) - y(k)$.

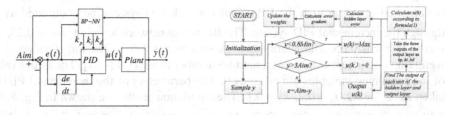

Fig. 3. Principle of PID controller based on BP-NN   **Fig. 4.** Program chart

(3) Calculate the input and output of neurons in each layer in the neural network, and take the outputs of the output layer k_p, k_i, k_d as the three parameters of the PID controller.

(4) Calculate the PID controller output $u(k)$ according to the formula (1).

(5) Learn the neural network, adjust the weighting coefficient online, and achieve the adaptive adjustment of PID controller parameters.

(6) $k = k + 1$, return to step (1).

When the ambient illumination is less than set $0.8Min$ (minimum), indicating the metro enters underground tunnel, the light output is the maximum. When the ambient illumination is greater than $3Aim$ (desired value), indicating the metro is out of the tunnel, the light output luminance is zero. The program chart is shown in Fig. 4.

5   MATLAB Simulation Results

In this simulation, since the controlled object belongs to a nonlinear discrete system, we take the approximate mathematical model [15]:

$$y(k) = \frac{\alpha(k) y(k-1)}{1 + y^2(k-1)} + u(k-1) \ . \tag{13}$$

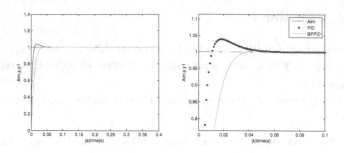

Fig. 5. Simulation results

In the formula, $\alpha(k) = 1.2\left(1 - 0.8e^{-0.1k}\right)$. The inputted target signal $Aim(k) = 1.0$,
is outputted as y through the BP neural network PID system, and outputted as $y1$
through the incremental PID system. The BP neural network learning rate is 0.25. The
inertia factor is 0.05. The neural network structure is 4-5-3. The initial value of the
weighting coefficient is taken as a random number in interval [-1,1]. The initial values
of the remaining variables are taken as 0. The parameters of the incremental PID are
taken as $k_p = 0.2$, $k_i = 0.2$, $k_d = 0.1$. The simulation results are shown in Fig. 5. Fig.
5(b) is a magnification of Fig. 5(a). Aim represents the inputted desired signal. $y1$
represents the output signal of the incremental PID control. y represents output signal
of the PID control with a BP neural network. The figures show that the outputs of the
incremental PID control and the PID control with a BP neural network both can
converge in 0.1 second. However, the overshot of the former is 4%, while that of the
latter is almost zero.

6   The Hardware Test and Analysis

The structure diagram of the metro constant illumination controller we developed is
shown in Fig. 6. "BH1750" is an illumination sensor and its highest precision is 0.12
LUX. "HS0038" is an infrared receiving head. "POWER" represents the power
module which provides a voltage of 5V for CPU and other parts of the controller.
"TLC5615" is a 12-bit D/A converter. "LED LAMP" represents the controlled light.
The real product of the controller is shown in Fig. 7. Via a illumination sensor, the
controller first samples interior illumination intensity, which is then processed by the
BP-NN PID control algorithm in MCU PIC18F2620, and at last the digital-analog
converter (D/A) module outputs a corresponding voltage to control the output of
LEDs to tune the internal illumination in metro.

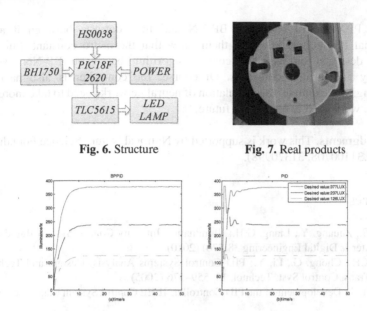

Fig. 6. Structure   Fig. 7. Real products

Fig. 8. Hardware test results

We have conducted three contrast tests on the controller with the incremental PID control algorithm ($k_p = 0.8$, $k_i = 0.8$, $k_d = 0.2$) and the BP neural network PID control algorithm (three PID parameters got by calculating the output of BP neural network real-timely; $\alpha = 0.05, \eta = 0.25$). The desired illumination values in the three tests are set as 377LUX, 237LUX and126LUX. In the beginning of the tests, the controllers are put in total darkness. For more precise test results, the controller's function of tuning brightness promptly in darkness is shielded. Fig. 8 is a collection of 100 results (collected every 0.5 seconds). Fig. 8(a) shows the output of PID controller with BP-NN and Fig. 8(b) shows the output of incremental PID controller. The figures show that the outputs of both controllers are close to a stable state at the 10th second. But the output of the latter experienced an overshot of 5LUX and an oscillatory of 20LUX before convergence. The main reason of the difference is that PID control based on BP-NN can tune the PID controller parameters real-timely while the parameters of conventional PID controller are fixed and cannot work well for nonlinear systems.

7   Conclusion

This paper presents the design of the metro constant illumination controller. In this controller, comfortable illumination can be set by infrared remote control. The controller adopts a new algorithm which combines traditional PID control and a BP neural network, and can tunes real-timely the three PID parameters. It overcomes the shortcomings of conventional nonlinear PID controller and achieve effective constant illumination control in a moving metro. This paper also presents the simulation and test results of PID control based on BP-NN and the contrasts between it and the conventional PID control. Both of them show that the metro constant illumination controller designed on the PID control algorithm based on BP-NN has good adaptability and strong robustness. Of course, the controller is also have some shortcomings, for example, the calculation of neural networks need to take more   CPU time. More work need to be done in future.

Acknowledgments. This work is supported by National Nature Science Foundation of China (No.31100708, 31370998).

References

1. Xing, S., Zhuang, Y., Liang, G.H.: Illumination Intensity Control System Based on PID. Computer & Digital Engineering 38(570) (2010)
2. Ang, K.H., Chong, G., Li, Y.: PID Control Systems Analysis, Design, and Technology. IEEE Trans. Control Syst. Technol. 13, 559–576 (2005)
3. Bennett, S.: Development of the PID Controller. IEEE Control System Magazine 13, 58–66 (1993)
4. Ohnishi, Y., Gravel, T.K.: A New Type Neural Network PID Control for Nonlinear Plants Control. IEEE Trans. on Neural Networks 11(4), 495–506 (2003)
5. Shahrokhi, M., Fanaei, M.A.: Comparison of Four Adaptive PID Controllers. Scientia Iranica 7, 129–136 (2000)
6. Jia, Q., Guo, J.Y., Zhao, X.F.: Application of Neural Network in the Boiler Combustion Control, Instrumentation user, Tianjin, China (2010)
7. Hecht, R.: Theory of Back-propagation Neural Networks. In: IEEE Proceedings of the International Conference on Neural Networks, vol. 1, p. 593 (1989)
8. Martins, G.F., Coelho, M.A.N.: Application of Feed-forward Artificial Neural to Improve Process Control of PID-based Control Algorithms. Computers and Chemical Engineering 24, 853–858 (2000)
9. Junghui, C., Huang, T.C.: Applying Neural Networks to On-line Updated PID Controllers for Nonlinear Process Control. Journal of Process Control 14, 211–230 (2004)
10. Li, G.Y.: Neural Fuzzy Control Theory and Application. Publishing House of Electronics Industry, Beijing (2009)
11. Gao, S.X., Cao, S.F., Zhang, Y.: Research on PID Control Based on BP Neural Network and Its Application. In: 2nd International Asia Conference on Informatics in Control, Automation and Robotics, pp. 91–94 (2010)
12. Liao, F.F., Xiao, J.: Research on Self-tuning of PID Parameters Based on BP Neural Networks. Journal of System Simulation 17(7), 1711–1713 (2005)
13. Wu, H.P., Ao, Z.G., Wang, G., Ao, W.Q.: PID Real-time Control for Parameter On-line Adjusting Based on BP Neural Network. Computer Knowledge and Technology 5(19), 5245–5246 (2009)
14. Huang, Y.R., Qu, L.G.: The PID Controller Parameter Tuning and Implementation. Science Press, Beijing (2010)
15. Liu, J.K.: Advanced PID Control and MATLAB Simulation, 3rd edn. Publishing House of Electronics Industry (2011)

A Study of Adaptive Neural Network Control System

Heng Zhong, Dingyuan Li, and Kun Tu

School of Mechatronic Engineering and Automation, Shanghai University,
Shanghai, 200072, P.R. China
Zhongheng_2008@126.com

Abstract. In a hydraulic support system of heavy equipment, the oil pressure is required to be a constant value. Due to the disturbances come from the external environment and the running process, the hydraulic support system would not be stable; therefore, we here present a closed-loop feedback system, which has an adaptive neuronal network control system to make the oil pressure stable by controlling the rotate speed of the hydraulic pump motor. In this hydraulic system, the response and stability are the key factors to judge the control method is good or not. In our simulation, it shows that this adaptive neural network control system can meet the design requirements. It has good response and stability characteristics.

Keywords: Neural Network, Steepest-descent Method.

1 Introduction

Adaptive control system works by measuring the input value and output value from the actual conditions. Due to the dynamic characteristic of adaptive control system, the error of the output value can be reduced; so the control characteristics can maintain at optimum performance and the output value can meet the anticipated result [1]. Neural network is a non-linear system which may be relatively limited for an individual neuron, but many functions can be realized if a large number of neurons compose a network [2].

Adaptive control system has a relatively strong robustness and neural network is good at self-learning function; so adaptive neural network control has a great advantage due to the integration of both. Adaptive neural network control system can enhance fault tolerance, real-time control and robustness. Adaptive neural network control system can make uncertain and complex nonlinear control achieved more effectively [3].

This paper is in the research and design of the hydraulic support system of heavy equipment which needs enough supported oil for working effectively. So it requires us to ensure that the oil pressure can be maintained at a stable value. In actual practice, however, the hydraulic oil pressure control system would not be stable due to the presence of interference; therefore keeping the stability in the process of adjusting is very important for this control system. As we have already analyzed the advantages of adaptive neural network control system, we design such an adaptive

M. Fei et al. (Eds.): LSMS/ICSEE 2014, Part II, CCIS 462, pp. 9–17, 2014.

neural network control system to make the oil pressure stable by controlling the rotate speed of the hydraulic pump motor. In our simulation, it shows that this adaptive neural network control system can competent on the design requirements. It has good response and stability characteristics[8,9,11,12].

2   Adaptive Neural Network Control System

2.1   Introduction of Hydraulic Support System

The structure of this hydraulic support system of heavy equipment is shown in Fig.1. The motor makes oil up through the pipeline to become oil film layer. The thickness of oil film layer can reflect the size of oil pressure value. Supporting a stable oil film layer is a necessary condition for equipment work normally [4].

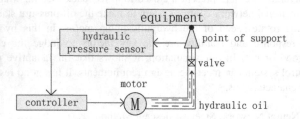

Fig. 1. Structure of this hydraulic support system

The adaptive neural network control system designed in this paper is based on DSP. DSP has powerful digital signal processing capability and excellent embedded control function, so it is widely used in industrial control situations and it also provides support of the hardware in this system. As Fig.1 shows, the control system contains three parts: control unit, actuator, and oil pressure signal acquisition unit. Because of the system requires real-time acquisition and oil pressure signals processing, we choose DSP as the controller[14,15].

2.2   Design of Adaptive Neural Network

For the large hydraulic control system, the key is the oil pressure should be stable by the controlling the rotate speed of the hydraulic pump motor. As the oil pressure is not easy to be collected directly, the oil pressure value can be converted to a corresponding voltage value. According to the corresponding relationship between oil pressure value X(Kg) and voltage value Y(V), we got 22 groups of sampling data. They are shown in the following table 1.

Table 1. Data of oil pressure values and voltage values

pressure value X voltage value Y					
X,Y	15.01, 2.98	14.93,2.84	14.83,2.7	14.74,2.58	14.63,2.43
X,Y	14.54,2.29	14.46,2.17	14.37,2.03	14.28,1.89	14.18,1.75
X,Y	14.07,1.58	13.97,1.45	13.88,1.18	13.69,1.03	13.59,0.9
X,Y	13.51,0.77	13.41,0.62	13.31,0.48	13.23,0.36	13.15,0.25
X,Y	13.07,0.14				

By Matlab simulation with the least squares method, we find the curve of oil pressure values and voltage values is a first-order system. Fig.2 is the graph of this first-order system[13,18,19].

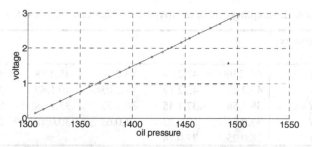

Fig. 2. Curve of oil pressure values and voltage values

We need oil pressure value be stable at 14(Kg); so according to Fig.2, the voltage value should be 1.5V. Through the adaptive neural network control system, oil pressure can reach a stable value. In this paper, the design of this control system has two neural networks: NN1 and NN2. As Fig.3 shows, NN1 is a controller and NN2 is an identifier. The design of this system is shown in Fig.3.

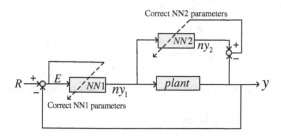

Fig. 3. Design of adaptive neural network system

Neural networks NN1 and NN2 have different functions in this system. According to the simulation result of this system, we choose the hidden layer neurons number is 5 which can effectively guarantee the accuracy and stability. The internal structures of NN1 and NN2 are showed in Fig.4.

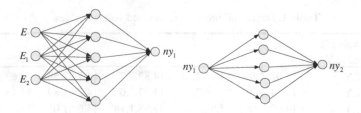

Fig. 4. Internal structures of NN1 and NN2

In Fig.4, E is the current error; E_1 is the first sampling error; E_2 is the second sampling error. Each learning time will have 100 samples, and we can change this number. By changing the frequency value ny_1(Hz), we can collect a corresponding voltage value y(V), and 22 groups of sampling data are shown in table 2.

Table 2. Sampling data of the frequency values and voltage values

frequency value ny_1 voltage value Y					
ny_1,Y	48, 2.98	47,2.84	46,2.7	45,2.58	44,2.43
ny_1,Y	43,2.29	42,2.17	41,2.03	40,1.89	39,1.75
ny_1,Y	38,1.58	37,1.45	36,1.32	35,1.18	34,1.03
ny_1,Y	33,0.9	32,0.77	31,0.62	30,0.48	29,0.36
ny_1,Y	28,0.25	27,0.14			

We find the curve of voltage values and frequency values is a first-order system in simulation with the least squares method. Fig.5 is the graph of this system.

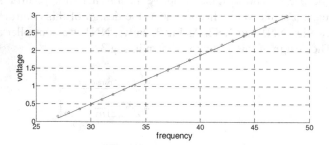

Fig. 5. Curve of frequency values and voltage values

The formula for this curve is shown below:

$$y = 0.1377ny_1 - 3.6283 \qquad (1)$$

Due to we need oil pressure value be stable at 14Kg, the voltage value should be 1.5V; so the initial value of R should be set to 1.5V. As Fig.5 shows, when the voltage value is required to achieve 1.5V, the frequency should be stable at 37Hz. Due to there are a lot of interferences from external environment, frequency may be unstable. But the neural network control system has a good generalization ability, and it can effectively solve this problem[16,17].

2.3   Learning Process of Adaptive Neural Network System

Adaptive neural network learning is a supervised learning process, so we need some target input and output samples for training. We can use random values as initial weights values at the beginning of the training; the target input samples are used as input values to get the output values of the network. The actual output value is compared with the target output value to calculate the error; then this system modifies the weights with the steepest descent method to reduce the error. Finally, the error almost has no decline in the network training [5]. Fig.6 is a block diagram of the adaptive neural network learning process.

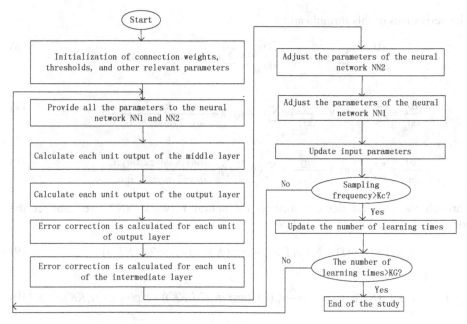

Fig. 6. Learning process of the adaptive neural network system

The basis function of this neural network system is $(e^x - e^{-x})/(e^x + e^{-x})$. K_C is sampling frequency and KG is the number of learning times; both of the two numbers are 500. These numbers can be adjusted according to simulation result. The inter-node input of neural network NN2 is:

$$y = 0.1377 ny_1 - 3.6283 \qquad (2)$$

Through the calculation of basis function, the inter-node output is:

$$nyz2(i,KK)=(\exp(nyzi2(i,KK))-\exp(-nyzi2(i,KK)))/(\exp(nyzi2(i,KK))+\exp(-nyzi2(i,KK))) \qquad (3)$$

The output value of NN2 through the intermediate node is:

$$ny_2(KK) = ny_2(KK) + nyz2(i,KK)w_{22}(i); \qquad (4)$$

W_{21} is the weight from input layer of NN2; W_{22} is the weight from output layer of NN2; q_2 is the function threshold of NN2; k_1 and k_2 are learning coefficients of NN1 and NN2. Due to the derivative for the output value of NN2 will be used to NN1, k_2 should be larger than k_1.

The squared error formula for NN2 is:

$$J = \frac{1}{2}\sum(y-ny_2)^2 \qquad (5)$$

The derivatives of this formula are:

$$\frac{\partial J}{\partial w_{22}} = \sum(y-ny_2)(-\frac{\partial ny_2}{\partial w_{22}}) = \sum(y-ny_2)(-nyz2) \qquad (6)$$

$$\frac{\partial J}{\partial q_2} = \sum(y-ny_2)(-\frac{\partial ny_2}{\partial q_2}) = \sum(y-ny_2)(-\frac{\partial ny_2}{\partial nyz2}\frac{\partial nyz2}{\partial nyzi2}\frac{\partial nyzi2}{\partial q_2}) \qquad (7)$$

$$\frac{\partial J}{\partial w_{21}} = \sum(y-ny_2)(-\frac{\partial ny_2}{\partial w_{21}}) = \sum(y-ny_2)(-\frac{\partial ny_2}{\partial nyz2}\frac{\partial nyz2}{\partial nyzi2}\frac{\partial nyzi2}{\partial w_{21}}) \qquad (8)$$

Through the steepest descent method for neural network NN2, we can get the parameters corrections shown below:

$$w_{22}(i) = w_{22}(i) - k_2(y(KK) - ny_2(KK))(-nyz2(i,KK)) \qquad (9)$$

$$q_2(i)=q_2(i)-k_2(y(KK)-ny_2(KK))(-w_{22}(i))/(\exp(nyzi2(i,KK))+\exp(-nyzi2(i,KK)))^2 \qquad (10)$$

$$w_{21}(i)=w_{21}(i)-k_2(y(KK)-ny_2(KK))(-w_{22}(i))ny_1(KK)/(\exp(nyzi2(i,KK))+\exp(-nyzi2(i,KK)))^2 \qquad (11)$$

Thus, the parameters of NN2 are corrected constantly. The process of NN1 is similar with NN2, and the inter-node input of neural network NN1 is:

$$nyzi1(i,KK) = w_{11}(1,i)E + w_{11}(2,i)E_1 + w_{11}(3,i)E_2 + q_1(i) \qquad (12)$$

Through the calculation of basis function, the inter-node output of NN1 is:

$$nyz1(i,KK)=(\exp(nyzi1(i,KK))-\exp(-nyzi1(i,KK)))/(\exp(nyzi1(i,KK))+\exp(-nyzi1(i,KK))) \qquad (13)$$

The output value of NN1 through intermediate node is:

$$ny_1(KK) = ny_1(KK) + nyz1(i, KK)w_{12}(i) \tag{14}$$

The squared error formula for NN1 is:

$$G = \frac{1}{2}\sum(y - R)^2 \tag{15}$$

The derivatives of this formula are:

$$\frac{\partial G}{\partial w_{12}} = \sum(y-R)(-\frac{\partial y}{\partial w_{12}}) \approx \sum(y-R)(-\frac{\partial ny_2}{\partial w_{12}}) = \sum(y-R)(-\frac{\partial ny_2}{\partial ny_1}\frac{\partial ny_1}{\partial w_{12}}) \tag{16}$$

$$\frac{\partial G}{\partial q_1} = \sum(y-R)(-\frac{\partial y}{\partial q_1}) \approx \sum(y-R)(-\frac{\partial ny_2}{\partial q_1}) = \sum(y-R)(-\frac{\partial ny_2}{\partial ny_1}\frac{\partial ny_1}{\partial nyz1}\frac{\partial nyz1}{\partial nyzi1}\frac{\partial nyzi1}{\partial q_1}) \tag{17}$$

$$\frac{\partial G}{\partial w_{11}} = \sum(y-R)(-\frac{\partial y}{\partial w_{11}}) \approx \sum(y-R)(-\frac{\partial ny_2}{\partial w_{11}}) = \sum(y-R)(-\frac{\partial ny_2}{\partial ny_1}\frac{\partial ny_1}{\partial nyz1}\frac{\partial nyz1}{\partial nyzi1}\frac{\partial nyzi1}{\partial w_{11}}) \tag{18}$$

Through the steepest descent method for neural network NN1, we can get the parameters corrections shown below:

$$w_{12}(i) = w_{12}(i) - k_1(y(KK) - R)YY(nyz1(i, KK)) \tag{19}$$

$$q_1(i) = q_1(i) - k_1(y(KK) - R)YYw_{12}(i) / (\exp(nyzi1(i, KK)) + \exp(-nyzi1(i, KK)))^2 \tag{20}$$

$$w_{11}(1,i) = w_{11}(1,i) - k_1(y(KK) - R)YYEw_{12}(i) / (\exp(nyzi1(i, KK)) + \exp(-nyzi1(i, KK)))^2 \tag{21}$$

$$w_{11}(2,i) = w_{11}(2,i) - k_1(y(KK) - R)E_1w_{12}(i)YY / (\exp(nyzi1(i, KK)) + \exp(-nyzi1(i, KK)))^2 \tag{22}$$

$$w_{11}(3,i) = w_{11}(3,i) - k_1(y(KK) - R)E_2w_{12}(i)YY / (\exp(nyzi1(i, KK)) + \exp(-nyzi1(i, KK)))^2 \tag{23}$$

In the adaptive neural network system, the data propagate backward step by step through the input layer, the hidden layer, and the output layer. Network weights can be modified along the direction of reducing the error. The error could be smaller during the learning procedure. Finally, the error almost has no decline in the network training [6,7,10].

2.4   Simulation Verification

According to the above method, Fig.7 is the error curve of system output values in the MATLAB simulation environment. The horizontal axis represents the number of learning times; the vertical axis represents the error values. Finally, E value is almost stable at 0. The simulation result shows that the adaptive neural network system can achieve the stability requirements of this hydraulic support system.

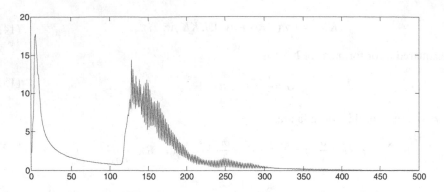

Fig. 7. Internal structures of NN1 and NN2

3   Conclusions

In this paper, an adaptive neural network control system based on DSP is designed for the hydraulic support system. This paper discusses the working process of adaptive neural network control system in detail by the steepest descent method. We have made a comprehensive explanation to the adjustment steps of all the relevant parameters. The design process of the system is described in detail. We have analysed the rationality and advantages of the adaptive neural network control system, and learned that it is suitable for the hydraulic support system.

Finally, the error is very small and can fully meet the requirements of this system. The error simulation test shows that this adaptive neural network control system has good fault-tolerance and robustness abilities.

References

1. Liu, L.: Uncertain Nonlinear Systems with Adaptive Neural Network Control. Liaoning University (2013)
2. Liu, H.: Neural Network Based System Identification. Xi'an Electronic Science and Technology University (2007)
3. Wang, W., Su, S., Xianxian, G.: Research of Adaptive Control Based of Neural Network. Computer Simulation 22(8), 132–135 (2005)
4. Zhao, Y.: Telescope Hydraulic Control System Based on DSP. Application of Electronic Technique 39(3), 23–26 (2005)
5. Chen, M.: MATLAB Neural Network Theory and Examples. Tsinghua University (2013)
6. Piche, S.W.: Steepest Descent Algorithms for Neural Network Controllers and Filters 5(2), 198–212 (1994)
7. Kan, J.: Self-Tuning PID Controller Based on Improved BP Neural Network. In: Second International Conference on Intelligent Computation Technology and Automation, vol. 1, pp. 95–98. IEEE Press, Changsha (2009)
8. Tayebi, A.: Reference Adaptive Iterative Learning Control for Linear Systems 20(9), 475–489 (2006)

9. Cigdem, A.-U., Berna, D.: A self-adapitive local search algorithm for the classical vehicle routing problem. Expert Systems with Applications 38(7), 8990–8998 (2011)
10. Narendra, K.S., Annaswamy, A.M.: Robust adaptive control in the presence of bounded disturbances. IEEE Transactions on Automatic Control (4), 306–315 (1986)
11. Hunt, K.J., Sbarbaro, D., Zbikowski, R., Gawthrop, P.J.: Neural networks for control system-asurvey. Automatica (28), 1083–1112 (1992)
12. Parthasarathy, K.: Identification and control for dynamical systems using neural netwokrs. IEEE Transaction on Neural Networks (1), 4–27 (1990)
13. Yu, Z.X., Du, H.B.: Adaptive neural tracking control for stochastic nonlinear systems with time-varying delay. Journal of Control Theory and Applications (2), 1808–1812 (2011)
14. Tong, S.C., Li, Y.M., Zhang, H.G.: Adaptive neural network decentralized back stepping output-feedback control for nonlinear large-scale systems with time delays. IEEE Transactions on Neural Networks (22) 1073–1086 (2011)
15. Fathi, F.: A greenhouse control with feed-forward and recurrent neural networks. Simulation Modelling Practice and Theory 15(8), 1016–1028 (2007)
16. Dong, X., Mei, W.: Design of an expert system based on neural network ensembles for missile fault diagnosis. In: Proceedings of 2003 IEEE International Conference on Robotics, vol. (2), pp. 903–908 (2003)
17. Martin, R., Heinrich, B.: A Direct Adaptive Method for Faster Back-propagation Learning. In: Ruspini, H. (ed.) Proceedings of the IEEE International Conference on Neural Networks, pp. 586–591 (1993)
18. Patrick, P.: Minimisation Method for Training Feed forward Neural Network. Neural Network (7), 145–163 (1994)
19. Behera, L., Kumar, S., Patnaik, A.: On adaptive learning rate that guarantees convergence infeed forward networks. Neural Networks 17(5), 1116–1125 (2006)

Design of Fuzzy Logic Controller
Based on Differential Evolution Algorithm

Li Shuai[1] and Sun Wei[2]

[1] School of Mechatronic Engineering and Automation, Shanghai University,
Shanghai, 200072
lishuaicumt@126.com
[2] College of Information & Electrical Engineering, China University of Mining & Technology,
Xuzhou, 221008
sw3883204@163.com

Abstract. In order to overcome the deficiency of fuzzy control algorithm, an adaptive fuzzy logic controller is proposed. In this method, the differential evolution algorithm (DE) was employed to optimize parameters of fuzzy controller: quantitative factor and proportional factor , they were designed as individuals of DE population, and evaluated using the fitness function provided until the termination condition was fulfilled. Then the selected parameter values were sent back to fuzzy logic controller. Simulation results concerning two-tank system show that the DE optimized fuzzy controller has good adaptability, as well as it's effectiveness, which provides a new approach to improve fuzzy control system.

Keywords: Fuzzy Logic Control, Differential Evolution Algorithm, Optimization.

1 Introduction

In1974, the British scholar E. H. Mamdani applied fuzzy logic control to boilers and steam engines controls, which indicated the birth of fuzzy control theory. Now fuzzy logic controllers have been widely applied to industrial process controls, and show to be a more accurate and efficient method. Although both fuzzy control theory and technology have been developed rapidly, there are still some shortcomings such as how to choose control rules, as well as how to adjust quantitative factors and proportional factor and so on. Some authors [1], [2] presented adaptive fuzzy controllers, which can be adjusted automatically by changing certain parameters such as shape and location of suitable membership functions. Pintu Chandra Shill et al [3] used QGA to optimize fuzzy control rules and membership function. However, the performance of the fuzzy controller mostly depends on human experience, and requires tedious trial and error processes, and it can't quickly converge to the optimal values.

Differential Evolution (DE) was proposed by R. Storn and K. Price in 1995[4]. As a new evolutionary computing technique, DE has some good properties, and it is simple, robust yet efficient in solving the global optimization problem. J. Vesterstrom

M. Fei et al. (Eds.): LSMS/ICSEE 2014, Part II, CCIS 462, pp. 18–25, 2014.

made a comparative study of differential evolution, evolutionary algorithms and particle swarm optimization on numerical benchmark problems, the results show that, DE outperforms PSO and other evolutionary algorithms[5].

In this paper, we proposed an optimized fuzzy logic controller based on differential evolution algorithm, through the quantitative factor and proportional factor to enhance the optimal design of fuzzy controller performance, and applied it to two-tank system.

2   Differential Evolution Algorithm

Differential Evolution (DE) is an evolutionary algorithm, which optimize problems by cooperation among individuals within populations and competition to. To apply a DE algorithm, we should generate a population, in which all individual initial values are chosen at random within bounds set by the user. Each individual is real-coded-dimensional vector depending on the problems. In the evolutional process, vectors in every generation undergo evolution through natural selection. Every vector crossover with a donor vector generated by mutating to form a trial vector, if cost function of the trial vector is less than that of the old ones, the trial vector replace the old to form next generation. The process will not end unless the termination condition is satisfied[6].Usually, the performance of a DE algorithm depends on three variables: the population size NP, the mutation scaling factor F, and the crossover rate CR. Fig.1 shows steps of the DE algorithm[7].

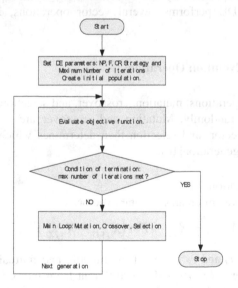

Fig. 1. Steps of the differential evolution algorithm

3   Fuzzy Logic Control Based on Differential Evolution Algorithm

3.1   The Design of Individual

Differential evolution is a parallel direct search method, it utilizes NP vectors, each of dimension D, which is the number of decision variable in the optimization problem. The DE vector $x_{i,j}$ is:

$$x_{i,j}(i = 1,2,...,NP; j = 1,2,...,D) \tag{1}$$

where i is the individual index, j is dimension index.

In this paper, we choose quantitative factors k_e, k_{ec} and proportional factor k_u to constitute the individual x [8]:

Fig. 2. The structure of individual

where k_e and k_{ec} denote quantitative factor of inputs, k_u denotes proportional factor of output. All these are based on real-coded.

The initial population is selected randomly in the bounds defined for each variable x. These bounds are specified by the user according to the optimization problem. After initialization, DE performs several vector operations, in a process called evolution[9].

3.2   Differential Evolution Operations

There are three DE operations: mutation, crossover, and selection. The initial value of vector x is selected randomly. Mutation and crossover are used to generate new vectors called trial vector, and selection then determines which of the vectors will survive into the next generation[10].

(a) Mutation Operation
Mutation for each vector creates a mutant vector:

$$V_{i,G+1} = X_{i,G} + F * \left(X_{r1,G} - X_{r2,G} \right) \tag{2}$$

where the indexes r_1 and r_2 represent the random and mutually different integers generated within range *[1, NP]* and also different from index i. F is a real number (*F∈[0,2]*) which controls the amplification of the difference vector ($x_{r1,G}$- $x_{r2,G}$).

(b) Crossover Operation
Through crossover operation, the target vector is mixed with the mutated vector, to yield the trial vector, following scheme:

$$U_{i,G+1} = \left(u_{1i,G+1}, u_{2i,G+1}, ..., u_{Di,G+1} \right) \tag{3}$$

Where

$$u_{ji,G+1} = \begin{cases} V_{ji,G+1} & \begin{array}{l} if\,(rand(j) \le CR) \\ or\,(j = mbr(i)), \end{array} \\ X_{ji,G} & \begin{array}{l} if\,(rand(j) > CR) \\ or\,(j \ne mbr(i)) \end{array} \end{cases}$$

for $i=1,2,...,NP$, $j=1,2,...,D$, $rand(j)$ is selected uniformly randomly within $[0,1]$, $mbr(i)$ represents a random integer within $[1,D]$, which is responsible for the trial vector containing at least one parameter from the mutant vector. $CR\in[0,1]$, which is a crossover parameter presenting the probability of creating parameters for trial vector from a mutant vector.

(c) Selection Operation

The selection operation selects the vector which will survive to be a member of the next generation. Which vector will be selected depending on the fitness value to evaluate the performance of the controller, it is necessary to define a fitness function. The target of our controller is to evaluate the dynamic and static characteristic of control system, such as rapid response, short adjusting time, small overshoot and small stable error etc., so the following fitness function is proposed:

$$x_{i,G+1} = \begin{cases} u_{i,G+1}, & f(u_{i,G+1}) < f(x_{i,G}) \\ x_{i,G}, & f(u_{i,G+1}) \ge f(x_{i,G}) \end{cases} \tag{4}$$

for $f(x)$ is the fitness function.

3.3   The Fitness Function

To evaluate the performance of the controller, it is necessary to define a fitness function. The target of our controller is to evaluate the dynamic and static characteristic of control system, such as rapid response, short adjusting time, small overshoot and small stable error etc., so the following fitness function is proposed:

$$f_i = at_s + b|\sigma| + \sum \left(ce^2 + du^2 \right) \tag{5}$$

for t_s represents regulating time. σ is the overshoot. e indicates the difference between the desired value and the actual value. u indicates the output of fuzzy controller, and a, b, c and d are constant coefficients.

3.4   The Algorithmic Step

The proposed overall tuning process works in following ways[11][12]:

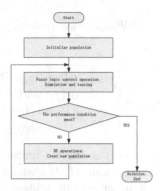

Fig. 3. Flowchart of DE based fuzzy control system

4   Simulation Results and Comparative Analysis

In this section, we applied the proposed method to the two-tank system [13]. Fig. 4 gives the structure. The experimental device includes of two water tanks T_1 and T_2, reservoir, junction valve V_1, leak valves V_2, V_3 and V_4, as well as execution structures and sensors. Water is draw from the reservoir by pump P_1, and through V_3 into water tank T_1, then through V_1 into water tank T_2, finally through V_2 go back to the reservoir. The leak valve V_4 is usually closed. We take the tank T_2 liquid level h_2 as the controlled variable.

Based on the mass conservation condition and Bernoulli law, the two-tank system model is:

$$A\frac{dh_1}{dt} = Q - Q_{12}, A\frac{dh_2}{dt} = Q_{12} - Q_{20} \tag{6}$$

for $Q_{12} = \mu_1 S \mathrm{sgn}(h_1 - h_2)(2g|h_1 - h_2|)^{1/2}$ is the water speed from T_1 to T_2, $Q_{20} = \mu_2 S(2gh_2)^{1/2}$ is the water speed from T_2 to the reservoir, $sgn(z)$ is the sign function of z, A is the cross sectional area of the tank, S for the pipe cross-sectional area, g is the acceleration due to gravity, μ_1, μ_2 are the flow coefficient, the values are as follows: $A = 6.3585 \times 10^{-3} \, m^2$, $S = 6.3585 \times 10^{-5} \, m^2$, $\mu_1 = 0.083$, $\mu_2 = 0.1133$.

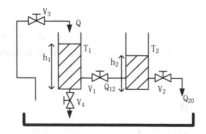

Fig. 4. The structure of two-tank system

For double tank water level control system, the aim is making liquid level of tank T2 reach the set value. Tank 1 and 2's initial liquid level is set to 0. To design fuzzy controller for double tank water level control system, we choose the error signal e(the difference between desired and actual value of h_2) and the error derivative signal de as input variables, and the control signal u (water speed Q) as the output variable. Five fuzzy sets were used for each input/output variable, like [NM NS ZE PS PM]. Then DE is used to optimize k_e, k_{ec} and k_u. In the optimal process, each vector is set as 3 dimensions and adopted the real-coded (look Fig. 2). DE population scale is NP=30, and mutation probability is F=0.6, crossover probability is CR=0.5, the maximum evolution generation is 30. The fitness function parameters are a=1, b=3, c=2, d=2. The sampling period is ts=0.01s, and The setting value of Tank 2's liquid level is : h_{20}=200mm for t=0~500s, h_{20}=350mm for t=501~1000s, h_{20}=160mm for t=1001~1600s. Fig. 5 shows the performance comparison curves.

From the simulation results, we know that, fuzzy controller has better performance than the PID controller. When the fuzzy controller is optimized by the DE algorithm, the output of the system became much better than the fuzzy controller, both in overshoot and in rapid respond time. So the method proposed in this paper is more effective.

To verify the anti-interference capability, we open the V_4 valve to add a disturbance at t=40s when the system is stable, allowing water to flow from the leak valve for some time, and then close the valve. It cost 30s for the system became stable, as shown in Fig. 6. So, it has good performance to overcome disturbance. After 30 generation, the best fitness value is 0.54784, Fig. 7 shows the fitness curve. Table 1 gives the comparison of the parameter values before and after optimization.

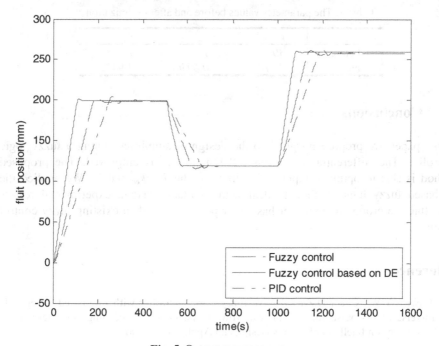

Fig. 5. Output response curves

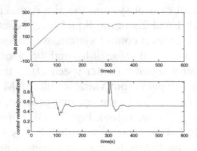

Fig. 6. The curves of output and control variable with disturbance

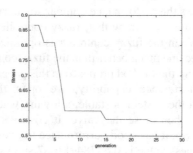

Fig. 7. Fitness function curve

Table 1. The parameter values before and after optimization

	k_e	k_{ec}	k_u
before	30	1	1.2
after	32.106	0.69367	1.9472

5   Conclusions

In this paper, we propose insight into the design of simplified tuning a fuzzy logic controller. The differential evolution (DE) algorithm is employed. The proposed method is able to optimize input and output weights k_e, k_{ec} and k_u. We applied the DE-based fuzzy logic method to deal with two-tank system, experimental results show that the proposed controller has better performance than existing fuzzy control system.

References

1. Huiwen, Y.W., Chen, D.Z.: Adaptive fuzzy logic controller with rule-based changeable universe of discourse for a nonlinear MIMO system. In: Proceedings of International Conference on Intelligent Systems Design and Applications (2005)

2. Jou, J.M., Chen, P.-Y., Yang, S.-F.: An adaptive fuzzy logic controller: its VLSI architecture and applications. IEEE Transactions on Very Large Scale Integration (VLSI) Systems 8, 52 (2000) ISSN:1063-8210
3. Shillt, P.C., Akhand, M.A.H., Muraset, K.: Fuzzy Logic Controller for an Inverted Pendulum System using Quantum Genetic Optimization. In: Proceedings of 14th International Conference on Computer and Information Technology (ICCIT), pp. 22–24 (2011)
4. Storn, R., Price, K.: Differential evolution—A simple and efficient heuristic for global optimization over continuous spaces. J. Global Optimize 11, 341–359 (1997)
5. Vesterstrom, J., Thomsen, R.: A comparative study of differential evolution, particle swarm optimization, and evolutionary algorithms on numerical benchmark problems. In: Proceedings of the IEEE Congress on Evolutionary Computation, pp. 1980–1987 (2004)
6. Cheng, S.-L., Hwang, C.: Optimal Approximation of Linear Systems by a Differential Evolution Algorithm. Transactions on Systems, Man, and Cybernetics—Part A: System and Humans 31(6) (November 2001)
7. Roger, L.S.T., Srinivas, M., Rangaiah, G.P.: Global optimization of benchmark and phase equilibrium problems useing differential evolution
8. Guo, X., Zhou, Y., Gong, D.: Optimization of fuzzy sets of fuzzy control system based on hierarchical genetic algorithms. In: Proceedings of IEEE TENCON 2002 (2002)
9. Brest, J., Zumer, V., Maucec, M.S.: Self-adaptive differential evolution algorithm in constrained real-parameter optimization. In: 2006 IEEE Congress on Evolutionary Computation, pp. 16–21 (2006)
10. Liu, J.L.: On setting the control parameter of the differential evolution method. In: Proc. 8th Int. Conf. Soft Computing (MENDEL 2002), pp. 11–18 (2002)
11. Shillt, P.C., Akhand, M.A.H., Muraset, K.: Fuzzy logic controller for an inverted pendulum system using quantum genetic optimization. In: Proceedings of 14th International Conference on Computer and Information Technology (ICCIT 2011), pp. 22–24 (2011)
12. Wonohadidjojo, D.M., Kothapalli, G., Hassan, M.Y.: Position control of electro-hydraulic actuator system using fuzzy logic controller optimized by particle swarm optimization. International Journal of Automation and Computing 10(3), 181–193 (2013)
13. Cheong, F., Lai, R.: Designing a hierarchical fuzzy logic controller using the differential evolution approach. Applied Soft Computing 7, 481–491 (2007)

Batch-Wise Updating Neuro-Fuzzy Model Based Predictive Control for Batch Processes

Qinsheng Li, Li Jia, and Tian Yang

Shanghai Key Laboratory of Power Station Automation Technology,
Department of Automation, College of Mechatronics Engineering and Automation,
Shanghai University, 200072 Shanghai, China

Abstract. In order to guarantee the control performance of the batch processes when uncertainties and disturbances exist, a neuro-fuzzy model (NFM) predictive controller based on batch-wise model modification is developed. Model modification along batch-axis is used to improve the accuracy of neuro-fuzzy model, and predictive control along time-axis can guarantee the optimal control consequence, which lead to superior tracking performance and better robustness compared with the conventional quadratic criterion based iterative learning control (Q-ILC) approach. An illustrative example is presented to verify the effectiveness of the investigated approach.

Keywords: batch processes, neuro-fuzzy system, model predictive control, batch-to-batch model modification.

1   Introduction

For deriving the maximum benefit from batch processes, optimal control of batch processes is very significant. The key of optimal control depends on obtaining an accurate model of batch processes, which can provide accurate predictions. Usually, developing first principle models of batch processes is time consuming and effort demanding [1].To overcome the problems, process input–output data based models, such as neural networks and fuzzy systems have been investigated to model nonlinear processes owing to their ability to approximate a nonlinear function to any arbitrary accuracy [2-6]. However, optimal control obtained from off-line process model is often suboptimal when applied to the real process because of the plant-model mismatches.

Exploiting the repetition of batch processes, run-to-run control uses results from previous batches to iteratively compute the optimal operating conditions for each batch and iterative learning control (ILC) has been widely used in the batch-to-batch control of batch processes. Zafiriou and co-workers proposed an approach for modifying the input sequence from batch-to-batch to deal with plant-model mismatch [7]. Lee and co-workers presented the quadratic criterion based iterative learning control (Q-ILC) approach for tracking control of batch processes based on a linear time-varying tracking error transition model [8]. Xiong and Zhang presented a

M. Fei et al. (Eds.): LSMS/ICSEE 2014, Part II, CCIS 462, pp. 26–35, 2014.

recurrent neural network based ILC scheme for batch processes where the filtered recurrent neural network prediction errors from previous batches are added to the model predictions for the current batch and optimization is performed based on the updated predictions [9]. Based on NFM and Q-ILC, Jia et al. proposed an integrated iterative learning control strategy with model identification and dynamic R-parameter to improve tracking performance and robustness for batch processes [10].

Considering conventional Q-ILC is an open-loop optimal control approach, which is difficult to against non-repetitive disturbances, and motivated by the previous works, a neuro-fuzzy model predictive control strategy (called as NF-MPC) for batch process is proposed. This strategy features the ability to obtain the optimal control profile based on online one-step prediction and adjust the neuro-fuzzy model parameters batch-to-batch. As a result, model uncertainties can be handled to improve the control performance. Simulation results show that by updating the model from batch to batch, the control profile demonstrates good tracking performance and robustness.

2   Batch-Wise Updating NFM for Batch Processes

2.1   Neuro-Fuzzy Model

Via integrating the TS-type fuzzy logic system and RBF neural network into a connection structure, the neuro-fuzzy system (NFS) is more suitable for online nonlinear systems identification and control. Similar to paper [10], the multiple input and single output (MISO) system is chosen:

$$R^j : IF \ X \ is \ A_j \ THEN \ y \ is \ w_j, \ j = 1, 2, \cdots N, \quad X = \left[x_1, x_2, \cdots x_M \right]^T$$

where R^j denotes the j-th fuzzy rule, N is the total number of fuzzy IF-THEN rules, X is the input variable of the system with M dimensions, y is the output variable of the j-th fuzzy rule, A_j is the fuzzy set defined on the corresponding universe and w_j is the j-th consequence of fuzzy rule.

The following nonlinear function is used as the consequence of fuzzy rule

$$w_j(k_t) = a_0^j + \sum_{i=1}^{M} a_i^j x_i(k_t) + \sum_{q=1}^{M} \sum_{p=1}^{M} b_{q,p}^j x_p(k_t) x_q(k_t) . \tag{1}$$

The NFS consists of four layers. The first layer is the input layer. The second layer is the membership function layer that receives the signals from the input layer and calculates the membership of the input variable. The third layer is the rule layer and the last layer is the output layer. The output of the whole neuro-fuzzy system is then given by:

$$y = \frac{\sum_{j=1}^{N} w_j \mu_j(X)}{\sum_{j=1}^{N} \mu_j(X)} = \Phi(X)W \ . \tag{2}$$

Where $\mu_j(X) = \exp\left(-\sum_{i=1}^{M} \frac{(x_i - c_{ji})^2}{\sigma_j^2}\right)$, c_{ji} and σ_j are center and width,

respectively, $\Phi(X) = \dfrac{[\mu_1(X), \mu_2(X), \cdots \mu_N(X)]}{\sum_{j=1}^{N} \mu_j(X)}$ And $W = [w_1, w_2 \cdots w_N]^T$.

Define that k is the batch index and t is the discrete batch time, the neuro-fuzzy model is employed to build the nonlinear relationship between input variable and product quality variable of batch processes as follow:

$$\hat{y}_k(t) = \Phi(X_k(t))W_k \ . \tag{3}$$

where $X_k(t) = [U_k(t-1), U_k(t-2), y_k(t-1), y_k(t-2)]$,

$W_k = [w_k(1), w_k(2) \cdots w_k(N)]$ are model adjustable parameters, and $\Phi(X_k(t))$ is a matrix decided by $X_k(t)$.

2.2   Model Modification

From literature [10], we define the object function as eq. (4), which should be minimized by the new updated weight W_k.

$$\min \ J(W_k) = \left\| \Phi(\bar{U}_k)W_k - \bar{Y}_k \right\|_{P_1}^2 + \left\| \Delta W_k \right\|_{P_2}^2 \ . \tag{4}$$

where

$$\Delta W_k = W_k - W_{k-1}$$

$$\bar{U}_k = (U_{k-winnum+1}^T, U_{k-winnum+2}^T \cdots U_k^T)^T$$

$$\bar{Y}_k = (Y_{k-winnum+1}^T, Y_{k-winnum+2}^T \cdots Y_k^T)^T$$

$$P_1 = p_1 \cdot I_{\min\{k, winnum\} \cdot T}$$

$$P_2 = p_2 \cdot I_{\min\{k, winnum\} \cdot T}$$

Where $winnum$ denotes the size of sliding window.

In this paper, $p_2 = \alpha k^2$, p_1 and α are both positive real numbers.

Let $\dfrac{\partial J}{\partial \mathbf{W_k}} = 0$, then we get eq. (5).

$$\mathbf{W}_k = \left(\mathbf{P_1}\mathbf{\Phi}^T(\bar{\mathbf{U}}_k)\mathbf{\Phi}(\bar{\mathbf{U}}_k) + \mathbf{P_2}\right)^{-1} \cdot \left(\mathbf{P_1}\mathbf{\Phi}^T(\bar{\mathbf{U}}_k)\bar{Y}_k + \mathbf{P_2}\mathbf{W}_{k-1}\right) .$$ (5)

Similar to paper [10], under the assumption that $\mathbf{\Phi}(\bar{\mathbf{U}}_k)$ and $\mathbf{W}_k$ are both bounded, and from eq. (5), we can obtain

$$\Delta \mathbf{W}_k = \mathbf{W}_k - \mathbf{W}_{k-1}$$

$$= \frac{p_1}{\alpha k^2}\left(\frac{p_1}{\alpha k^2}\mathbf{\Phi}^T(\bar{\mathbf{U}}_k)\mathbf{\Phi}(\bar{\mathbf{U}}_k) + I_N\right)^{-1} \times \mathbf{\Phi}^T(\bar{\mathbf{U}}_k)(\bar{Y}_k - \mathbf{\Phi}(\bar{\mathbf{U}}_k)\mathbf{W}_{k-1})$$

$$\lim_{k \to \infty} \Delta \mathbf{W}_k = 0$$

This is to say, model-plant mismatch can be eliminated by yielding the new updated weight W_k from batch to batch.

3   NF-MPC for Batch Process

Set u_k and y_k are, respectively, the input variable of the product qualities and product quality variable of batch processes, U_k is the input sequence at the k-th batch, $\hat{y}_k(t)$ is the predicted values of product quality variable at time t of the k-th batch, NFM model is employed to predict the values of product quality variable step by step online. In every batch, this procedure is repeated to let the product qualities asymptotically converge to the desired product qualities $y_d(t_f)$ at the batch end time t_f.

Owing to the model-plant mismatches, the offset between the measured output and the model prediction is termed as model prediction error defined by

$$\hat{e}_k(t) = y_k(t) - \hat{y}_k(t) .$$ (6)

For employing one-step prediction of NFM, meanwhile, minimizing model prediction error to enhance control performance, the following objective function is chosen for NF-MPC.

$$\min J(U_k(t+1)) = \left\|y_d(t_f) - y_k^p(t+1)\right\|_{\mathbf{Q}}^2 + \left\|\hat{e}_k(t)\right\|_{\mathbf{Q}}^2 + \left\|\Delta U_k(t+1)\right\|_{\mathbf{R}}^2 .$$ (7)

s.t.

$$y_k^p(t+1) = \hat{y}_k(t+1) + \hat{e}_k(t)$$
$$\Delta U_k(t+1) = U_k(t+1) - U_k(t)$$
$$u^{low} \le u_k(t) \le u^{up}$$
$$y^{low} \le y_k(t) \le y^{up}$$

where u^{low} and u^{up} are the lower and upper bounds of the input trajectory, y^{low} and y^{up} are the lower and upper bounds of the final product qualities, $\mathbf{Q}$ and $\mathbf{R}$ are defined as $\mathbf{Q} = q \times \mathbf{I}_n$ and $\mathbf{R} = r \times \mathbf{I}_n$ where r and q are the factors.

In summary, the following describes the steps to update the control sequences step by step using the above-mentioned approach at k-th batch:

Step 1 Build NFM based on historical batch operation data. Initialize $U_k(1)$ and parameters Q, R.

Step 2 Update the weight of NFM according to eq.(4), then get updated $U_k(t)$ according to eq.(7), the corresponding output $y_k(t)$ is measured. set t=1

Step 3 Calculate $\hat{y}_k(t)$ and $\hat{y}_k(t+1)$ by eq.(3), next, compute the model prediction error $\hat{e}_k(t)$ by eq. (6).

Step 4 The objective function shown in eq. (7) is minimized and an updated input sequences $U_k(t+1)$ is updated.

Step 5 If t<t_f ,set $t = t+1$ and go to Step 2

4   Convergence Analysis

In this section, a rigorous theorem has been used to prove that the tracking error $e_k(t) = y_d(t_f) - y_k(t)$,under the NF-MPC law, can converge to a bound, as $t \to t_f$.

Theorem 1. Let $e_k(t)$ be the tracking error sequece based on the NFM under the NF-MPC law,if there is no model prediction error, namely $\hat{e}_k(t) = y_k(t) - \hat{y}_k(t) = 0$, and under the assumption that the terminal tracking error is bounded, namely $\|e_k(t_f)\| = \|y_d(t_f) - y_k(t_f)\| \le \varepsilon$, ε is small positive constant, then

$$\lim_{t \to t_f} \|e_k(t)\| \le \varepsilon$$

Proof   As $U_k(t+1)$ is the optimal control trajectory of the optimization problem, if $U_k^{\bullet}(t+1) \ne U_k(t+1)$, then the following inequality is true:

$$J(U_k(t+1)) \le J(U_k^{\bullet}(t+1)) . \tag{7}$$

Set $U_k^{\bullet}(t+1) = U_k(t)$, from ineq. (7), we obtain

$$J(U_k(t+1)) \le J(U_k(t)) = \left\| y_d(t_f) - y_k(t) \right\|_Q^2 + \left\| U_k(t) - U_k(t) \right\|_R^2 = \left\| e_k(t) \right\|_Q^2 \tag{8}$$

Next, under the assumption of no model prediction error, the following inequality holds.

$$J(U_k(t+1)) = \left\| y_d(t_f) - y_k(t+1) \right\|_Q^2 + \left\| U_k(t+1) - U_k(t) \right\|_R^2 \ge \left\| e_k(t+1) \right\|_Q^2 . \tag{9}$$

From inequalities (8) and (9), we get

$$\left\| e_k(t+1) \right\|_Q^2 \le J(U_k(t+1)) \le \left\| e_k(t) \right\|_Q^2 . \tag{10}$$

Therefore, the limits of $\left\| e_k(t) \right\|$ exist as $t \to t_f$, namely

$$\lim_{t \to t_f} \|e_k(t)\| = \left\| e_k(t_f) \right\| \le \varepsilon$$

This completes the proof.   Q.E.D

Remark 1.   From ineq. (10), It is clear that $e_k(t)$ is convergent along with time-axis, and $e_k(t)$ decreases to zero as $t \to \infty$, namely $\lim_{t \to \infty} e_k(t) = 0$. In fact, the time interval within each batch is finite for batch process, hence t_f can not be infinite, and $e_k(t)$ converges to a bounded region at the batch end. Furthermore, we can derive that the bound of end tracking error $e_k(t_f)$ depends on batch run length t_f at k-th batch.

5   Examples

This case study is a typical nonlinear batch reactor, in which a first-order irreversible exothermic reaction $A \xrightarrow{k_1} B \xrightarrow{k_2} C$ takes place [10, 11, 12]. This process can be described as follow:

$$\dot{x}_1 = -4000\exp(-2500/T)x_1^2$$
$$\dot{x}_2 = 4000\exp(-2500/T)x_1^2 - 6.2\times10^5\exp(-5000/T)x_2$$

where T denotes reactor temperature, x_1 and x_2 are, respectively, the reactant concentration.

In this simulation, the reactor temperature is first normalized using $T_d = (T - T_{min})/(T_{max} - T_{min})$, in which T_{min} and T_{max} are respectively 298 (K) and 398 (K). T_d is the control variable which is bounded by $0 \le T_d \le 1$ and $x_2(t)$ is the output variable. The control objective is to manipulate the reactor temperature T_d to control concentration of B at the end of the batch $x_2(t_f)$.

The independent random signal with uniform distribution between [0, 1] are used to simulate 30 input-output data for training purpose and another 20 input-output data for testing purpose. And the robustness of the proposed method is evaluated by introducing Gaussian white noise with 0.002^2 standard deviation to the output variable of the process. In the simulation, the model of six fuzzy rules is chosen. The following parameters in batch process are assumed: $r = 1, q = 10$, $U_k(1) = 0$, and $y_d(t_f) = 0.6100$. Then 3% noisy is put into the quality output of the 5th batch, and fig.1 to fig.5 show the strong stability of the control strategy in minimizing the effect of the noisy. It can be observed from fig4 and fig5 that the investigated control strategy have faster convergence rate along batch-axis and better tracking performance against disturbances compared with conventional Q-ILC method.

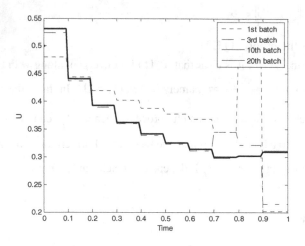

Fig. 1. Control trajectory

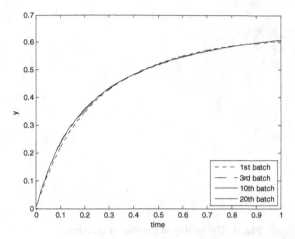

Fig. 2. Product quality trajectory

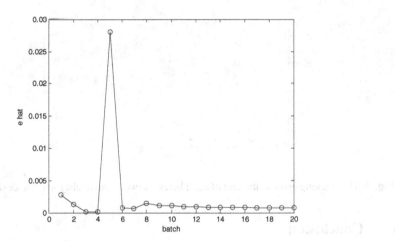

Fig. 3. The error between the model and the plant

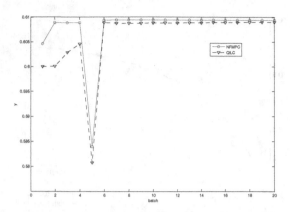

Fig. 4. The quality at the end of each batch

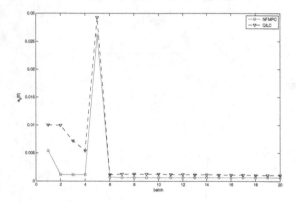

Fig. 5. The tracking error at the end of each batch between the quality and the desired quality

6   Conclusion

In this paper, a neuro-fuzzy model predictive control with batch-to-batch model updating mechanism is proposed. The results on a simulated batch reactor show that the superior tracking performance and better robustness have been obtained compared to conventional Q-ILC method. In future works, we will investigate how to combine time-wise NFMPC with batch-wise Q-ILC strategy, which can integrate time-wise information and batch-wise information into one uniform frame, and will get perfect control performance.

Acknowledgments. Supported by National Natural Science Foundation of China (61374044), Shanghai Science Technology Commission (12510709400), Shanghai Municipal Education Commission (14ZZ088), Shanghai talent development plan.

References

1. Zhang, J.: Batch-to-batch optimal control of a batch polymerisation process based on stacked neural network models. Chemical Engineering Science 63, 1273–1281 (2008)
2. Cybenko, G.: Approximation by superposition of a sigmoidal function. Math. Control Signal Systems 2, 303–314 (1989)
3. Girosi, F., Poggio, T.: Networks and the best approximation property. Biological Cybernetics 63, 169–179 (1990)
4. Park, J., Sandberg, I.W.: Universal approximation using radial-basis-function networks. Neural Comput. 3(2), 246–257 (1991)
5. Wang, L.X.: Fuzzy systems are universal approximators. In: Proc. IEEE Internat. Conf. on Fuzzy Systems, pp. 1163–1170 (1992)
6. Tian, Y., Zhang, J., Morris, A.J.: Modeling and optimal control of a batch polymerization reactor using a hybrid stacked recurrent neural network model. Ind. Eng. Chem. Res. 40, 4525–4535 (2001)
7. Zafiriou, E., Adomaitis, R.A., Gattu, G.: An approach to run-to-run control for rapid thermal processing. In: Proceedings of the American Control Conference, pp. 1286–1288 (1995)
8. Lee, J.H., Lee, K.S., Kim, W.C.: Model-based iterative learning control with a quadratic criterion for timevarying linear systems. Automatica 36(5), 641–657 (2000)
9. Xiong, Z., Zhang, J.: A batch-to-batch iterative optimal control strategy based on recurrent neural network models. J. Process Contr. 15, 11–21, 19 (2005)
10. Jia, L., Yang, T., Qiu, M.: An integrated iterative learning control strategy with model identification and dynamic R-parameter for batch processes. Journal of Process Control 23, 1332–1341 (2013)
11. Xiong, Z., Zhang, J., Wang, X., Xu, Y.: Run-to-run iterative optimization control of batch processes based on recurrent neural network. In: Yin, F.-L., Wang, J., Guo, C. (eds.) ISNN 2004. LNCS, vol. 3174, pp. 97–103. Springer, Heidelberg (2004)
12. Ray, W.H.: Advanced Process Control. McGraw-Hill, New York (1981)

A Novel Learning Algorithm
for Pallet Grouping Technology

Weitian Lin[1], Zhigang Lian[1], Bin Jiao[1], Xingsheng Gu[2], and Wei Xu[3]

[1] Shanghai DianJi University, 200240 Shanghai, China
[2] Key Laboratory of Advanced Control and Optimization for Chemical Processes of Ministry of Education, East China University of Science and Technology, 200237 Shanghai, China
[3] Shanghai Electric Group Co., Ltd. Central Academe, 200070 Shanghai, China
linwt@sdju.edu.cn, {lllzg,binjiaocn}@163.com, xsgu@ecust.edu.cn, xuwei0729@gmail.com

Abstract. The Pallet Grouping Problem (PGP) is defined as minimizing the number of pallets for placing all materials in a collection and distribution center. A key to solving the PGP is to tackle the Pallet Loading Problem (PLP). The Pallet Loading Problem aims to maximize the number of identical rectangular boxes placed within a rectangular pallet. All boxes have identical rectangular dimensions and, when placed, must be located completely within the pallet. In this paper, a novel pallet grouping technology is proposed and a new learning algorithm, namely Learning Only from Excellence, (LOE) is presented for solving the pallet loading problem. Simulation results show that compared with the conventional Genetic Algorithm (AG) for two pallet loading problems with different scales, the new learning algorithm is proved to be more efficiently.

Keywords: pallet grouping, collection and distribution, learning algorithm, genetic algorithm.

1   Introduction

The Pallet Grouping Problem (PGP) has direct impact on the enterprise logistic transportation speed, thus the production efficiency. If the pallet grouping size is too big, i.e. all materials are contained in one pallet, then the workshop will not be supplied with materials on time, unless some materials arrive too early, implying storage at the shop floor, further, take up workshop site resources, affect production and processing. If the pallet grouping size too small, i.e. one small portion of materials on one pallet, which will then make the whole transportation so busy that the transportation speed will be slowed down, the efficiency is reduced and big cost is incurred.

The pallet grouping problem is a kind of scheduling problem. How to optimize the number of pallets placing all given materials is the so called Pallet Loading Problem (PLP). More specifically, the Pallet Loading Problem (PLP), as an optimization problem, aims to place products into boxes on a rectangular pallet, in order to

M. Fei et al. (Eds.): LSMS/ICSEE 2014, Part II, CCIS 462, pp. 36–45, 2014.

optimize its utilization without overlapping. Dyckhoff [1] shows that the problem can be classified as 2/B/O/C (Two-dimensional, Selection of items, One object, Identical items). So the problem extends to a special case of cut and packing problems.

The PLP widely exists in materials and goods distribution chain. An increase in the number of boxes packed into the pallet will lead to a decrease in the logistics cost [2]. According to Hodgson [3], the pallet loading problem is generally classified into two types including the Manufacturer's Pallet Loading Problem (MPLP) and Distributor's Pallet Loading Problem (DPLP). The difference is that the MPLP considers boxes with identical dimensions while the DPLP deals with boxes of different dimensions. It's noted that the boxes are packed in horizontal layers in both cases.

In this paper we consider the MPLP, where given a fixed pallet volume V, load-bearing W, the problem is to arrange the maximum number of identical boxes $\{(v_1, w_1),(v_2, w_2)$ $(v_3, w_3),\ldots, (v_n, w_n)\}$ into the pallet (V,W). In addition the loading problem this paper researched also constrained in time, namely the loading on the material must timely be sent to the destination. The paper consider Weight, Volume and Time constraints for the pallets.

Some intelligent algorithms have been proposed to solve NP-hard problems[4][5][6], especially the MPLP, For example, Lau et al. [7] presented a hybrid approach based on heuristic and genetic algorithms (GA), for solving the profit-based multi-pallet loading problem which was mathematically formulated as a nonlinear integer programming problem. Kocjan and Holmström [8] described an Integer Programming model for generating stable loading patterns for the Pallet Loading Problem under several stability criteria. Within a tree-search structure [9,10], new algorithms have been developed for the pallet loading problem. For some special cases of the pallet loading problems with upper bounds, Martins and Dell [11], Ribeiro and Lorena [12] and Pureza and Morabito [13] came up with several solution. Martins and Dell [14] presented new bounds, heuristics, and an exact algorithm for the Pallet Loading Problem. Yaman and Şen [15] studied the contents of a pre-determined number of mixed pallets so as to minimize the total inventory holding and backlogging costs of its customers over a finite horizon. The problem was formulated as a mixed integer linear program, and valid inequalities were incorporated to strengthen the formulation. Further, techniques based in identified structures shown as G4 [16] and L [17,18] were also reported. Some meta-heuristic methods were also reported, such as the Tabu Search Genetic Algorithms [19,20].

In this paper, a novel pallet grouping technology is proposed and a new learning algorithm, which is defined as *Learning Algorithm Only from Excellent~Pbest,Gbest and Gpbest*, for short LAOE~PGpG is developed to solve the pallet loading problem. The core of this novel algorithm is the learning methods including ML1 and ML2, which represent point-learning and period-learning respectively. Further, the manufacturing problem is shown to be a particular case of the linear programming 0-1 model.

2   Problem Model

For a better understanding of the problem, a general example is introduced as follows.

In a shop-building factory, all materials needed for the production work are collected, distributed and transported timely by the scheduling center. Details are listed as follows (it is assumed that pallets used and transportation destinations are the same.)

Let all kinds of materials placed on the i-th pallet be $x_{i1}, x_{i2}, \cdots, x_{in}$ and the number of materials are $n_{i1}, n_{i2}, \cdots, n_{in}$ respectively. Time to collect and distributing the j-th materials in the i-th pallet is d_{ij}. The model for the proposed objective problem is expressed as follows:

$$\min \quad z = i$$

$$\text{st} \quad \begin{cases} \sum_{j=1}^{n} n_{ij} x_{ij} w_{ij} \leq W \\ \sum_{j=1}^{n} n_{ij} x_{ij} v_{ij} \leq V \\ d_{j2} < D_j (j = 1, 2, \ldots, n) \\ d_{j2} > d_{j1} (j = 1, 2, \ldots, n) \\ \sum_{j=1}^{n} n_{ij} x_{ij} < N_j (j = 1, 2, \ldots, n) \\ x_1, x_2, \cdots, x_n \text{ are nonnegative number} \end{cases} \quad (1)$$

In the above equation (1), $d_{j2} < D_j$ suggests that the time of collecting and distributing is less than the due date. $d_{j2} > d_{j1}$ means that the material x_j collection complete time is greater than beginning time.

3   Algorithm Development

3.1   The Learning Algorithm

Suppose that the searching space is D-dimensional, and m particles form the colony. The i-th particle represents a D-dimensional vector $X_i (i = 1, 2, \cdots, m)$. It means that the i-th particle locates at $X_i = (x_{i1}, x_{i2}, \cdots, x_{iD})(i = 1, 2, \cdots, m)$ in the searching space. We could calculate the fitness of the particles by putting its position into a designated objective function. Denote the best knowledge represented by Ibest of the i-th particle as $P_{best} = (i_{i1}, i_{i2}, \cdots, i_{iD})$, the best knowledge represented by Gbest of the colony as $G_{best} = (g_1, g_2, \cdots, g_D)$ and the best knowledge searched by the time of the k-th running of the colony as $G_{bes}^{pre}(k) = (g_{p1}, g_{p2}, \cdots, g_{pD})$ respectively. If the particles learn from P_{best}, G_{best} and $G_{bes}^{pre}(k)$, the algorithm is defined as *Learning Only from*

Excellence (LOE), The algorithm could be performed according to the following equations.

$$X_{teacher}(k) = G_{best}(k) \cup P_{best}(k) \cup G_{best}^{pre}(k).$$ (2)

$$X_i(k+1) = X_i(k) \oplus X_{teacher}(k).$$ (3)

where k is the times of iteration; $X_{teacher}(k)$ is the superior particle obtained via $X_i(k)$ learning from $G_{best}(k)$ and $G_{best}(k)$. Due to signal coding or natural number coding in the algorithm, $(x_{i1}(k), x_{i2}(k), \cdots, x_{iD}(k))$ of $X_i(k)$ are signal or natural number; $\oplus$ represents learning; $\cap$ represents choosing best particles in the next generation. The ending condition is either the maximum iteration number is reached or the pre-set value is achieved.

3.2   Encoding

While the algorithm is applied to optimize different problems, the encoding scheme can be different. For the production scheduling problem, the integer encoding is used as it is a combinatorial problem and the solution space is discrete. In this paper, for the Pallet Grouping Problem (PGP), binary encoding scheme is adopted.

0-1 encoding: Making an instance with N kinds of materials, then a series of sequence is generated as follows:

0-2

$$se(sequence) = [\underbrace{1, 0, \cdots, 0, 1, 0}_{N}].$$ (4)

In equation (4), it has different meanings. 1 and 0 indicate whether or not a material is presented in a pallet.

A new model, namely LOE model, is introduced in details as follows, including the encoding and learning method.

3.3   LOE Learning Method

For the LOE optimizing model, several learning methods are illustrated with charts for twelve kinds of materials.

In ML1and ML 2 shown above, two learning points are chosen for an operation. Some points in ① are chosen and copied to immature particle ②, which form a novel particle ②. Then some other effective learning points are chosen to match with the novel particle ②. Then, new particles are formed for the next operation. The operation process are illustrated in Fig. 1(a,b), in which the operated position is marked with a line.

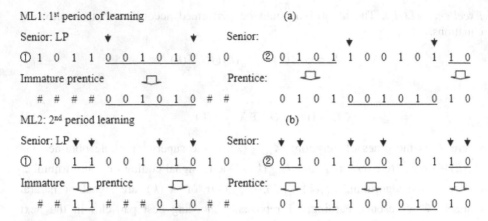

Fig. 1. The LOE learning method for PGP

ML3 is like as ML1 and ML 2 shown above, three learning points are chosen for an operation. Some points in ① are chosen and copied to immature particle ②, which form a novel particle ②. Then some other effective learning points are chosen to match with the novel particle ②. Then, new particles are formed for the next operation. The operation process is no cumbersome.

3.4   Mutation Operation in GA

Details on the mutation scheme in GA can be found in a number of literatures.

For the mutation M1 shown above in one section, one *insert point* is chosen for moving and inserting operation, and *one section* in ① is chosen and moving to insert after *insert point*, which forms a novel particle ②. The operation process is illustrated in Fig. 2(a), in which the operated position is marked with a line.

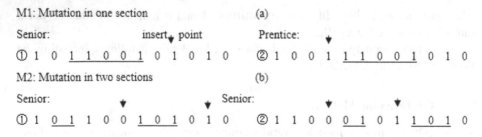

Fig. 2. Mutation method for GA optimizing PGP

For the mutation M2 shown above in two sections, two *insert points* are chosen for moving and inserting operation, and two *sections* in ① are chosen and moving to insert after *insert point* respectively, which form a novel particle ②. The operation process is illustrated in Fig. 2(b), in which the operated positions are marked with two line.

4   Experiment

To illustrate the effectiveness and performance of LOE for the pallet loading problem (PLP), two representative instances taken from Internet based on practical data have been used. The comparisons are made between LOE and the standard GA.

Test 1: Suppose that ten kinds of materials are provided for production work, for which the needed material delivery is named as XQWLT, and the list of due date XQWLT =[2,4,6,8,10,12,14,16,18,20]; quantity of demand items is name as XQWLN, and the needed number of materials' XQWLN, =[12,16,9,19,47,63,51, 52,87,32]; collected material procurement/finish dates are named as WLT and the due date of completing the materials collection WLT =[1,2,4,6,9,10,13,15,16,18]. 10 item weight and volume are named as WLW and WLV respectively, weight and volume of ten materials are listed as WLW =[57,64,8,21,46,17,97,32,72,99] and WLV =[11,4,30,13,35,21,36,28,15,25]. The upping values of the pallet's weight W and volume V are 1000 and 2000.

Test 2: Suppose that twenty kinds of materials are provided for production work,for which the due date is set as XQWLT =[2,3,4,7,9,10,11,12,13,14,16,17,18,19,20,23, 25,26,29,30], demand number of materials is XQWLN =[12,16,9,19,7,13,21,18,8, 20,10,4,30,25,20,19,5,10,26,18] and the due date of completing materials collection is WLT =[1,3,4,5,6,7,9,11,12,13,15,16,17,18,20,23,24,25,28,29]. In details, weight and volume of ten materials are listed as WLW =[83,120,139,51,140,70,84,120,77, 67,166,62,88,156,75,79,190,80,161,89] and WLV =[132,69,148,70,161,81,177,69, 156,65,146,143,77,90,181,60,136,82,184,64]. Also, the upping values of the pallet's weight W and volume V are 1000 and 2000.

Remark: Min, Average and Max stand for minimum, Average and maximum fitness value. EGN is the maximum generation number. Size is the size of population for GA and LOE. Run defines the number of running circles. Goods represent number of materials. P: Gp:G Percent represents the ratio of present individual historic best, historic global best from previous several generations to present generation and present global best, namely the ratio of $X_{teacher}(k)$. L1,L2,L3 give three different learning methods. Also, for the GA algorithm, m is the mutation rate, M1,M2 are the mutation methods and C1,C2,C3 are the three different crossing operations can been seen in paper [21], which are like L1,L2,L3 learning methods respectively. There is no the same parameter in two different algorithms, so in this paper their best or average solution is used to compare their performance.

The experimental result is shown in Table 1.

Table 1. The comparisons of LOE and GA for test 1 and test 2

algorithm		Min/Average/Max			
			Learning method		
	Problem	P: Gp:G Percent	L1	L2	L3
LOE EGN=1000 Size=120 Run=10	Test1: Goods: 10	50:25:25	32/32.8/33/0	31/32.6/34/1	31/32.5/34/1
		25:50:25	32/32.8/34/0	**31**/32.6/34/1	**31/32.5**/33/1
	Test2: Goods: 20	50:25:25	49/50.6/52/0	49/50.2/51/0	49/51.7/52/0
		25:50:25	**48/49.6**/51/1	**48**/50.7/52/1	49/50.5/52/0
	Problem	Mutation Percent		Crossover	
		Pm / mutation	C1	C2	C3
GA EGN =1000 Size=120 Run=10	Test1: Goods: 10	0.1 / M1	33/34/35/0	33/33.9/35/0	**32/33.5**/34/0
		0.08 / M2	33/33.9/35/0	33/33.7/35/0	33/33.6/34/0
	Test2: Goods: 20	0.1 / M1	50/51.1/52/0	**49**/51.4/53/0	50/51.2/53/0
		0.08 / M2	**49/50.8**/52/0	50/51.1/53/0	50/51.2/52/0

From Table 1, we can conclude that the LOE outperforms the GA for the pallet loading problem. Firstly, comparing the results between LOE and GA, we find that the best fitness values are obtained by LOE in both tests 1 and test 2, which are marked with bold letters. Secondly, for LOE the better fitness value can be obtained when Learning method L3 is chosen and the P: Gp:G Percent is equaled to 25:50:25 in test 1, and in test 2, the better solution is obtained when L1 is chosen and P: Gp:G Percent is 25:50:25. Comparing the results using the GA, we can find the better fitness value can be obtained when C3 is used for Crossover and for M1, Pm =0.1 in Test 1. For Test 2, the better solution is obtained when C1 and M2 are used, where Pm=0.08.

From Fig.3, it is obvious that the convergence rate of LOE is faster than the GA. LOE finally gives the better fitness value than GA algorithm in both test 1 and test 2. Also, the average fitness value found by LOE are better than GA for optimizing both test 1 and test 2. Finally, LOE is better than GA according to average value.

In summary, LOE has shown to be an effective and stable algorithm for PLP. Compared with the GA algorithm, the novel algorithm is more efficient and effective.

Associated with the example of Table 2, from table 2 part, the solutions have been described below: First batch is two palettes. Pallet 1: 1 unit of material 1 and 6 units of material 2. Pallet 2: 11 units of material 1; Second batch is one palette. Pallet 3: 10 units of material 2.

Pallet Grouping and Dispatching Problem of issues discussed in this article are not just the loading problem, which is related to delivery to the destination on time constraints limit. Loads much be distributed better onto pallets by time as much as possible, no delay is needed and of course it is better not to advance dispatching. Maybe advanced delivery to destination is not needed, and there is no place to pile up.

This is not a simple discussion   about   how to   load all   materials into the pallet using the least. Instead, our objective is to use minimum pallets and optimal deliver material to the destination, no delay is needed and of course it is better not to advance dispatching.

Finally, the convergence of the two algorithms for optimizing the PLP are shown in Fig. 3.

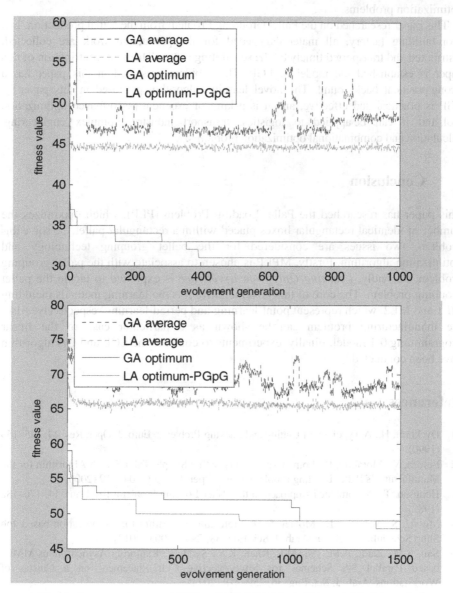

Fig. 3. Convergence curve of GA and LA-PGpG for problem test 1 and test 2

In the actual production, if material distribution is science, it will greatly reduce the number of pallet and times of back and forth transportation, reduce costs, and improve efficiency. The problem this paper presented has a strong practical background; it can be applied to logistics and transport industry. The novel learning algorithm proposed in this paper is a kind of excellent performance optimization tool, and it can be applied to complex engineering calculation, combinational optimization problems.

This paper researched on the Pallet Grouping Problem from the actual production. In a shop-building factory, all materials needed for the production work are collected, distributed and transported timely by the scheduling center. The main contribution of this paper is established the model of PGP. The problem presented in this paper has a strong practical background. The novel learning algorithm proposed in this paper for PGP is practical and effective, and it is a kind of excellent performance optimization tool, and it can be applied to logistics, transport industry, complex engineering calculation and combinational optimization problems.

5   Conclusion

This paper has researched the Pallet Loading Problem (PLP), which maximizes the number of identical rectangular boxes placed within a rectangular pallet. To solve this problem, two issues are considered for the pallet grouping technology and optimization algorithm. Firstly, MPLP is chosen to associate with the pallet grouping problem. Secondly, *Learning Only from Excellence* is exploited to tackle the pallet grouping problem. The core of this novel algorithm is the learning methods including ML1 and ML2, which represent point-learning and period-learning respectively. Also, the manufacturing problem can be shown as a particular case of the linear programming 0-1 model. Finally, experiments to compare the LOE and GA algorithm have been conducted.

References

1. Dyckhoff, H.: A Typology of Cutting and Packing Problems. Euro. J. Oper. Res. 44, 145–159 (1990)
2. Pureza, V., Morabito, R.: Some Experiments with a Simple Tabu Search Algorithm for the Manufacturer's Pallet Loading Problem. Com. Oper. Res. 33, 804–819 (2006)
3. Hodgson, T.: A Combined Approach to the Pallet Loading Problem. IIE Trans. 14, 176–182 (1982)
4. Molaei, S., Vahdani, B., Molaei, S.: A Molecular Algorithm for an Operation-based Job Shop Scheduling Problem. Arab. J. Sci. Eng. 38, 2993–3003 (2013)
5. Sait, S.M., Zaidi, A.M., Ali, M.I., Khan, K.S., Syed, S.: Exploring Asynchronous MMC-based Parallel SA Schemes for Multiobjective Cell Placement on a Cluster of Workstations. Arab. J. Sci. Eng. 36, 259–278 (2011)
6. Fadaei, M., Zandieh, M.: Scheduling a Bi-objective Hybrid Flow Shop with Sequence-Dependent Family Setup Times using Metaheuristics. Arab. J. Sci. Eng. 38, 2233–2244 (2013)

7. Lau, H.C.W., Chan, T.M., Tsui, W.T., Ho, G.T.S., Choy, K.L.: An AI Approach for Optimizing Multi-pallet Loading Operations. Exp. Sys. App. 36, 4296–4312 (2009)
8. Kocjan, W., Holmström, K.: Computing Stable Loads for Pallets. Euro. J. Oper. Res. 207, 980–985 (2010)
9. Bhattacharya, R., Roy, R., Bhattacharya, S.: An Exact Depth-first Algorithm for the Pallet Loading Problem. Euro. J. Oper. Res. 110, 610–625 (1998)
10. Alvarez-Valdez, R., Parrel, F., Tamarit, J.M.: A Branch-and-cut Algorithm for the Pallet Loading Problem. Com. Oper. Res. 32, 3007–3029 (2005)
11. Martins, G.H.A., Dell, R.F.: The Minimum Size Instance of a Pallet Loading Problem Equivalence Class. Euro. J. Oper. Res. 179, 17–26 (2007)
12. Ribeiro, G.M., Lorena, L.A.N.: Lagrangean Relaxation with Clusters and Column Generation for the Manufacturer's Pallet Loading Problem. Com. Oper. Res. 34, 2695–2708 (2007)
13. Pureza, V., Morabito, R.: Some Experiments with a Simple Tabu Search Algorithm for the Manufacturer's Pallet Loading Problem. Com. Oper. Res. 33, 804–819 (2006)
14. Martins, G.H.A., Dell, R.F.: Solving the Pallet Loading Problem. Euro. J. Ope. Res. 184, 429–440 (2008)
15. Yaman, H., Şen, A.: Manufacturer's Mixed Pallet Design Problem. Euro. J. Ope. Res. 186, 826–840 (2008)
16. Scheithauer, G., Terno, J.: The G4-heuristic for the Pallet Loading Problem. J. Oper. Res. Soc. 47, 511–522 (1996)
17. Lins, L., Lins, S., Morabito, R.: An L-approach for Packing (l,w)-rectangles into Rectangular and L-shaped Pieces. J. Oper. Res. Soc. 54, 777–789 (2003)
18. Birgin, E.G., Morabito, R., Nishihara, F.H.: A Note on an L-approach for Solving the Manufacturer's Pallet Loading Problem. J. Oper. Res. Soc. 56, 1448–1451 (2005)
19. Alvarez-Valdez, R., Parre, F., Tamarit, J.M.: A Tabu Search Algorithm for Pallet Loading Problem. OR Spectrum 27, 43–61 (2005)
20. Kanga, K., Moonb, I., Wang, H.: A Hybrid Genetic Algorithm with a New Packing Strategy for the Three-Dimensional Bin Packing Problem. App. Math. Com. 219, 1287–1299 (2012)
21. Lian, Z.G., Gu, X.S., Jiao, B.: A Dual Similar Particle Swarm Optimization Algorithm for Job-shop Scheduling with Penalty. In: Proceedings of the 6th World Congress on Control and Automation, pp. 7312–7316. IEEE Press, New York (2006)

Optimization of Data Query and Display Algorithm for a Wireless Monitoring and Visualization System

Jiqiu Chen[1], Jingqi Fu[1], and Weihua Bao[2]

[1] School of Mechatronics Engineering and Automation, Shanghai University,
Shanghai, 200072, China
[2] Shanghai Automation Instrumentation Co.,Ltd.
191 Guangzhouxi Road, Shanghai, China
a4052@163.com, jqfu@staff.shu.edu.cn, bwh@saic.sh.cn

Abstract. As wireless sensor networks are widely used in industry, the wireless monitoring system for sniffing data packets from wireless sensor networks has been realized. Many novel methods has been utilized to efficiently store data and provide reliable query and display functions in real-time listening, such as adjustment of storage structure and introduction of memory module. This paper designs and realizes a wireless monitoring and visualization system, which can accurately collect data from the communication among a variety of wireless terminal devices in real-time through changing original data storage structure, improving data query algorithms and optimizing data display algorithms.

Keywords: Wireless sensor networks, sniff, query, display, algorithm.

1 Introduction

Currently, wireless sensor networks (WSN) has been widely used in a number of areas because of its low cost, low power consumption and so on. Militarily, the US Science Application International Corporation adopted WSN to build an electronic perimeter defense system. In precision agriculture, Intel establishes the world's first wireless vineyard in Oregon. In addition, the Institute of Computing Technology, Chinese Academy of Science, makes a safety monitoring system using WSN for the National Palace Museum, which contributes WSN technology in the field of civil security [1].

The current development of WSN shows the following characteristics: First, the number of data and the size of wireless communication surge; Second is the instability of communication links which is caused by the complexity of WSN deployment environment and the resource-constrained sensor nodes; Thirdly, more and more users are not satisfied with just a simple record of the original data and repeated display.

WSN-based wireless monitoring system can monitor wireless communication and effectively store, manage and use these real-time data. For a wireless monitoring system, there are several major difficulties, namely how to accurately receive wireless

M. Fei et al. (Eds.): LSMS/ICSEE 2014, Part II, CCIS 462, pp. 46–55, 2014.

data, the second is how to efficiently store and manage data, the third is how to efficiently query and display data. In this paper, a wireless sensor network monitoring and visualization system (WSNMVS) has been designed and implemented. This system can achieve data communications between wireless devices for real-time monitoring, and supports for a variety of monitoring equipment. It also can receive information simultaneously from multiple devices, which means several different channels can be monitored simultaneously; Meanwhile, WSNMVS has two modes: online and offline mode. The former supports for the database. On this basis, the paper changes the original data storage structure, improves data query algorithms and optimizes the data display algorithm, thus greatly improving the efficiency of data storage, query and display.

In the remaining chapters, the second part will show some relevant research; the third part describes the design and implementation of WSNMVS, including the optimization of data storage structure and data query and display algorithms; some practical results of real-work will show in Part 4; the last part is a brief conclusion for this paper.

2   Related Works

There are several similar wireless monitoring and visualization systems for WSN, for example CoMaDa[2], SpyGlass[3], M-DAD[4], SNAMP[5], NSSN[6], CROWD[7], Mote-VIEW[8], Z-Monitor[9], etc. Brief description will be presented for some representative systems.

CoMaDa is an adaptive framework with graphical support for configuration, management, and data handling tasks for WSN, which can provide supports for management tasks of different types of hardware and the data visualization. CoMaDa consists in a modular fashion, making it easy for users to expand according to their own needs.

SNAMP is a multi-sniffer and multi-view visualization platform for WSN. It uses a variety of listening node for wireless data collection, avoiding data bottlenecks. And SNAMP has a variety of visual module: topology analysis module, sensor data module, packet analysis module, and network management module.

NSSN is a network monitoring and packet sniffing tool for WSN. It can automatically search the channel of the target WSN, parsing and displaying ZigBee, 6LoWPAN and other protocols. And NSSN provides functions like network troubleshooting and performance energy analyzing. In addition, NSSN also supports remote data monitoring.

These systems mentioned above have their own advantages and characteristics, and also drawbacks. Some only support specific hardware, some support only one wireless listening devices, some are not good at storing and managing large amounts of data, and others' efficiency of query and display is not high. All these are achieved or improved in WSNMVS.

3   System Design and Implementation of WSNMVS

3.1   Architecture of WSNMVS

The overall architecture of WSNMVS is shown in Fig.1, consisting of four main parts: the target WSN, wireless monitoring terminal, PC client and the database server.

The target WSN is a running wireless sensor networks in industry field, which should support the IEEE 802.15.4 protocol in physical layer and MAC layer and Zigbee protocol in network layer support. Then the target WSN will become a monitoring goal for WSNMVS.

Wireless terminals contain a variety of wireless listening devices, mainly including wireless gateways, wireless handheld devices, etc. These wireless terminals can receive wireless data from the whole 16 channels in 2.4G band (2400 ~ 2483.5MHz), then they send those data to the PC client through a wired (including serial, USB, Ethernet, etc.) or wireless (Wi-Fi) way.

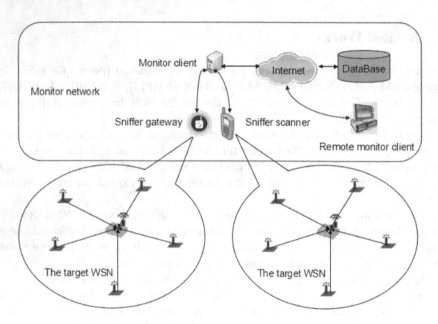

Fig. 1. The architecture of WSNMVS

The PC clients can be divided into local clients and remote clients, which receive data from the wireless terminal monitor and store it to a data storage (including database or local temporary files, etc.). The data in storage will be queried, analyzed and finally visualized to users.

The database server is responsible for data manage and user permissions, providing online query and analysis capabilities.

3.2   Software Design of WSNMVS

The software design of WSNMVS is divided into two parts: the embedded wireless terminal and PC clients. The design and implementation of the PC clients program will be discussed in detail.

The entire PC clients consist the following four modules: graphical user interface (GUI) module, controller module, sniffer module and storage module, Fig.2 shows the interaction between the four modules.

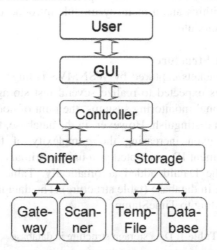

Fig. 2. Architecture of monitor client

The flexibility of software is improved by using the MVC pattern on structure, separating visual interface, control strategies and data model.

The GUI module actually contains two modules: one is the interface module, the other is the graphics module. The former responds to user input, the latter displays data controlled by the controller module. The visualization is also charged by this module.

The control module receives all kinds of messages from sniffer module, GUI modules and storage module, and makes the final decision to notify the appropriate modules to respond. The control module also contains a dozen strategy modules, each responsible for their own decision-making functions. Here is the major strategy module: sniffer selection module, channel selection module, node selection module, protocol configuration module, display configuration module, display filtering module, data detailing module, monitor operating module.

The sniffer module is in charge of unifying interface to use a variety of monitoring equipment or temporary replacement of monitoring equipment without affecting the other modules. Each sub-module under the unified interface is responsible for one listening device.

The wireless monitoring data is saved and managed by the storage module, the interface of which also need   to be unified to use a variety of storage devices.

3.3   Optimization for the Query and Display Algorithm of WSNMVS

To achieve the goal of capturing real-time data from WSN, the reliable monitoring equipment, good data storage structure, efficient query and display algorithms are necessary. The WSNMVS has the most functions which is essential to the traditional monitoring systems, such as topology structure and data display, and also it supports a variety of listening devices for sniffing at the same time. WSNMVS improves the query and display efficiency because of the improved storage structure, the optimized query and display algorithms and a memory module introduced between the display module and the storage module.

3.3.1   Modify Storage Structure

Due to the number of packets captured by WSNMVS is huge, a simple and reliable architecture of storage is expected to realize several fast storage and complex logic queries. In the conventional monitoring systems, the joint of node ID and TimeStamp is the prime key for data distinguish. However, in the database, this kind of prime key is not quite efficient, even increasing the complexity of the storage structure. Therefore the auto increment ID is adopted as a unique primary key, namely that each time a data received the DataId added automatically. Table 1 shows the storage structure of the raw data in database (table structure). The data in storage first ordered by DeviceId, then according to TimeStamp.

Table 1. Data storage structure in database

Field	Type	Null	Key	Default	Extra
DataId	int(10) unsigned	NO	PRI	NULL	Auto_increment
DeviceId	varchar(25)	NO		NULL	
TimeStamp	bigint(25)	NO		NULL	
DataLength	tinyint(3) unsigned	NO		NULL	
RawData	varbinary(255)	NO		NULL	

3.3.2   Introduction of the Memory Module

Whatever data storage pattern (memory, temporary files or databases) it is, it takes much time to extract data through query command from storage. Especially when much query command is executed frequently in a short time, the extraction operation will be repeated, however, in fact a large part of which is duplicated. Because usually it is the index range of query changed rather than the query itself.

In order to eliminate the factor that affects the efficiency, a memory module has been introduced between the display module and the display filter module. The capability of the memory module is that when a query command inputted by a user, whether to extract data from the storage module relies on the memory module. Three judging criteria are as followed: first, query commands with or without changes; second, whether the new data comes; last, the range of data display is or not beyond the scope of the data storage buffer.

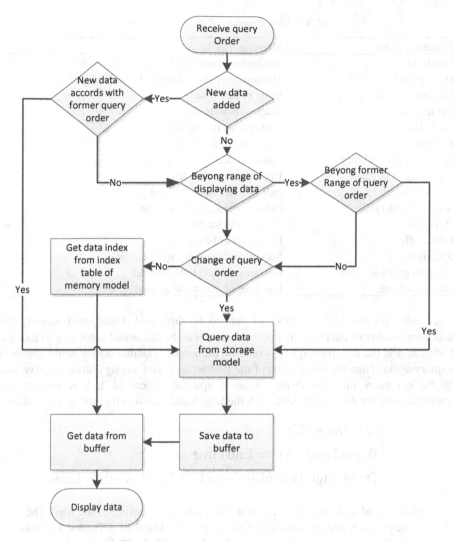

Fig. 3. Flow chart of data query and display

As long as there is one of the three criteria is met, the query command in the memory module will be updated. Meanwhile the table of index data will be built, by which the data will be extracted from the buffer to the storage module. Otherwise, instead of executing the real query operation, the scope of the table of index data, will be changed and then the data will be got directly from the buffer. Fig.3 shows the entire flow of the data query and display.

3.3.3   Improve the Algorithm of Query and Display

In order to ensure fluency of data display, you need the appropriate algorithms of query and display. Firstly, a query structure according to the storage structure should be defined, as shown in Table 2.

Table 2. Data structure of query parameter

Parameter Name	Meaning
DataLimit	Maximum number of data to be displayed
DeviceLimit	Maximum number of device to be displayed
DataIndex	Index of first data
DeviceIndex	Index of first device
BeginTime	First time to receive data
EndTime	Last time to receive data
DataLengthDownLimit	Lower limit of data length
DataLengthUpLimit	Upper limit of data length
SelectedIndex	Index of data to be selected
SelectedDeviceIndex	Index of device to be selected
DeviceID	Table of device ID
DeviceIdIndex	Device ID to be queried
SpecificIndex	Specific index of data to be queried
QueryBeginTime	First time of data to be queried
QueryEndTime	Last time of data to be queried

Secondly, the maximum number of data to be displayed (DataLimit), namely the maximum number of data seen by user each time, can be calculated according to the size of screen. And the data from query should be numbered (DataIndex) for easily obtained by query again. Then the index value of the first term of data among current display area will be got every time the display zone is updated (where id, t, l, is respectively representatively for the current device ID, the length and timestamp of the received data):

$$\begin{cases} id \in DeviceID \\ BeginTime <= t <= EndTime \\ DataLengthDownLimit <= l <= DataLengthUpLimit \end{cases} \quad (1)$$

It will be found and marked in the storage. Finally, according to the mark the next data will be got fast, and the number of data equals to DataLimit. Figure 4 shows the comparison of the performance between old and new query and display algorithms.

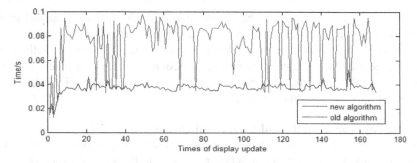

Fig. 4. Compare of performance between old and new query display algorithm

Figure 4 shows the time required for query and display in each time. It prettily proves the query efficiency has been significantly improved and more stable after the query and display algorithm optimized. This means that every query command needs less than 40 milliseconds for response, which makes the instantaneity of the system better.

3.4   Eal-Time Running Effect

After the sniffer has been selected, WSNMVS will automatically connect to the monitoring equipment, and prepare for everything. When sniffing starts, those listening device will transfer the wireless data to the PC client, which will be stored in the corresponding storage. The function of real-time monitoring in multiple different channels is supported by WSNMVS. Fig.5 shows the real-time raw data packets captured from target WSN by WSNMVS.

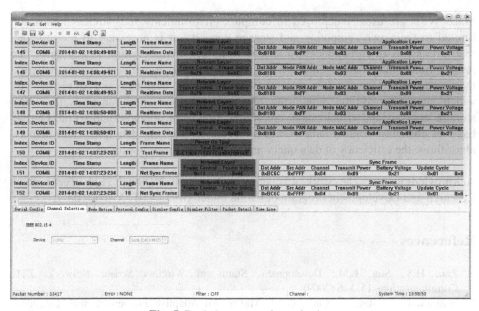

Fig. 5. Real-time network monitoring

Besides the raw data display showed above, WSNMVS also supports the timeline display. Unlike grouping data by the device ID in main interface, the timeline module orders data by time, which shows the sequence between multiple devices, but also can, identify the communication error clearly. As shown in Fig.6.

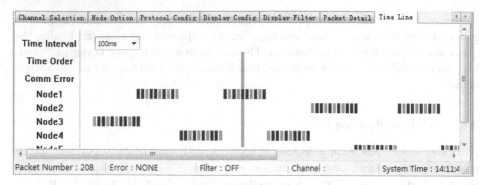

Fig. 6. Time line module for communication diagnosis

4   Conclusion

Since WSN are usually deployed in complex environments and wireless nodes are limited to their own computing and storage capacity, it is necessary to monitor the real-time status of communications and the interactive information to guarantee the proper functioning of the entire WSN. In this paper, a wireless sensor network monitoring and visualization system (WSNMVS) is designed and implemented, which supports real-time monitoring of the target WSN and functions like data storage, analysis and visualization, communication error diagnose and so on. On the basis of changing original data storage structure, optimizing data query and display algorithms, the efficiency of data query and display has been improved.

Acknowledgments. The work was supported by the National High Technology Research and Development Program of China under Grants 2011AA040103-7.

References

1. Zhu, H.S., Sun, L.M.: Development Status of Wireless Sensor Network. ZTE Communications 15, 1–6 (2009)
2. Schmitt, C., Freitag, A., Carle, G.: CoMaDa: An Adaptive Framework with Graphical Support for Configuration, Management, and Data Handling Tasks for Wireless Sensor Networks. In: 9th International Conference on Network and Service Management, pp. 211–218. Zurich (2013)
3. Buschmann, C., Pfisterer, D., Fischer, S., Fekete, S.P., Kroller, A.: SpyGlass: A Wireless Sensor Network Visualizer. ACM SIGBED Rev. 2, 1–6 (2005)
4. Hammoudeh, M., Newman, R., Dennett, C., Mount, S.: Inductive as a support of deductive data visualisation in Wireless Sensor Networks. In: ISCC 2009: IEEE Symposium on Computers and Communications, pp. 277–280, Sousse (2009)
5. Yang, Y., Xia, P., Huang, L., Zhou, L., Xu, Y., Li, X.: SNAMP: A Multisniffer and Multi-view Visualization Platform for Wireless Sensor Networks. In: Proc. the 2006 1ST IEEE Conference on Industrial Electronics and Applications, pp. 1–4. IEEE Computer Society, Washington, DC (2006)

6. Zhong, H., Sun, L.M.: NSSN: A network monitoring and packet sniffing tool for wireless sensor networks. In: 8th International Wireless Communications and Mobile Computing Conference, pp. 537–542. IEEE Press, Limassol (2012)
7. Jeong, Y.S., Chung, Y.J., Park, J.H.: Visualisation of efficiency coverage and energy consumption of sensors in wireless sensor networksusing heat map. IET Communications 5, 1129–1137 (2011)
8. Turon, M.: MOTE-VIEW: A Sensor Network Monitoring and Management Tool. In: 2nd IEEE Workshop on Embedded Networked Sensors, pp. 11–17. IEEE Computer Society, Washington, DC (2005)
9. Koubaa, A., Chaudhry, S., Gaddour, O., Chaari, R., Al-Elaiwi, N., AlSoli, H., Boujelben, H.: Z-monitor: Monitoring and Analyzing IEEE 802.15.4-based Wireless Sensor Networks. In: Proc. the 6th IEEE LCN Workshop on Network Measurements, in Conjunction with LCN 2011, Bonn, Germany (2011)

A Simple Human Learning Optimization Algorithm

Ling Wang[1], Haoqi Ni[1], Ruixin Yang[1], Minrui Fei[1], and Wei Ye[1,2]

[1] Shanghai Key Laboratory of Power Station Automation Technology, School of Mechatronics Engineering and Automation, Shanghai University, Shanghai, 200072, China
[2] Institute of Computer Science, University of Munich, Munich, 80538, Germany
`{wangling,nhq12721182,yrx816,mrfei,yeweiysh}@shu.edu.cn`

Abstract. This paper presents a novel Simple Human Learning Optimization (SHLO) algorithm, which is inspired by human learning mechanisms. Three learning operators are developed to generate new solutions and search for the optima by mimicking the learning behaviors of human. The 0-1 knapsack problems are adopted as benchmark problems to validate the performance of SHLO, and the results are compared with those of binary particle swarm optimization (BPSO), modified binary differential evolution (MBDE), binary fruit fly optimization algorithm (bFOA) and adaptive binary harmony search algorithm (ABHS). The experimental results demonstrate that SHLO significantly outperforms BPSO, MBDE, bFOA and ABHS. Considering the ease of implementation and the excellence of global search ability, SHLO is a promising optimization tool.

Keywords: human learning optimization, meta-heuristic, global optimization, learning operators, optimization algorithm.

1 Introduction

The computational drawbacks of existing derivative-based numerical methods such as complex derivatives, sensitivity to initial values, and the large amount of enumeration memory required have forced researchers to rely on meta-heuristic algorithms to solve complicated optimization problems, such as Genetic Algorithms [1], Ant Colony Optimization [2], Particle Swarm Optimization [3], Harmony Search [4], and Fruit Fly Optimization Algorithms [5]. To effectively and efficiently solve hard optimization problems, new powerful meta-heuristics inspired by nature, especially by biological systems, must be explored, which is a hot topic in evolutionary computation community now [6].

Many human learning activities are similar to the search process of meta-heuristics. For instance, when a person learns how to play Sudoku, he or she repeatedly studies and practices to master and improve new skills and evaluate his or her performance for guiding the following study while meta-heuristics iteratively generate new solutions and calculate the corresponding fitness values for adjusting the following search. In most activities human can solve problems by random learning, individual learning, and social learning. For the example of learning Sudoku again, a person may

M. Fei et al. (Eds.): LSMS/ICSEE 2014, Part II, CCIS 462, pp. 56–65, 2014.

randomly learn due to lack of prior knowledge or exploring new strategies (random learning), learn from his or her previous experience (individual learning) and learn from his or her friends and related books (social learning). Inspired by this simple learning model, a simple human learning optimization algorithm is proposed.

The rest of the paper is organized as follows. Section 2 introduces the idea, operators and implementation of SHLO in detail. Then, the presented SHLO is applied to tackle a set of 0-1 knapsack problems to evaluate its performance in Section 3. Finally, Section 4 concludes the paper.

2   Simple Human Learning Optimization Algorithm

2.1   Initialization

The binary-coding framework is adopted in SHLO, and consequently an individual in SHLO is represented by a binary string as Eq. (1),

$$x_i = \begin{bmatrix} x_{i1} & x_{i2} & \cdots & x_{ij} & \cdots & x_{iM} \end{bmatrix}, x_{ij} \in \{0,1\}, 1 \leq i \leq N, 1 \leq j \leq M \tag{1}$$

where x_i denotes the *i-th* individual, N is the size of the population, and M is the dimension of solutions. Each bit of a binary string is initialized as "0" or "1" randomly, which stands for a basic element of the knowledge or skill that people want to learn and master.

2.2   Learning Operators

2.2.1   Random Learning Operator

At the beginning of learning, people usually learn at random as there is no prior knowledge of problems. In the following studying, due to forgetting, only knowing partial knowledge of problems and other factors, individuals cannot fully replicate previous experience and therefore they still learn with a certain randomness. To emulate these phenomena of randomness in human learning, a simplified random learning operator is developed for SHLO as Eq. (2).

$$x_{ij} = Rand(0,1) = \begin{cases} 0, & 0 \leq rand() \leq 0.5 \\ 1, & else \end{cases} \tag{2}$$

where *rand()* is a stochastic number between 0 and 1.

2.2.2   Individual Learning Operator

Individual learning is defined as the ability to build knowledge through individual reflection about external stimuli and sources [7]. It is very common that people use their own experience and knowledge to avoid mistakes and improve their performance during the process of study. To mimic individual learning of human in SHLO, an individual knowledge database (IKD) is used to store personal best experience as Eq. (3-4).

$$IKD = \begin{bmatrix} ikd_1 \\ ikd_2 \\ \vdots \\ ikd_i \\ \vdots \\ ikd_N \end{bmatrix}, 1 \le i \le N \tag{3}$$

$$ikd_i = \begin{bmatrix} ikd_{i1} \\ ikd_{i2} \\ \vdots \\ ikd_{ip} \\ \vdots \\ ikd_{iL} \end{bmatrix} = \begin{bmatrix} ik_{i11} & ik_{i12} & \cdots & ik_{i1j} & \cdots & ik_{i1M} \\ ik_{i21} & ik_{i22} & \cdots & ik_{i2j} & \cdots & ik_{i2M} \\ \vdots & \vdots & & \vdots & & \vdots \\ ik_{ip1} & ik_{ip2} & \cdots & ik_{ipj} & \cdots & ik_{ipM} \\ \vdots & \vdots & & \vdots & & \vdots \\ ik_{iL1} & ik_{iL2} & \cdots & ik_{iLj} & \cdots & ik_{iLM} \end{bmatrix}, 1 \le p \le L \tag{4}$$

where ikd_i denotes the IKD of person i, L is the pre-defined number of solutions saved in the IKD, and ikd_{ip} stands for the p-th best experience of person i.

When SHLO conducts individual learning, it generates new solutions based on the knowledge in the IKD, which is operated as Eq.(5)

$$x_{ij} = ikd_{ipj} \tag{5}$$

2.2.3   Social Learning Operator

Although a person could learn and solve problems on his or her own experience, i.e. through individual learning, the learning process may be very slow and inefficient if problems are complicated. In the social environment, people can learn from a collective experience through social learning to further develop their ability [8, 9]. In this context, people directly or indirectly transfer knowledge and skills, and hence the efficiency and effectiveness of learning will be improved from experience share [10, 11]. For emulating this efficient learning strategy, the social knowledge data (SKD) is used to reserve the knowledge of the population as Eq. (6)

$$SKD = \begin{bmatrix} skd_1 \\ skd_2 \\ \vdots \\ skd_q \\ \vdots \\ skd_H \end{bmatrix} = \begin{bmatrix} sk_{11} & sk_{12} & \cdots & sk_{1j} & \cdots & sk_{1M} \\ sk_{21} & sk_{22} & \cdots & sk_{2j} & \cdots & sk_{2M} \\ \vdots & \vdots & & \vdots & & \vdots \\ sk_{q1} & sk_{q2} & \cdots & sk_{qj} & \cdots & sk_{qM} \\ \vdots & \vdots & & \vdots & & \vdots \\ sk_{H1} & sk_{H2} & \cdots & sk_{Hj} & \cdots & sk_{HM} \end{bmatrix}, 1 \le q \le H \tag{6}$$

where H is the size of the SKD and skd_q is the q-th solution in SKD.

Based on the knowledge in the SKD, SHLO can perform social learning as Eq. (7) to generate better solutions in the search process.

$$x_{ij} = sk_{qj} \qquad (7)$$

In summary, SHLO uses the random learning operator, individual learning operator and social learning operator to yield new solutions and search for the optima based on the knowledge stored in the IKD and SKD just like human learning and improving skills by these three learning forms, which can be integrated and operated as Eq. (8)

$$x_{ij} = \begin{cases} Rand(0,1), & 0 \leq rand() \leq pr \\ ik_{ipj}, & pr < rand() \leq pi \\ sk_{qj}, & else \end{cases} \qquad (8)$$

where pr is the probability of random learning, and the values of (pi-pr) and (1-pi) represents the probabilities of performing individual learning and social learning, respectively.

2.3   Updating Operation

After all individuals generate new candidate solutions, the fitness of each individual is evaluated according to the pre-defined fitness function which is used to update IKD and SKD for the following search, just like people evaluate their performance of new practices to summarize and update their experience for leaning better in the following steps. For the updating of the IKD, the new generated solution will be stored in the IKD if its fitness value is better than the worst one in the IKD or the current number of solutions in the IKD is less than the pre-defined value. For the updating of SKD, the best solution of the current generation will be saved in the SKD if its fitness value is superior to that of the worst one in the SKD or the current number of solutions in the SKD is less than the pre-defined number. Note that the SKD updates no more than one solution at each iterative step, which can keep a better diversity of the algorithm to avoid the premature.

SHLO runs the learning operators and updates the IKD and SKD repeatedly till it finds the optima of problems or the termination criterions are met. The procedure of SHLO can be illustrated as Fig. 1.

3   Experimental Results and Discussions

To evaluate the performance of the algorithm, SHLO, as well as other four binary-coding optimization algorithms, i.e. binary PSO (BPSO) [12], modified binary differential evolution (MBDE) [13], binary fruit fly optimization algorithm (bFOA) [14] and adaptive binary harmony search algorithm (ABHS) [15], was applied to solve 0-1 knapsack problems (0-1 KPs). For a fair comparison, the recommended parameters of BPSO, MBDE, bFOA and ABHS were used to tackle these problems. As there is no adaptive strategy in the original bFOA and MBDE which significantly spoils their performance on high-dimensional problems, the adaptive strategy is introduced into these two algorithms and the parameters are set based on a parameter study. The parameters of all the algorithms are listed in Table 1. As the benchmark problems are the "single-objective" problems, the sizes of the IKD and SKD are both set to 1 based on trails and error to enhance search efficiency and reduce the cost of computation.

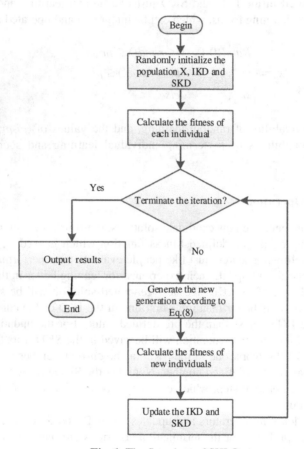

Fig. 1. The flowchart of SHLO

3.1   0-1 Knapsack Problems

Knapsack problems have been studied intensively in the last few decades, attracting both theorists and practitioners. From a practical point of view, knapsack problems can model many application problems such as capital budgeting, cargo loading and cutting stock [16].

Table 1. Parameters settings of SHLO, BPSO, MBDE, bFOA and ABHS

Algorithm	Parameters
SHLO	$pr = \dfrac{5}{M}$, $pi = 0.85 + \dfrac{2}{M}$
BPSO [12]	c_1=1.5, c_2=1.5, w_{min}=0.1, w_{max}=0.9, V_{max}=4, V_{min}=-4
MBDE [13]	F_1=0.5, F_2=0.005, CR_{max}=0.8, CR_{min}=0.2
bFOA [14]	$S = 3$, $L = 3$, $b_{max} = 30$, b_{min}=6
ABHS [15]	$C = 15$; $PAR = 0.2$; $HMS = 30$; $NGC = 20$;

* M is the dimensionality of the solution.

Given a set of n items and each item j having an integer profit p_j as well as an integer weight w_j, the 0-1 knapsack problem (0-1 KP) is defined to choose a subset of items such that their overall profit is maximized while the overall weight does not exceed a given capacity, which can be formulated as Eq. (9)

$$Max\, f(x) = \sum_{j=1}^{N} p_j x_j$$

$$s.t. \begin{cases} \sum_{j=1}^{N} w_j x_j \leq C \\ x_j = 0 \ or \ 1, j = 1,2,...,N \end{cases} \tag{9}$$

where the binary decision variable x_j is used to indicate whether item j is included in the knapsack or not. Without loss of generality, 0-1 KPs assume that all profits and weights are positive and all weights are smaller than the capacity C.

Note that 0-1 KPs are constrained problems, and thus the penalty function as Eq. (10) is adopted to deal with infeasible solutions of which the total weight exceeds the limit C. No heuristic strategy of KPs is introduced in this paper to avoid the influence on the real performance of the algorithm,

$$MaxF(x) = f(x) - \lambda \times \max(0, c)$$

$$c = \sum_{j=1}^{N} w_j x_j - C \tag{10}$$

where λ, called the penalty coefficient, is a big constant which guarantees that the fitness of the best infeasible solution is poorer than that of the worst feasible solution.

A set of 0-1 knapsack problems were devised to validate SHLO as well as BPSO, MBDE, bFOA and ABHS. The numbers of items were set to 100, 200, 400, 600, 800, 1000, 1200, and 1500, and two instances of each scale were generated to test the performance of the algorithms more exactly. The weight w_j and the profit p_j were generated according to [16], i.e. from 5 to 20 and from 50 to 100, respectively. The weight capability C was set to 1000, 2400, 4000, 6000, 8000, 10000, 14000, and 16000, respectively. The population size and maximum generation of all the algorithms were set to 100, 10000 and 200, 40000 for the instances less and no less than 1000 items, respectively. The experimental results are presented in Table 2 and Table 3.

Table 2. The results of SHLO, BPSO, MBDE, bFOA and ABHS on low-dimensional 0-1 knapsack problems

	Algorithm	Best	Mean	Worst	Std
	SHLO	**6526**	**6526.0**	**6526**	**0.000**
	bFOA	6526	6525.9	6524	0.889
Kp100.1	ABHS	**6526**	**6526.0**	**6526**	**0.000**
	MBDE	6526	6524.2	6523	1.962
	BPSO	6526	6525.6	6522	1.265
	SHLO	**6824**	**6824.0**	**6824**	**0.000**
	bFOA	6824	6823.8	6823	0.731
Kp100.2	ABHS	**6824**	**6824.0**	**6824**	**0.000**
	MBDE	6824	6823.2	6822	0.872
	BPSO	6824	6822.9	6822	1.039
	SHLO	**14999**	**14999.0**	**14999**	**0.000**
	bFOA	14999	14998.0	14993	1.944
Kp200.1	ABHS	14999	14998.6	14997	0.843
	MBDE	**14999**	**14999.0**	**14999**	**0.000**
	BPSO	14999	14998.8	14997	0.632
	SHLO	**14791**	**14791.0**	**14791**	**0.000**
	bFOA	14791	14786.4	14780	4.879
Kp200.2	ABHS	14791	14787.7	14784	3.613
	MBDE	**14791**	**14791.0**	**14791**	**0.000**
	BPSO	**14791**	**14791.0**	**14791**	**0.000**
	SHLO	**27100**	27099.1	**27095**	1.524
	bFOA	27100	27094.9	27091	3.381
Kp400.1	ABHS	27097	27096.0	27086	4.879
	MBDE	27099	27095.4	27092	2.119
	BPSO	27100	27097.5	27092	3.126
	SHLO	27099	26448.7	26209	7.969
	bFOA	26657	26454.2	26253	1.643
Kp400.2	ABHS	26859	26425.5	26237	8.679
	MBDE	27099	26453.3	26092	2.119
	BPSO	**27099**	**26461.8**	**26250**	**3.512**
	SHLO	**40216**	**40216.0**	**40216**	**0.000**
	bFOA	40216	40212.4	40202	4.648
Kp600.1	ABHS	40216	40210.3	40204	8.644
	MBDE	40216	40212.7	40204	4.596
	BPSO	**40216**	**40216.0**	**40216**	**0.000**
	SHLO	**39602**	**39601.2**	**39594**	17.769
	bFOA	39602	39601.0	39597	12.236
Kp600.2	ABHS	39602	39599.5	39531	23.282
	MBDE	39602	39600.8	39583	11.155
	BPSO	39602	39600.3	39587	12.284

Table 3. The results of SHLO, BPSO, MBDE, bFOA and ABHS on high-dimensional 0-1 knapsack problems

	Algorithm	Best	Mean	Worst	Std
	SHLO	53855	53851.8	53837	5.473
	bFOA	53855	53845.8	53832	7.451
Kp800.1	ABHS	53850	53844.1	53822	4.838
	MBDE	53850	53841.9	53829	6.045
	BPSO	**53855**	**53851.9**	**53843**	**3.510**
	SHLO	**52705**	**52701.5**	**52692**	**8.697**
	bFOA	52703	52683.8	52667	13.312
Kp800.2	ABHS	52695	52691.3	52690	6.883
	MBDE	52692	52690.7	52689	1.528
	BPSO	52705	52698.3	52688	7.329
	SHLO	**66882**	**66857.4**	**66844**	**13.082**
	bFOA	66867	66837.7	66829	14.930
Kp1000.1	ABHS	66855	66840.2	66814	18.520
	MBDE	66860	66849.4	66828	9.009
	BPSO	66853	66830.8	66801	15.803
	SHLO	**66905**	**66899.2**	**66898**	**23.122**
	bFOA	66891	66867.4	66849	17.097
Kp1000.2	ABHS	66900	66887.1	66830	33.084
	MBDE	66905	66895.3	66893	16.429
	BPSO	66853	66841.6	66823	11.393
	SHLO	**86823**	**86820.7**	**86805**	**5.559**
	bFOA	86805	86796.8	86776	10.497
Kp1200.1	ABHS	86811	86710.5	86703	42.393
	MBDE	86812	86796.9	86776	10.775
	BPSO	86823	86810.2	86798	9.578
	SHLO	**86715**	**86702.8**	**86698**	**8.927**
	bFOA	86701	86700.2	86694	1.304
Kp1200.2	ABHS	86705	86698.8	86677	4.083
	MBDE	86701	86694.7	86686	6.506
	BPSO	86715	86702.2	86698	7.430
	SHLO	**102657**	**102622.4**	**102551**	**28.982**
	bFOA	102608	102598.0	102586	29.541
Kp1500.1	ABHS	102619	102575.9	102534	27.843
	MBDE	102603	102583.2	102563	12.674
	BPSO	102602	102523.1	102473	41.089
	SHLO	**104860**	**104851.8**	**104748**	**30.506**
	bFOA	104860	104833.2	104742	36.573
Kp1500.2	ABHS	104840	104820.8	104761	29.835
	MBDE	104831	104816.3	104754	19.655
	BPSO	104824	104801.6	104770	23.384

As can be seen in Table 2 and Table 3, SHLO finds the best numerical results on 14 out 16 instances and is only inferior to BPSO on Kp400.2 and Kp800.1, bFOA on Kp400.2, and MBDE on Kp400.2, respectively. For the instances in which the items are less than 200, all the algorithms can find the best-known solutions and achieve satisfactory results. When items increase to 1500, BPSO, ABHS and MBDE cannot reach the best-known values any more. Based on the ranking results given in Table 4, it is clear that SHLO outperforms BPSO, MBDE, bFOA and ABHS on the 0-1 knapsack problems.

Table 4. The ranks of SHLO, BPSO, MBDE, bFOA and ABHS on 0-1 knapsack problems

	SHLO	bFOA	ABHS	MBDE	BPSO
Kp100.1	1	3	1	5	4
Kp100.2	1	3	1	4	5
Kp200.1	1	5	4	1	3
Kp200.2	1	5	4	1	1
Kp400.1	1	5	3	4	2
Kp400.2	4	2	5	3	1
Kp600.1	1	4	5	3	1
Kp600.2	1	2	5	3	4
Kp800.1	2	3	4	5	1
Kp800.2	1	5	3	4	2
Kp1000.1	1	4	3	2	5
Kp1000.2	1	4	3	2	5
Kp1200.1	1	4	5	3	2
Kp1200.2	1	3	4	5	2
Kp1500.1	1	2	4	3	5
Kp1500.2	1	2	3	4	5
Average	1.25	3.50	3.56	3.25	3.00

4   Conclusion

Inspired by the mechanisms of human learning, this paper presents a novel meta-heuristic algorithm, named simple human learning optimization (SHLO), in which three learning operators, i.e. the random learning operator, individual learning operator, and social learning operator are developed by mimicking human learning behaviors to generate new solutions and search for the optimal solution of problems. To evaluate the performance of the proposed algorithm, low-dimensional and high-dimensional 0-1 KP benchmarks are adopted as benchmark problems to test SHLO. For a fair comparison, other four binary-coding optimization algorithms, i.e. BPSO, MBDE, bFOA, and ABHS, are also used to solve the benchmark problems with the recommended parameters. The experimental results demonstrate that SHLO outperforms BPSO, MBDE, bFOA and ABHS.

Acknowledgments. This work is supported by National Natural Science Foundation of China (Grant No. 61304031, 61374044 & 61304143), and Innovation Program of Shanghai Municipal Education Commission (14YZ007).

References

1. Goldberg, D.E.: Genetic Algorithms in Search, Optimization, and Machine Learning. Addison Wesley, Boston (1989)
2. Colorni, A., Dorigo, M., Maniezzo, V.: Distributed Optimization by Ant Colonies. In: Proceedings of First European Conference on Artificial Life. MIT Press/Bradford Books, Paris (1991)
3. Kennedy, J., Eberhart, R.: Particle Swarm Optimization. In: Proceedings of IEEE International Conference on Neural Network vol. 4, pp. 1942–1948. IEEE Press (1995)
4. Lee, K.S., Geem, Z.W.: A New Meta-heuristic Algorithm for Continuous Engineering Optimization: Harmony Search Theory and Practice. Computer Methods in Applied Mechanics and Engineering 194, 3902–3933 (2005)
5. Pan, W.T.: A New Fruit Fly Optimization Algorithm: Taking the Financial Distress Model as an Example. Knowledge-Based Systems 26, 69–74 (2012)
6. Fister, J.I., Yang, X.S., Fister, I.: A Brief Review of Nature-Inspired Algorithms for Optimization. Elektrotehnski Vestnik 80, 1–7 (2013)
7. Forcheri, P., Molfino, M.T., Quarati, A.: ICT Driven Individual Learning: New Opportunities and Perspectives. Educational Technology & Society 3, 51–61 (2000)
8. Ickes, W., Gonzalez, R.: "Social" Cognition and Social Cognition: From the Subjective to the Intersubjective. Small Group Research 25, 294–315 (1994)
9. Ellis, A.P.J., Hollenbeck, J.R., Ilgen, D.R., Porter, C.O.L.H., West, B.J., Moon, H.: Team learning: Collectively connecting the Dots. Journal of Applied Psychology 88, 821–835 (2003)
10. Andrews, K.M., Delahaye, B.L.: Influences on Knowledge Processes in Organizational Learning: The Psychosocial Filter. Journal of Management Studies 37, 797–810 (2002)
11. McEvily, S.K., Chakravarthy, B.: The Persistence of Knowledge-based Advantage: An Empirical Test for Product Performance and Technological Knowledge. Strategic Management Journal 23, 285–305 (2002)
12. El-Maleh, A.H., Sheikh, A.T., Sait, S.M.: Binary Particle Swarm Optimization (BPSO) Based State Assignment for Area Minimization of Sequential Circuits. Applied Soft Computing 13, 4832–4840 (2013)
13. Wu, C.Y., Tseng, K.Y.: Topology Optimization of Structures Using Modified Binary Differential Evolution. Structural and Multidisciplinary Optimization 42, 939–953 (2010)
14. Wang, L., Zheng, X., Wang, S.: A Novel Binary Fruit Fly Optimization Algorithm for Solving the Multidimensional Knapsack Problem. Knowledge-Based Systems 48, 17–23 (2013)
15. Wang, L., Yang, R., Xu, Y.: An Improved Adaptive Binary Harmony Search Algorithm. Information Sciences 232, 58–87 (2013)
16. Zou, D., Gao, L., Li, S., Wu, J.: Solving 0-1 knapsack problem by a novel global harmony search algorithm. Applied Soft Computing 11, 1556–1564 (2011)

An Improved Non-dominated Sorting Genetic Algorithm for Multi-objective Optimization Based on Crowding Distance

Tian-liang Xia and Shao-hua Zhang

Key Laboratory of Power Station Automation Technology,
School of Mechatronic Engineering and Automation,
Shanghai University, Shanghai 200072, China
{xiatianliang123,eeshzhan}@126.com

Abstract. An improved non-dominated sorting genetic algorithm (INSGA) is introduced for multi-objective optimization. In order to keep the diversity of the population, a modified elite preservation strategy is adopted and the evaluation of solutions' crowding degree is integrated in crossover operations during the evolution. The INSGA is compared with the NSGA-II and other algorithms by applications to five classical test functions and an environmental/economic dispatch (EED) problem in power systems. It is shown that the Pareto solution obtained by INSGA has a good convergence and diversity.

Keywords: genetic algorithm, crowding distance, elite preservation, multi-objective optimization.

1 Introduction

Multi-objective optimization is an integral part of optimization activities and has a tremendous practical importance, since almost all real-world optimization problems are ideally suited to be modeled using multiple conflicting objectives [1]. As such, how to optimize these objectives simultaneously becomes a very important issue.

Over the past decades, a number of multi-objective evolutionary algorithms (MOEAs) have been developed and practically applied [2-4]. Among these algorithms, the genetic algorithm (GA) is one of the most commonly applied evolutionary optimization approaches. Especially, the non-dominated sorting genetic algorithm (NSGA) and its later improved version (NSGA-II) proposed in [5] are widely applied in real-life multi-objective optimization problems. In NSGA-II, a fast non-dominated sorting approach and elite preservation strategy is introduced to improve the efficiency of the algorithm, and crowding distance comparison between solutions maintains diversity of the population. However, the NSGA-II does not assure non-elite solutions be part of the population, which will suppress the population diversity and global convergence of the algorithm.

In this paper, we use a modified elite preservation strategy. It assures non-elite solutions be part of the population. In addition, the evaluation of solutions' crowding

M. Fei et al. (Eds.): LSMS/ICSEE 2014, Part II, CCIS 462, pp. 66–76, 2014.

degree is integrated in crossover operations during the evolution. By comparing the crowding distance of parent solutions in crossover operation, offspring solution is mainly formed by parents with better diversity. From the simulation results, we illustrate that the Pareto solutions obtained by the proposed algorithm are more uniformly distributed and the algorithm have a good global convergence.

2   Multi-objective Optimization Problem (MOP)

Assume that a multi-objective optimization problem is consisted of n decision variables, m optimization objectives, p inequality constraints and q equality constraints. A typical MOP can be formulated as follows:

$$\min_{x \in D} F(x) = [f_1(x), f_2(x), \cdots, f_m(x)]$$
$$s.t. \quad g_i(x) \leq 0, \quad i = 1, 2, \cdots, p$$
$$h_j(x) = 0, \quad j = 1, 2, \cdots, q \tag{1}$$
$$x = (x_1, x_2, \cdots, x_n) \in D$$
$$x_i^{\min} \leq x_i \leq x_i^{\max}, i = 1, 2, \cdots, n$$

Here, D is the decision region defined by decision variables' lower boundary $x_i^{\min}$ and upper boundary $x_i^{\max}$. To aid descriptions, we introduce the following definitions regarding multi-objective optimization.

Definition 1. Solution $x = \{x_1, \cdots, x_n\}$ is said to dominate solution $y = \{y_1, \cdots, y_n\}$, denoted as $x \succ y$, if and only if

$$\forall i \in \{1, 2, \cdots, m\}, f_i(x) \leq f_i(y),$$
$$and \quad \exists j \in \{1, 2, \cdots, m\}, f_j(x) < f_j(y). \tag{2}$$

Reciprocally, solution y is said to be dominated by solution x, denoted as $y \prec x$.

Definition 2. Solution x^* is said to be a Pareto optimum only if there is no such a solution $x \in D$ existed that makes $x \succ x^*$, where $x^* = \{x_1^*, x_2^*, \cdots, x_n^*\}$ and $x = \{x_1, x_2, \cdots, x_n\}$. All Pareto optima constitute a Pareto optimal set, denoted as X^*.

Definition 3. Pareto-optimal front is defined as $PF = \{F(x^*) | x^* \in X^*\}$, i.e., the mapping of Pareto-optimal set in the objective space.

3   Improved Non-dominated Sorting Genetic Algorithm (INSGA)

3.1   Crowding Distance

Crowding distance is used to estimate the density of solutions surrounding a particular solution in the population. The solution with a relatively large crowding distance in the same Pareto front indicates that it has a relatively low density of solutions surrounded and a better diversity. Contrarily, less crowding distance indicates worse diversity of the solution. As shown in figure 1, the crowding distance calculation requires sorting the population according to the value of each objective function in ascending order. Then, for each objective function, the boundary solutions (solutions with the smallest and largest objective values) are assigned an infinite distance value. All other intermediate solutions are assigned a distance value equal to the absolute normalized difference in the function values of two adjacent solutions. The overall crowding distance value is calculated as the sum of individual distance values corresponding to each objective. Each objective function is normalized before calculating the crowding distance.

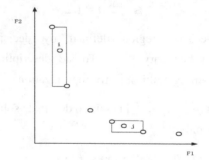

Fig. 1. Calculation of crowding distance

The calculation of solution i's crowding distance is as follows:

$$d(i) = \sum_{j=1}^{m} \frac{\Delta f_{i,j}}{f_{j\max} - f_{j\min}} \quad . \tag{3}$$

where $\Delta f_{i,j}$ is the absolute normalized difference in j^{th} objective function values of solution $i's$ two adjacent solutions, m is the number of optimization objectives, $f_{j\max}$, $f_{j\min}$ is the maximum and minimum value of jth objective function in the same Pareto front, respectively.

3.2   Elite Preservation Strategy

Elite preservation strategy is commonly adopted in evolutionary algorithm to ensure the elite solutions in the population can be selected in the next generation, and it is a common practice to copy the elite solutions to next generation directly, just as presented in NSGA-II. By observing the NSGA-II algorithm experimental test, we found that the population becomes all non-dominated solutions composed after about 30 iterations, which means there are no other solutions but elite ones are involved and participate in evolution operations. This may restrain the diversity of the population and lead the algorithm more likely to converge to local optimum. As such, we adopt a modified elite preservation strategy. First, the solutions of the population are clustered according to their non-domination ranks and form a number of sub-populations. By pre-assigning a series of ratios corresponding to these sub-populations, a regular number of solutions in each sub-population will be selected to form the new population. It makes non-elite solutions be part of the population and improves the diversity of the population. The ratio assignment equation is as follows [8]:

$$N(i) = \frac{P(1-r)}{1-\sin r^k} \sin r \qquad (4)$$

where $N(i)$ is the maximum number of solutions that can be selected in sub-population with non-domination rank equal to i. P is the size of the population, k represents the number of non-domination rank that exists in the combined population with size of $2P$, $r(0<r<1)$ is a user define parameter. For $P=100$, $k=8$, $r=0.65$, the distribution diagram of $N(i)$ is shown in figure 2.

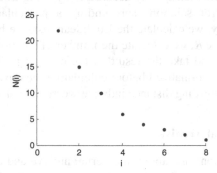

Fig. 2. Distribution curve of $N(i)$

As we can see from figure 2, this method makes non-elite solutions be part of the population and maintains the diversity of the population. Some modifications should be noted.

1) We can determine the number of solutions that can be selected in each sub-population with non-domination rank equal to i by Equation (4). When the size of distributed population (the sum of $N(i)$) is less than P, we fill the deficit with solutions with higher non-domination rank.

2) If $N(i)$ exceeds the number of solutions in non-domination rank i, we make that extra portion redistribute to the solutions with lower non-domination rank.

We can deduce from Equation (4) that the number of non-dominated solutions in the population must be less than P. In order to search for more elite solution, we define a parameter NI to denote the number of iterations of using this new elite preservation strategy. The algorithm adopts the modified strategy from the beginning to iteration NI, after that the original elite preservation strategy presented in [7] is adopted.

3.3   Crossover

The algorithm use a $n{\times}L$ matrix as solution expression by binary encoding where n is the number of variables and L is the encoding length. The window crossover operator as mentioned in [6] is used for the binary crossover. It works by randomly selecting two parents from the mating pool and then randomly selecting a window size. The entries within the window portion are exchanged between the two parents to generate offspring. To compare the fitness of two parent solutions, it is very common by taking the solutions' non-domination rank or function value as the measurement. However, it ignores the distribution property of solutions and decreases the diversity of the algorithm to some extent. Therefore, we compare the non-domination rank of solutions during the comparison in the first place, the one with higher rank gets more portion be presented in the offspring. If the two parent solutions have the same non-domination rank, we compare their crowding distance.

Because of the distribution of population is irregular during the iteration, it cannot reflect the solution's distribution density accurately by using Equation (4). To get an estimate of the density of solutions surrounding a particular solution i in the population in a better way, we calculate the Euclidean distance between solution i to others. By given a set value R, we calculate the number of solutions whose distance to solution i is less than R, and take the result as the crowding distance of solution i. Each objective function is normalized before calculating the crowding distance. Note that lower values of the crowding distance indicate worse diversity of population.

3.4   Update of External Archive

The non-dominated solutions are stored in external archive and crowding distance is used to maintain diversity. Initially, this archive is empty. As the evolution moves, good solutions enter the archive. When there is a new solution obtained by evolution in an iteration of the algorithm, the following operations will be applied to update the external archive: 1) If the new solution is dominated by any solution in the external archive, it will be discarded; 2) If there are solutions in the external archive dominated by the new solution, remove them from the archive while the new solution will be added into the archive; 3) If the new solution is non-dominated with all the solutions in the external archive, it will be added into the archive. Finally, when the external archive reaches its maximum allowed capacity, the one with the lowest crowding distance is removed from the archive and add the new solution into the archive.

3.5   Main Process of the INSGA

Step 1: Initialization. A random parent population is created. The iteration count of *gencnt* is set to 0 and the maximum iteration number is *maxgen*.

Step 2: *gencnt=gencnt+1*. If *gencnt>maxgen*, the iteration stops, otherwise, calculate the objective function value of each solution.

Step 3: Sorting the population based on the non-domination. Each solution is assigned a fitness value equal to its non-domination rank.

Step 4: Crossover; Mutation.

Step 5: Update the external archive. If *gencnt=1*, then generate the archive within the current population.

Step 6: Selection. If *gencnt<NI*, the modified elite preservation strategy is adopted, otherwise, use the original one. Jump to step 2.

4   Experiments and Discussions

4.1   Metrics

There are some metrics that can be used to quantify the quality of the Pareto front obtained by the algorithms. The following metrics will be used in this paper: Spacing and Maximum spread.

4.1.1   Spacing
The goal is to measure the spread (distribution) of non-dominated solutions throughout the Pareto front. It can be evaluated by Equation (5):

$$SP = \sqrt{\frac{1}{ns-1}\sum_{i=1}^{ns}(\overline{d} - d_i)^2} \ .$$ (5)

where $d_i = \min_j (\sum_{k=1}^{m}|f_k^i - f_k^j|)$ $i, j = 1, 2, 3, \cdots, ns, i \neq j$. $\overline{d} = \frac{1}{n}\sum_{i=1}^{ns}d_i$, ns is the number of solutions in the Pareto front, d_i is the total distance between solution i to its adjacent solutions on every objective dimension and m is the number of objectives. A lower value of *SP* indicates that the algorithm is able to maintain a better spread of solutions.

4.1.2   Maximum Spread
This metric is to evaluate maximum extension covered by the non-dominated solutions in the Pareto front. In a two-objective problem, the maximum spread corresponds to the Euclidean distance between the two farther solutions.

$$MS = \sqrt{\sum_{k=1}^{m} (|\max_{i=1}^{ns} f_k^i(x) - \min_{i=1}^{ns} f_k^i(x)|)^2} \quad . \tag{6}$$

One should note that higher values of *MS* indicate better performance.

4.2   Comparative Numerical Study

4.2.1   Test Functions

In order to test the performance of algorithm, we used 5 classical test functions described in Table 1 as benchmark functions. The algorithm is compared with NSGA-II, both binary coded. We have used 10 bits to code each decision variable. The population size of *popsize*=100, maximum iteration number of *maxgen*=500. The crossover probability of P_c=0.9 and a mutation probability of P_m=0.1. The size of external archive is 200, *NI*=200. The test runs 20 times on each case.

Table 1. Test problems used in the study

Problem	Objective functions	Variable bounds
ZDT1	$f_1(x) = x_1$ $f_2(x) = g(x)[1 - \sqrt{x_1/g(x)}]$ $g(x) = 1 + \dfrac{9}{n-1}\sum_{i=2}^{n} x_i$	$[0,1]^n$ $n=30$
ZDT2	$f_1(x) = x_1$ $f_2(x) = g(x)[1 - (x_1/g(x))^2]$ $g(x) = 1 + \dfrac{9}{n-1}\sum_{i=2}^{n} x_i$	$[0,1]^n$ $n=30$
ZDT3	$f_1(x) = x_1$ $f_2(x) = g(x)[1 - \sqrt{x_1/g(x)} - x_1 \sin(10\pi x_1)/g(x)]$ $g(x) = 1 + \dfrac{9}{n-1}\sum_{i=2}^{n} x_i$	$[0,1]^n$ $n=30$
ZDT4	$f_1(x) = x_1$ $f_2(x) = g(x)[1 - \sqrt{x_1/g(x)}]$ $g(x) = 1 + 10(n-1) + \sum_{i=2}^{n} [x_i^2 - 10\cos(4\pi x_i)]$	$0 \le x_1 \le 1$ $-5 \le x_i \le 5$ $i = 2,3,\cdots,n$ $n = 10$
KUR	$f_1(x) = \sum_{i=1}^{n-1} [-10\exp(-0.2\sqrt{x_i^2 + x_{i+1}^2})]$ $f_2(x) = \sum_{i=1}^{n} (|x_i^{0.8}| + 5\sin x_i^3)$	$[-5,5]^n$ $n=3$

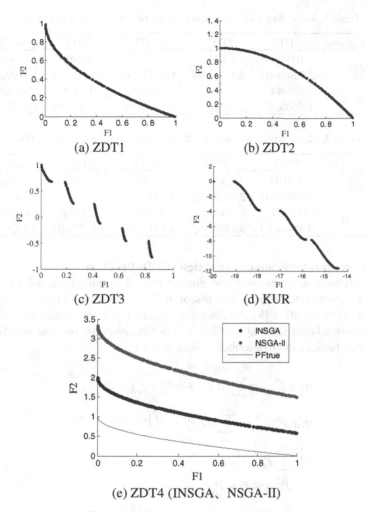

Fig. 3. Pareto front obtained by INSGA

We can see that the Pareto fronts all converge to global optimum on ZDT1, ZDT2, ZDT3 and KUR from Fig. 3. ZDT4 is very difficult to optimize because of its local optimums, both algorithms didn't converge to the global optimum. But we can see that the front obtained by INSGA is better than NSGA-II from Fig. 3 (e).

Table 2 shows that INSGA performs better than NSGA-II in all test problems. INSGA also has a better robustness than NSGA-II. From Table 3 we can see that the numerical difference of mean values of *MS* is very small, but INSGA performs more stable with lower standard deviation.

We also compared INSGA with several other evolutionary algorithms (MOPSO, m-DNPSO, MOPSO-CDLS, CSS-MOPSO, MOPSO-CDR) mentioned in literature [7]. On most test problem, INSGA performs better in diversity metric *SP* than other algorithms, only CSS-MOPSO in ZDT1 and m-DNPSO, MOPSO-CDR in ZDT4 are better than INSGA.

Table 2. Mean (first rows) and standard deviation (second rows) of *SP*

Test Functions	ZDT1	ZDT2	ZDT3	ZDT4	KUR
INSGA	0.0030	0.0031	0.0032	0.0051	0.0643
	2.7246e-04	2.3501e-04	1.0173e-03	7.0686e-04	1.5103e-03
NSGA-II	0.0044	0.0047	0.0055	0.0055	0.0659
	1.5600e-04	2.4387e-04	9.1802e-03	9.7126e-04	1.6117e-03

Table 3. Mean (first rows) and standard deviation (second rows) of *MS*

Test Functions	ZDT1	ZDT2	ZDT3	ZDT4	KUR
INSGA	1.4100	1.4112	1.9658	2.2048	12.9023
	5.6785e-03	2.9165e-03	3.5873e-03	2.0236e-01	6.4018e-03
NSGA-II	1.4171	1.4146	1.9451	2.2064	12.8966
	9.4946e-03	8.7616e-03	9.1377e-02	3.1652e-01	7.1529e-03

4.2.2   Environmental/Economic Dispatch (EED) Problem

With an increasing concern over the environmental pollution caused by thermal power plants, environmental/economic dispatch (EED) problem in power systems has drawn much more attention [9]. A reasonable dispatch scheme would not only result in great economic benefit, but also reduce the pollutants emission. The simplified bi-objective EED model can be described as follows:

$$\min F_{fuel} = \sum_{i=1}^{ng} (a_i P_i^2 + b_i P_i + c_i).$$

$$\min F_{emission} = \sum_{i=1}^{ng} (\alpha_i P_i^2 + \beta_i P_i + \gamma_i). \tag{7}$$

$$s.t. \sum_{i=1}^{ng} P_i = P_D.$$

$$P_{i\min} \leq P_i \leq P_{i\max}.$$

Where ng is the number of committed generation units; a_i, b_i and c_i are the fuel cost coefficients of ith unit; α_i, β_i and γ_i are coefficients of the ith unit's emission characteristics; P_i is the output of ith unit; P_D is the total load demand; P_{imin} and P_{imax} is the minimum and maximum output limit(MW) of ith unit, respectively.

To validate the feasibility of the proposed INSGA algorithm for the EED problems, we apply it to a 6-unit test system. The study period is 1 hour. The fuel cost, emission coefficients and output limits of the units are given in Table 4.

Both algorithms are binary coded. We have used 30 bits to code each decision variable which is the generation output of units. Solutions in the population are repaired using priority list (PL) [10] based on either fuel cost or emission coefficients (with equal probability) to satisfy the load demand equality constraint. The population size of *popsize*=100, *maxgen*=200, P_c=0.9 and P_m=0.1. The size of external archive is 100, *NI*=100. The test runs 20 times for each algorithm.

As we can see from Table 5, INSGA performs better than NSGA-II with more uniformly distributed and better spread Pareto solutions. INSGA is also more stable than NSGA-II.

Table 4. Test results of *SP* and *MS* obtained by INSGA and NSGA-II

Algorithm	SP				MS			
	Max	Min	Ave	Std	Max	Min	Ave	Std
INSGA	1.429	0.684	0.997	0.161	203.80	193.05	200.62	3.00
NSGA-II	1.790	0.931	1.102	0.181	211.99	151.41	198.27	11.63

Table 5. Fuel cost, emission coefficients and output limits of units

		G_1	G_2	G_3	G_4	G_5	G_6
Fuel Cost	a_i	0.15247	0.10587	0.02803	0.03546	0.02111	0.01799
	b_i	38.53973	46.15916	40.39655	38.30553	36.32782	38.27041
	c_i	756.7989	451.3251	1049.3251	1243.5311	1658.5696	1356.6592
Emission	α_i	0.00419	0.00419	0.00683	0.00683	0.00461	0.00461
	β_i	0.32767	0.32767	-0.54551	-0.54551	-0.51116	-0.51116
	γ_i	13.8593	13.8593	40.2669	40.2669	42.8953	42.8953
Output	P_{imin}	10	10	40	35	130	125
limits	P_{imax}	125	150	250	210	325	315

5   Conclusion

We have proposed an improved non-dominated genetic algorithm based on crowding distance. By adopting a modified elite preservation strategy, it improves the diversity of the population and global search ability of the algorithm. We introduce the crowding distance comparison into crossover operation to maintain solutions' dispersed distribution. According to the numerical test results, we believe that the INSGA is able to converge to the true Pareto front efficiently with a better distribution of solutions.

References

[1] Deb, K.: Multi-objective optimization. Search Methodologies, pp. 403–449. Springer US (2014)
[2] Deb, K.: Multi-objective optimization using evolutionary algorithms. John Wiley & Sons, Chichester (2001)
[3] Fonseca, C.M., Fleming, P.J.: Genetic Algorithms for Multi-objective Optimization: Formulation Discussion and Generalization. In: ICGA, vol. 93, pp. 416–423 (1993)
[4] Zitzler, E.: Evolutionary algorithms for multi-objective optimization: Methods and applications. Shaker, Ithaca (1999)

[5] Deb, K., Pratap, A., Agarwal, S., et al.: A fast and elitist multi-objective genetic algorithm: NSGA-II. IEEE Transactions on Evolutionary Computation 6(2), 182–197 (2002)

[6] Valenzuela, J., Smith, A.E.: A seeded memetic algorithm for large unit commitment problems. Journal of Heuristics 8(2), 173–195 (2002)

[7] Santana, R.A., Pontes, M.R., Bastos-Filho, C.J.A.: A multiple objective particle swarm optimization approach using crowding distance and roulette wheel. In: Ninth International Conference on Intelligent Systems Design and Applications, ISDA 2009, pp. 237–242. IEEE (2009)

[8] Wang, L.: Research on Genetic Algorithms for Multi-Objective Optimization Algorithms. Wuhan University of Technology, Wuhan (2006)

[9] Zahavi, J., Eisenberg, L.: An application of the Economic-environmental power dispatch. IEEE Transactions on Systems, Man and Cybernetics 7(7), 523–530 (1977)

[10] Senjyu, T., Shimabukuro, K., Uezato, K., et al.: A fast technique for unit commitment problem by extended priority list. IEEE Transactions on Power Systems 18(2), 882–888 (2003)

Comprehensive Analysis
of Cooperative Particle Swarm Optimization
with Adaptive Mixed Swarm

Jing Jie*, Beiping Hou , Hui Zheng, and Xiaoli Wu

College of Automation & Electronic Engineering,
Zhejiang University of Science and Technology,
310023, No. 318, Liuhe Road, Hangzhou City, P.R. China
Jjing277@sohu.com

Abstract. Inspired by the collective intelligence of natural mixed flocking, the paper develops a mixed swarm cooperative search model for particle swarm optimization(MCPSO). Firstly, makes some analysis about the hinting principles and search mechanism behind the natural mixed flocking, and proposes the construction of mixed swarm for optimization. Secondly, introduces the mixed swarm into PSO and researches the main search behaviors of MCPSO, including coarse search and fine search, cooperative search and learning. Finally, the proposed MCPSO was applied to some well-known benchmarks. The experimental results and relative analysis show mixed swarm cooperative search mechanism can greatly benefit the global optimization performance of PSO.

Keywords: swarm intelligence, particle swarm optimization, mixed swarm, cooperative search.

1 Introduction

Swarm intelligence is a new fruit inspired by the social behavior lying in certain biologic systems in nature, for example: bird flocking, fish schooling or ant colonizing etc... Generally, the individual in those biologic systems is of low intelligence and is small or weak in nature, but a swarm of them as a whole will emerge high collective intelligence and is robust or rival with startling complex behavior. Particle swarm optimization(PSO) is a notable optimization algorithm based on swarm intelligence. Since the original version of PSO is introduced in 1995 by Kennedy and Eberhart[1], it has attracted lots of researchers from different background around the world, and has been applied successfully in many areas recently [2-6]. However, like general evolutionary algorithms, PSO also suffers from the premature convergence problem, especially for the large scale and complex problems.

---

* Corresponding author.

M. Fei et al. (Eds.): LSMS/ICSEE 2014, Part II, CCIS 462, pp. 77–86, 2014.

Many techniques have been proposed to improve the original PSO. Shi and Eberhart introduced a parameter of inertia weight to control the search of PSO[7]; Ratnaweera developed PSO with time-varying acceleration coefficients to balance the local and the global search ability[8]. Cui improved PSO using fitness landscape [9]. Jie proposed a knowledge-based cooperative particle swarm optimization to keep the balance of the exploration and the exploitation of the algorithm[10]; Van den Bergh[11] also used multiple swarms to optimize different components of the solution vector cooperatively. Besides that, many researches have been done to analyze the behavior and the convergent ability of PSO. Clerc[2] made an analysis of the stability properties of the algorithm based on a simplification of the original PSO, and proposed a set of coefficients to control the system's convergent tendencies. Later, Kadirkamanathan[12] carried out a stability analysis of the particle dynamic based on the passive theorem and Lyapunov stability, and concluded that the PSO system is controllable and observable. He further pointed out the primary aim of PSO is optimization, not only maintaining its system stability.

In order to develop new valid model of PSO, we try to go back and seek help from some natural biological flocking here. In nature, it's not uncommon to find that birds of several species flocking come together to be a mixed-species flocking. The reason may be that such flocking can make different species defend predators or detect forages better. Inspired and attracted by its existing mode and collective intelligence in nature, we try to develop an initial and simple idea about the cooperative PSO in [13]. Here, we go on improving the idea about the mixed swarm cooperative search model, and make some comprehensive analysis about its hinting principles, the search mechanism and the optimization performance.

The remaining of the paper is organized as the follows. Section 2 presents some inspiration from mixed swarm in nature. Section 3 describes the construction of the mixed swarm for optimization. Section 4 introduces the details of MCPSO, including the computation model, the coarse search and the fine search, the cooperative search and learning. Some experimental results and some comparison analysis about the proposed models are provided in section 5. Finally, Section 6 concludes with some remarks

2   Inspiration from Mixed Swarm in Nature

Mixed flocking is often observed in tropical forests, especially in the non-breeding season or migration dates. No doubt, the phenomena are the result of nature selection. The main reason may be that the mixed flocking can increases the number of eyes and ears available to detect predators, defend attackers or find food sources. Obviously, such mixed flocking can make full use of different ability of each species and enhance the existence ability of those species as a whole. For example: nearsighted gleaning birds such as Red-eyed Vireos can be permitted by farsighted sallies like Yellow-margined Flycatchers to move in their groups on the tropical wintering grounds; Downy Woodpeckers generally can admit chickadees and titmice in their foraging. Apparently, farsighted sallies or Downy Woodpeckers are superior to nearsighted gleaning birds or titmice in the viability in such mixed species. The former seem to lose some prey to the latter, but the lost can be compensated by the improving safety due to the latter's early detection of approaching dangers.

The hinting principles behind the mixed-species flocking firstly is cooperation or social symbiosis that leads every species to benefit from acting as a whole, whether in detecting food or dangers, and then is competition for sources each other. As far as the optimization is concerned, the mixed-species flocking not only provides a proper mode to construct the searching swarm, but also can help us develop a global optimization model based on its social behavior principles.

3 Mixed Swarm for Optimization Search

Inspired from mixed flocking in nature, we construct a mixed swarm for optimization search. Here, the optimization problem is one kind of non-linear programming problem that can be described as the following:

$$\min f(X) \quad X \in \Omega \subseteq R^D; \Omega = [a,b]^D . \tag{1}$$

Where, $f(X)$ is a multi-modal function while X represents any one solution matching to the function; Ω is the solution space.

Considering the optimization search, the mixed swarm is constructed by two kinds species which are referred to as exploration species (S_1) and exploitation species(S_2) here. Each species consists of a group of individuals, each individual is provided with a position to represent a solution to the optimization problems.

In order to describe the dynamic relationship between the two species of the mixed swarm we introduce the following definitions first.

Definition 1. Supposed the best and the worst positions of S_j are $X_{best}^{S_j}(t)$, $X_{worst}^{S_j}(t)$ respectively in the tth iteration, its activity territory can be defined as:

$$X_T^{S_j}(t) = [X_{\min}^{S_j}(t), X_{\max}^{S_j}(t)]^D . \tag{2}$$

Here,

$X_{\min}^{S_j}(t) = \{x_{\min,d}^{S_j}(t) \mid d = 0,1,\cdots,D-1\}$ and $x_{\min,d}^{S_j}(t) = \min(\min\{x_{worst,d}^{S_j}(t)\}, \min\{x_{best,d}^{S_j}(t)\})$

$X_{\max}^{S_j}(t) = \{x_{\max,d}^{S_j}(t) \mid d = 0,1,\cdots,D-1\}$ and $x_{\max,d}^{S_j}(t) = \max(\max\{x_{worst,d}^{S_j}(t)\}, \max\{x_{best,d}^{S_j}(t)\})$

Obviously, $X_T^{S_j}(t)$ indicates the max super-space overcastted by the exploration species $S_i(t)$.

Definition 2. Supposed the activity territory of S_j is $X_T^{S_j}(t)$, its activity radius can be computed by the following:

$$r_T^{S_j}(t) = \frac{1}{2}(X_{\max}^{S_j}(t) - X_{\min}^{S_j}(t)) . \tag{3}$$

In the mixed swarm, the two species take on different tasks with different ability. S_1 (Exploration species) acts on the role of the farsighted sallies in nature and is designated to go on the coarse exploration in the solution space, which generally can find out the potential territory of a better solution; while S_2 (the exploitation species) mainly undertakes the fine exploitation in the local territory which has been located by the exploration species S_1, just as the nearsighted gleaning birds, who generally follow the farsighted sallies to food. With the same aim to find out the global optimum, the two species are cooperative and learn from each other during the search, which can greatly benefit the tradeoff of global search and local search in the optimization.

4   Mixed Swarm Cooperative Search Model for PSO

Here, we introduce the mixed swarm search mechanism into SPSO and try to develop a valid MCPSO. Its main search behaviors can be discussed in this section.

4.1   Computation Model of MCPSO

In SPSO, a swarm of particles have been defined to represent the potential solutions of an optimization problem. In order to search an optimum, each particle begins with an initial position randomly and flies through the D-dimensional search space. The flying behavior of each particle can be described by its velocity and position, so the update equations about the velocity and position of the individual in mixed swarm can be formulated as the follows:

$$v_{id}^{S_j}(t+1)=wv_{id}^{S_j}(t)+c_1r_1(p_{id}^{S_j}(t)-x_{id}^{S_j}(t))+c_2r_2(p_{gd}^{S_j}(t)-x_{id}^{S_j}(t)). \tag{4}$$

$$x_{id}^{S_j}(t+1)=x_{id}^{S_j}(t)+v_{id}^{S_j}(t+1) \tag{5}$$

Where, S_j (j=1,2) just stands for the species S_1 or S_2.

4.2   Adaptive Coarse Search and Fine Search

Though the update equation of velocity is the same for the particles in the deferent species, the search behaviors of the particles in the two species are basically different from each other. The main difference essentially lies in the length of the search step and the search area of the particles in the two species.

As we know, the velocity of each particle generally should be limited to avoid flying out the solution space. A bigger boundary value means the particle can choose a bigger step to search, while a smaller value indicates the particle only search with a smaller step. So we manipulate the two species to go on the coarse search and the fine search through adopting different velocity limitation to their particles.

Since the exploration species S_1 is expected to make the coarse search in a broader area, the velocity of its particles is designated in the limitation of $[-V_{max}^{S_1}(t), V_{max}^{S_1}(t)]^D$, and $V_{max}^{S_1}(t)$ is set as the dynamic partition of the search boundary during the search:

$$V_{max}^{S_1}(t) = \frac{MaxIteration - t}{MaxIteration} \cdot (b-a).$$ (6)

Obviously, the velocity limitation $V_{max}^{S_1}(t)$ can adaptively permit the particles of S_1 to search with a relative bigger step over time, and make a coarse search in a broader area.

As a helper of exploration species, the exploitation species S_2 mainly undertakes the fine exploitation in the territory of S_1. So, the search step of its particles just depends on the territory diameter of S_1 in the tth iteration, which is limited in $[-V_{max}^{S_2}(t), V_{max}^{S_2}(t)]^D$, $V_{max}^{S_2}(t)$ can be computed as the following:

$$V_{max}^{S_2}(t) = R(X_{max}^{S_1}(t) - X_{min}^{S_1}(t)).$$ (7)

Observing the equation above, it's easy to know $V_{max}^{S_2}(t)$ can be modified adaptively with the change of the territory of S_1, and its value is always more smaller than the value of $V_{max}^{S_1}(t)$, which just ensure the exploitation species can make a good fine search among the territory of S_1.

Here, we denote the update operators of the mixed swarm, the exploration species and the exploitation species as $O_U^{S_M}$, $O_C^{S_1}$ and $O_F^{S_2}$ respectively.

4.3   Cooperative Search and Learning

After the initialization, the two species of the mixed swarm go on their search for some times in parallel. The exploration species act on the coarse search and try to discover new potential area with a bigger step, while the exploitation species keep the fine search in the current best area found by the exploration species with a smaller step. During their search, the two species exchange information each other. If one species discovers the other side detecting new better position, its particles will modify their search direction at a special time. In this means, the two species in the mixed swarm not only are cooperative, but also compete to each other during the search.

In order to ensure the exploitation species to make enough fine searches, we introduce a parameter to control the cooperation between the two species, which is denoted as cooperation gap T_0. Once each cooperation gap T_0 is reached (that means $t=kT_0$, $k=1,2,...K$), the cooperation will begin, and the two species will obtain the useful information from each other to decide where to go on their following search.

Denoting $P_g^{S_1}(t)$ and $P_g^{S_2}(t)$ as the best positions found by the exploration species and the exploitation species respectively in the current iteration, $F(P_g^{S_1}(t))$ and $F(P_g^{S_2}(t))$ are the fitness value of the two positions. If $F(P_g^{S_1}(t)) < F(P_g^{S_2}(t))$, it indicates S_2 has captured one better position omitted by S_1, so all the particles of S_1 should modify their search direction and go back to learn from S_2. The learn operation can be implemented by the following:

$$v_{id}^{S_1}(t)=wv_{id}^{S_1}(t-1)+c_1r_1(p_{id}^{S_1}(t-1)-x_{id}^{S_1}(t-1))+c_2r_2(p_{gd}^{S_2}(t-1)-x_{id}^{S_1}(t-1)). \tag{8}$$

If $F(P_g^{S_1}(t)) > F(P_g^{S_2}(t))$, it indicates S_2 fails to find out a new better position around the history territory of S_1, so S_2 should give up its history search area and fly to the new territory of S_1. At this moment, we can choose its best position $X_{best}^{S_1}(t)$, the worst position $X_{worst}^{S_1}(t)$ and the center position $X_{center}^{S_1}(t)$ as three guide information to generate S_2. Around the three positions above, three parts of particles will be generated by normal distribution respectively to form the exploitation species S_2, and the standard deviation is decided by the territory radius of S_1 in the current iteration:

$$\sigma^{S_2}(t) = r_T^{S_1}(t). \tag{9}$$

Denote the above learning operation of S_1 as $O_L^{S_1}$, the reproducing operator of S_2 as $O_R^{S_2}$, and the cooperation operation of the mixed swarm S_M as $O_{CP}^{S_M}$. After that, the two species will go on the search of other T_0 iterations to make coarse search and fine search independently each other until a satisfied position is found or a termination criterion is matched.

4.4   Stagnation Avoiding Operation

Based on the above analysis, it's easy to know that MCPSO can keep the tradeoff of the exploration and the exploitation validly through the coarse search and the fine search of mixed swarm. However, one phenomenon should be pointed out and deserved to pay more attentions, which is the stagnation of the particles. With the search going on, all of particles tend to converge to the best position found by the mixed swarm in a sense, and the velocity of each particle is about to reduce slowly until the particle stagnates around the best position in the end. The most unfortunate thing is that the best position where most particles stagnate is a local optimum, but the particles have no any energy to make another potential search, what just means a fail search happens. In order to avoid the stagnation phenomenon of the particles and a fail search, the simplest method is to observe the dynamic velocity of the best particle in the mixed swarm. If the velocity of the best particle in S_M is smaller than the designated value ε(a small positive value), the best particle can be regards as being stagnated and its velocity will be reinitialized with a proportion to the interzone $[-V_{max}, V_{max}]^D$. After that, the best particle will get new energy and guide others particles to search into another area, which will contribute to the global optimization of MCPSO greatly. Here, the operation is referred to as the stagnation avoiding operation and is denoted as $O_{SA}^{S_M}$.

4.5   Pseudocode of MCPSO

MCPSO main
```
{ t=0;  Initialize S_M through O_I^{S_M} = O_I^{S_1} + O_I^{S_2}
   While (not (termination criteria)) do
      {Update S_M through O_U^{S_M} = O_C^{S_1} + O_F^{S_2} ;
         If (t %T_0 = =0)
         {Carry out O_{CP}^{S_M} on S_M;
            {If F(P_g^{S_1}(t)) < F(P_g^{S_2}(t))
               Let P_g^{S_1}(t) ← P_g^{S_2}(t) and do O_{CP}^{S_M} = O_L^{S_1} + O_F^{S_2} ;
               Else reproduce S_2 and do O_{CP}^{S_M} = O_C^{S_1} + O_R^{S_2} ;}
         If (V_g^{S_M}(t)<ε) { carry out O_{SA}^{S_M} }
         t ++;
      }
}
```

5   Experimental Studies

In this section, we choose four multimodal function optimization problems to demonstrate the validity of MCPSO. The optimization problems are listed in the following table.

Table 1. Test Functions F_1-F_4

Func.	Name	Equations
F_1	Rosenbrock Function	$F_3(\mathbf{x}) = \sum\limits_{i=1}^{D-1} (100(x_i^2 - x_{i+1})^2 + (x_i - 1)^2)$
F_2	Rastrigrin Function	$F_4(\mathbf{x}) = \sum\limits_{i=1}^{D} (x_i^2 - 10\cos(2\pi x_i)+10)$
F_3	Griewank Function	$F_5(\mathbf{x}) = \dfrac{1}{4000}\sum\limits_{i=1}^{D} x_i^2 - \prod\limits_{i=1}^{D} \cos(\dfrac{x_i}{\sqrt{i}})+1$
F_4	Ackley Function	$F_6(\mathbf{x}) = 20 + e - 20\exp(-0.2\sqrt{\dfrac{1}{D}\Sigma_{i=1}^{D} x_i^2}) - \exp(\dfrac{1}{D}\Sigma_{i=1}^{D}\cos 2\pi x_i)$

A series of experiments have been done to make some performance analysis about MCPSO. In all cases, the parameters of MCPSO have been set as the following: the swarm size N=60 and each sub-swarm size M is 30. Inertia weight w is decreased linearly from 0.9 to 0.4 over time, the two learning factors $c_1=c_2=1.49$, the cooperation interval $T_0=10$. All the statistic results are achieved by 25 runs independently.

5.1 Search Ability Analysis of MCPSO

In order to observe the performance of MCPSO in different hyperspaces, we use the algorithm to solve F_1-F_4 with 10-D, 20-D, 30-D, 100-D, 500-D and 1000-D respectively, and get the following box-plots of the statistics results in 25 runs .

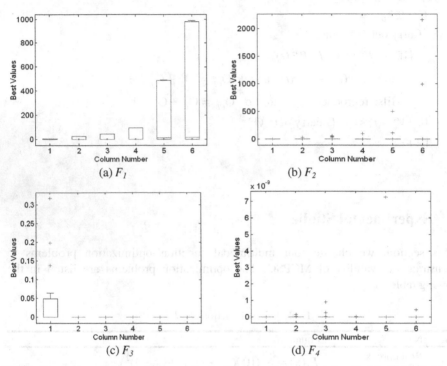

Fig. 1. Box-plots of the statistics results in 25 runs when optimizing F_1 (a), F_2(b), F_3(c) and F_4(d)

In Fig.1, the column numbers from 1 to 6 indicate the 10-D, 30-D, 50-D, 100-D, 500-D and 1000-D respectively, and the figure in each column describes the distribution of the 25 statistic results in each hyperspace. Observing the Fig.(a)-(d), it' easy to know the best results come from the optimization of the Ackley function F_4. Though MCPSO has one or two distinct outlier in several columns denoted by the red "+", the values of the outliers all are near to the optimum(0). The worst case happens on the Rosenbrock function(F_1). According to Fig.1.(a), it's obvious that the box in each column is more and more higher with the dimension increasing, that means MCPSO has more and more troubles in optimizing the function F_1 with the dimension increasing.Seen Fig.1.(b), MCPSO is very robust and almost reach to the optimum (0) in 25 runs for Rastrigin function (F_2) in the 10-D hyperspace. With the dimensions increasing, one or two outliers appear in each column, and the value of the outliers is more and more far from the global optimum, especially for 500-D and 1000-D, which means MCPSO encounters the premature convergence in large scale hyperspace. Fig.1.(c) show us MCPSO performs very well for F_3 with 30-D, 50-D, 100-D, 500-D and 1000-D. In each cases, MCPSO always can capture the global optimum with a high precision in 25 runs. As we known, Griewank function is more difficult to optimize in lower dimension hyperspace than in higher dimension hyperspace, so we find that the performance of MCPSO isn't robust for F_3 in 10-D. observing the box-plots in the first column of Fig.1.(c), three red "+" indicate MCPSO falls into the local optimum for 3 times in the search,   but the other values all fall into a relative compact extent [0, 0.05]. Moreover, the medial value in 25 runs(denoted by the red "–" in the box-plots) is superposed to the bottom of the box-plots, that indicates half of the 25 values are converge to the global optimum "0" of the function. All above show MCPSO also is a rival technique for complex optimization.

5.2   Performance Comparison of MCPSO with MSPSO

In this section, we design the following experiment to make a comparison analysis between MCPSO and the standard PSO based on multiple sub-swarms (MSPSO). Here, MSPSO adopts two sub-swarms to cooperative search just as MCPSO, but the two sub-swarms obey the same search operation and the same update mode as the standard PSO. Except for the above points, the other components all are the same for the two algorithms. The two algorithms are used to solve F_1- F_4 with 100 dimensions one by one.·

Table 2 lists the statistics results got by MCPSO and MSPSO in 25 runs, including the best values, the worst values, the mean values and the standard errors(Std). Comparing each item of the two algorithms, it's easy to know that MCPSO outperforms MSPSO greatly. Different from MSPSO, MCPSO take advantage of exploitation species as a helper of exploration species to conduct the fine search. During the fine search, the velocity of each particle can be modified adaptively according to the current territory radius of the exploration species. Through the coarse search of the exploration species and the fine search of the exploitation species, and their cooperative search, MCPSO can keep the balance of the global search and the local search truly, so it can present an outstanding global performance than MSPSO.

Table 2. Comparisons of MCPSO with MSPSO

Algorithm	Items	F_1	F_2	F_3	F_4
MSPSO	Best	4.291e+001	3.087e+001	1.316e-004	2.425e+000
	worst	2.645e+003	7.768e+001	5.755e-002	7.649e+000
	Mean	2.979e+002	5.565+001	1.120e-002	5.591e+000
	Std	5.074e+002	1.028e+001	1.389e-002	1.261e+000
MCPSO	Best	4.043e-019	4.392e-013	0.000e+000	5.887e-016
	worst	4.752e+001	5.945e+001	8.537e-005	6.994e-006
	Mean	2.758e+001	8.654e+000	7.016e-006	3.666e-007
	Std	2.278e+001	2.425e+001	1.944e-005	1.375e-006

6   Concluding Remarks

The paper develops a cooperative search model for PSO based on mixed swarm, and makes comprehensive analysis about its hinting principles, the search mechanism and the optimization performance. The proposed MCPSO was applied to some well-known benchmarks. Some comparison analyses have been made. According to the relative statistics results, MCPSO is superior to the compared algorithm in most of the functions. All the results show MCPSO is a robust global optimization technique for the complex optimization problems. **However,** some failures also appear when MCPSO is used to optimize the large scale functions in the hyperspace with higher dimension. The future work should focus on how to improve the convergent speed of MCPSO in the final search and achieve higher accurate solutions for the large scale optimizations.

Acknowledgments. This work is supported by Natural Science Foundation of China under Grant No. 61203371, and Planned Science and Technology Project of Zhejiang province under Grant No. 2011C31G2130052.

References

1. Kennedy, J., Eberhart, R.C.: Particle Swarm Optimization. In: Proc. IEEE Conference on Neural Networks, Perth, pp. 1942–1948. IEEE Press, New York (1995)
2. Clerc, M., Kennedy, J.: The Particle Swarm–Explosion, Stability, and Convergence in a Multidimensional Complex Space. IEEE Trans. Evol. Comput. 6(1), 58–73 (2002)
3. Liang, J.J., Qin, A.K., Suganthan, P.N., Baskar, S.: Comprehensive Learning Particle Swarm Optimizer for Global Optimization of Multimodal Functions. IEEE Trans. Evol. Comput. 10, 281–295 (2006)
4. Coello, C.A.C., Pulido, G.T., Lechuga, M.S.: Handling multiple objectives with particle swarm optimization. IEEE Trans. Evol. Comput. 8(3), 256–279 (2004)
5. Xiao, R.B., Chen, W.M., Chen, T.G.: Modeling of Ant Colony's Labor Division for the Multi-Project Scheduling Problem and Its Solution by PSO. J. Comput. Theor. Nanos. 9, 223–232 (2012)
6. Zeng, J.C., Jie, J., Cui, Z.H.: Particle Swarm Optimization. Science Press, Beijing (2004)
7. Shi, Y., Eberhart, R.C.: A modified particle swarm optimizer. In: Proc. Conference on Evolutionary Computation, pp. 69–73. IEEE Press, Piscataway (1998)
8. Ratnaweera, A., Halgamuge, S.K., Watson, H.C.: Self-Organizing Hierarchical Particle Swarm Optimizer With Time-Varying Acceleration Coefficients. IEEE Trans. Evol. Comput. 8, 240–255 (2004)
9. Cui, Z.H., Cai, X.J., Shi, Z.Z.: Using Fitness Landscape to Improve the Performance of Particle Swarm Optimization. J. Comput. Theor. Nanos. 9, 223–232 (2012)
10. Jie, J., Zeng, J.C., Han, C.Z., Wang, Q.H.: Knowledge-based Cooperative Particle Swarm Optimization. Appl. Math. Comput. 205, 861–873 (2008)
11. Van den Bergh, F., Engelbrecht, A.P.: A cooperative approach to particle swarm optimization. IEEE Trans. Evolutionary Computation 8, 225–239 (2004)
12. Kadirkamanathan, V., Selvarajah, K., Fleming, P.J.: Stability Analysis of the Particle Dynamics in Particle Swarm Optimizer. IEEE Trans. Evol. Comput. 10, 245–255 (2006)
13. Jie, J., Wang, W.L., Xu, X.L., Zhang, J.: Mixed-Flocking-Based Cooperative Particle Swarm Optimization. Advan. Sci. Let. 9, 795–800 (2012)

Imperialist Competitive Algorithm with Trading Mechanism for Optimization

Shuaiqun Wang[1], Aorigele[2], Jingyi Luo[3,4], and Shangce Gao[2,4,5]

[1] Department of Computer Science, Tongji University, Shanghai, China
[2] Faculty of Engineering, University of Toyama, Toyama, Japan
[3] Faculty of Information Science and Electrical Engineering,
Kyushu University, Fukuoka, Japan
[4] College of Information Sciences and Technology,
Donghua University, Shanghai, China
[5] Engineering Research Center of Digitized Textile & Fashion Technology,
Ministry of Education, Donghua University, Shanghai, China

Abstract. International trade is the exchange of capital, goods, and services across different countries. Trading has been explored by economists to be an important mechanism for maintaining development. In an imperialistic country, trading makes imperialists capture resources from colonies, meanwhile providing technologies or cultures for colonies to develop themselves. Inspired by this economic phenomenon, this paper transplants the trading mechanism to imperialist competitive algorithm (ICA) and proposes an improved ICA with import and export mechanisms (IICA). IICA is designed to alleviate the problem of slow convergence without significantly impairing the parallel competitive feature of ICA. It is characterized by allowing the imperialist to capture useful aspects of colonies to enhance itself, and meanwhile making colonies learn advanced components from their imperialist. In this way, the trading mechanism enables imperialists and colonies to strengthen interactions during them. The performance of IICA is validated on 23 benchmark functions. Its high performance is confirmed by comparing with other ICA variants.

1 Introduction

In recent years, swarm intelligence has received increasing interests and regarded as powerful methods for the difficult optimization problems. Among them, imperialist competitive algorithm (ICA) [1] is inspired by the behavior of imperialists which attempt to control and possess colonies. It starts with a randomly generated population of countries composed of imperialists and colonies. ICA divides its population into some sub-populations (empires) and searches for the optimal solution through two processes: assimilation and competition. Imperialists are the best candidate solutions, while colonies are divided among those imperialists according to their objective function values. Assimilation operation moves colonies towards their imperialists and competition operation removes a colony from the weakest empire and adds it to another empire. These operations lead to

M. Fei et al. (Eds.): LSMS/ICSEE 2014, Part II, CCIS 462, pp. 87–98, 2014.

the search for better solutions in ICA. Eventually, the population converges to certain areas of the search space. Until now, ICA has been successfully applied on various complex problems, such as pattern classification [2, 3], scheduling problems [4–7] and machinery design problems [8, 9].

The distinct advantage of ICA is its inherent parallel evolutionary mechanism which allows all empires to interact via competition with each other. To improve the search performance of ICA, the scholars have put forward various improved algorithms. For instances, Bahrami [10] utilized chaotic mapping to decide the moving direction of the colony in assimilation operator. Zhang [11] randomly selected a part of colonies to update their position after moving. Lin [12] proposed the perturbed ICA and used artificial imperialists to replace the weaker ones in order to enhance the information interaction between the empires. To a certain extent, the above improved ICA variants increased the local optimization ability of the algorithm. In original ICA, the imperialist improves its power only used exchanging position with its colony, and the convergence speed is therefore becoming slow. Inspired by this phenomenon, we tried to develop ICA by enhancing the information interaction between the imperialist and their colonies, through considering the trading mechanism between two countries. Compared with other variants of ICA, the proposed algorithm (IICA) can utilize the effective information from their colonies to speed up the imperialist moving towards the optimal solution. Meanwhile, the colonies by trading with their imperialist accelerate the speed of approaching imperialist. Thus, the trading mechanism enables imperialists and colonies to strengthen interactions during them, accelerating the convergence speed of the algorithm.

2   Imperialist Competition Algorithm with Trading Mechanism

International trade is exchanging activities of goods and services between different countries and regions. International trade is bound to stimulate innovation mechanism of the enterprize, promote the technical progress and the development of a country's economy. Inspired by such social phenomenon, colonies which trade with economically powerful imperialist are expected to speed up their own economic developments. On the other hand, although the power of colony is weaker than its imperialist, there must exist advanced products or resources in colonies. Therefore, within the process of trading, imperialist can also promote the development of its economy. The algorithmic flowchart of IICA is shown in Fig. 1 which is described in details in the following.

Initialize Empire: The goal of optimization is to find the optimal solution of the problem. An appropriate format for representing a solution must be determined. For the D dimensional optimization problem, a country can be represented as: $country = [p_1, p_2, ..., p_D]$, The cost of a country is used the cost function f to estimate: $Cost = f(country) = f(p_1, p_2, ..., p_D)$.

The initial population size is N_{pop}. N_{imp} most powerful countries are selected as imperialists and the rest $N_{col}(N_{col} = N_{pop} - N_{imp})$ of the countries are as-

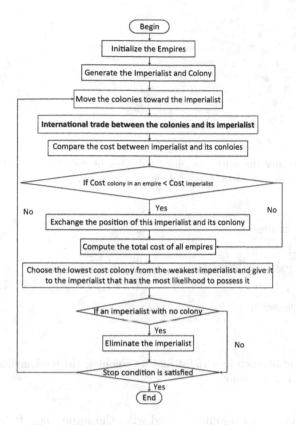

Fig. 1. Flowchart of the proposed algorithm

signed to these empires according to the power of imperialists as their colonies. To assign the colonies among imperialists proportionally, normalized cost of m-th imperialist is defined by:

$$C_m = c_m - max\{c_i\}, i \in 1, 2, ..., N_{imp} \tag{1}$$

where C_m is the normalized cost of m-th imperialist and c_m denotes the cost of m-th imperialist. The normalized power for this imperialist is defined by

$$p_m = \left| \frac{C_m}{\sum_{i=1}^{N_{imp}} C_i} \right| \tag{2}$$

The normalized power of an imperialist reveals the approximate number of colonies that should be possessed by this imperialist. Thus the initial number of colonies of m-th empire will be $NC_{mcol} = round\{p_m \cdot N_{col}\}$, where NC_{mcol} is the initial number of colonies of m-th empire and N_{col} is the total number of colonies. To generate each empire, we randomly choose NC_{mcol} colonies and give them to each imperialist. Fig. 2(a) shows the initial population of each empire

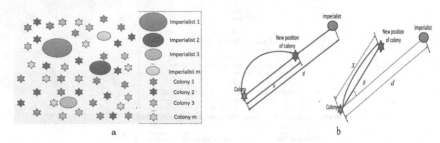

Fig. 2. (a) Generating the initial empires; (b) Moving colonies toward their relevant imperialist in a randomly distance and deviated direction

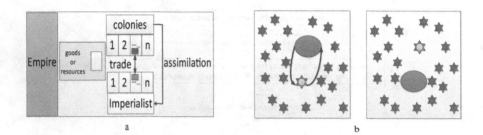

Fig. 3. (a) Trading between imperialist and its colonies; (b) Exchanging the positions of the imperialist and a colony

including imperialist and colonies labeled with the same color. It is obvious that bigger imperialists have greater number of colonies and vice verse.

Colonies Assimilation: Each imperialist has their own colonies and attempts to develop their colonies. This process is called assimilation and is shown in Fig. 2(b), in which the colony moves toward the imperialist by x units. The new position of colony is shown in a darker color and bigger icon. During this process, colonies of an empire move toward the imperialist. $x \sim U(0, \beta \times d), \beta > 1$ is a random variable with uniform distribution.

The original direction of movement is the vector from colony to imperialist. To search wider area around the imperialist, a random amount of deviation to the direction of movement is added. Fig. 2(b) illustrates the new direction and $\theta \sim U(-\gamma, \gamma)$ angle is a random number with uniform distribution, where γ is a parameter that adjust the deviation from the original direction.

Trading Mechanism: Each empire takes possession of many colonies and the colonies are weaker than their imperialist. Nevertheless both colony and imperialist are countries which are composed of many factors, for example, natural resources, commodities and cultures, etc. Not all factors of colonies are weaker than those of their imperialist. Thus the imperialists trade with its colony to make them become more powerful. Inspired by this thought, an improved imperialist competition algorithm is proposed as shown in Fig. 3(a). In this figure, the

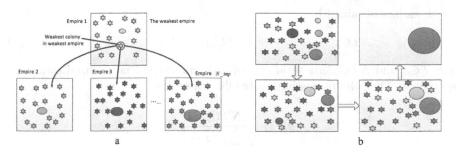

Fig. 4. (a)Imperialistic competition; (b)Empire convergence

rectangle represents a factor in a country and small blue and red rectangle stands for a part of this factor which trades between imperialist and colony. In the original variants of ICA, the colonies of an empire are assimilated by its imperialist and the imperialist is developed only based on exchanging position. However, in the proposed algorithm, the imperialist also develops itself by trading with its colonies.

Exchanging Positions of the Imperialist and a Colony: While colonies moving toward the imperialist, one of them may reach to a position with lower cost than that of imperialist. In this case, the imperialist and the colony switch their position. Then the algorithm will continue by the new imperialist and colonies start moving toward position of new imperialist. Fig. 3(b) shows the position exchange between the imperialist and the best colony which is shown in a yellow color and has lower cost than its imperialist.

Total Cost of Empire: The power of the entire empire is mainly determined by its imperialist. The total cost of m-th empire is defined as:

$$TC_m = Cost(imperialist_m) + \xi \cdot mean\{Cost(colonies\ of\ empire_m)\} \quad (3)$$

where TC_m is total cost of m-th empire and ξ is a positive number between 0 and 1. Different values show the corresponding weights that the cost of imperialist is in the total cost. The small value for ξ causes the cost of the entire empire to mainly depend on the cost of imperialist. As the value of ξ increases, the influence of the colony over the cost will be increased. In the experiments, the value of ξ is set to be 0.1.

Imperialist Competition: In ICA, all the empires are trying to occupy colonies from other empires and control them. Empire competition will lead to weakest empire lose their colonies which will be allocated to other empires. In the process of competition, each empire is likely to have the colonies. However, the more powerful empire will have more opportunities to occupy the colonies and will be getting stronger. As the competition, the weakest empire will lose all of the colonies and be collapsed finally. Fig. 4(a) shows the process of the imperialist competition. The population is composed of N_{imp} empires. Possession probability of each empire is computed based on its total power. Normalized total cost of m-th empire obtained by

$$NTC_m = TC_m - max\{TC_i\}, i = 1, 2, ..., N_{imp} \tag{4}$$

where TC_m and NTC_m are total cost of mth empire and normalized total cost respectively. Then the possession probability of each empire is

$$p_m = \left| \frac{NTC_m}{\sum_{i=1}^{N_{imp}} TC_i} \right| \tag{5}$$

It is obvious that $p_1 + p_2 + ... + p_{N_{imp}} = 1$. The colonies are divided among empires based on their possession probability. Based on the computation speed, a new method is used in algorithm that has less computational effort than Roulette Wheel selection. At the beginning, we form vector P as $P = [p_1, p_2, ..., p_{N_{imp}}]$. Then a vector is created with the same size as P and its elements must be uniformly distributed random numbers between 0 and 1.

$$\boldsymbol{R} = [r_1, r_2, ..., r_{N_{imp}}], r_1, r_2, ..., r_{N_{imp}} \sim U(0, 1) \tag{6}$$

Then vector D is obtained by

$$\boldsymbol{D} = \boldsymbol{P} - \boldsymbol{R} = [D_1, D_2, ..., D_{N_{imp}}] = [r_1, r_2, ..., r_{N_{imp}}]$$
$$= [p_1 - r_1, p_2 - r_2, p_3 - r_3, ..., p_{N_{imp}} - r_{N_{imp}}] \tag{7}$$

Based on vector D, we will give the weakest colony in weakest empire to an empire with maximum index in D.

Eliminating the Powerless Empires and Convergence: In imperialist competition, weaker empires will collapse and their colonies will be divided among others. Many criteria are made to eliminate the powerless empires. In this article, an empire collapsed when it loses all of its colonies. When all empires have collapsed except the most powerful one, the imperialist competition will be terminated and all colonies will be under control of the existed empire. In such case, the algorithm stops and finds the optimal solution. Fig. 4(b) shows this procedure.

3   The Pseudocode of Trading Mechanism

The pseudocode of implementing the trading mechanism between imperialist and colony is shown in Algorithm 1.

Algorithm 1– **Implementation of the trading mechanism**

Begin-learning: Input a imperialist Old_{imp} and one of its colonies Old_{col};
set $\beta = rand()$; $New_{imp} = Old_{imp}, New_{col} = Old_{col}$;
For i=1:n
set k is the randomly chosen dimension;
$Old_{cost.imp} = cost(Old_{imp})$;
$Old_{cost.col} = cost(Old_{col})$;

$New_{imp}(k) = \beta * Old_{imp}(k) + (1 - beta) * (Old_{col}(k));$
$New_{col}(k) = \beta * Old_{col}(k) + (1 - beta) * (Old_{imp}(k));$
$New_{cost.imp} = cost(New_{imp});$
$New_{cost.col} = cost(New_{col});$
if $New_{cost.imp} < Old_{cost.imp};$ $Old_{imp} = New_{imp};$ **end-if;**
if $New_{cost.col} < Old_{cost.col};$ $Old_{col} = New_{col};$ **end-if;**
end-For;
End-learning: Output improved imperialist New_{imp} and colony $New_{col};$

4 Experimental Studies

4.1 Experiment Settings

In this section, a set of 23 benchmark functions, adopted from Yao and Liu [13], is tested in this experiment to evaluate the relative strength and weakness of IICA. The lower bound and upper bound of variant, the number of dimension and the optimal value of each function are also listed in function table. Functions $f_1 \sim f_7$ are unimodal and $f_8 \sim f_{13}$ are multimodal functions with many minima, $f_{14} \sim f_{23}$ are multimodal functions with less minima.

In this experiment, IICA compares with original ICA [1] with the same initial parameters and operating steps except the proposed trading mechanism. For the two ICA methods, the number of countries and the number of imperialists are set to 88 and 8 respectively. In the process of trading, the number of exchanging goods is very important and decides the performance of the algorithm. In optimal algorithm, the goods in trading mechanism map to dimensions in function. Fig. 5 shows convergence plot of unimodal function f_1 and multimodal function f_9 (30 dimensions) with different number of exchanging dimensions. From f_1, it is clear that the bigger number of exchanging dimensions for unimodal function is more excellent convergent result. The balance of execution efficiency and the optimization results of the algorithm should be considered. So the number of exchanging dimensions is set to 5 for unimodal functions. For multimodal function f_9, different numbers of exchanging dimensions lead to a greater difference

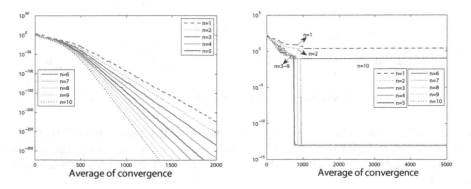

Fig. 5. Convergence plot of unimodal function f_1 and multimodal function f_9

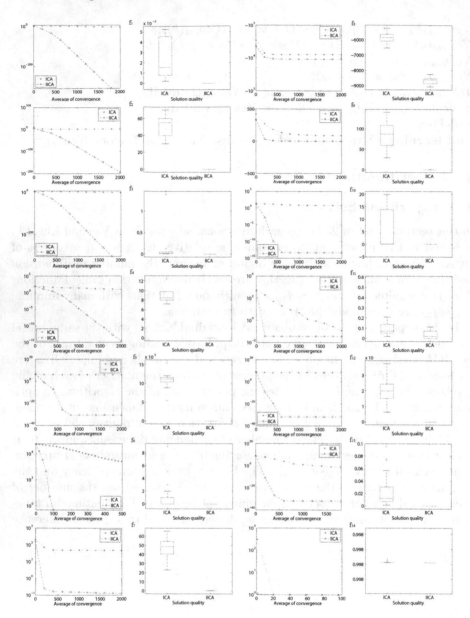

Fig. 6. Convergence plot of benchmark functions ($f_1 \sim f_{14}$)

between the results. But the number of exchanging dimensions from 3 to 9 (removing 6) has the approximate results. So the number of exchanging dimensions for multimodal functions with 30 dimensions is set to 3 and with less minima is set to 1.

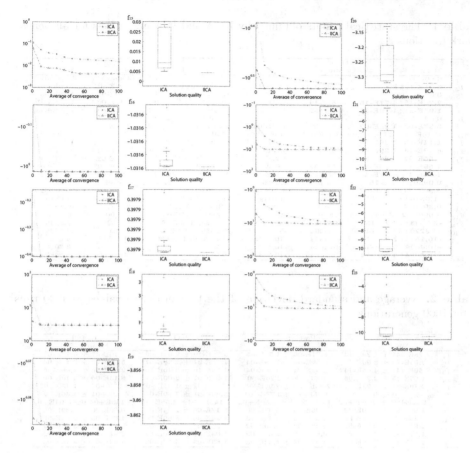

Fig. 7. Convergence plot of benchmark functions ($f_{15} \sim f_{23}$)

4.2   Performance Comparison

For validation of the proposed IICA, experiments have been performed on similar functions belonging to different categories. Table 1 lists the statistical results obtained by IICA and ICA algorithms for 23 benchmark functions. It is obviously observed that IICA with trading mechanism can prompt the algorithm to reach the optimal solution speedily and get better results than ICA. For each test function, 30 runs are performed and the average outcome is plotted in the figure for 2000 generations with 30 dimensions and for 100 generations with less dimensions.

From Table 1, for the unimodal functions $f_1 \sim f_7$, 3 out of 7 functions has been found out the optimal values. The results of the remaining unimodal functions are also superior to ICA. The progress of mean best solution with the number of generations is shown in Fig. 6. From the figure, it can be easily visualized that IICA performs much better than ICA in terms of convergence speed and quality of solution. Multimodal functions have been used to check the performance of algorithm

Table 1. The statistical results obtained by IICA and ICA algorithms for all benchmark functions

Func.	IICA			ICA		
	Mean ± Std	Min	Max	Mean ± Std	Min	Max
f_1	0±0	0	0	5.74E-02±1.13E-01	3.40E-03	5.59E-01
f_2	3.21E-195±0	1.09E-196	1.54E-194	5.20E+01±1.40E+01	3.00E+01	7.00E+01
f_3	0±0	0	0	1.68E-01±4.30E-01	7.00E-03	1.39E-00
f_4	2.26E-15±4.77E-15	3.82E-19	1.51E-14	8.94E-00±1.56E-00	7.19E-00	1.25E+01
f_5	1.97E-31±1.10E-31	7.40E-32	2.96E-31	1.08E+06±2.54E+05	5.40E+05	1.55E+06
f_6	0±0	0	0	9.67E-01±1.88E+00	0	9
f_7	3.37E-02±1.25E-02	1.36E-02	4.81E-02	4.25E+01±1.32E+01	2.30E+01	6.81E+01
f_8	-1.10E+04±2.59E+02	-1.15E+04	-1.08E+04	-7.91E+03±6.71E+03	-8.87E+03	-6.96E+03
f_9	1.14E-13±5.48E-14	0	2.27E-13	8.60E+01±2.59E+01	2.89E+01	1.40E+02
f_{10}	4.62E-14±7.29E-15	3.91E-14	6.75E-14	6.03E+00±8.14E+00	1.34E-02	2.00E+01
f_{11}	3.55E-02±3.81E-02	0	1.12E-01	1.16E-01±1.17E-01	1.93E-02	5.78E-01
f_{12}	1.59E-32±9.43E-34	1.57E-32	2.08E-32	1.99E+08±7.81E+07	6.30E+07	3.76E+08
f_{13}	1.35E-32±5.57E-48	1.35E-32	1.35E-32	2.32E-02±2.42E-02	1.30E-03	9.53E-02
f_{14}	0.9980±3.43E-16	0.9980	0.9980	0.9980±2.58E-12	0.9980	0.9980
f_{15}	4.30E-03±9.31E-07	4.30E-03	4.30E-03	1.47E-02±9.90E-03	5.00E-03	2.86E-02
f_{16}	-1.0316±1.11E-15	-1.0316	-1.0316	-1.0316±4.24E-07	-1.0316	-1.0316
f_{17}	0.3979±0	0.3979	0.3979	0.3979±5.59E-07	0.3979	0.3979
f_{18}	3.0000±2.05E-15	3.0000	3.0000	3.0000±4.15E-06	3.0000	3.0000
f_{19}	-3.8628±2.39E-15	-3.8628	-3.8628	-3.8625±1.40E-03	-3.8628	-3.8549
f_{20}	-3.3220±3.02E-14	-3.3220	-3.3220	-3.2556±7.07E-02	-3.3201	-3.1298
f_{21}	-10.1532±2.80E-13	-10.1532	-10.1532	-8.6936±2.16E+00	-10.1517	-4.6715
f_{22}	-10.4029±5.32E-13	-10.4029	-10.4029	-9.2144±1.90E+00	-10.4013	-3.7028
f_{23}	-10.5364±6.50E-12	-10.5364	-10.5364	-9.0160±2.36E+00	-10.5289	-2.8685

Table 2. Average and standard deviation of the best objective values over 30 runs after 1000 generations

Func.	IICA Mean ± Std	Perturbed ICA Mean ± Std	ICAAI Mean ± Std	ICACI Mean ± Std
f_1	3.76E-145 ± 7.29E-145	8.31E-06 ± 1.30E-05	3.76E-10 ± 2.00E-09	2.10E-07 ± 9.70E-07
f_2	2.93E-79 ± 3.90E-79	3.56E-04 ± 7.48E-04	1.10E-07 ± 2.38E-07	5.08E-05 ± 2.10E-04
f_3	2.50E-144 ± 4.04E-144	2.69E-04 ± 4.50E-04	1.53E-10 ± 6.40E-10	2.84E-06 ± 7.00E-06
f_4	1.09E-05 ± 2.29E-05	6.61E-00± 2.20E-00	1.99E-01 ± 2.00E-01	8.13E-00 ± 3.10E-00
f_5	2.79E-31 ± 1.27E-31	2.30E+02 ± 2.96E+02	1.00E+02 ± 1.31E+02	1.27E+02 ± 1.45E+02
f_6	0 ± 0	1.96E-01 ± 3.77E+01	3.00E-01 ± 7.90E-01	4.62E+01 ± 1.46E+02
f_8	-1.12E+04 ± 2.88E+02	-1.140E+04 ± 2.80E+02	-1.142E+04± 2.56E+02	-1.143E+04 0± 3.04E+02
f_9	1.99E-01 ± 4.20E-02	5.95E-00 ± 3.03E-00	5.17E-00 ± 2.94E-00	6.00E-00 ± 2.89E-00
f_{10}	3.94E-14± 7.76E-14	1.20E-03 ± 1.30E-03	4.14E-06 ± 8.60E-06	1.06E-03 ± 2.00E-03
f_{11}	2.91E-02 ± 3.33E-02	2.28E-02 ± 3.00E-02	1.23E-02 ± 1.70E-02	3.81E-02 ± 3.70E-02
f_{12}	1.57E-32 ± 2.89E-48	6.91E-03 ± 2.60E-02	1.04E-02 ± 3.20E-02	6.91E-03 ± 2.60E-02
f_{13}	1.35E-32 ± 2.89E-48	1.81E-03 ± 4.04E-03	1.83E-03 ± 4.16E-03	1.83E-03 ± 4.16E-03

from locating global optima and local optima. The main reason for introducing these functions is to test the characteristics of the algorithm in continuous and multimodal search space. The whole multimodal functions are divided into two classes based on their functional behavior. The multimodal functions $f_8 \sim f_{13}$ are the combinations of multimodal and unimodal functions. The proposed algorithm IICA compared with ICA for these functions is also shown in Fig. 6. It is obvious that IICA performs much better than ICA due to advance searching strategy that effectively exploits and explores the search space without depending on the silhouette of the functions. Function $f_{14} \sim f_{23}$ with few local minima appear to be simple and similar to that of the unimodal functions. But the major difference is their convergence trend. From Fig. 7, It is not difficult to find that the convergence graph consists of standstill point in comparison to the unimodal functions. All functions except f_{15} find the optimal solution. ICA get stacked for longer number of generations, whereas IICA appear to converge speedily.

Table 2 lists average and standard deviation of the best objective values over 30 runs after 1000 generations. For unimodal functions $f_1 \sim f_6$, the mean values within limited generations obtained from IICA is much better than PICA, ICAAI and ICACI [14]. It accounts for trading mechanism used in ICA can accelerate the speed of convergence. For multimodal funcitons $f_8 \sim f_{13}$, the result of f_8 obtained from IICA is inferior to PICA, ICAAI and ICACI and there is no significant difference between the four methods for f_{11}, the rest of the multimodal functions with IICA are superior to other three variants. It draws a conclusion that trading mechanism can avoid falling into local optimum.

5 Conclusions

A new ICA variant called IICA has been developed. The IICA is characterized by allowing the imperialist to capture useful aspects from colonies to enhance itself, and meanwhile making colonies learn advanced components from the imperialist. In this way, the trading mechanism enables imperialists and colonies to strengthen interactions during them. The significance of IICA mainly lies on two aspects. First, in terms of algorithmic design, trading is a noteworthy phenomenon. By now, as the idea that trading is closely related to crossover evolution and learning has been accepted by more and more economists, it is worthwhile verifying if the trading mechanism is helpful to swarm intelligence techniques. In this research, such attempt is demonstrated by applying trading on ICA. Second, in terms of performance, the proposed IICA manages to alleviate slow convergence and keep the parallel competitive feature of ICA as well.

Acknowledgment. This work is partially supported by the National Natural Science Foundation of China (Grants No. 61203325), Shanghai Rising-Star Program (No. 14QA1400100), "Chen Guang" project supported by Shanghai Municipal Education Commission and Shanghai Education Development Foundation (No. 12CG35), Ph.D. Program Foundation of Ministry of Education of China (No. 20120075120004), the Fundamental Research Funds for the Central Universities (No. 2232013D3-39).

References

1. Atashpaz-Gargari, E., Lucas, C.: Imperialist competitive algorithm: An algorithm for optimization inspired by imperialistic competition. In: IEEE Congress on Evolutionary Computation, CEC 2007, pp. 4661–4667. IEEE (2007)
2. Karami, S., Shokouhi, S.B.: Application of imperialist competitive algorithm for automated classification of remote sensing images. International Journal of Computer Theory & Engineering 4(2) (2012)
3. Mousavi Rad, S., Akhlaghian Tab, F., Mollazade, K.: Application of imperialist competitive algorithm for feature selection: A case study on bulk rice classification. International Journal of Computer Applications 40 (2012)
4. Shokrollahpour, E., Zandieh, M., Dorri, B.: A novel imperialist competitive algorithm for bi-criteria scheduling of the assembly flowshop problem. International Journal of Production Research 49(11), 3087–3103 (2011)

5. Karimi, N., Zandieh, M., Najafi, A.: Group scheduling in flexible flow shops: A hybridised approach of imperialist competitive algorithm and electromagnetic-like mechanism. International Journal of Production Research 49(16), 4965–4977 (2011)
6. Forouharfard, S., Zandieh, M.: An imperialist competitive algorithm to schedule of receiving and shipping trucks in cross-docking systems. The International Journal of Advanced Manufacturing Technology 51(9-12), 1179–1193 (2010)
7. Behnamian, J., Zandieh, M.: A discrete colonial competitive algorithm for hybrid flowshop scheduling to minimize earliness and quadratic tardiness penalties. Expert Systems with Applications 38(12), 14490–14498 (2011)
8. Lucas, C., Nasiri-Gheidari, Z., Tootoonchian, F.: Application of an imperialist competitive algorithm to the design of a linear induction motor. Energy Conversion and Management 51(7), 1407–1411 (2010)
9. Coelho, L.D.S., Afonso, L.D., Alotto, P.: A modified imperialist competitive algorithm for optimization in electromagnetics. IEEE Transactions on Magnetics 48(2), 579–582 (2012)
10. Bahrami, H., Faez, K., Abdechiri, M.: Imperialist competitive algorithm using chaos theory for optimization (cica). In: 2010 12th International Conference on Computer Modelling and Simulation (UKSim), pp. 98–103. IEEE (2010)
11. Zhang, Y., Wang, Y., Peng, C.: Improved imperialist competitive algorithm for constrained optimization. In: International Forum on Computer Science-Technology and Applications, IFCSTA 2009, pp. 204–207. IEEE (2009)
12. Lin, J.L., Cho, C.W., Chuan, H.C.: Imperialist competitive algorithms with perturbed moves for global optimization. Applied Mechanics and Materials 284, 3135–3139 (2013)
13. Yao, X., Liu, Y., Lin, G.M.: Evolutionary programming made faster. IEEE Transactions on Evolutionary Computation 3(2), 82–102 (1999)
14. Lin, J.L., Tsai, Y.H., Yu, C.Y., Li, M.S.: Interaction enhanced imperialist competitive algorithms. Algorithms 5(4), 433–448 (2012)

An Improved Artificial Fish Swarm Algorithm
and Application

Xinyuan Luan [*], Biyao Jin, Tingzhang Liu [**], and Yingqi Zhang

School of Mechatronic Engineering and Automation, Shanghai Key Laboratory of Power Station Automation Technology, Shanghai University, Shanghai 200072
liutzh@staff.shu.edu.cn

Abstract. An improved Artificial Fish Swarm Algorithm (AFSA) based on Hooke-Jeeves (HJ) algorithm is proposed and improved AFSA is applied to design lamps of changeable color temperature and high luminous efficacy in this paper. The disadvantage of AFSA stochastic moving without a definite purpose is improved by HJ algorithm, owing to HJ's great ability of local searching. Accuracy of solution is improved by the adaptive weight. The improved AFSA is verified through an example of how to search for the most luminous efficacy of LED mixing color. The white, red, green and blue LEDs are chosen to design LED lamp samples. LED proportions of 5000K color temperature among those LEDs are optimized by AFSA and new AFSA in the Matlab. The obtained results indicate that improved AFSA is faster and higher accuracy. After LED lamps are tested by integrating sphere, the results show that the difference between the actual value and simulation calculation value is tiny, the new AFSA is effective. The improved AFSA provides a new efficient calculation method of LED proportions. Compared with the traditional manual calculation LED proportions, new method not only saves a significant amount of time, but also achieves higher luminous efficacy for lamps. All this shows that the new method is effective and has high practical value.

Keywords: LED, AFSA, HJ, color mixing, optimization.

1 Introduction

Because the Light Emitting Diode (LED) has advantages of long lifetime, energy saving, high efficiency and environment protection, LED lamps are widely used on various occasions. With the rapid development of the LED lighting industry, the color temperature issue attracts attention of many researchers. Color temperature has effect on the nocturnal change in core temperature and melatonin in human body [1]. Lighting source color temperature has influence on visual performance in tunnel and road area [2]. Color temperature affects user's emotional feelings as well [3]. The right color

---

[*] This work is supported by National Science and Technology Infrastructure Program under Grant 2011BAE34B01.
[**] Corresponding author.

M. Fei et al. (Eds.): LSMS/ICSEE 2014, Part II, CCIS 462, pp. 99–110, 2014.

temperature relaxes consumer and makes them feel comfortable; the inappropriate color temperature may reduce consumer's ability of distinguishing things. To solve this problem, a flexible color temperature light source is need. Compared to the traditional light source, LED light source has advantages in changing the color temperature by mixing different proportion brightness of different color LEDs. The mixing color method of theoretical calculation has been shown in some papers [4-8]. The white, amber and blue LEDs are selected to mix color in the paper [7], the luminous efficacy are almost 80lm/w. The red, green and blue LEDs are selected in the paper [8] and the luminous efficiency are about 24.13lm/W. Using both solutions in paper [7] and paper [8] can achieve the same color temperature at the same power consumption level, but the former luminous efficiency is higher than the latter; the former is brighter than the latter. Therefore, there are lots of methods to implement the same color temperature and luminous efficiency may be different. It is a very worthwhile study to find out higher luminous efficacy mixing color method.

AFSA [9] is a swarm intelligence optimization algorithm, which is inspired by the behavior of fish. Through the simulation of fish's behavior, such as preying, swarm, random and following, the optimal solution is found in the target space. AFSA uses target function as algorithm evaluation function and finds out the appropriate solution quickly. Therefore, AFSA can quickly search for the optimal LED brightness proportion at a certain color temperature. However, AFSA has its shortcomings: the solution accuracy is not high and the convergence rate at the later stage of calculation is slow.

On the other hand, Hooke-Jeeves (HJ) algorithm [10] as the traditional algorithm has strong local search ability, but the global convergence ability is poor. HJ algorithm alternates searching n-dimensions coordinate directions, and changes one of the coordinates every time, then finds a new start point.

How to improve AFSA by HJ algorithm and how to optimize LED brightness proportion by AFSA and improved AFSA will be discussed in this article.

2   Mixing Color Method

The Fig.1 is a CIE 1932 chromaticity diagram. The diagram shows the all possible hues and their relationship to the green, red and blue indicated by the vertices of the triangle. In CIE xyY color space, x and y are the chromaticity coordinates, and Y is lumen value, which we can often see in the LED datasheet.

A special color can be mixed by XYZ (that corresponds to RGB) is known as the CIE XYZ color space. X, Y and Z are tristimulus values, which come from CIE standard observer functions [6].

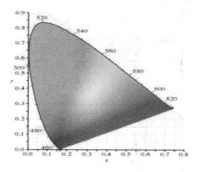

Fig. 1. The CIE 1931 color space chromaticity diagram

The relation between CIE xyY color space and CIE XYZ color space can be described as the formula (1) [11].Here i means LED kind. If LED kinds are n $(n>2)$, mixing color formula can be written as formula (2) shown.Chromaticity value (x_{mix}, y_{mix}) and tristimulus value Y_{mix} are given as the target value , than formula (3) can be derived from formula (1) and formula (2).

Normally, LED kinds selection finished, the LED kinds' number is constant, and the LED combinations' chromaticity values (x_i, y_i) are also constant. Then, formula (3) is transformed into a kind of linear equations of higher degree, and tristimulus values Y_i can be solved from the equations. If Y_{mix} is variable, obviously, Y_i may have many kinds of solutions. Y_{mix} is bigger, the luminous efficacy is higher. It converts the problem about Y_i solutions to the problem about maximum for Y_{mix}.

$$\begin{cases} x_i = \dfrac{X_i}{X_i + Y_i + Z_i} \\ y_i = \dfrac{Y}{X_i + Y_i + Z_i} \\ z_i = (1 - x_i - y_i) \end{cases} \Leftrightarrow \begin{cases} X_i = \dfrac{Y_i}{y_i} x_i \\ Y_i = Y_i \\ Z_i = \dfrac{Y_i}{y_i}(1 - x_i - y_i) \end{cases} \tag{1}$$

$$\begin{cases} X_{mix} = \sum_{i=1}^{n} X_i \\ Y_{mix} = \sum_{i=1}^{n} Y_i \\ Z_{mix} = \sum_{i=1}^{n} Z_i \end{cases} \tag{2}$$

$$
\begin{cases}
x_{mix} = \dfrac{\displaystyle\sum_{i=1}^{n} \dfrac{x_i Y_i}{y_i}}{\displaystyle\sum_{i=1}^{n} \dfrac{x_i Y_i}{y_i} + \sum_{i=1}^{n} Y_i + \sum_{i=1}^{n} \dfrac{Y_i}{y_i}(1 - x_i - y_i)} \\[4ex]
y_{mix} = \dfrac{\displaystyle\sum_{i=1}^{n} Y_i}{\displaystyle\sum_{i=1}^{n} \dfrac{x_i Y_i}{y_i} + \sum_{i=1}^{n} Y_i + \sum_{i=1}^{n} \dfrac{Y_i}{y_i}(1 - x_i - y_i)} \\[4ex]
Y_{mix} = \displaystyle\sum_{i=1}^{n} Y_i
\end{cases}
\tag{3}
$$

3   LED Down Light Lamp Design

Considering the high color rendering index of lamp, red, green and blue LED are selected in the lamp [8]. Normally, white LED luminous efficacy is higher than red, green and blue LED's luminous efficacy. In order to improve the luminous efficacy of lamp, white LED is added. In other words, WRGB mode replaces the RGB mode. A down light lamp is designed which includes 10pcs white, 6pcs red, 5pcs green and 4pcs blue Osram LED , the parameters of four different color LED are shown in Table1. In Table1, x and y are the chromaticity coordinates; Lumen/c means lumen/current.

5000K is set as target color temperature. So, the formula (3) is transformed into Eq. (4). Y_w, Y_r, Y_g and Y_b are white, red, green and blue LED lumen value. Those are linear equations with 4 variables. Through the Eq. (4), the relationship among Y_w, Y_g and Y_{mix} is shown in Fig. 2. The best Y_{mix} is found out by traditional manual calculation firstly, and values are 2448.01.This value will be used as setting target to test optimization algorithm. The Eq. (4) is just an optimal problem with restrictions.

Table 1. LED group chromatic coordinate and luminous

LED	Lumen/c	x	y	x/y	$(1-x-y)/y$
White	156/350mA	0.37	0.42	0.8810	0.5000
Red	60/400mA	0.7006	0.2993	2.3408	0.0003
Green	93/350mA	0.1547	0.8059	0.1920	0.0489
Blue	28/350mA	0.1208	0.0705	1.7135	11.4709
Mix	Y_{mix}	0.3451	0.316	0.9815	0.8629

$$\begin{cases} -0.0593Y_w - 1.1879Y_r + 0.2362Y_g + 3.1810Y_b = 0 \\ -0.1629Y_w + 0.1747Y_r - 0.5637Y_g + 3.9869Y_b = 0 \\ Y_{mix} = Y_w + Y_r + Y_g + Y_b \end{cases}$$

$$\begin{aligned} \max \quad & Y_{mix}(Y_w, Y_r, Y_g, Y_b) \\ st. \quad & Y_w \in [0,1560] \\ & Y_r \in [0,300] \\ & Y_g \in [0,465] \\ & Y_b \in [0,112] \end{aligned} \qquad (4)$$

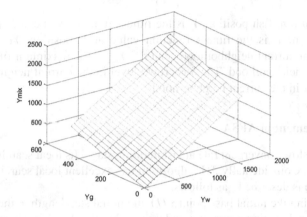

Fig. 2. Relationship among Y_w, Y_g and Y_{mix}

4   Improved AFSA

4.1   AFSA Analysis

Artificial Fishes (AFs) are randomly generated in the AFSA. The maximum value is found out by AFs' preying, swarm and following behavior[9]. Preying behavior can be described as formula (5).The part $(X_i\text{-}X_j)/\ //X_i\text{-}X_j//$ indicates AFs moving direction. The constant Step and random variable decide the moving distance. The $X_{i/next}$ is updated by random variable also in some status, this help AFSA escape from the local optimal solution. Swarm and following behavior can be described as formula (6) and formula (7).They help AFSA locate the optimal solution. The *prey ()* function just is Preying behavior. From the formula (5) - (7) analysis we can see that the updated step of AFSA has effect on the accuracy of the optimal solution and convergent rate.

$$X_{i \mid next} = \begin{cases} X_i + \text{Random (Step)} \dfrac{X_j \cdot X_i}{\|X_j - X_i\|} & \text{if } (Y_i < Y_j) \\ X_i + \text{Random (Step)} & \text{else} \end{cases} \quad (5)$$

$$X_{i \mid next} = \begin{cases} X_i + \text{Random (Step)} \dfrac{X_c \cdot X_i}{\|X_c - X_i\|} & \text{if}(\dfrac{Y_c}{n_f} > \delta Y_i) \\ prey() & \text{else} \end{cases} \quad (6)$$

$$X_{i \mid next} = \begin{cases} X_i + \text{Random (Step)} \dfrac{X_{max} - X_i}{\|X_{max} - X_i\|} & \text{if}(\dfrac{Y_{max}}{n_f} > \delta Y_i) \\ prey() & \text{else} \end{cases} \quad (7)$$

Here, X_i is current fish position. Y_i is the fitness function value at X_i position. X_j is next fish position. Y_j is the fitness function value at X_j position. Yc is the fitness function value at current neighborhood center. Y_{max} is the maximum fitness function value at current neighborhood. The δ is crowd level in the current neighborhood. The nf is AF number in the current neighborhood.

4.2   Improvement of AFSA

The AFSA has slow convergent rate at last iteration stage. Its local search ability needs to be enhanced . Coincidentally ,HJ   algorithm has excellent local search ability.

HJ algorithm is described   as follows.

Step1. Defining the initial base point $x\ (1)$, the initial step length d, the jth coordinate direction e_j, the calculation accuracy r and the acceleration factor α. Coordinate system is n-dimensional coordinate. Let $y\ (1) = x\ (1)$, $k=j=1$.

Step2. If $f(y^{(j)}+de_j) < f(y^{(j)})$, the direction is correct; let $y^{(j+1)}=y^{(j)}+de_j$, and go to Step3. If $f(y^{(j)}+de_j) \geq f(y^{(j)})$,   the direction is not correct. In this case, if $f(y^{(j)}-de_j) < f(y^{(j)})$, let $y^{(j+1)}=y^{(j)}-de_j$, and go to Step3. If $f(y^{(j)}-de_j) \geq f(y^{(j)})$, let $y^{(j+1)}=y^{(j)}$.

Step3. If $j<n$, let $j =j + 1$, repeat Step2. If $j=n$, and $f(y^{(n+1)}) \geq f(x^{(k)})$, go to Step5;otherwise,   if $f(y^{(n+1)}) < f(x^{(k)})$, go to Step4.

Step4. Let $x^{(n+1)}=y^{(n+1)}$, and let $y^{(1)}=x^{(k+1)} + \alpha(x^{(k+1)}-x^{(k)})$. Let $k =k + 1$. Reset $j=1$, and go to Step2.

Step5. If $d \leq r$, the computation is end, $x^* \approx x^{(k)}$. Otherwise, let $d = d/2$, $y^{(1)}=x^{(k)}$, $x^{(k+1)}=x^{(k)}$, $k = k + 1$, $j=1$, and repeat Step2.

A new variable $(1- 0.5*(gen/GEN)^2-o)$ is applied to AFSA to improve the accuracy of AFSA. The gen is iteration generation. The GEN is the setting maximum iteration generation. The o is the adjustment coefficient, $o \in (0\ 1]$. At the last iteration stage, the updated step value becomes smaller, and the accuracy of solution is higher. Thus, formula (6) and formula (7) are replaced by formula (8) and formula (9).

$$X_{ilnext} = \begin{cases} X_i + (1-0.5\times(\frac{gen}{GEN})^2 \cdot o) * Random(Step)\frac{X_c - X_i}{\| X_c - X_i \|} & if(\frac{Y_c}{n_f} > \delta Y_i) \\ prey() & else \end{cases} \quad (8)$$

$$X_{ilnext} = \begin{cases} X_i + (1-0.5\times(\frac{gen}{GEN})^2 \cdot o) * Random(Step)\frac{X_{max} - X_i}{\| X_{max} - X_i \|} & if(\frac{Y_{max}}{n_f} > \delta Y_i) \\ prey() & else \end{cases} \quad (9)$$

$Y_{mix\ target}$ is defined as actual best solution; Y_{mix} is defined as best solution of optimization algorithm. The ε is defined as accuracy sign, and it is one of ending conditions.

$$| Y_{mix\ target} - Y_{mix} | <= \varepsilon \quad (10)$$

4.3   New AFSA Flow

New AFSA process is described as below:

Step1. Initialization: N pieces of AF are generated randomly, $x_i = \{x_1, x_2, \dots , x_n\}$.Initializing N as 100、 ε as10^{-4} 、 vision field as 20、 moving step as 4、 crowd level as 0.618、 maximum trying number as 100 and maximum iterations as 60.

Step2. Fitness function value of every AF is calculated and compared with each other . Maximum fitness function value is got and written in the bulletin board.

Step3. AF implements swarm, following and preying behavior. Fitness function value of every AF is calculated and compared with the value in the   bulletin board.

Step4. Updating the bulletin board. If the fitness function value is better than the value in the bulletin board, the latter is replaced by the former. Otherwise, the value of the bulletin boards remains the same.

Step5. Determining whether to update AF. If the updated condition (iteration number is even) is satisfied, every previous AF is updated by new one   based on HJ algorithm, $x_j = \{x_1, x2, \dots , x_n\}$. HJ algorithm start point is just old point(AF),which uses pattern search to find out a new point (AF)that fitness value is bigger than old point (AF).

Step6. Judging whether it reaches the maximum number of iterations or satisfies other ending conditions, if it satisfies one of ending conditions, it stops. Otherwise, it goes to Step3, and enters next loop.

5   Results and Discussion

The AFSA and the new AFSA both are implemented 30 times in the MATLAB R2010a, simulation results are shown in Table 2. Convergence is defined as: algorithm reaches the setting accuracy. From Table 2, several conclusions can be drawn. The convergence rate of the new AFSA is twice as high as the original AFSA; the new AFSA's average iterations of reaching the setting accuracy are less than half of the original AFSA. The new AFSA can converge in 4 iterations, but the old algorithm requires at least 10 iterations.

Table 2. Simulation results

Algorithm	AFSA	New AFSA
The rate of convergence at 30 times	50%	100%
Average iterations at convergence	39	16
Minimum iterations at convergence	10	4

In the Fig. 3, blue points are original AFs; red triangles are new AFs which are optimized by HJ algorithm once. Every New AF fitness value is bigger than original one. This is advantageous to the convergence rate. Here Y_g, Y_w and Y_{max} mean green, white LED and optimal solution lumen value correspondingly.

The new AFSA's fastest iteration results are shown in the Fig. 4. The new AFSA reaches the accuracy at 4th iteration. Here, Gen means generation.

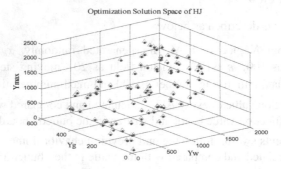

Fig. 3. New AFSA optimal solution space

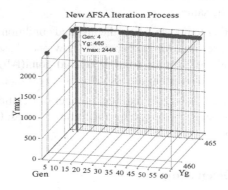

Fig. 4. The changing of the new AFSA optimal solution

The original AFSA's fastest iteration results are shown in the Fig. 5, and the original AFSA reaches the target at 10th iterations. In the Fig. 6, blue dots are optimal values. The whole optimal process is represented by dots moving. After optimal solution moves to the 4th position, it keeps the same value until end. This means new AFSA gets optimal solution at 4th iteration.

These results demonstrate that the improved methods are effective. New AFSA's performance is better than original AFSA's.

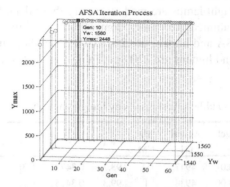

Fig. 5. Original AFSA optimal solution iteration process

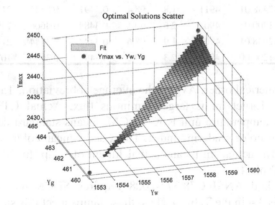

Fig. 6. The optimal solution of new AFSA moving process

From the optimal solution, the ratio between 4 kinds LED is calculated. Meanwhile the LED proportions at 5000K color temperature are calculated by manual, and lumens are 1200lm. These parameters will be stored in flash of microcontroller to control LED drivers (see Table 3).

Table 3. Lumen ratio

LED	Best solution	Total lm	Ratio
Y_w	1560.00	1560	100.00%
Y_r	312.67	360.00	86.85%
Y_g	456.00	456.00	100%
Y_b	110.34	112.00	98.52%
Y_{mix}	2448.01	2497.00	98.04%

Where, Y_w, Y_r, Y_g, Y_b and Y_{mix} are simulation calculation lumen value; Best solution means the optimal solution of new AFSA; Ratio, Total lumen was divided by simulation best solution lumen value (see Table 3).

Several LED down light lamps are made to verify these data. Both two kinds LED proportions are programmed into microcontroller. One kind is that lumens are calculated by new AFSA and lumens are 2448.01lm .Another kind is that lumens are calculated by manual and lumens are 1200lm .Table 4 is the test results of three lamps by integrating sphere.

Table 4. Test results by integrating sphere

NO.	Target CT	Target lm	Actual CT	SD CM	Ra	x	y	Actual lm	lm/W
1	5000K	2448.01	4947	0.8	98.7	0.3474	0.3608	2110.20	97.21
		1200.00	4938	1	99.3	0.3475	0.3593	1077.60	95.72
2	5000K	2448.01	4845	3.2	98.8	0.3511	0.368	2086.92	100.75
		1200.00	4843	3.4	99.2	0.351	0.3662	1067.23	99.62
3	5000K	2448.01	4911	1.8	98.4	0.3491	0.3643	2138.78	100.19
		1200.00	4908	2.6	99.1	0.3488	0.3666	1090.92	98.79
Avg.	5000K	2448.01	4901	1.9	98.6	0.3492	0.3644	2111.97	99.38
		1200.00	4896	2.3	99.2	0.3491	0.3640	1078.58	98.04

Here, NO. is numerical order. Avg. is the average abbreviation. Target CT is target color temperature. Target lm is target luminous flux. Actual CT is actual color temperature. Color temperature unit is K. SDCM, means standard deviation of color matching. Ra means color rendering index. The x and y mean color coordinate in CIE xyY color space. The actual lm means actual lumen of lamp. lm/W is the luminous efficacy unit (see Table 4).

The USA Standard ANSI C78.377-2008 requires SDCM is less than 7. The maximum SDCM is 3.4 in the Table 4.Thus, these lamps meet this standard.

The diffuse board makes the light of lamp look more uniform and soft, but it decreases luminous efficacy and lets color coordinate drift slightly also. At same time, reflective sheeting color has effect on the color coordinate; Reflection efficiency of reflective sheeting has effect on luminous efficacy. And the LED driver not only drives LED but also consumes power. These are reasons that Actual average lumens of lamps are only 86% of target lumens.

When the average lumens are 2111.97lm, the color temperature is 4901K. The color temperature value is almost same as the value when the average lumens are 1078.58lm. But luminous efficacy of former is higher than the latter's. In the group which target lumens are calculated by new AFSA, the maximum luminous efficacy is 100.75lm/W. In another group witch target lumens are calculated by manual, the maximum luminous efficacy is 99.62lm/W. New AFSA not only increases total lumens, but also improves luminous efficacy.

2700K and 5000K color temperature actual effects are shown the Fig.7. The diffuse board is in the front bezel and diffuses light. The white paper part is reflective sheeting in the lamp.99.38 lm/W lamp luminous efficacies are obtained. These luminous efficacies are quite satisfactory, comparing with 24.13lm/W luminous efficacies in paper [8].

Actually, to achieve the target color temperature, 4 kinds or more kinds of LEDs can be selected to mix color based on this method. With LED kind's increment, it's increasingly difficult to get maximum brightness by manual solving equations; time complexity has increased dramatically. On the other hand, new AFSA is heuristic search algorithm, decreases time complexity sharply. The advantage of new AFSA is fairly obvious.

Fig. 7. The comparison of different color temperature

6   Conclusions

In this paper, some methods are adopted to improve AFSA, and improved AFSA is applied to design lamps of changeable color temperature and high luminous efficacy. The HJ algorithm and the adaptive weight are introduced into AFSA. The solution space of AFSA is optimized by the HJ algorithm; accuracy of solution is improved by the adaptive weight. Simulation results demonstrate that new AFSA is effective. Some lamps are designed based on simulation results. After these lamps are tested by integrating sphere, the results indicate that simulation results meet designed target.

Compared with original AFSA, improved AFSA gets faster convergence rate and higher accuracy. The improved AFSA provides a new efficient calculation method of LED proportions. Compared with the traditional manual calculation LED proportions, new method not only saves a significant amount of time, but also achieves higher luminous efficacy for lamps.

All this shows that the new method is effective and has high practical value.

References

1. Morita, T., Takura, H.: Effects of Lights of Different Color Temperature on the Nocturnal Changes in Core Temperature and Melatonin in Humans. Applied Human Science 15(5), 243–246 (1996)
2. Zhang, Q.W., Chen, Z.L., Hu, Y.K., Yang, C.Y.: Study on the Influence of Lighting Source Color Temperature on Visual Performance in Tunnel and Road Lighting. China Illuminating Engineering Journal 19(2), 24–29 (2008)

3. Kim, I.T., Choi, A.S., Jeong, J.W.: Precise Control of a Correlated Color Temperature Tunable Luminaire for a Suitable Luminous Environment. Building and Environment 57, 302–312 (2012)
4. Lin, Y., Ye, L., Liu, W.J., Lv, Y.J.: Optimization Algorithm of Correlated Color Temperature for LED Light Sources by Dichotomy. Acta Optica Sinica 29(10), 2791–2794 (2009)
5. Dai, C.H., Yu, J.L., Yu, J., Yin, C.Y.: Uncertainty Analysis of the Color Temperature and the Correlated Color Temperature. Acta Optica Sinica 25(4), 547–552 (2005)
6. Harris, A.C., Weatherall, I.L.: Objective Evaluation of Color Variation in The Sand-burrowing Beetle Chaerodes Trachyscelides White (Coleoptera: Tenebrionidae) by Instrumental Determination of CIE LAB Values. Journal of the Royal Society of New Zealand (The Royal Society of New Zealand) 20(3), 253–259 (1990)
7. Osram: Color Detection for Multi-color LED Systems Like Brilliant Mix. OSRAM-OS Application Guide-Brilliant Mix Feedback Loop_v1_1_4_2012. 1-12 (2012)
8. Yan, L.Q., Yang, W.Q., Li, S.Z., Cheng, B., Zhang, J.H.: Dynamic Color Temperature White Lighting Source Based on Red Green and Blue Light Emitting Diode. Acta Optica Sinica 31(5), 0523004-1–0523004-7 (2011)
9. Li, X.L., Shao, Z.J., Qian, J.X.: An Optimizing Method Based on Autonomous Animals: Fish swarm Algorithm. Systems Engineering-Theory & Practice 11, 32–38 (2002)
10. Zheng, X.P., Chen, Z.Q.: Back Calculation of Source Strength and Location of Toxic Gases Releasing Based on Pattern Search Method. China Safety Science Journal 20(5), 29–34 (2010)
11. Smith, T., Guild, J.: The C.I.E. Colorimetric Standards and Their Use. Transactions of the Optical Society 33(3), 73–134 (1931-1932)

A Novel Deterministic Quantum Swarm Evolutionary Algorithm

Xikun Wang[1], Lin Qian[1,2], Ling Wang[1,*], Muhammad Ilyas Menhas[3],
Haoqi Ni[1], and Xin Du[1]

[1] Shanghai Key Laboratory of Power Station Automation Technology,
Shanghai University, Shanghai 200072,China
[2] Shanghai Power Construction Testing Institute, Shanghai 200031, China
[3] Department of Electrical Engineering, Mirpur University of Science and Technology,
Mirpur A.K., Pakistan
wangling@shu.edu.cn

Abstract. This paper presents a novel deterministic quantum swarm
evolutionary (DQSE) algorithm based on the discovery of the drawback of the
standard quantum swarm evolutionary (QSE) algorithm, in which a
deterministic search strategy, inspired by the nature of qubit-based evolutionary
algorithms and the characteristics of qubits, is proposed to avoid the misleading
of search and strengthen the global search ability. The experimental results
show that the developed DQSE outperforms the quantum-inspired evolutionary
algorithm, the quantum-inspired evolutionary algorithm with NOT gate and
QSE in terms of the search accuracy and the convergence speed, which
demonstrates that DQSE is an effective and efficient optimization algorithm.

Keywords: quantum evolutionary algorithm, particle swarm optimization,
quantum swarm evolutionary algorithm, Q-bit, qubit.

1 Introduction

Inspired by the concept and principles of quantum computing, Han et al. proposed the
quantum evolutionary algorithms [1, 2] which provided links between quantum
computing and evolutionary algorithms. After that, quantum-inspired evolutionary
algorithms (QEAs) have been studied and applied to a variety of optimization
problems, such as multidimensional knapsack problems [3], fault diagnosis [4],
traveling salesman problems [5], clustering [6] and network design problems [7].
Recently hybrid QEAs have been a main research direction of QEAs for improving
the performance, and various hybrid QEAs are developed, such as the quantum ant
colony optimization algorithms [8, 9], genetic quantum evolutionary algorithm [10],
immune quantum evolutionary algorithms [11, 12] and quantum swarm evolutionary
algorithms [13-15]. Among these hybrid QEAs, quantum swarm evolutionary (QSE)
algorithms employ the search mechanism of particle swarm optimization (PSO) to

---

* Corresponding author.

M. Fei et al. (Eds.): LSMS/ICSEE 2014, Part II, CCIS 462, pp. 111–121, 2014.

update the quantum angles automatically, and therefore the robustness and the search ability of the algorithm are enhanced and it is easy to be implemented. The previous works show that QSE outperforms the standard QEA on 0-1 knapsack problems, traveling salesman problems and multiuser detection problems [13-15]. However, as quantum observing is also used to generate binary solutions in QSE, which would mislead the search direction, the performance of QSE on complicated problems may be very poor. To make up for it, this paper presents a novel deterministic quantum swarm evolutionary (DQSE) algorithm in which a deterministic search strategy is proposed to improve the global search ability of QSE.

The rest part of the paper is organized as follows. The standard QEA is first introduced in Section 2 for understanding DQSE better. Section 3 describes the presented deterministic quantum swarm evolutionary algorithm in detailed. In Section 4, the simulation results and the comparisons with QEAs and QSE are given and analyzed. Finally, Section 5 concludes the paper.

2   Quantum-inspired Evolutionary Algorithms

2.1   Initialization

Unlike other evolutionary algorithms using the classical representation approaches such as binary, numeric or symbolic coding, QEAs use Q-bits. One Q-bit is defined with a pair of complex number $[\alpha, \beta]^T$, which is characterized as $|\psi\rangle = \alpha|0\rangle + \beta|1\rangle$, where $|0\rangle$ and $|1\rangle$ are the quantum states. α and β are complex numbers that specify the probability amplitudes of the corresponding states; α^2 denotes the probability that the Q-bit will be found in the "0" state and β^2 gives the probability that the Q-bit will be found in the "1" state. For an M-dimensional individual q of QEAs, it can be defined as Eq. (1)

$$q = \begin{bmatrix} \alpha_1 \dots \alpha_j \dots \alpha_M \\ \beta_1 \dots \beta_j \dots \beta_M \end{bmatrix} 1 \le j \le M \quad . \tag{1}$$

where M is the length of Q-bit chromosome, α_j and β_j are the corresponding probability amplitudes of the j-th Q-bit and satisfy the normalization condition $|\alpha_j|^2 + |\beta_j|^2 = 1$. All the Q-bits are set to $\frac{1}{\sqrt{2}}$ which means that each Q-bit chromosome is initialized with the linear superposition of all possible states with the same probability.

2.2   Quantum Observing

To solve optimization problems, the corresponding solutions of Q-bit chromosomes are needed. In QEAs, a conventional binary solution is constructed by observing Q-bits. For a bit p_{ij} of the binary individual p_i, a random number r is generated. If $|\alpha_{ij}|^2 > r$, then set p_{ij} with "0" ; otherwise set p_{ij} with "1", i.e.

$$p_{ij} = \begin{cases} 1 & if \ r > |\alpha_i|^2 \\ 0 & otherwise \end{cases} . \tag{2}$$

Then the population P of binary solutions can be generated by observing the states of the current Q-bit individuals.

2.3   Quantum Rotation Gate

By using the corresponding binary solutions, the fitness of quantum individuals can be calculated and adopted to evaluate the performance. Then the quantum rotation gate is used to update quantum individuals in the QEA and leads the algorithm close to the best solution gradually, which is defined as follows:

$$\begin{bmatrix} \alpha'_{ij} \\ \beta'_{ij} \end{bmatrix} = \begin{bmatrix} \cos(\theta_{ij}) & -\sin(\theta_{ij}) \\ \sin(\theta_{ij}) & \cos(\theta_{ij}) \end{bmatrix} \begin{bmatrix} \alpha_{ij} \\ \beta_{ij} \end{bmatrix} = U(\theta_{ij}) \begin{bmatrix} \alpha_{ij} \\ \beta_{ij} \end{bmatrix} . \tag{3}$$

$$\theta_{ij} = \Delta\theta_{ij} \cdot S(\alpha_{ij}, \beta_{ij}) . \tag{4}$$

where θ_{ij} is the quantum rotation angle, $S(\alpha_{ij}, \beta_{ij})$ is the sign of θ_{ij} that determines the direction, and $\Delta\theta_{ij}$ is the magnitude of the rotation angle which is determined according to the lookup table [1]. With the updating of the quantum angle, $|\alpha_i|^2$ or $|\beta_i|^2$ approaches to 1 or 0, that is, the Q-bit chromosome converges to a single state, and finally the optimal solution can be found. More details of QEAs can be found in [1, 2].

3   Deterministic Quantum Swarm Evolutionary

QEAs only use the information of the optimal individual to guide the search, and therefore the algorithm is likely to be trapped in the local optimum when solving complex problems. In addition, the rotation angle, as the main updating strategy, determines the optimization performance in QEAs, which is given based on the empirical values. Note that it is not balanced for QEAs to update the "1" state and "0" state which may spoil the search ability of algorithms for some problems, and it is very difficult to set new quantum angle rotation rules for QEAs in a new application. To make up for it, QSE presents a simply but efficient approach for quantum angle rotation, and as the local best information, as well as the global best information, is used and therefore the search ability of QSE is enhanced. However, the search direction of QSE may be misled due to the characteristic of quantum observing which will be discussed in the following section, thus a new deterministic updating approach is developed and used in the proposed DQSE.

3.1   Initialization

For simplification, the quantum individual of DQSE uses the quantum angle as the coding scheme, and the population Q of DQSE can be represented as Eq.(5)

$$Q = \begin{bmatrix} \theta_{11}, \theta_{12}, \theta_{13}, ..., \theta_{1M} \\ \theta_{21}, \theta_{22}, \theta_{23}, ..., \theta_{2M} \\ ... \\ \theta_{i1}, \theta_{i2}, \theta_{i3}, ..., \theta_{iM} \\ ... \\ \theta_{N1}, \theta_{N2}, \theta_{N3}, ..., \theta_{NM} \end{bmatrix} . \tag{5}$$

where $\theta_{ij} \in [0, \frac{1}{2}\pi]$ is the rotation angle, $\cos\theta_{ij} = |\alpha_{ij}|$, $\sin\theta_{ij} = |\beta_{ij}|$, and $\sin^2\theta_{ij} + \cos^2\theta_{ij} = 1$. Obviously, θ_{ij} can be used to replace $|\alpha_{ij}|$ and $|\beta_{ij}|$.

3.2   Quantum Angle Updating

In the standard QSE, the search mechanism of PSO is introduced to search for the optimal quantum angle to solve optimization problems, which can be represented as Eq. (6-7):

$$v_{ij} = \omega \times v_{ij} + c_1 \times r_1 \times (\theta p_{ij} - \theta_{ij}) + c_2 \times r_2 \times (\theta g_j - \theta_{ij}) . \tag{6}$$

$$\theta_{ij} = \theta_{ij} + v_{ij} . \tag{7}$$

where ω is inertia factor; c_1 and c_2 are constants; r_1 and r_2 are random numbers between 0 and 1; v_{ij}, θ_{ij}, and θp_{ij}, are the velocity, the current quantum angle, and the corresponding individual best quantum angle, respectively; θg_j is the global best quantum angle value.

Compared with QEAs, QSE updates the quantum angle based on the evolution strategy of PSO easily and efficiently. However, due to the characteristic of quantum observing, the search direction of the quantum angle may be misled in QSE. For instance, consider $\cos^2\theta_{ij} = 0.01$, which indicates the corresponding binary value is "0" with probability of 0.01 and "1" with probability of 0.99, respectively. If the finally observed bit is "0" and the corresponding binary individual is the optimal solution, the quantum angle of this dimension of the population will be attracted to move to the current angle value, which means that more and more "1" will be generated by QSE in the following quantum observing operation process while we actually need the algorithm to move to "0" state. This misleading may seriously spoil the performance of QSE.

Note that in qubit-based evolutionary algorithms including QEAs and QSE, the updating operation is used to rotate the quantum angle and thus each Q-bit converges to a single state gradually, i.e. approaching either 1 or 0. In DQSE or QSE, the quantum angle is used, and therefore the goal of updating operation is to lead quantum angles to 0 or $\pi/2$, i.e. approaching 0 or 1. Inspired by this nature of qubit-based evolutionary algorithms, a novel deterministic velocity updating strategy is developed and used in DQSE to remedy the misleading and improve search efficiency as Eq. (8-10)

$$v_{ij} = \omega \times v_{ij} + c_1 \times r_1 \times \left(D\theta p_{ij} - \theta_{ij} \right) + c_2 \times r_2 \times \left(D\theta g_j - \theta_{ij} \right) \quad . \tag{8}$$

$$D\theta p_{ij} = \begin{cases} 0 & if \ Pbest_{ij} = 0 \\ \dfrac{\pi}{2} & if \ Pbest_{ij} = 1 \end{cases} \quad . \tag{9}$$

$$D\theta g_j = \begin{cases} 0 & if \ Gbest_j = 0 \\ \dfrac{\pi}{2} & if \ Gbest_j = 1 \end{cases} \quad . \tag{10}$$

where $Pbest_{ij}$ and $Gbest_j$ are the corresponding binary values of the local best solution and global best solution, respectively.

However, considering the characteristics of Q-bits as well as quantum observing operation and avoiding the premature of the algorithm, the converging states of quantum angles are improved as Fig.1 where γ is a constant angle, and consequently the deterministic velocity updating operation used in DQSE is modified as follows:

$$v_{ij} = \omega \times v_{ij} + c_1 \times r_1 \times \left(D\theta p_{ij} - \theta_{ij} \right) + c_2 \times r_2 \times \left(D\theta g_j - \theta_{ij} \right) \quad . \tag{11}$$

$$D\theta p_{ij} = \begin{cases} \gamma & if \ Pbest_{ij} = 0 \\ \dfrac{\pi}{2} - \gamma & if \ Pbest_{ij} = 1 \end{cases} \quad . \tag{12}$$

$$D\theta g_j = \begin{cases} \gamma & if \ Gbest_j = 0 \\ \dfrac{\pi}{2} - \gamma & if \ Gbest_j = 1 \end{cases} \quad . \tag{13}$$

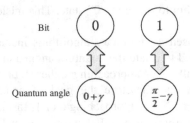

Fig. 1. The modified relationship between bits and quantum angles in DQSE

3.3   Implementation of NQSE

In summary, the implementation of DQSE is described as follows:

Step 1: Initialize the population Q, and each quantum angle is random generated between 0 and $\pi/2$.

Step 2: Generate the corresponding binary population P by performing quantum observing as Eq. (2), calculate the fitness value of each individual with the binary solutions, and set the initial local best solutions and the initial global best solution.

Step 3: Compute the velocity of quantum angles according to Eq. (11-13), and then update quantum angles as Eq. (7).

Step 4: Generate new binary population P by executing the quantum observing operator as Eq. (2) and calculate the fitness of new individuals.

Step 5: Update the local best solutions and the global best solution according to the fitness values.

Step 6: If the termination condition is satisfied, output the best binary solution; otherwise goto Step 3.

4   Parameter Study

Obviously, the parameter γ affects the search ability of DQSE. To briefly observe the performance of DQSE and set a fair γ, a parameter study of γ is performed. Two functions among the 15 benchmark functions listed in Table 1, i.e. f_4 and f_{14}, are used for testing DQSE with different γ, i.e. 0, 0.01 , 0.05 , 0.08 , 0.1 , 0.11 , 0.12 , 0.13 , 0.15 , 0.20 and 0.25. The population size, Q-bit length of each decision variable, and number of generations are set to be 30, 32, and 3000, respectively. The recommended parameters of PSO are used, that is, ω =0.7289, c_1=1.42 and c_2=1.47. DQSE was run 50 times independently for each function. The results of the parameter study are given in Table 2 including the success rate (SR), the best solution (Best), the average solution (Ave), the average generation of finding the optimal solution (AveG), and the minimum generation of finding the optimal solution (MinG).

Table 2 shows that DQSE works well with γ in a fair range. However, it is not surprised that a too small or too big value of γ significantly spoils the performance of DQSE as the algorithm loses the capability of escaping from the local optima and the advantage of Q-bits when γ is equal or very close to "0" while DQSE degrades to a random search when γ is equal or very close to "$\pi/2$". Based on the results of two benchmark functions, $\gamma = 0.11\pi$ is recommended and adopted in this paper.

5 Experimental Results and Analysis

To evaluate the performance of DQSE, we ran DQSE on the 15 benchmark functions, as well as the standard QSE [13], the standard QEA [2], and an improved QEA, i.e. the QEA with NOT gate (NQEA) [10]. Each algorithm with the recommended parameters was run 50 times independently on all the functions. The results are presented in Table 3.

Table 1. Benchmark functions

	Functions	Type
f_1	Sphere Model	unimodal
f_2	Schwefel's Poblem 2.22	unimodal
f_3	Schwefel's Poblem 1.2	unimodal
f_4	Rosenbrock's Valley	multimodal
f_5	Schaffer F6	multimodal
f_6	Shubert Function	multimodal
f_7	F1 Problem	unimodal
f_8	Rastrigin's Function	multimodal
f_9	Ackley's Function	multimodal
f_{10}	F2 Problem	multimodal
f_{11}	Six Hump Camel Back Function	multimodal
f_{12}	Branin's Function	multimodal
f_{13}	Levy F5	multimodal
f_{14}	Glankwahmdee Function	multimodal
f_{15}	Freudenstein-roth Function	multimodal

Table 3 shows that DQSE outperforms QSE, QEA and NQEA on 15, 14, and 12 functions and is inferior to QSE, QEA and NQEA on 0, 1, 3 functions, respectively. However, although the SR results of DQSE are poorer than those of NQEA on f_{14} and f_{15}, the average values of DQSE are superior to those of NQEA, which indicates that DQSE can efficiently avoid being trapped in the local optima. The performance of QSE is even worse than that of QEA due to the misleading. Compared with QSE, DQSE has much better results which demonstrates that the proposed deterministic search strategy can fix the misleading problem.

Table 2. Parameter study of γ

F	γ	SR(%)	Best	Ave	AveG	MinG
	0	0	4.8281E-4	0.25502874	/	/
	0.01π	0	6.2015E-4	0.15092999	/	/
	0.05π	8	0	0.05446532	1115	241
	0.08π	34	0	0.02457259	874	129
	0.10π	46	0	0.01028581	639	128
f_4	0.11π	70	0	7.5550E-5	784	86
	0.12π	60	0	1.2516E-7	720	170
	0.13π	62	0	1.2485E-7	755	311
	0.15π	24	0	3.2377E-6	1770	503
	0.20π	0	1.7227E-6	1.1919E-4	/	/
	0.25π	0	5.8238E-5	0.00416987	/	/
	0	2	0	0.88311280	6	6
	0.01π	4	0	0.69021723	52	31
	0.05π	16	0	0.16990252	38	17
	0.08π	18	0	0.16977962	33	22
	0.10π	20	0	0.05555525	57	32
f_{14}	0.11π	20	0	1.6659E-4	70	50
	0.12π	12	0	1.9696E-4	118	75
	0.13π	12	0	5.2872E-4	231	132
	0.15π	8	0	4.5201E-4	357	175
	0.20π	0	1.1472E-5	0.00186329	/	/
	0.25π	0	1.8890E-4	0.19118453	/	/

Table 3. Results of QEA, NQEA, QSE, DQSE on benchmark functions

		SR (%)	Best	Ave	AveG	MinG
	QEA	88	0	1.24454E-6	528	41
f_1	NQEA	100	0	0	93	53
	QSE	2	0	0.01315277	122	122
	DQSE	100	0	0	59	13
	QEA	98	0	7.8125E-5	422	46
f_2	NQEA	100	0	0	101	44
	QSE	2	0	0.05087223	1331	1331
	DQSE	100	0	0	56	27
	QEA	40	0	0.00241363	75	59
f_3	NQEA	100	0	0	120	57
	QSE	4	0	0.959670410	953	882
	DQSE	100	0	0	81	24
	QEA	0	1.49011E-8	0.040532328	/	/
f_4	NQEA	62	0	0.008534873	395	55
	QPSO	0	2.38424E-7	0.048355591	/	/
	DQSE	70	0	7.555E-5	784	86

Table 3. (*continued*)

f_5	QEA	48	-1	-0.99533627	1046	46
	NQEA	68	-1	-0.99689090	215	58
	QSE	2	-1	-0.99172843	4	4
	DQSE	42	-1	-0.99455897	692	56
f_6	QEA	0	-186.7193	-172.378978	/	/
	NQEA	20	-186.7308	-182.077938	1701	460
	QSE	0	-186.7279	-175.330374	/	/
	DQSE	38	-186.7308	-184.402919	725	70
f_7	QEA	2	-38.8503	-38.6553702	57	57
	DQEA	18	-38.8503	-38.7609278	987	125
	QSE	0	-38.8448	-38.6744453	/	/
	NQSE	34	-38.8503	-38.8263129	898	140
f_8	QEA	14	-80.7066	-80.3395065	281	78
	NQEA	12	-80.7066	-80.6559051	767	182
	QSE	4	-80.7066	-79.9507025	400	324
	DQSE	36	-80.7066	-80.6768844	723	21
f_9	QEA	82	0	0.002101440	404	56
	NQEA	100	0	0	108	58
	QSE	10	0	0.448213776	1105	3
	DQSE	100	0	0	56	18
f_{10}	QEA	0	-1.1283	-1.125200744	/	/
	NQEA	6	-1.1511	-1.130163501	344	74
	QSE	2	-1.1511	-1.128282123	2968	2968
	DQSE	6	-1.1511	-1.131805166	238	120
f_{11}	QEA	36	-1.031628	-1.030175749	851	84
	NQEA	96	-1.031628	-1.031624569	979	85
	QSE	2	-1.031628	-0.995954155	2840	2840
	DQSE	100	-1.031628	-1.03162812	318	16
f_{12}	QEA	48	0.39789	0.4041765031	182	35
	NQEA	64	0.39789	0.3994956086	250	44
	QSE	24	0.39789	0.4169677564	856	77
	DQSE	74	0.39789	0.3987495188	232	10
f_{13}	QEA	4	-176.1375	-162.1123380	1459	323
	NQEA	78	-176.1375	-173.9238963	1293	93
	QSE	0	-176.1322	-153.7435019	/	/
	DQSE	94	-176.1375	-176.1359707	556	71
f_{14}	QEA	10	0	0.0342618884	471	76
	NQEA	22	0	0.0014913356	122	85
	QSE	0	8.1057E-6	0.2711418264	/	/
	DQSE	20	0	1.6659E-4	70	50
f_{15}	QEA	18	0	2.0294479787	753	92
	NQEA	26	0	1.7870134E-4	142	78
	QSE	2	0	12.361067861	18	18
	DQSE	20	0	1.2085406E-4	137	65

6   Conclusions

In this paper the drawback of QSE is pointed out. To make up for it, a novel deterministic quantum swarm evolution algorithm is presented in which a deterministic search strategy is developed to avoid the misleading of quantum angle rotation, inspired by the nature of qubit-based evolutionary algorithms and the characteristics of qubits. The performance of DQSE is evaluated and compared with QEA, NQEA and QSE on benchmark functions. The results show that DQSE outperforms QEA, NQEA and QSE in terms of the search accuracy and the convergence speed, which demonstrates that the presented DQSE is an efficient optimization tool.

Acknowledgments. This work was supported by National Natural Science Foundation of China (Grant No. 61304031, 61374044 & 61304143), and Innovation Program of Shanghai Municipal Education Commission (14YZ007).

References

1. Han, K.H., Kim, J.H.: Genetic Quantum Algorithm and Its Application to Combinatorial Optimization Problem. In: IEEE International Conference on Evolutionary Computation, pp. 1354–1360. IEEE Press, La Jolla (2000)
2. Han, K.H., Kim, J.H.: Quantum-Inspired Evolutionary Algorithm for a Class of Combinatorial Optimization. IEEE Trans. Evolutionary Computation 6(6), 580–593 (2002)
3. Wang, L., Wang, X.T., Fei, M.R.: A Novel Quantum-Inspired Pseudorandom Proportional Evolutionary Algorithm for the Multidimensional Knapsack Problem. In: GEC 2009, pp. 545–552 (2009)
4. Wang, L., Niu, Q., Fei, M.R.: A Novel Quantum Ant Colony Optimization Algorithm and Its Application to Fault Diagnosis. Transactions of the Institute of Measurement and Control 30(3-4), 313–329 (2008)
5. You, X.M., Liu, S.: Quantum Computing-Based Ant Colony Optimization Algorithm for TSP. In: 2009 2nd International Conference on Power Electronics and Intelligent Transportation System, vol. 3, pp. 359–362 (2009)
6. Xiao, J., Yan, Y.P., Zhang, J., et al.: A Quantum-Inspired Genetic Algorithm for K-means Clustering. Expert Systems With Applications 37, 4966–4973 (2010)
7. Lin, D.Y., Waller, S.T.: A Quantum-Inspired Genetic Algorithm for Dynamic Continuous Network Design Problem. Transportation Letters-the International Journal of Transportation Research 1(1), 81–93 (2009)
8. Wang, L., Niu, Q., Fei, M.: A Novel Quantum Ant Colony Optimization Algorithm. In: Li, K., Fei, M., Irwin, G.W., Ma, S. (eds.) LSMS 2007. LNCS, vol. 4688, pp. 277–286. Springer, Heidelberg (2007)
9. Zhang, Y., Liu, S.H., Fu, S., et al.: A Quantum-Inspired Ant Colony Optimization for Robot Coalition Formation. In: Control and Decision Conference, CCDC 2009, pp. 626–631. IEEE Press, Guilin (2009)
10. Wang, L., Tang, F., Wu, H.: Hybrid Genetic Algorithm Based on Quantum Computing for numerical Optimization and Parameter Estimation. Applied Mathematics and Computation 171(2), 1141–1156 (2005)

11. Li, X., Qian, L.H.: A Modified Quantum-Inspired Evolutionary Algorithm Based on Immune Operator and Its Convergence. In: Fourth International Conference on Natural Computation, ICNC 2008, pp. 136–140. IEEE Press, Jinan (2008)
12. Jiao, L.C., Li, Y.Y., Gong, M.G., et al.: Quantum-Inspired Immune Clonal Algorithm for Global Optimization. IEEE Transactions on Systems, Man, and Cybernetics, Part B: Cybernetics 38(5), 1234–1253 (2008)
13. Wang, Y., Feng, X.Y., Huang, Y.X., et al.: A Novel Quantum Swarm Evolutionary Algorithm and Its Applications. Neurocomputing 70(4-6), 633–640 (2007)
14. Yang, Y., Tian, Y.F., Yin, Z.F.: Hybrid Quantum Evolutionary Algorithms Based on Particle Swarm Theory. In: Industrial Electronics and Applications, pp. 1–7. IEEE Press, Singapore (2006)
15. Huang, Y.R., Tang, C.L., Wang, S.: Quantum-Inspired Swarm Evolution Algorithm. In: IEEE International Conference on Computational Intelligence and Security Workshops, pp. 208–211. IEEE Press, Harbin (2007)

Application of the Improved Quantum Genetic Algorithm

Yufa Xu, Xiaojuan Mei, Zhijun Dai, and Qiangqiang Su

Electrical Department, Shanghai Dianji University, Shanghai 200000

Abstract. The paper put forward an improved QGA in order to solve the shortcomings of traditional QGA in optimizing the multimodal function, like lower convergence speed, easier to local optimum. The improved QGA will adjust the rotation angle of quantum gate according to the dynamic evolutionary process, so it can speed up the convergence of the multimodal function. In order to avoiding the individual evolved to local optimum, so mutation was introduced. Lastly, through the typical complex multimodal function test to prove the validity of the improved QGA.

Keywords: Quantum Genetic Algorithm(QGA), Multimodal Function, Quantum Gate, Quantum Mutation.

1   Introduction

QGA was dated from 1990s [1].Narayanan and the others put forward the concept of QGA firstly. In essence, QGA is belongs to GA. QGA just introduce the concepts of quantum computing into GA. Like QGA introduces the parallelism of quantum computing into GA, so the speed of GA increased, QGA introduces the Many Universes concept of quantum computing into GA, so the scope of GA increased and the efficiency of GA is improved.

When QGA was put forward, it came into notice at once, so there were big achievements of QGA. Reference [3] improves the coding method of QGA and through testing the high dimensional function, improved QGA is better than traditional QGA. Reference [4] a chromosome of the standard QGA is regarded as a node and the chromosome population is seen as a network. Improved QGA chromosomes are divided into some sub-groups, when updating chromosomes, an optimal chromosome in locality or in other sub-groups is chosen based on a certain probability as the evolution target for each chromosome. The new network structure of the chromosome is favorable to the diversity of individual chromosomes. Reference [5] puts forward an improved QGA that adjust the quantum gate rotation angle dynamically and optimize complex function. Reference [6] put forward an improved QGA based on catastrophe and local search, and give a new algorithm—MarQ. The improved QGA solve the problem of precocity, local search and so on. Reference [7] adopts a new quantum rotation gate-H_ε to updating the population. It can avoid the algorithm fall into local search and improve the ability of global search. Reference [8] put forward single real number encoding instead of real number encoding to improving the efficiency of QGA

M. Fei et al. (Eds.): LSMS/ICSEE 2014, Part II, CCIS 462, pp. 122–128, 2014.

and reduce the elapse time. And simplify rotation matrix into add and subtract step in order to reduce elapse time. The paper put forward an improved QGA in order to solve the shortcomings of traditional QGA in optimizing the multimodal function, like lower convergence speed、 easier to local optimum. The improved QGA will adjust the rotation angle of quantum gate according to the dynamic evolutionary process, so it can speed up the convergence of the multimodal function. In order to avoiding the individual evolved to local optimum, so mutation was introduced. Lastly, through the typical complex multimodal function test to prove the validity of the improved QGA.

2   Quantum Genetic Algorithm (QGA)

Quantum genetic algorithm is a kind of algorithm that combines GA and quantum computing. It is based on q-bit and some concepts of quantum mechanics [9]. Q-bit is the smallest information storage unit. On the basis of the expression of quantum bit, put the application of the probability amplitude into chromosome coding, then a chromosome is a not sure number and it can express some different superposition state[10].This is the reason that we said quantum computing has parallelism. In the process of evolution, quantum revolving door realizes the evolutionary operation and makes it evolve in the direction of optimal solution.

2.1   Quantum Bit (Q-bit) [5, 9]

Q-bit is the smallest storage unit in quantum computer and it is similar to binary bit 0 and 1. However, unlike ordinary computer, q-bit can not only be in 0 state and 1 state, but also can in any superposition state between 0 and 1. The basic unit can be presented as $|0>$、 $|1>$. Then any superposition state can be presented as:

$$|\varphi>=\alpha|0>+\beta|1>$$ (1)

Where α、 β are complex numbers that specify the probability amplitudes of $|0>$、 $|1>$. And α、 β must meet:

$$|\alpha|^2 + |\beta|^2 = 1$$ (2)

Where $|\alpha|^2$ gives the probability that the q-bit will be found in state 0, $|\beta|^2$ gives the probability that the q-bit will be found in the state 1.

2.2   Quantum Revolve Gate [7,8,11]

Quantum revolve gate realize the process of evolution. So, the design of revolve gate is the most important operation to QGA. Quantum revolve gate can be presented as:

$$\begin{pmatrix} \alpha_i' \\ \beta_i' \end{pmatrix} = U_R \begin{pmatrix} \alpha_i \\ \beta_i \end{pmatrix} = \begin{pmatrix} \cos(\theta_i) & -\sin(\theta_i) \\ \sin(\theta_i) & \cos(\theta_i) \end{pmatrix} \begin{pmatrix} \alpha_i \\ \beta_i \end{pmatrix} \tag{3}$$

Where $\left(\alpha_i',\beta_i'\right)$ is the probability amplitudes of i bit after updating, (α_i,β_i) is the probability amplitudes of i bit before updating, θ_i is the angle, $\theta_i = S(\alpha_i,\beta_i)*\Delta\theta_i$, $S(\alpha_i,\beta_i)$ is the direction of rotation.

In the traditional QGA, angle usually adopts fixed strategy. So this strategy is confirmed and it can not be change according to the question. The strategy as follows:

x_i : the current chromosome's i-bit; $best_i$; the current best chromosome's i-bit ; $f(x)$: fitness function; $\Delta\theta_i$: the size of rotation angle.

Table 1. The choice of angle strategy

x_i	$best_i$	$f(x) > f(best)$	$\Delta\theta$	$S(\alpha_i,\beta_i)$			
				$\alpha_i\beta_i > 0$	$\alpha_i\beta_i < 0$	$\alpha_i = 0$	$\beta_i = 0$
0	0	False	0	0	0	0	0
0	0	True	0	0	0	0	0
0	1	False	0.01*pi	1	-1	0	1、-1
0	1	True	0.01*pi	-1	1	1、-1	0
1	0	False	0.01*pi	-1	1	1、-1	0
1	0	True	0.01*pi	1	-1	0	1、-1
1	1	False	0	0	0	0	0
1	1	True	0	0	0	0	0

The rotation angle can be choice according to the table 1. The strategy can be as follows: firstly, compare the current fitness function $f(x)$ with the current best fitness function $f(best_i)$, then ,if $f(x) > f(best)$, change the q-bit and let it evolve to the direction of x_i ; otherwise change q-bit and let it evolve to the direction of best.

3   Improved Quantum Genetic Algorithm(IQGA)

3.1   Dynamic Quantum Rotation Gate

From the table 1, we can see, $\Delta\theta$ is a fixed value and it can not be changed according to the need of procedure. In the fact, $\Delta\theta$ belongs to [0.001*pi, 0.05*pi] and it is concluded by experience without some theoretical guidance [5]. Therefore, this strategy has shortcomings. If the value of $\Delta\theta$ is small, the precision will be high, but the speed of convergence will be slow; if the value of $\Delta\theta$ is too large, the speed will be increased, but it will cause to "early-maturing". So the better strategy should modulate the angle according to the need of process. The specific strategies are as follows: firstly, at the beginning of evolution set a large angle, usually it is $\Delta\theta_{max}$ =0.05*pi; secondly, decrease the angle gradually with the increase the evolution. There has:

$$\Delta\theta = \frac{\Delta\theta_{max}}{N} \tag{4}$$

Where N is the size of population. We know, in the begin of the evolution, individual evolution direction and optimal solution has a big gap, so it is good to use big angle called "rough search" and it can accelerate the search speed; with the evolution becomes big, individual evolution direction and optimal solution draws near, so it is good to use small angle called " fine search" and it can increase the search precision. That is to say, combine "rough search" in the beginning and "fine search" in the process can not only speed up convergence, but also solve the problem of early-maturing. It can be better to evolution.

3.2   Quantum Mutation

From the formula (4) above, when N is big, $\Delta\theta$ will change a little. That is to say, with the increase of the size of population, the evolution of individual will be standstill. It is a fatal mistake for multimodal function optimization, so in case of individual being caught in local extremum, the paper introduces the quantum mutation. The aim of quantum mutation is when size of population goes to a certain times like 200 times adding a little disturbance to disturb its evolution direction that can stop evolution fall into local optimization. The mutation method as follows: exchange the probability amplitude (α_i, β_i), then the direction of evolution will be reversal about 180 and change the direction of evolution and avoid the local optimization.

3.3   The Workflow of Improved Quantum Genetic Algorithm

(1)   Population initialized. All the probability amplitude initializes with $(1/\sqrt{2}, 1/\sqrt{2})$, and every state is equiprobable;
(2)   Give one measurement of the population and record the value, that is, binary coding;
(3)   Give the value fitness assessment, and record the best individual and fitness;
(4)   Judge the precision meet the demand (precision must $<10^{\wedge}(-4)$). If the conditions meet, then end, otherwise continue.
(5)   Use formula (4) and (5) to update rotation angle and gain a new population;
(6)   Judge whether the individual need mutation. If needed, then introduce mutation operation; otherwise continue (7);
(7)   Let t=t+1, turn to (2) and continue cycling.

4   Typical Function Test

Test the property of the improved algorithm through doing the minimum value of typical function:

(1) De Jong function

$$F_1 = 100(x_1^2 - x_2)^2 + (1 - x_1)^2 \qquad \begin{cases} -2.048 \le x_1 \le 2.048 \\ -2.048 \le x_2 \le 2.048 \end{cases}$$

This is a pathological single peak function and it is difficult to convergence. In theory, it has the minimum value f(1,1)=0.

(2) Six peak hunchback function

$$F_2 = (4 - 2.1x_1^2 + \frac{1}{3}x_1^4)x_1^2 + x_1 x_2 + (-4 + 4x_2^2)x_2^2 \qquad \begin{cases} -3 \le x_1 \le 3 \\ -3 \le x_2 \le 3 \end{cases}$$

This function has 6 local minimum value and the smallest is f(-0.0898,0.7126)=f(0.0898,0.7126)=-1.031628.

(3) Schaffer function

$$F_3 = 0.5 + \frac{\sin^2 \sqrt{x_1^2 + x_2^2} - 0.5}{[1 + 0.001(x_1^2 + x_2^2)]^2} \qquad \begin{cases} -100 \le x_1 \le 100 \\ -100 \le x_2 \le 100 \end{cases}$$

This function has a minimum value f(0,0)=0 in its domain.

(4) Shubert function

$$F_4 = \sum_{j=1}^{5} j\cos[(j+1)x_1 + j] * \sum_{j=1}^{5} j\cos[(j+1)x_2 + j] \qquad \begin{cases} -10 \le x_1 \le 10 \\ -10 \le x_2 \le 10 \end{cases}$$

This is a multiple hump function. In its domain, it has 760 minimal values and the smallest one is f=-186.731.

The test result are presented in table 2: the initial value are set as follows ; the maximum number of iterations N=500, population size sizepop=50, chromosome q-bit coding is 20. The maximum angle of improved QGA delta=0.05*pi, the range of mutation is (0,50).

Table 2. The result of testing typical function

function	Algorithm	Average convergence number	Average value of function	Optimal value of function
F1	QGA	92	0.0000992	0
	ImprovedQGA	40	0.00000615	
F2	QGA	100	1.03002	-1.031628
	ImprovedQGA	50	1.03221	
F3	QGA	158	0.000732	0
	ImprovedQGA	90	0.000089	
F4	QGA	146	-186.29512	-186.731
	ImprovedQGA	62	-186.57132	

The testing result is presented in table 2. From table 2, we can see, the improved QGA is better than traditional QGA in convergence number and accuracy of computing. The performance of the algorithm is considered from convergence speed and accuracy. From the table 2, we can see improved QGA has a higher speed of convergence. Compared function from experience with the real value, we can see, improved QGA is nearer to real value, so the accuracy of improved QGA is improved. In a word, the improved QGA is better in the speed of convergence and computing accuracy. It verifies that the improved QGA's efficiency.

5   Conclusion

The paper put forward an improved QGA and it makes the convergence speed and computing accuracy in some extent. Through the Matlab simulation of 4 typical functions, it proves that the improved QGA is better. However, the paper only does a little improve on quantum angle and add mutation to QGA. QGA is a new algorithm relatively, so it has many to research and development. Besides, because of the introduction of mutation and the complex of procedure, so the elapsed time increased. How to decrease the elapsed time on the basis of not slowing down convergence speed and computing accuracy is a direction in the future.

References

1. Liang, C., Bai, H.: Research progress of quantum genetic algorithm. Application Research of Computers 29, 2402–2404 (2012) (in Chinese)
2. Narayanan, A., Moore, M.: Quantum-inspired genetic algorithm. In: Proc. of IEEE International Conference on Congress on Evolutionary Computation, pp. 6–66 (1996)
3. Abs, A.V., Cruz, M.B.R., Vellasco, M.A.C.P.: Quantum-inspired evolutionary algorithm applied to numerical optimization problems. In: Evolutionary Algorithm: Computation (CEC), Rio de Janeiro, Brazil, pp. 1–6 (2012)
4. Guo, J., Sun, L., Wang, R., Yu, Z.: An improved quantum genetic algorithm. In: Third International Conference on Genetic and Evolutionary Computing, pp. 14–17 (2009) (in Chinese)
5. Zhang, Z.: An improved quantum genetic algorithm. Computer Engineering 36, 18–182 (2010) (in Chinese)
6. Liu, H.: The optimization of reactive power in power system based on the improved quantum genetic algorithm. Southwest Jiaotong University (2006) (in Chinesse)
7. Zhu, X., Zhang, X.: The optimization of continuous function based on improved quantum genetic algorithm research. Computer Engineering and Design 28, 5195–5197 (2002) (in Chinese)
8. Fang, C.: The improved quantum genetic algorithm and its application on the distribution of the optimization of thermal power unit. North China Electric Power University (2011) (in Chinese)
9. Han, K., Kim, J.H.: Genetic Quantum Algorithm and its Application to Combinatorial Optimization Problem. Press by IEEE
10. Wang, L., Wu, H.: Hybrid quantum genetic algorithm and its performance analysis. Control and Decision 20, 157–158 (2005) (in Chinese)
11. Zhang, G., Jin, W.: The improved quantum genetic algorithm and its application. Journal of Southwest Jiaotong University 38, 718–722 (2003) (in Chinese)
12. Shi, F., Wang, H.: 30 cases analysis of intelligent algorithm. Beihang University, Beijing (2010) (in Chinese)

Strategy Analysis
of an Evolutionary Spectrum Sensing Game[*]

Dongsheng Ding[**], Guoyue Zhang, Donglian Qi[**], and Huhu Zhang

College of Electrical Engineering, Zhejiang University,
Hangzhou 310027, P.R. China
{donsding,zhangguoyue,qidl}@zju.edu.cn,
donsding@126.com

Abstract. Evolutionary game has been shown to greatly improve the spectrum sensing performance in cognitive radio. However, as selfish users are shortsighted for the long-term profits, they are not willing to collaborate to sense. In this paper, we propose an evolutionary spectrum sensing game to improve the long-term spectrum utilization. The new spectrum sensing model takes advantage of the long-term effect of the future actions on the current actions by using the concept of present value (PV) in repeated game. The collaboration conditions of two strategies, i.e., tit-for-tat and grim strategy are discussed. It is proved that the grim strategy can enhance secondary users' sensing positivity greatly, and so is the overall spectrum efficiency. Finally these new developments are illustrated in our experiments.

Keywords: Cognitive radio, evolutionary game, present value, spectrum sensing.

1 Introduction

Cognitive radio (CR) is a dynamic spectrum access soft technology [1], which means that secondary users (SUs) can identify whether the licensed spectrum is empty or not by spectrum sensing. If primary users (PUs) are not using the licensed spectrum, the SUs can utilize this vacant spectrum to increase the throughput of CR to its full potential. Game theory provides a theoretical framework that studies the process of how to cooperatively sense the licensed spectrum, which has attracted much attention recently [1--9].

In [2--7], it has been shown that the performance of spectrum sensing can be improved through spectrum sensing game modeling. In [2], light weight cooperation in sensing based on hard decisions was proposed to reduce the sensitivity requirement. It was shown in [3] that cooperative sensing could reduce the detection time of the PU and increase the overall agility. How to choose proper SUs for

---

[*] This work was supported by National High Technology Research and Development Program of China under Grant 2014AA052501, National Natural Science Foundation under Grant 61371095 and 51177146, and Fundamental Research Funds for the Central Universities under Grant 2014QNA4011.

[**] Corresponding author.

M. Fei et al. (Eds.): LSMS/ICSEE 2014, Part II, CCIS 462, pp. 129–139, 2014.
© Springer-Verlag Berlin Heidelberg 2014

cooperation was investigated in [4]. The design of sensing slot duration to maximize SUs' throughput under certain constraints was studied by [5]. Two energy-based cooperative detection methods using weighted combining were proposed in [6]. The spatial diversity was introduced in [7] to improve spectrum sensing capabilities of centralized cognitive radio networks. The main reason is that the time a SU spent on spectrum sensing can be reduced greatly by sharing sensing results. Recently it was verified in [8, 9] that, the proposed evolutionary framework can achieve a higher throughput than the case where SUs sense individually without cooperation. Among these developments, a fully cooperative scenario is assumed that all SUs voluntarily cooperate to sense and share the sensing results.

However, not all SUs are willing to share their results. The spectrum sensing game, by its very nature is non-cooperative. Given a required detection probability to protect the PU from interference, SUs are willing to sense the licensed spectrum for a higher immediate throughput. For SUs who do not take part in sensing, they can overhear the sensing results and have more time for their own data transmission. If none of them take time to sense the licensed spectrum, all of the users (include the PU) would not get a higher throughput than it obtained by themselves. On the contrary, even if all users succeed in cooperating to sense, the licensed spectrum sharing may be unable to complement the cost of sensing. Therefore, instead of seeking the current maximization of payoffs, SUs need to predict the long-term payoff according to different strategies.

In this paper, we utilize the concept of present value (PV) in repeated game [10] to describe the long-term effect on the current actions from the future actions. The payoffs of SUs are PVs of future possible throughputs, rather than the immediate throughputs. We establish it as an evolutionary spectrum sensing game (ESSG). Two common strategies, i.e., tit-for-tat and grim strategy are discussed. It is proved that the grim strategy can enhance SUs' sensing positivity greatly, and so is the overall spectrum efficiency. Our main contributions are divided to three aspects as follows.

(i) To authors' best knowledge, this is the first to propose ESSG by use of the concept of PV.

(ii) The conditions of cooperation using the tit-for-tat strategy and the grim trigger strategy in ESSG are provided. The grim trigger strategy is tested as a suitable strategy for cooperation in ESSG.

(iii) Our model is tested with a satisfactory performance.

The remainder of the paper is organized as follows. The spectrum sensing model of CR is introduced in Section 2. In Sections 3, ESSG is proposed, where the strategy analysis are provided for two strategies, i.e., tit-for-tat and grim strategy. Simulation and performance is discussed in Section 4. Finally Section 5 concludes this paper and provides some future works.

2   Spectrum Sensing Model

Consider a CR network with a PU and K SUs, where each SU can take spectrum sensing and sharing, and data transmission. The licensed spectrum is divided into K

sub bands, and each SU operates exclusively in one of the K sub bands when the PU is absent. The transmission time is slotted into intervals of length T_s. Once the PU become active, SUs within their transmission ranges can sense the PU jointly.

The received signal $r(t)$ of a SU can be expressed by

$$r(t) = \begin{cases} hs(t) + \omega(t), H_1, \\ \omega(t), H_0. \end{cases} \tag{1}$$

where the hypotheses H_1, H_0 denotes the PU is present or not. The channel gain is h from the PU to SUs; $s(t)$ is the signal from the PU, which is assumed to be an i.i.d. random process with zero mean and σ_s^2 variance; and $\omega(t)$ is an additive circularly symmetric Gaussian noise with zero mean and σ_ω^2 variance. $s(t)$ and $\omega(t)$ are assumed to be independent.

The spectrum is sensed in a SU by use of an energy detector [11]. The test statistics $T(r)$ is defined as

$$T(r) = \frac{1}{N}\sum_{t=1}^{N}|r(t)|^2 \tag{2}$$

where N is the number of received samples.

Assume the PU performs a complex PSK signal, the probability density function (PDF) of $T(r)$ can be approximated by $N\left(\sigma_\omega^2, \frac{\sigma_\omega^4}{N}\right)$ and $N\left((\gamma+1)\sigma_\omega^2, \frac{(2\gamma+1)\sigma_\omega^4}{N}\right)$ under H_0, H_1 respectively, where $\gamma = \frac{|h|^2\sigma_s^2}{\sigma_\omega^2}$ is the received signal-to-noise ratio (SNR) of the PU under H_1.

Definition 1. [8] The probability of detecting the presence of the PU under H_1 is defined as the detection probability P_d; the probability of detecting the presence of the PU under H_0 is defined as the false alarm probability P_f.

$$P_d(\lambda) = \frac{1}{2}erfc\left(\left(\frac{\lambda}{\sigma_\omega^2} - \gamma - 1\right)\sqrt{\frac{N}{2(2\gamma+1)}}\right) \tag{3}$$

$$P_f(\lambda) = \frac{1}{2}erfc\left(\left(\frac{\lambda}{\sigma_\omega^2} - 1\right)\sqrt{\frac{N}{2}}\right) \tag{4}$$

where λ is the threshold of the energy detector and $erfc(\cdot)$ is the complementary error function.　　　　　　　　　　　　　　　　　　　　　　　　　　　□

Given a target detection probability $\bar{P}_d$, the threshold λ can be derived and the false alarm probability P_f can be written as

$$P_f(\bar{P}_d, N, \gamma) = \frac{1}{2}erfc\left(\sqrt{(2\gamma+1)}erfc^{-1}(1 - 2\bar{P}_d) + \gamma\sqrt{\frac{N}{2}}\right) \tag{5}$$

When a SU is sensing the licensed spectrum, its data transmission cannot be performed. If the sampling frequency is f_s and the frame duration is T_s, the time duration for data transmission can be represented by $T_s - \delta(N)$, where $\delta(N) = \frac{N}{f_s}$ is the time spent in sensing spectrum.

Definition 2. [9] When the PU is absent, in those time slots where no false alarm is generated, the average throughput of a SU is defined as R_{H_0}; when the PU is present, but not detected by SUs, the average throughput of a SU is defined as R_{H_1}.

$$R_{H_0} = \frac{T_s - \delta(N)}{T_s}(1 - P_f)C_{H_0} \tag{6}$$

$$R_{H_1} = \frac{T_s - \delta(N)}{T_s}(1 - P_d)C_{H_1} \tag{7}$$

where C_{H_1}, C_{H_0} is the data rate of the SU under H_1 and H_0 respectively.   □

If the probability of the absence of the primary user is denoted by P_{H_0}, the total throughput of a SU is represented by

$$R(N) = P_{H_0}R_{H_0}(N) + \left(1 - P_{H_0}\right)R_{H_1}(N) \tag{8}$$

In dynamic spectrum access, the target detection probability $\bar{P}_d$ required by the PU is very close to 1. Due to the interference from the PU to SUs, the second term can be omitted because it is much smaller than the first term.

$$\tilde{R}(N) \approx P_{H_0}R_{H_0}(N) = \frac{T_s - \delta(N)}{T_s}(1 - P_f)P_{H_0}C_{H_0} \tag{9}$$

Before data transmission, the SUs need to sense the PU's activity. Two kinds of actions can be made by the SUs. The first is that the SUs can cooperate to sense and share the results. The opposite is not to serve for the common goal and act by maximizing own throughputs selfishly. Before take such actions, there are always a cooperative strategy and a defecting strategy. SUs can be labeled, according to their choice of strategy, as either cooperators or defectors.

3   Evolutionary Spectrum Sensing Game

In evolutionary spectrum sensing game (ESSG), actions of the SUs are based on the belief when the game is played repeatedly. In [8, 9], the mixed strategies may change between generations based on the comparison between the current payoffs for SUs and the average payoff. In our ESSG, we use the concept of present value (PV) in repeated game to describe the long-term effect of the future actions on the current actions. The collaboration conditions of two strategies, i.e., tit-for-tat and grim strategy are discussed. It is proved that the strategy of grim strategy can enhance SUs' sensing positivity greatly.

Firstly, we can model the spectrum sensing model as a spectrum sensing game (SSG).

Definition 3. In a SSG with K SUs, the set of players is denoted by $T = \{p_1, \cdots, p_K\}$. Each player p_i can choose one of two actions in $A = \{C, D\}$, where C is the contribute sensing (cooperator) and D is the refuse to contribute sensing (defector). The payoff of each player is the throughput of SUs under different strategies. □

Definition 4. Assume the set $T_c = \{p_1, \cdots, p_J\}$ is the J SUs who cooperate to sense. The false alarm probability of the cooperative sensing among set T_c with a fusion rule 'RULE' and a target detection probability $\bar{P}_d$ is defined by $P_f^{T_c} = P_f(\bar{P}_d, N, \{\gamma_i, i \in T_c\}, RULE)$. The payoff of a cooperator $p_j \in T_c$ is defined as $\tilde{U}_{C,p_j}$; the payoff of a defector $p_i \notin T_c$ is defined as $\tilde{U}_{D,p_i}$.

$$\tilde{U}_{C,p_j} = P_{H_0}\left(1 - \frac{\delta(N)}{|T_c|T_s}\right)(1 - P_f^{T_c})C_{p_j}, |T_c| \in [1, K] \tag{10}$$

$$\tilde{U}_{D,p_i} = \begin{cases} P_{H_0}(1 - P_f^{T_c})C_{p_i}, |T_c| \in [1, K] \\ 0, |T_c| = 0 \end{cases} \tag{11}$$

where $|T_c|$ is the number of contributors, C_{p_j} is the data rate of p_j under H_0. □

Given a $\bar{P}_d$ for T_c, the target detection probability P_{d,p_j} of each SU can be obtained by solving the following equation.

$$\bar{P}_d = \sum_{k=\lceil\frac{1+|T_c|}{2}\rceil}^{|T_c|} \binom{|T_c|}{k} \bar{P}_{d,p_j}^k \left(1 - \bar{P}_{d,p_j}\right)^{|T_c|-k} \tag{12}$$

We assume each contributor takes the same responsibility, $\bar{P}_{d,p_j}, p_j \in T_c$ are the same. Similar to (5), we have

$$P_{f,p_j} = \frac{1}{2}erfc\left(\sqrt{(2\gamma_{p_j}+1)}\,erfc^{-1}\left(1 - 2\bar{P}_{d,p_j}\right) + \sqrt{\frac{N}{2|T_c|}}\gamma_{p_j}\right) \tag{13}$$

In this paper, we use the majority rule [14] as the fusion rule 'RULE' in Definition 4, that is

$$P_d = Pr\{at\ leat\ half\ users\ in\ T_c\ report\ H_1|H_1\}$$

$$P_f = Pr\{at\ leat\ half\ users\ in\ T_c\ report\ H_1|H_0\}$$

Definition 5. An ESSG is defined as $G = \{T, A, S, U\}$, where the set of players T and the action set A are defined in Definition 3. The number of SUs following strategies s_j is n_j. The population profile is $x = \{x_j\}$ and $x_j = \frac{n_j}{K}$. The strategy set is S.

Consider a two-player game. Let $P = 1 - P_f^{T_c}, T = T_c, B_i = 1 - P_{f,p_i}, D_i = P_{H_0}C_{p_i}, i = 1, 2$ and $\tau = \frac{\delta(N)}{T}$, the payoff matrix can be written as Table 1.

Table 1. Payoff matrix

	SU 1 Cooperate	SU 1 Defect
SU 1 Cooperate	$D_1 P\left(1 - \dfrac{\tau}{2}\right), D_2 P\left(1 - \dfrac{\tau}{2}\right)$	$D_1 B_1 (1 - \tau), D_2 B_1$
SU 1 Defect	$D_1 B_2, D_2 B_2 (1 - \tau)$	$0, 0$

Since this game is not the prisoners' dilemma, we try mixed strategies to solve. Let x_1 and x_2 are the probabilities of SU 1, SU 2 taking action C. If SU 1 choose C, the expected payoff is

$$\tilde{U}_{S_1}(C, x_2) = D_1 P\left(1 - \frac{\tau}{2}\right) x_2 + D_1 B_1 (1 - \tau)(1 - x_2) \tag{14}$$

$$\tilde{U}_{S_1}(C, C) = D_1 P\left(1 - \frac{\tau}{2}\right) x_1 x_2 + D_1 B_1 (1 - \tau) x_1 (1 - x_2) + D_1 B_2 (1 - x_1) x_2 \tag{15}$$

Similarly, If SU 2 choose C, we can obtain the expected payoff accordingly. The replicator dynamics of SU 1 and SU 2 are expressed as the followings [9].

$$\dot{x}_1 = x_1 (1 - x_1) D_1 [B_1 (1 - \tau) - E_1 x_2] \tag{16}$$

$$\dot{x}_2 = x_2 (1 - x_2) D_2 [B_2 (1 - \tau) - E_2 x_1] \tag{17}$$

where $E_1 = B_2 + B_1 (1 - \tau) - P\left(1 - \frac{\tau}{2}\right)$ and $E_2 = B_1 + B_2 (1 - \tau) - P\left(1 - \frac{\tau}{2}\right)$.

We assume that the SNR in each sub band within the same licensed spectrum band is the same, $\gamma_{S_1} = \gamma_{S_2}, C_{S_1} = C_{S_2}$. The steady-state of (16) and (17) is defined as the evolutionary stable strategy (ESS), a detail analysis you can refer to [8, 9].

In dynamic spectrum access, if none of them take time to sense the licensed spectrum, all of the users (include the PU) would not get a higher throughput than it obtained by themselves. On the contrary, even if all users succeed in cooperating to sense, the licensed spectrum sharing may be unable to complement the cost of sensing. Therefore, the current maximization of payoffs is unreasonable. Since the data transmission of SUs is a long-term process, we can describe the effective payoffs using the present valve (PV).

Definition 5. [12] PV is the sum that a player is willing to accept currently instead of waiting for the future payoff, i.e., accept smaller payoff today that will be worth more in the future, similar to making an investment in the current period that will be increased by a rate r in the next period.   □

If the payoff is 1 in the next time, the payoff that a player is willing to accept will be $\frac{1}{1+r}$ now. Actually there is a probability p that the game will stop, the payoff that a player is willing to accept will be $\frac{1-p}{1+r} \triangleq \delta$, where $\delta \in [0,1]$ is the discounted factor. If the expected payoff in the next time is X, the PV of the next round game is δX. Assume the current payoff is 1, the PV of the infinite game is $PV = 1 + \delta + \delta^2 + \cdots = \frac{1}{1-\delta}$.

In repeated games, contingent strategies are frequently used to model the sequential nature of the relationship that users can adopt strategies that depend on behavior in preceding plays of the games. Most contingent strategies are trigger strategies. Two common trigger strategies are the tit-for-tat (TFT) and the grim trigger strategy [10, 12]. This paper only consider the case $p = 0$, that is the infinite game.

Definition 6. In an ESSG, TFT means choosing, in any specified period of game, the action chosen by your rival in the preceding period of play. When playing TFF, you cooperate with your rival if she cooperated during the most recent play of the game and defect (as punishment) if your rival defected. This punishment phase lasts only as long as your rival continues to defect; you will return to cooperation one period after she chooses to do so.   □

Theorem 1. In an ESSG, both sides take TFT and tend to cooperate if the discount factor satisfies

$$\delta > \frac{1 - P_{f,s_j}}{1 + P_{f,s_j}}, j = 1,2 \tag{18}$$

Proof. If both sides take TFT, the PV of strategy C is

$$PV_{Cooperate} = \frac{1}{1 - \delta} D_1 P \left(1 - \frac{\tau}{2}\right) \tag{19}$$

The PV of strategy D is the sum of the payoffs

$$PV_{Cheat} = D_1 B_2 + \delta D_1 B_1 (1 - \tau) + \frac{\delta^2}{1 - \delta} D_1 P \left(1 - \frac{\tau}{2}\right)$$

To promote one SU to cooperate, the PV of strategy C is preferable for each SU, we have $PV_{Cooperate} > PV_{Cheat}$, that is (18). So far, the proof is completed.   □

It is shown in Theorem 1 that if $\delta > 1$, P_{f,s_j} will go to zero. So TFT is impractical. SUs have a strong desire to cheat for a high payoff as increasing N, since the cost of sensing will be increased with a large N.

Another strategy which can promote cooperation is the grim trigger strategy, which is more harsh strategy than TFT.

Definition 7. In an ESSG, the grim strategy entails cooperating with your rival such time as she defects from cooperation; once a defection has occurred, you punish your rival (by choosing the defect strategy) on every play for the rest of the game.   □

The punishment for a SU who chooses not to sense is more serious using the grim trigger strategy than that using TFT.

Theorem 2. In an ESSG, both sides take the grim trigger strategy and tend to cooperate if the discount factor satisfies $\delta \geq \frac{1}{2}$.

Proof. If both sides take the grim trigger strategy, the PV of strategy C is

$$PV_{Cooperate} = \frac{1}{1-\delta} D_1 P \left(1 - \frac{\tau}{2}\right) \tag{21}$$

If a SU chooses the strategy D, it means that it will be punished to sense alone forever. Thus the PV of the strategy C is

$$PV_{Cheat} = D_1 B_2 + \frac{\delta}{1-\delta} D_1 B_1 (1-\tau) \tag{22}$$

To promote one SU to cooperate, the PV of strategy C is preferable for each SU, we have $PV_{Cooperate} > PV_{Cheat}$.

$$\delta > \frac{1-\left(1+P_{f,s_j}\right)\left(1-\frac{\tau}{2}\right)}{\tau}, j = 1,2 \tag{23}$$

As the increase of N, P_{f,s_j} will go to zero and the right term of (23) can reach the maximum $\frac{1}{2}$. □

4   Simulation and Performance

The simulation parameters of ESSG are set as follows. The PU's signal is assumed to be baseband QPSK modulated, where the sampling frequency is $f_s = 1MHz$ and the time duration is $T = 20ms$. The probability of PU's absent is $P_{H_0} = 0.9$ and the required target detection probability $\bar{P}_d = 0.95$. The SNR $\gamma_{s_j} = -12dB$.

Firstly, we do not use the concept of PV. The algorithm of ESSG is shown in Table 2. The initial values are set to $x = 0.8, C = 1$. The comparison between the cases $x_1 = x_2 = 1$ and $x_1 = x_2 = ESS$ is shown in Figure 1, where the evolutionary stable strategy is denoted as ESS.

When τ is smaller than 0.1, the cost of spectrum sensing increases with τ. It is shown that two SUs are willing to sense. However, when τ increases larger than 0.2, the sensing probability of each SU decreases and they tend to defect. The worst is that the throughput decreases at the same time, which is shown in Figure 1. The maximum difference between cooperating completely and cooperating at ESS happens at $\tau = 1$. And the throughput of each SU is decreased too much. The main reason is that each SU only considers the current payoff in each round game.

Table 2. The game algorithm

STEP 1 Parameters initialization.
STEP 2 Compute payoffs (10) and (11) for m circles.
STEP 3 Compute expected payoffs (14) and (15).
STEP 4 Update strategies s_i (16) and (17).
STEP 5 If ESS is achieved, STOP; else go back to STEP 2.

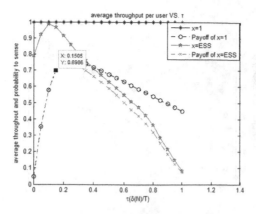

Fig. 1. The average throughput and probability

Now, we use the concept of PV to extend ESSG. To testify the grim trigger strategy can promote cooperation, the game algorithm in Table 2 is adopted with the same parameters. Note the payoffs in STEP 2 are replaced by (21) and (22). The simulation results are shown in Figures 2, 3 and 4. The ESS of ESSG is denoted as $x = ESS'$.

In Figure 2, the sensing probability increases faster at ESS' in ESSG than the result at ESS. The throughput of each SU is improved, especially when τ is close to 1. When a SU choose the strategy D in current period, the other one will choose D forever. So, each SU tends not to take the adventure to wait for the others' sensing result.

The sensing positivity is increased when δ is increased from 0.5 to o.53 in Figure 3. As SUs consider the effect of future actions, the throughput of each SU is increased, especially when τ closes to 1. As shown in Figure 4, x keeps stable when δ is increased to 0.6. It means that, when the cost of sensing is large, both SUs will keep cooperating to share for reducing individual cost. The throughput is just the same as they always choose to cooperate.

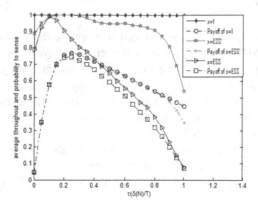

Fig. 2. The average throughput and probability $\delta = 0.5$

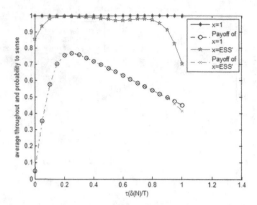

Fig. 3. The average throughput and probability $\delta = 0.53$

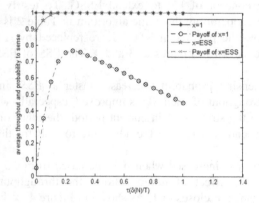

Fig. 4. The average throughput and probability $\delta = 0.6$

5   Conclusion

In this paper, the concept of PV is introduced to improve the evolutionary spectrum sensing game. The SUs not only consider the current payoff, but also the current effect from future payoff. Two strategies, i.e., tit-for-tat and grim strategy are discussed. It is proved that the strategy of grim strategy can enhance secondary users' sensing positivity greatly, and so is the overall spectrum efficiency. The simulation results show that the interaction of the two SUs using the grim trigger strategy can increase the throughput of each SU greatly.

How to utilize the global information to promote the SUs to sense remains a tedious work. Our simulation results show the improvement was obvious. The future work is to investigate the general case of p and find other suitable strategies.

References

1. Masonta, M.T., Mzyece, M., Ntlatlapa, N.: Spectrum decision in cognitive radio networks: A survey. IEEE Communications Surveys & Tutorials, vol 15(3), 1088–1107 (2013)
2. Mishra, S.M., Sahai, A., Brodersen, R.W.: Cooperative sensing among cognitive radios. In: IEEE International Conference on Communications, vol. 4, pp. 1658–1663 (2006)
3. Ganesan, G., Li, Y.: Cooperative spectrum sensing in cognitive radio. IEEE Transactions on Wireless Communications 6(6), 2204–2222 (2007)
4. Peh, E., Liang, Y.C.: Optimization for cooperative sensing in cognitive radio networks. In: IEEE Wireless Communications and Networking Conference, pp. 27–32 (2007)
5. Liang, Y.C., Zeng, Y., Peh, E.C., Hoang, A.T.: Sensing-throughput tradeoff for cognitive radio networks. IEEE Transactions on Wireless Communications 7(4), 1326–1337 (2008)
6. Visser, F.E., Janssen, G.J., Paweczak, P.: Multinode spectrum sensing based on energy detection for dynamic spectrum access. In: Vehicular Technology Conference, pp. 1394–1398 (2008)
7. Ganesan, G., Li, Y., Bing, B., Li, S.: Spatiotemporal sensing in cognitive radio networks. IEEE Journal on Selected Areas in Communications 26(1), 5–12 (2008)
8. Wang, B., Liu, K.R., Clancy, T.C.: Evolutionary game framework for behavior dynamics in cooperative spectrum sensing. In: IEEE Global Telecommunications Conference, pp. 1–5 (2008)
9. Wang, B., Liu, K.R., Clancy, T.C.: Evolutionary cooperative spectrum sensing game: how to collaborate? IEEE Transactions on Communications 58(3), 890–900 (2010)
10. Antoniou, J., Papadopoulou, V., Vassiliou, V., Pitsillides, A.: Cooperative user–network interactions in next generation communication networks. Computer Networks 54(13), 2239–2255 (2010)
11. Lutz, E., Cygan, D., Dippold, M., Dolainsky, F., Papke, W.: The land mobile satellite communication channel-recording, statistics, and channel model. IEEE Transactions on Vehicular Technology 40(2), 375–386 (1991)
12. Dixit, A.K., Skeath, S., Reiley, D.: Games of strategy. Norton, New York (1999)

Particle Swarm Optimization Based on Shannon's Entropy for Odor Source Localization

Nanqi Li, Qiang Lu, Yang He, and Jian Wang

School of Automation, Hangzhou Dianzi University,
Hangzhou, 310018, China
lvqiang@hdu.edu.cn

Abstract. This paper proposes the particle swarm optimization based on Shannon's entropy to deal with the problem of odor source localization. First, a measurement model by which the robots can always observe a position is briefly described. When the detection events occur, the position of the odor source lies in the vicinity of the observed position with a higher probability. When the non-detection events occur, the position of the odor source does not lie in the vicinity of the observed position with a higher probability. Second, on the basis of the measurement model, the posteriori probability distribution on the position of the odor source is established where the detection events and non-detection events are taken into account. Third, each robot can understand the search environment by using Shannon's entropy which can be calculated in terms of the posteriori probability distribution on the position of the odor source. Moreover, each robot should move toward the direction of the entropy reduction. By means of this principle, the particle swarm optimization algorithm is introduced to plan the movement of the robot group. Finally, the effectiveness of the proposed approach is investigated for the problem of odor source localization.

1 Introduction

In nature, the creatures usually make use of the odor to find food and attract mates [1–3]. For example, the moths can find mates in the air; the lobsters can seek food under water; and the rats can avoid predators on land. With the rapid development of the chemical sensor technologies, robots can be used to imitate these creatures to conduct some searching tasks under the extreme environments [4–6], such as the detection of harmful gases leak, the search of fires source, and military missions, to name a few. These tasks can be formulated as the problem of odor source localization. In the last two decades, the problem of odor source localization has been widely studied based on a single robot in the science and engineering field. Correspondingly, the various solutions such as chemotaxis [14, 15], anemotaxis [5] and infotaxis [16] have been proposed.

Recently, how to use the multiple mobile robots to locate the odor source has received much attention from researchers due to a major benefit over a single robot, i.e. a wider detection range, which can enable the robot group to better capture the time-varying plume [7–9, 12]. As a swarm intelligence technique,

M. Fei et al. (Eds.): LSMS/ICSEE 2014, Part II, CCIS 462, pp. 140–148, 2014.
© Springer-Verlag Berlin Heidelberg 2014

the particle swarm optimization (PSO) algorithm has been well applied in co-ordinating the multiple mobile robots to search for the odor source [12, 13]. For example, Jatmiko et al. (2007) [13] proposed the charged PSO algorithm (CPSO) to coordinate the multiple mobile robots where two types of robots (neutral and charged robots) are used to search for the odor source. On the basis of the CPSO algorithm, Jatmiko et al. (2007) [13] further gave two wind utility algorithms: one is the WUI-45 algorithm while the other is the WUII algorithm. For the WUI-45 algorithm and the WUII algorithm, the wind information is simply employed to guide the movement direction of the robot group. By analyzing the PSO algorithm, Lu and Han (2011) [11] put forth a probability particle swarm optimization with information-sharing mechanism (PPSO-IM) algorithm to control the robot group. It is worth mentioning that the PPSO-IM algorithm make use of swarm information to model the probability distribution on the position of the odor source such that the robot group is guided to search for the appropriate range with a higher probability. It is worthwhile to note that the the aforementioned PSO algorithms mainly employ the concentration information to predict the position of the odor source and then to plan the movement direction of the robot group. However, the concentration cues pointing toward the location of the source are not always available because mixing in a flowing medium breaks up regions of high concentration into random and disconnected patches. In order to deal with this issue, a method based on Shannon's entropy was proposed to guide a single robot to locate the odor source [16]. Specifically, according to the priori knowledge about the odor dispersion, at each time, the robot would move towards the reduction direction of entropy. However, there exist two issues for this method based on Shannon's entropy. On one hand, this method requires priori knowledge, that is to say, the robot needs to know the odor dispersion model before the search task. If the odor dispersion model is not appropriate for the real odor distribution, this method will not obtain a better search result. On the other hand, this method is only used to control a single robot rather than the multi-robot system, which results in that this method cannot be used to coordinate the multi-robot system. Therefore, how to deal with two issues is the motivation of the current study.

To sum up, we will deal with the problem of odor source localization based on Shannon's entropy by using a multi-robot system. Firstly, we will briefly describe a measurement model. By this measurement model, an observed position can always be available. When the detection events occur, i.e., the robot detects the odor cues, the observed position is in the vicinity of the real position of the odor source with a higher probability. When the non-detection events occur, i.e., the robot group loses the odor clues, the observed position is not in the vicinity of the real position of the odor source with a higher probability. Secondly, we will establish the posterior probability distribution on the position of the odor source based on the measurement model. Thirdly, we will design the particle swarm optimization based on Shannon's entropy, which means that the robot group always moves towards the direction of the entropy reduction. Finally, the effectiveness of the particle swarm optimization based on Shannon's entropy is investigated for the problem of odor source localization.

2 Particle Swarm Optimization Based on Shannon's Entropy

In this section, we will first describe the measurement model. Then, we will establish the posteriori probability distribution on the position of the odor source. Finally, we will plan the movement of the robot group based on the particle swarm optimization algorithm.

2.1 Measurement Model

The measurement model of filament has been established in [7]. Here, we make a brief introduction about the measurement model. Movement process of a single filament can be given by

$$\dot{x}(t) = \mu(t) + n(t) \tag{1}$$

where $x(t)$ denotes the position of filament at time t; $\mu(t)$ is the mean wind velocity at $x(t)$; $n(t)$ is a random process which satisfies a standardized normal distribution. Suppose the filament released from the odor source at t_1, and then the filament position at t_k $(t_k > t_1)$ is

$$x(t_1, t_k) = \int_{t_1}^{t_k} \mu(\tau)d\tau + \int_{t_1}^{t_k} n(\tau)d\tau + x_0(t_1) \tag{2}$$

where $x_0(t_1)$ denotes the real position of the odor source at time t_1. Further, we need to discretize the movement process of filament

$$\int_{t_1}^{t_k} \mu(\tau)d\tau \approx \sum_{i=t_1}^{t_k} \mu(i)\Delta t \tag{3}$$

Let

$$v(t_1, t_k) = \sum_{i=t_1}^{t_k} \mu(i)\Delta t \tag{4}$$

and

$$w(t_1, t_k) = \int_{t_1}^{t_k} n(\tau)d\tau \tag{5}$$

Assume that the position of the odor source is stationary as

$$x_0(t_k) = x_0(t_1) \tag{6}$$

The equation (2) can be simplified as

$$x(t_1, t_k) = x_0(t_k) + v(t_1, t_k) + w(t_1, t_k) \tag{7}$$

Since time t_l is unknown, it can be regarded as the start time of the search process. Hence,

$$\overline{v}(t_k) = \frac{1}{k} \sum_{t_1=0}^{t_{k-1}} v(t_1, t_k) \tag{8}$$

Similarly

$$\overline{w}(t_k) = \frac{1}{k} \sum_{t_1=0}^{t_{k-1}} w(t_1, t_k) \tag{9}$$

where $\overline{w}(t_k)$ is a Gaussian process with zero mean and $\frac{1}{k} \sum_{t_1=0}^{t_{k-1}} (t_k - t_1)\sigma^2$ variance. In particular, $v(t_1, t_k)$ is an effective movement distance which should satisfy $\|v(t_1, t_k) - x_i(t_k)\|_2 < \beta$, $\|.\|$ is 2-norm. β is a control parameter which is significant for the quality of data obtained. If the inequality $\|v(t_1, t_k) - x_i(t_k)\|_2 < \beta$ is not satisfied, moving distance of filament can be ignored. Therefore, parameters should be given to comply with the real search environment. $x_i(t_k)$ denotes the current position of the ith robot at time t_k. Next, the discrete-time index k can replace the t_k, and then the equation (7) can be simplified as

$$x(k) = x_0(k) + \overline{w}(k) + \overline{v}(k) \tag{10}$$

Defining our measurement model is

$$z^i(k) = x_0(k) + \overline{w}(k) \tag{11}$$

where $z^i(k) = x(k) - \overline{v}(k)$ can be regarded as a measurement of $x_0(k)$ at time k for the ith robot and $\overline{w}(k)$ is viewed as a measurement noise.

2.2 Posterior Probability Distribution

It should be pointed out that the search range is divided into m grids and d_l denotes the center position of the lth grid. The posterior probability distribution on the position of the odor source is based on the above measurement model. A likelihood function can be given by

$$L_{d_l} = e^{- \int_0^t Q(z^i(t')|d_l) dt'} \prod_{j=1}^{H} Q(z^i(t_j)|d_l) \tag{12}$$

with

$$Q(z^i(k)|d_l) = e^{-\|z^i(k) - d_l\|} \tag{13}$$

where H is the number of detection events on the search path; t_j is the corresponding time; t' is the time at which non-detection events occur; and $z^i(k)$ is the measurement value of the ith robot on current position at time k. The posterior probability distribution is defined by

$$P_t(d_l) = \frac{L_{d_l}}{\int L_x dx} \tag{14}$$

$P_t(d_l)$ is probability where the position of the odor source locates on d_l at time t, and the posterior probability distribution combining (12), (13) with (14) is obtained as

$$P_t(d_l) = \frac{\exp[-\int_0^t e^{-\|z^i(t')-d_l\|}\mathrm{d}t'] \prod\limits_{j=1}^{H} e^{-\|z^i(t_j)-d_l\|})}{\int (\exp[-\int_0^t e^{-\|z^i(t')-x\|}\mathrm{d}t'] \prod\limits_{j=1}^{H} e^{-\|z^i(t_j)-x\|})\mathrm{d}x} \tag{15}$$

The detection probability is closely correlated with the distance between the measurement result $z^i(k)$ of the robot and d_l. If the robot encounters the filaments and the distance between d_l and $z^i(k)$ is smaller, then the probability that d_l is the position of odor source is higher. On the contrary, if the robot can not detect the filaments and the distance between d_l and $z^i(k)$ is relatively smaller, then the probability that d_l is not the position of odor source is higher.

2.3   Particle Swarm Optimization Based on Shannon's Entropy

If a probability distribution function is given, denoted by $P_t(d_l)$, the Shannon's entropy is

$$s_t^i = -\sum_m P_t(d_l) \ln P_t(d_l) \tag{16}$$

where s_t^i is the ith robot's entropy at time t; m is the grid number of the search range; d_l is the center position of the grid and $l \in \{1,,2,\ldots,m\}$. Moreover, we can get a position that corresponds to the maximum probability among all grids for the ith robot as

$$x_l(k) = \arg\max(P_t(d_1), P_t(d_2), \ldots, P_t(d_m)) \tag{17}$$

where $x_l(k)$ is the center position corresponding to the maximum probability among all grids for the ith robot. Let $s^1, s^2, \ldots, s^n$ be the corresponding previous minimum entropies for each robot and then we have a position that corresponds to the global minimum entropy.

$$x_g(k) = \arg\min(s^1, s^2, \ldots, s^n) \tag{18}$$

where $x_g(k)$ is the position corresponding to the global minimum entropy. According to the idea of the particle swarm optimization algorithm, we have

$$\begin{aligned} v_i(k+1) &= v_i(k) + u_i(k) \\ x_i(k+1) &= x_i(k) + v_i(k+1) \end{aligned} \tag{19}$$

with

$$u_i(k) = (\omega - 1)v_i(k) + \alpha_1(x_l(k) - x_i(k))$$
$$+\alpha_2(x_g(k) - x_i(k)) \tag{20}$$

where ω, α_1, and α_2 are the same parameters with the ones used by the standard particle swarm optimization algorithm. $x_i(k)$ is the position of the ith robot while $v_i(k)$ is the velocity of the ith robot.

3   Simulation Results

In this section, we will adopt Farrell's odor model to build a simulation environment to verify the feasibility of our algorithm. The parameters of simulation environment are shown in Table 1, where C_1, C_2, C_3, C_4, C_5 and C_6 denote the different initial wind velocities and σ_1, σ_2, σ_3, σ_4, σ_5 and σ_6 refer to the different spectral densities. The dispersion velocity can be estimated by the different spectral density. The higher spectral density means the faster dispersion velocity. The parameters $v_{max}= 0.8$ m/s and $\omega_{max} =1.57$ rad/s are used to limit the maximum linear velocity and angular velocity of robots, respectively. The control parameter $\alpha=0.85$.

Table 1. The parameters of the Farrell's odor dispersion model

Variables	Values
Area (m×m)	100×100
Odor source position (m)	(10,0)
Q	5123.7618
K_x, K_y	10,10
Growth rate	0.001
Initial wind velocity (m/s)	$C_1 - C_6$
Spectral density	$\sigma_1 - \sigma_6$
Max linear velocity (m/s)	0.8
Max angular velocity(rad/s)	1.57

The six cases about spectral densities and velocities are shown in Table 2. For Case 4, the wind velocity is faster and dispersion velocity is also faster, which will result in the wider and fast fluctuant plume. Instead, for Case 2, the wind velocity and dispersion velocity are smaller, which will result in the narrower and gradually changed plume. A circle is predefined, which has the radius of 1 m. The real position of the odor source is viewed as the center of circle and the search task is finished when any robot enters the circle.

Fig. 1 shows odor source localization process of five robots. In Fig. 1(a), the initial positions of five robots are set at the right up corner in the search region

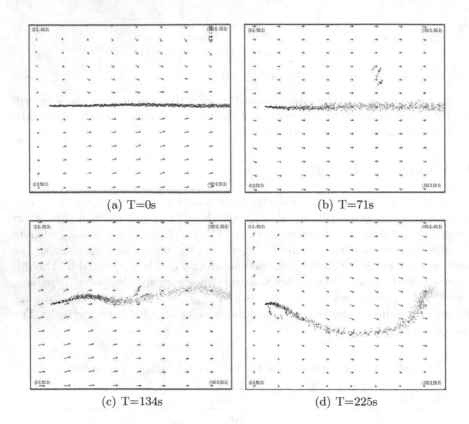

(a) T=0s

(b) T=71s

(c) T=134s

(d) T=225s

Fig. 1. The search process of five robots

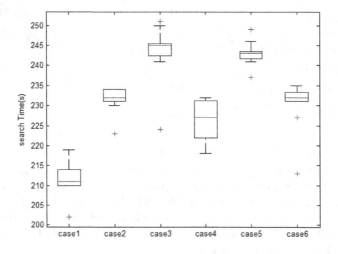

Fig. 2. The search time of five robots for six cases

Table 2. The six cases

Cases	The spectral density	The initial wind speed
Case 1	(4, 2)	(1, 0)
Case 2	(1, 1)	(0.6, 0)
Case 3	(7, 4)	(1.3, 0)
Case 4	(9, 7)	(2, 0)
Case 5	(6, 3)	(1.2, 0)
Case 6	(5, 2)	(1.1, 0)

Table 3. The success rate based 75 runs

Cases	Case 1	Case 2	Case 3	Case 4	Case 5	Case 6
Success rate	100	100	98.7	98.7	100	100

at $T=0$s. Each robot has different color. The black dots denote the filaments that form a plume. The blue arrows denote the wind speed and direction. Fig. 1(b) and Fig. 1(c) show the search process at $T = 71$ s and $T = 134$ s, respectively. In Fig. 1(d), the odor source is found at $T = 225$ s. The success rates obtained by the proposed algorithm is shown in Table 3. Fig. 2 shows the statistical results of six cases for the search time.

4   Conclusion

The problem of odor source localization has been addressed. The measurement model, which is designed based on wind information, has been described. By the measurement, a position is always observed in order to judge the real position of the odor source. Then, we have established the posteriori probability distribution on the position of the odor source. Next, in terms of Shannon's entropy, we have planned the movement of the robot group, where the idea from the PSO algorithm is introduced. Finally, the effectiveness of the particle swarm optimization based on Shannon's entropy has been investigated for the problem of odor source localization. Furthermore, the particle swarm optimization based on Shannon's entropy can be further improved and the new research result can be found in [10].

Acknowledgments. The research work was supported in part by the National Natural Science Foundation of China under Grants 61375104, 61203025, 61175093, and the Zhejiang Provincial Natural Science Foundation of China under Grant No. LQ13F030014.

References

1. Settles, G.S.: Sniffers: Fluid-Dynamic Sampling for Olfactorytrace Detection in Nature and Homeland Security. Journal of Fluids Engineering 127, 189–218 (2005)
2. Hayes, A.T., Martinoli, A., Goodman, R.M.: Distributed Odor Source Localization. IEEE Sensors Journal 2, 260–271 (2002)
3. Zimmer, R.K., Butman, C.A.: Chemical Signaling Processes in Themarine Environment. Biological Bulletin 198, 168–187 (2000)
4. Farrell, J.A., Murlis, J.X., Long, Z., Li, W., Card, R.T.: Filament-Based Atmospheric Dispersion Model to Achieve Short Time-Scale Structure of Odor Plumes. Environment Fluid Mechanics 2, 143–169 (2002)
5. Farrell, J.A., Pang, S., Li, W.: Plume Mapping via Hidden Markov Methods. IEEE Transactions on System, Man, Cybernetics: Part B, Cybernetics 33, 850–863 (2003)
6. Pang, S., Farrell, J.A.: Chemical Plume Source Localization. IEEE Transactions on System, Man, Cybernetics: Part B, Cybernetics 36, 1068–1080 (2006)
7. Lu, Q., Liu, S., Xie, X., Wang, J.: Decision Making and Finite-Time Motion Control for a Group of Robots. IEEE Transactions on Cybernetics 43, 738–750 (2013)
8. Lu, Q., Han, Q.-L., Liu, S.: A Finite-Time Particle Swarm Optimization Algorithm for Odor Source Localization. Information Sciences 277, 111–140 (2014)
9. Lu, Q., Han, Q.-L., Xie, X., Liu, S.: A Finite-Time Motion Control Strategy for Odor Source Localization. IEEE Transactions on Industrial Electronics 61, 5419–5430 (2014)
10. Lu, Q., He, Y., Wang, J.: Localization of Unknown Odor Source Based on Shannon's Entropy Using Multiple Mobile Robots. In: 40th Annual Conference of the IEEE Industrial Electronics Society. IEEE Press, New York (2014)
11. Lu, Q., Han, Q.-L.: A Probability Particle Swarm Optimizer with Information-sharing Mechanism for Odor Source Localization. In: 18th World Congress of the International Federation on Automatic Control, pp. 9440–9445. IFAC Press (2011)
12. Lu, Q., Luo, P.: A Learning Particle Swarm Optimization for Odor Source Localization. International Journal of Automation and Computing 8, 371–380 (2011)
13. Jatmiko, W., Sekiyama, K., Fukuda, T.: A PSO-Based Mobile Robot for Odor Source Localization in Dynamic Advection-Diffusion with Obstacles Environment: Theory, Simulation and Measurement. IEEE Computational Intelligence Magazine 2, 37–51 (2007)
14. Russell, R., Bab-Hadiashar, A., Shepherd, R., Wallace, G.: A Comparison of Reactive Chemotaxis Algorithms. Robotics and Autonomous Systems 45, 83–97 (2003)
15. Lytridis, C., Kadar, E.E., Virk, G.S.: A Systematic Approach to the Problem of Odour Source Localisation. Autonomous Robots 20, 261–276 (2006)
16. Vergassola, M., Villermaux, E., Shraiman, B.: 'Infortaxis' as a Strategy for Searching Without Gradients. Nature 445, 406–409 (2007)

Face Detection and Tracking Based on Adaboost CamShift and Kalman Filter Algorithm

Kun Chen, ChunLei Liu, and Yongjin Xu

Department of Automation, College of Mechatronics Engineering and Automation,
Shanghai University, Shanghai Key Laboratory of Power Station Automation Technology,
200072 Shanghai, China
chenkun10086@163.com, liuchunlei602A@126.com,
xuyj@mail.shu.edu.cn

Abstract. Face detection is an important component of the intelligent video surveillance system. Based on the MeanShift algorithm, we have developed into the CamShift algorithm. Although the traditional Camshift algorithm can track the moving object well, it has to set the tracking object by manually. Meanwhile it fails to track the object easily while the object is occluded and interfered by the same color obstructions. In order to solve the problem, according to the CamShift algorithm features, in this article, I will combine Adaboost, CamShift and Kalman filtering algorithm, which can be relied on to realize face detection and tracking automatically and accurately.

Keywords: Adaboost, CamShift, Kalman, face detection and tracking.

1 Introduction

In recent years, with the development of computer hardware technology, image sensor based on video surveillance system has developed rapidly [1]. Many of that facing the complex applications for background video surveillance system have appeared in large numbers. MeanShift algorithm is a non-parameter estimation algorithm, which uses the statistical characteristics, so it has strong robustness to noise. Moreover MeanShift has used the gradient optimization methods to reduce the time to search for matching features which makes it a strong real-time feature [2]. Based on MeanShift, Gary R.Bradski proposed CamShift (Continuously Adaptive Mean-Shift) algorithm, which overcomes the problem that the model could not be updated, etc. According to the color histogram, CamShift can adjust the window size automatically to fit the size of the tracked face in the image, which can effectively solve facial deformation and blocked problems. However, we find that CamShift has some problems, for example, we have to select the face manually, and it means inconvenient for us. In addition, it is easy to lose tracked objects facing with the same skin color and body [3]. For the above two issues, two different methods will be used to solve these two problems. For the former, Adaboost algorithm which based on the Harr classifier can be used to locate the position and extent of the face. For the latter, we need to introduce the state

M. Fei et al. (Eds.): LSMS/ICSEE 2014, Part II, CCIS 462, pp. 149–158, 2014.

estimator to predict the motion parameters of the target to reduce the misjudged rate. As we know, Kalman filter [4] is an optimal estimation error covariance in the minimum criteria, a small amount of calculation, real-time, and the actual motion parameters that can take advantage of constantly revised estimated value of the future to improve the state of motion estimation accuracy, timeliness and robustness. Therefore, we combine Camshift and Kalman filter to track faces that means automatic and accurate face tracking [5]. And we will firstly talk about the principle of algorithm and then try to do some experiments to confirm how this method works and its results.

2   Principle of Algorithm

2.1   CamShift Algorithm

The RGB color space is relatively sensitive to changes in illumination brightness. In order to reduce the impact of changes in the brightness of light tracking effect, Camshift algorithm will transform form RGB color space into HSV color space and then subsequent processing. So, Based on the target image color histogram model, Camshift algorithm will convert the image to color probability distribution, to initiate a search window size and position, and adaptively adjust the position and size of the search window based on the results obtained in the previous frame, and then locate the central location for the current image [6]. The flowing shows how the CamShift algorithm works:

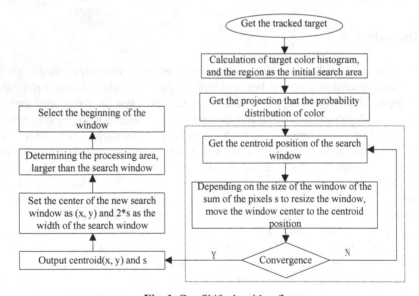

Fig. 1. CamShift algorithm flow

The following gives the flow of CamShift algorithm:

(1) Define the initial target and its regional. We define the point (x, y) as the pixel position in the search window. The point I(x, y) is the pixel value of (x, y) in projection map. So, we define M_{00} as the zero-order moment and M_{01}, M_{10} as the first moment:

$$M_{00} = \sum_x \sum_y I(x, y) \tag{1}$$

$$M_{01} = \sum_x \sum_y yI(x, y) \tag{2}$$

$$M_{10} = \sum_x \sum_y xI(x, y) \tag{3}$$

And then calculate the direction and size of the target track, and the second moment:

$$M_{20} = \sum_x \sum_y x^2 I(x, y) \tag{4}$$

$$M_{02} = \sum_x \sum_y y^2 I(x, y) \tag{5}$$

$$M_{11} = \sum_x \sum_y xyI(x, y) \tag{6}$$

$$(x_c, y_c) = (M_{10} / M_{00}, M_{01} / M_{00}) \tag{7}$$

And then define:

$$a = M_{20} / M_{00} - x_c^2 \tag{8}$$

$$b = 2(M_{11} / M_{00} - x_c y_c) \tag{9}$$

$$c = M_{02} / M_{00} - y_c^2 \tag{10}$$

(2) Adjust the search window size, and move the center of the search window to the center of mass. If the moving distance is greater than the preset fixed threshold, we recalculate the center of mass after adjustment until the window center and quality heart moving distance is less than a preset fixed threshold, or cyclic operation count reaches the maximum number of times, it is considered to meet the convergence conditions [7]. Compared to the current frame, we define the length and width of the search window for the next frame as l and w, namely:

$$1 = \sqrt{\frac{(a+c) + \sqrt{b^2 + (a-c)^2}}{2}} \tag{11}$$

$$w = \sqrt{\frac{(a+c) + -\sqrt{b^2 + (a-c)^2}}{2}} \tag{12}$$

2.2   Adaboost Algorithm

Adaboost algorithm is an iterative algorithm, and the core idea is to set the different classifiers for the same training, namely the weak classifiers, and then put these weak classifiers together to construct a stronger final classifier. The basic idea is: when the samples classified by classifier, we reduce the weight of samples [8]; otherwise, increase the weight of samples, so that the learning algorithm concentrate harder training samples for comparative study in the subsequent study, eventually getting the ideal classifier recognition rate and we design such a weak classifier for face detection:

$$h(x, f, p, \theta) = \begin{cases} 1, & \text{if } pf(x) < p\theta \\ 0, & \text{otherwise} \end{cases} \tag{13}$$

Where x is a sample, f (x) is a characteristic of a training sample, θ is a threshold, and p indicates the direction of inequality. Firstly, in order to understand the algorithm better, I will show you the flow of Adaboost algorithm as follows:

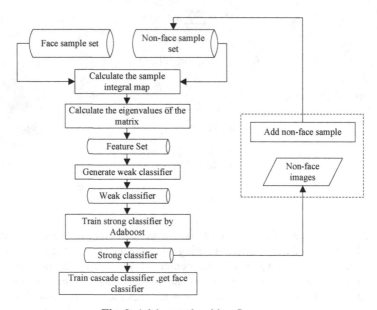

Fig. 2. Adaboost algorithm flow

The following gives the flow of Adaboost algorithm:

(1) Given samples image (x1, y1), (x2, y2), ... , (xn, yn), where yi=0 ,1 means non-face samples and face samples.
(2) Initialize the weights:

$$w_{1,i} = \frac{1}{2m}, \frac{1}{2l} \qquad (14)$$

Where m, l means the number of non-face samples and face samples.
(3) Normalized weights:

$$\frac{w_{t,i}}{\sum\limits_{j=1}^{n} w_{t,j}} \to w_{t,i} \qquad (15)$$

Where t=1, 2, ..., T
(4) Choose the best weak classifier:

$$\varepsilon_t = \min \sum_i w_i |h(x_i, f, p)| \qquad (16)$$

And set $h_t(x) = h(x, f_t, p_t, \theta_t)$
(5) Update weights : $w_{t+1,i} = w_{t,i} \varphi_t^{1-e_i}$, if the sample x_i is correctly classified and set $e_i = 0$ else $e_i = 1$, and $\varphi_t = \frac{\varepsilon_t}{1-\varepsilon_t}$
(6) Get the strong classifier:

$$c(x) = \begin{cases} 1, & \sum\limits_{t=1}^{T} \alpha_t h_t(x) \geq \frac{1}{2}\sum\limits_{t=1}^{T} \alpha_t \\ 0, & otherwise \end{cases} \qquad (17)$$

Where $A(k) = \begin{pmatrix} 1 & 0 & T & 0 \\ 0 & 1 & 0 & T \\ 0 & 0 & 1 & 0 \\ 0 & 0 & 0 & 1 \end{pmatrix}$

2.3   Kalman Algorithm

The kalman filter, also known as linear quadratic estimation, is an algorithm that uses a series of measurements observed over time, containing noise (random variations) and other inaccuracies, and produces estimates of unknown variables that tend to be more precise than those based on a single measurement alone [9].

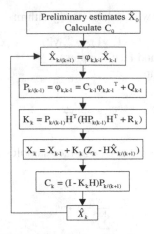

Fig. 3. Kalman algorithm flow

It contains two equations: prediction equation and observation equation:
Signal equation:

$$X(k) = A(k-1)X(k-1) + W(k) \qquad (18)$$

Observation equation:

$$Z(k) = H(k)X(k) + V(k) \qquad (19)$$

Where $X(k)$ is the state vector of the system and $Z(k)$ is the observation vector of the system at $t(k)$, $A(k-1)$ is the state transition matrix, $H(k)$ is the observation matrix. Dynamic noise $W(k)$ and observation noise $V(k)$ is uncorrelated as follows:

$$P(W(k)) \sim N(0, Q(k)) \qquad (20)$$

$$P(V(k)) \sim N(0, R(k)) \qquad (21)$$

For this part, to track face, three steps as follows:

(1) Initializing Kalman filter. Set the initial value for Kalman state parameters.
(2) State prediction. The position of the search window for the next frame will be predicted before searching.
(3) Status update. Updating the state of the filter.

During the tracking process, as the two adjacent frame images at a short interval of time (typically tens of milliseconds), the smaller target change the state of motion of face, a face can be approximated that the target uniform linear motion in the two frame interval.

Here, define T as the interval, $X(k) = (x(k), y(k), v_x(k), v_y(k))^T$ as the state vector, $Z(k) = (x(k), y(k))^T$ as the observation state vector, $x(k), y(k)$ as the position of the face in the center of the image, $v_x(k), v_y(k)$ as the speed of face moving.

So, define the state transition matrix of system:

$$H(k) = \begin{bmatrix} 1 & 0 & 0 & 0 \\ 0 & 1 & 0 & 0 \end{bmatrix} \tag{22}$$

And observation matrix:

$$H(k) = \begin{bmatrix} 1 & 0 & 0 & 0 \\ 0 & 1 & 0 & 0 \end{bmatrix} \tag{23}$$

In short, I plan to use Adaboost algorithm to find the face automatically not chose the tracking face manually. In this way, the workload will be reduced further, and make us feel more convenient. Meanwhile, by using the Kalman algorithm, we can distinguish the part which does not belong to the face but has the same color with face. Now, I will elaborate on the combination of the three methods [10].

3   Face Detection and Tracking System by Combining Adaboost, CamShift and Kalman

As mentioned above, the Adaboost algorithm will used to search the face from the image immediately the video system works. Of course, the video should contain face. And then Camshift tracking algorithm will be initialized automatically. Because Camshift algorithm is real-time tracking and non-rigid object, morph targets, and the rotation has better adaptability, but it does not make use of target direction of movement in space and velocity information, when there is a serious ambient occlusion or target motion during fast and easy to lose the target . So it will work well and CamShift algorithm is also a good algorithm. However, when the serious ambient happens, it seems worried. Therefore, we need to do something to improve it. Kalman algorithm is a good choice. As we know, Kalman filter is a sequence of linear dynamic systems minimum variance estimation algorithm that can accurately predict the position and velocity of the target, and a small amount of calculation, real-time computing. When tracking, according to the result of the Adaboost algorithm, we set $x(0)$, $y(0)$ as the center of the search window of CamShift algorithm, where $v_x(0) = 0$, $v_y(0) = 0$. By using CamShift algorithm, we calculate the position and the size of the tracking window. Define $Z(k)$ as the center of mass calculated by Camshift algorithm output window is a measure value of Kalman Filtering. We should correct the face position of prediction, and set the center position for the search window of the number of $(k+n)$ frame according to the center of mass: $X(k+n|k)$ by tracking human face through Kalman filter. By this way, we can realize the predictable Camshift tracking, which not only saves time and improves tracking efficiency, but also can overcome severe occlusion caused by defects in the target face losing. We have narrated a lot and I will let you know by doing some experiments.

4   Results and Analysis

Based on the platform of OpenCV, firstly, we test the Adaboost algorithm and find that it can get the human face quickly and automatically. The result achieves the excepted goals.

(a)   Detection with Adaboost   (b)   Detection without Adaboost

(c) Face detection for many people with Adaboost

Fig. 4. Face detection results

From the above we can see that the faces can be detected automatically and quickly. No matter the image contains one face or more faces, the results have achieved the expected results.

Then, when the face was blocked by some obstacles, we also need to track the face. Fig.5 shows the result by comparing the traditional CamShift algorithm with the algorithm of this paper. Picture a shows the CamShift algorithm that passing through the obstacle. Picture b shows the mixed method when dealing with the situation.

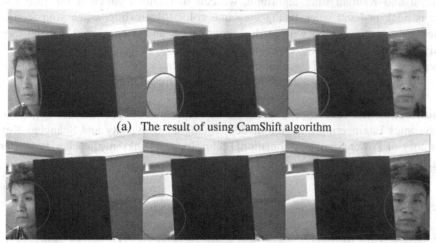

(a)　The result of using CamShift algorithm

(b)　The result of combining CamShift algorithm and Kalman filter

Fig. 5. Two different ways dealing with blocked

From above, we can see that if we use the traditional CamShift algorithm, you will not track the face if blocked. However, we are able to give the forecast estimates on the motion information by using Kalman filter algorithm. In this way, after seconds, the face will be tracked as fast as possible so that we will not lose the target. Now, let us make a thorough inquiry for this when the same color appears. Generally speaking, the neck and the hand will interference the result. We call it the Mixed algorithm. In order to express my idea, I will list it in this graph.

Table 1. Comparison of the two methods

Items	CamShift Algorithm	The Mixed Algorithm
Automatic detection	It can not	Yes, it can
Deal with obstacles	Losing the target	Track the target well
Processing speed	Fast	Slow

In all, by these experiments, the method proposed by this article can fulfill my requirements, and solve the occlusion problem effectively.

5   Conclusion

Based on Camshift tracking algorithm, we propose the algorithm that combining Adaboost, Camshift and Kalman algorithm to track the face. Maybe, when using the method, it can cause some bad in performance. However, it has more advantages. On the one hand, by mixing Camshift algorithm and Adaboost algorithm, it solves the trouble that the face should selected by yourself which becomes inaccurate with the passage of time. So, it is auto and accurate. On the other hand, by mixing CamShift algorithm and Kalman filter algorithm, we can distinguish which is the face and where is neck or hand when facing the same color. Meanwhile, when your face is blocked by something, you will not lose your target and it can search for the face quickly and accurately. The experimental results show that the method is high efficiency of the method of calculation, real-time to track moving human face, and it is good to achieve a rapid, automatic and accurate for face detection and tracking.

Acknowledgments. This work is supported by Department of Automation, College of Mechatronics Engineering and Automation, Shanghai University and Shanghai Key Laboratory of Power Station Automation Technology. Thanks my teacher Professor YongJin Xu.

References

1. Sorenson, H.W.: Kalman filtering: Theory and application. IEEE Press, New York (1985)
2. Viola, P., Jones, M.: Robust real-time face detection. Int. J. Computer Vision 57, 137–154 (2004)
3. Liu, S.R., Jiang, X.Y.: A moving object tracking algorithm based on modified CamShift/Kalman filter. Control Engineering of China (2010)
4. Kalman, E.: A new approach to linear filtering and prediction problems. Transaction of the ASMF: Journal of Basic Engineering, 35–45 (1960)
5. Wang, H.Z., Suter, D., Schindler, K.: Adaptive object tracking based on an effective appearance filter. IEEE Trans on Pattern Analysis and Machine Intelligence 9(29), 1661–1667 (2007)
6. Comanieiu, D., Ramesh, V., Meer, P.: Kemel-based object tracking. IEEE Trans. on Pattem Analysis and Maehine Intelligenee 25(5), 564–577 (2003)
7. Wei, Y., Zhang, T.W.: A new method for detection of moving targets in complex scenes. Jounal of Computer Research and Development 35(8), 724–728 (1998)
8. Xin, Y.H., Yang, W.H.: Pseudo-linear Kalman Filter Based Passive Location and Tracking Techniques by Two Infrared Stations. Journal of Xidian University 31(4), 505–508 (2004)
9. http://en.wikipedia.org/wiki/Kalman_filter
10. Zhu, C.H.: Face detection and tracking based on OpenCV. Computer Engineering and Applications 48(26), 157–161 (2012)

A Quick Method for Matching Object Subspaces Based on Visual Inspection

Wenju Zhou[1,2], Zixiang Fei[1], Huosheng Hu[1], Li Liu[2], and Jingna Li[2]

[1] School of Computer Science and Electronic Engineering, University of Essex,
Colchester, CO4 3SQ, UK
[2] School of Information and Electronic Engineering, Ludong University, Yantai 264025, China

Abstract. In visual inspection , the object image subspace should be segmented and matched, then the affine relationship is built between the template image and the sample image. But sometime the illumination is uneven on the surface of object image, it is difficult to obtain accurate position of the object subspace quickly. In this paper, a novel strategy is proposed to adopt discrete radial search paths instead of searching all points in the image. Therefore, the searching time can be reduced. In order to reduce the influence coming from the industrial environment, the paper proposes another method that is local energy level set segmentation, which can locate the object subspace quickly and accurately. The detection upon crown caps is as an example in the paper, then the detection effects and computing time are compared between several detection methods, and the mechanism of inspecting has been analyzed. The industrial applications are also given in the paper.

Keywords: visual inspection, object subspace, fast match, level set, local energy function.

1   Introduction

In visual inspection, the target image contour should be extracted firstly from the original image captured from the industrial produce line. Therefore, the target subspace should be located accurately so that the subspace can be detected in detail. In the process of the industrial production, the spatial position of the target images always keeps changing because of mechanical vibration, the expansion and contraction of the support structure, the delays of signal and so on. Thus, the mapping relation between the template subspace and the sample subspace generate deviation, which will make the increasing difficulty of the defect detection for the target image at the next steps. Eventually, the detection results and accuracy will be influenced. Therefore, the mapping relation should be built between the subspace of the standard image and the sample image by locating and segmenting the sample subspace fast and accurately.

Among many methods of image segmentations and image matches, the energy function method for segmenting image has attracted many researchers attention. The basic idea of energy function method is to use curves to express the target contour. The process of segmentation will be turned into the problem to finding the energy minimum. It can be realized by solving the corresponding Euler (Eider-Lagrange)

M. Fei et al. (Eds.): LSMS/ICSEE 2014, Part II, CCIS 462, pp. 159–168, 2014.

equation. Then, the method of image segmentation can be divided into two categories based on edge information [1] and based on region information [2, 3]. The model based on edge information uses part edge information to look for boundary of the target, whose representatives are Active Contour Model and Snake Model [4]. These models mainly depend on the levels set of the edge information, which are irrelevant with the distribution of the global brightness. Therefore, it has the ability to segment uneven illumination images. However, the segmentation model based on edge information is very sensitive to noise and boundary leaking problem will happen easily in weak boundary of the image. That is, the models based on region information are proposed, for example Piecewise Constant (PC) [5] model, which drives the evolution from active contour toward the target boundary through defining the region description. The whole region characteristics are used for realizing image segmentation. This method has greater robustness than the edge-based segmentation method. But it is usually difficult to define the description of the region of the image since the description of the region is based on uniform illumination hypothesis. It cannot be carried out effective segmentation when the brightness of the image is uneven. Vese Chan and Tsai [6] proposed Piecewise Smooth(PS) [7] model for segmenting uneven illumination images, which turns the image segmentation into looking for the optimization and approximation of the block smooth function. So, this model can overcome the influence of the inconsistencies in brightness in some case. However, it needs search point by point over the whole image with very heavy computation load. It is very difficult to be used on the real-time detection on industrial line. Currently, the application of the level set method is mainly reflected in target tracking [8], image restoration [9], image denoising [10], robustness analysis of image segmentation [3], and so on.

To overcome these disadvantages lie in segmentation algorithm in target subspace, this paper proposes the novel method which uses local energy and discrete paths based on level set method to match the target subspace with the template subspace. Our method can locate the object subspace quickly and accurately. The crown caps detection is as an example in the paper, then the detection effects and computing time are compared between several detection methods, and the mechanism of inspecting have been analyzed. The results show that our method can satisfy the real-time requirement in the industrial applications.

This paper is organized as follows: we first describe our propose method in section 2 include discrete paths level set method and local energy level set method. The experiments results and their comparing for these methods are presented in section 3. Section 4 shows the several applications on industrial line. Finally, in section 5 we describe the conclusion and future works.

2   Discrete Path Search Strategy Based on Local Energy and Level Set

2.1   Path Settings for the Level Set

In order to match the target subspace to the template subspace, firstly, the edge of the target subspace should be found. Due to the fact that the level set method can reduce the disturbance to search image edge of the uneven lighting illumination. Level Set (LS) was firstly established in 1988 by Osher and Sethian [11], they used this method based on thermodynamics equation for solving the problem about the flame shape change process. It is difficult for the traditional parametric to express describe the change of the flame shape since the flame shape is dynamics and uncertainty of topological structure. Therefore Osher and Sethian proposed the level set method to descript the interface whose movement is time-dependent. Their main idea is to move the curve (surface) as the zero Level Set embedded in the higher dimensional function. Level set method as a numerical technique to model for tracking the shape interface. Its advantages are in two aspects. The first one is that the evolution of the curve (surface) only need to be numerical calculated on Cartesian coordinates and does not need to get the parametric of the curve (surface). The calculation process is carried out on the fixed grid which accords the Euler (Euler) framework. Another advantage is that it is convenient to deal with the topology of the evolutionary curve (surface) changes and effectively avoid the hard problems in the parametric curve process. The zero level set is the middle part between the negative area and the positive area on images. Assuming the curved surface function is ϕ, the evolution curve at the time t is $C(t)$, then the zero level set is $\phi(C(t),t)=0$. According to the composite function of derivation rules, derivation of $\phi(C(t),t)=0$ can obtain:

$$\frac{\partial \phi}{\partial t} + \nabla \phi \cdot \frac{\partial C}{\partial t} = 0 \quad . \tag{1}$$

Where $\nabla \phi$ is gradient, it is same orientation as normal vector.

According to the curve evolution theory, the normal unit vector on the curve is

$$\bar{N} = -\frac{\nabla \phi}{|\nabla \phi|} \quad . \tag{2}$$

And the curve evolution equation is

$$\frac{\partial C}{\partial t} = F\bar{N} \quad . \tag{3}$$

By substituting(2), (3) into(1) , we get the level set evolution equation

$$\frac{\partial \phi}{\partial t} = F|\nabla \phi| \quad . \tag{4}$$

In order to solve the equation(4), a partial time-dependent differential equation is needed, which is the Hamilton-Jacobi expression [11]. The final result can be obtained by separation variables, that is, the variables should be processed in time domain and space domain. Apparently, it is difficult and complex to solve equation(4). So, the level set can be initialized as the Signed Distance Function (SDF) in engineering. The SDF calculation complexity is $O(nm)$ [12], where n denotes the points number in a grid and m is the number of grids. In the initial case, the calculation is very large. In order to accelerate the speed to generate SDF, Sethian[12] proposed the fast marching method (FMM), which sorts the all the neighborhood points using the complete binary tree. So, the computation complexity of distance function is reduced from $O(nm)$ to $O(n\log m)$. In order to further reduce the computational complexity, The level set $\phi(x,y,t)$ can be represented using the discrete grid forms. Assume the space length of a discrete grid is h, the time pace is Δt, then the level set of grid point (i,j) at the time n is $\phi(ih, jh, n\Delta t)$ can be abbreviated as ϕ_{ij}^n. Therefore, evolution equation (4) can be discrete as following

$$\frac{\phi_{ij}^{n+1} - \phi_{ij}^n}{\Delta t} = F_{ij}^n \left| \nabla_{ij} \phi_{ij}^n \right| \tag{5}$$

Where F_{ij}^n denotes the value of speed function at time n. The equation (5) above can be solved using Upwind Finite Differential Method (UFDM). Firstly, six operators are defined as first-order difference, first-order forward difference and first-order back difference

$$\phi_x^0 = \frac{1}{2h}(\phi_{i+1,j} - \phi_{i-1,j}) \qquad \phi_y^0 = \frac{1}{2h}(\phi_{i,j+1} - \phi_{i,j-1})$$

$$\phi_x^+ = \frac{1}{h}(\phi_{i+1,j} - \phi_{i,j}) \qquad \phi_y^+ = \frac{1}{h}(\phi_{i,j+1} - \phi_{i,j}) \tag{6}$$

$$\phi_x^- = \frac{1}{h}(\phi_{i,j} - \phi_{i-1,j}) \qquad \phi_y^0 = \frac{1}{h}(\phi_{i,j} - \phi_{i,j-1})$$

Substitute (6)into(5), we can get

$$\phi_{ij}^{n+1} = \phi^n + \Delta t(\max(F_{ij}^n, 0)\nabla^+ + \min(F_{ij}^n, 0)\nabla^-) \quad . \tag{7}$$

Where ∇^+ and ∇^- are defined as follow,

$$\nabla^+ = [\max(\phi_x^-, 0)^2 + \min(\phi_x^+, 0)^2 + \max(\phi_y^-, 0)^2 + \min(\phi_y^+, 0)^2]^{1/2}$$
$$\nabla^- = [\max(\phi_x^+, 0)^2 + \min(\phi_x^-, 0)^2 + \max(\phi_y^+, 0)^2 + \min(\phi_y^-, 0)^2]^{1/2} \tag{8}$$

The equation (7) is used to segment image, the segment speed function is

$$F = F_{prop} + F_{curv} + F_{adv} \quad , \tag{9}$$

Where $F_{prop} = V_0$ denotes the evolution speed on the length, $F_{curv} = -\varepsilon\kappa$ is the evolution speed on the curvature, $F_{adv} = \vec{U} \cdot \vec{V}$ is convection velocity on the level where $\vec{U} = (u(x,y,t), v(x,y,t))$. u is the gradient on x orientation and v is the gradient on y orientation. Thus, equation (7) can be rewritten as

$$\phi_{ij}^{n+1} = \phi^n + \Delta t[(\max(V_{0ij}, 0)\nabla^+ + \min(V_{0ij}, 0)\nabla^-) + (\max(u_{ij}^n, 0)\phi_x^- + \min(u_{ij}^n, 0)\phi_x^+)$$
$$+ (\max(v_{ij}^n, 0)\phi_y^- + \min(v_{ij}^n, 0)\phi_y^+) - \varepsilon\kappa_{ij}^n((\phi_x^0)^2 + (\phi_y^0)^2)^{1/2}] \tag{10}$$

The level set can be updated continuously by(10). The step apace Δt should be met Courant-Friedrichs-Levy (CEL) as follow. In this case, the grid space h has been given

$$F \cdot \Delta t \leq h \tag{11}$$

Since equation (10) is calculated for the whole area in the image, the computation is enormous when the image size is large. In order to reduce the computation, Chop [13] proposed the idea of Narrow Band in 1993, and Adalsteinsson et al. [14] given its specific realize in 1995. The method is able to rapid evolution to obtain the value of a level set. The main principle is only to update the narrow band around the shape of the level set region. Due to the less grid points, the computation is greatly reduced. Although Narrow Band can reduce the calculation, it still not satisfies the real-time application in the industrial field. Therefore, we propose a discrete path searching method, whose search is along several fixed lines but not full image. The computational complexity of the search becomes 1D from 2D. The search space is shrunk greatly, and the search efficiency is improved. Thus, the real-time search can be satisfied.

In this paper, crown cap image is as an application example for exploring the level set method. Since the cap shape is a circular image, the search paths can along the direction of the radius, the search method is shown in Figure 1.

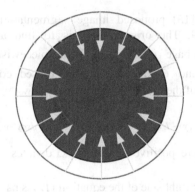

Fig. 1. Discrete paths and direction

As shown in Figure 1, the level set search is performed from the outer circle to inner circle. The search path is along the radius and the search direction is the same as the arrow direction. Because the arrow direction is almost orthogonally with the edge of the cap image, the search paths may be the shortest paths and the gradient may be the maximum along the paths. One search path can get one zero level set point. Thus, the number of level set points has been greatly reduced and the computing efficiency has been greatly enhanced.

Because the crown cap image is captured as it is being transferred on the produce line, the target and transfer equipment are often adhesive in the image, which is shown as Figure 2. In Figure 2, the cap left edge is together with the holding ratchet, which will cause inaccurate to locate the edge of the cap image. Therefore, this paper proposes the local energy level set method based on the discrete path level set search, which can accurately search caps edge according to the energy difference between the caps surface and the caps skirt.

Fig. 2. Crown cap image from industrial line

2.2 Local Energy Model

Mumford and Shah [15] proposed image segmentation model based on energy minimization in 1989. The original idea is to find a contour curve (I_0, C) to approximate a given image I, where I_0 is the piecewise smooth approximation of the original image, and C denotes smooth closed contour curves. The energy functional expression of the model is written as follow

$$E(I,C) = \int_{\Omega} |I(x,y) - I_0(x,y)|^2 dxdy + \mu \int_{\Omega \backslash C} |\nabla I(x,y)|^2 dxdy + v \cdot length(C) \quad . \tag{12}$$

Where μ and v are positive constants, Ω denotes image region, and $C \subset \Omega$ is the contour curve.

The first term on the right side of the equation (12) is named fidelity, which is used to express the similarity between the segment image and the original image. The second term denotes smoothness, which is used to ensure the smoothness of the segment regions. The third term is the constraint, which is used to constrain the length to reach the minimum. When the equation (12) obtains the minimum value on the left, the I and C on the right can get the desire results. However, the solution process for the equation (12) is very complicated, thus the Mumford-Shah model need be simplified in actual operation. If the Mumford-Shah model I_0 is simplified for the piecewise constant function, such as I_0 is the constant in each target area, Chan-Vese (CV) model [16] can be obtained. That is, CV is the simplified approximation of the Mumford-Shah model, which can be obtained by minimizing the energy function as following

$$E(u_1, u_2, C) = \mu \cdot Length(C) + \lambda_1 \cdot \int_{inside(C)} |I(x,y) - u_1|^2 dxdy + \lambda_2 \cdot \int_{outside(C)} |I(x,y) - u_2|^2 dxdy \quad . \tag{13}$$

Where μ, λ_1 and λ_2 are positive constants, they are usually constants as $\mu = \lambda_1 = \lambda_2 = 1$. u_1 and u_2 are the grey scale average at the outside and inside of the curve C. The first term of the energy function (13) is used to normalize the curve. The second and third together are as fidelity, their roles are to attract the curve to the target contour.

In order to obtain the minimum energy of the $E(u_1, u_2, C)$, the level set ideas is used, that is, the level set instead of the unknown evolution curve. When the points are inside of the curve, the level set is defined as $\varphi(x, y) > 0$. The points are outside of the curve, the level set is defined as $\varphi(x, y) < 0$. The points are on the curve, the level set is defined as $\varphi(x, y) = 0$. Thus, the equation (13) can be rewritten as

$$E(u_1, u_2, C) = \mu \int_\Omega \delta_\varepsilon(\phi(x, y)) |\nabla \phi(x, y)|^2 \, dxdy + \lambda_1 \cdot \int_\Omega |I(x, y) - u_1|^2 H_\varepsilon(\phi(x, y)) dxdy$$
$$+ \lambda_2 \cdot \int_\Omega |I(x, y) - u_2|^2 (1 - H_\varepsilon(\phi(x, y))) dxdy \tag{14}$$

Where $\delta_\varepsilon(z)$ and $H_\varepsilon(z)$ are Dirac and Heaviside. $\delta_\varepsilon(z)$ and $H_\varepsilon(z)$ are expanded to write as follows

$$\delta_\varepsilon(z) = \begin{cases} \dfrac{1}{2\varepsilon}[1 + \cos(\dfrac{\pi z}{\varepsilon})], & |z| \le \varepsilon \\ 0, & |z| > \varepsilon \end{cases} \tag{15}$$

$$H_\varepsilon(z) = \begin{cases} \dfrac{1}{2}(1 + \dfrac{z}{\varepsilon} + \dfrac{1}{\pi}\sin(\dfrac{\pi z}{\varepsilon})), & |z| \le \varepsilon \\ 1, & z > \varepsilon \\ 0, & z < -\varepsilon \end{cases} \tag{16}$$

The minimize result of the equation (14) can be solved by energy functional form of Euler-Lagrange. The following level set evolution equations can be obtained

$$\frac{\partial \phi}{\partial t} = \delta_\varepsilon(\phi)[\mu div(\frac{\nabla \phi}{|\nabla \phi|}) - \lambda_1(I - u_1)^2 + \lambda_2(I - u_2)^2], \tag{17}$$

$$\phi(0, x, y) = \phi_0(x, y) \qquad in \ \Omega, \tag{18}$$

$$\frac{\delta_\varepsilon(\phi)}{|\nabla \phi|}\frac{\partial(\phi)}{\partial \tilde{n}} = 0 \qquad on \ \partial\Omega. \tag{19}$$

Where the equation (18) is initial condition, and (19) is boundary. The grey scale u_1 and u_2 are updated by following

$$u_1(\phi) = \frac{\int_\Omega I(x, y) H_\varepsilon(\phi(x, y)) dxdy}{\int_\Omega H_\varepsilon(\phi(x, y)) dxdy},$$
$$u_2(\phi) = \frac{\int_\Omega I(x, y)(1 - H_\varepsilon(\phi(x, y))) dxdy}{\int_\Omega (1 - H_\varepsilon(\phi(x, y))) dxdy} \tag{20}$$

3   Experimental Results

In this paper, the results of the subspace matching experiments are compared among level set method, discrete path level set and local energy path level set method. In above matching experiments, the crown cap image is obtained on actual industrial line. According to the circular features of caps, the search path is pre-set to be along the radius of the caps' circle, which is shown in Figure 1. The experiments were carried out on an Intel i5 PC with 4G RAM. All the computations were performed with C#.

In order to compare the effectiveness and the efficiency of searching cap edge between different methods, the experiments in different paths is carried out. Figure 3 and Figure 4 are the crown cap images captured on the line detection system. Figure 3 shows the search of the crown caps seal side using discrete path level set method. Figure 4 shows the search of the crown caps surface side using discrete path level set method. The number of searching paths in two figures are 15, 30 and 60. As shown in figures, the edges of the seal side and surface side can be successfully grasped using discrete path level set method.

As show in Figure 3 and Figure 4, the skirts of the edge and the gripping device interference the searching for the edge of the cap, the fitted circles deviates actual cap seal circle and the surface circle. Although the error can be reduced by adding the number of search path, the error still persists there.

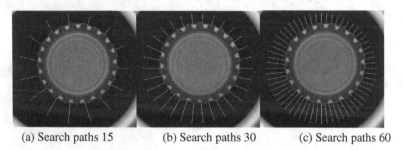

(a) Search paths 15   (b) Search paths 30   (c) Search paths 60

Fig. 3. Fitting the inner seal circle and edge circle

(a) Search paths 15   (b) Search paths 30   (c) Search paths 60

Fig. 4. Fitting the surface circle and edge circle

From equation(17), Figure 5 shows the results using the local energy discrete path level set method to search the edge of the bottle caps. The search paths in Figure 5 are the same as that of Figure 3 and Figure 4.

As shown in Figure 5, the upper images are the inner seals fitted and the down images are the surface circles fitted. The edge search results have been satisfactory as the number of search paths is 30.   So Figure 5 is only shown the experiments results when the number of search paths is 15 and 30. The search results and fitting with 60 paths are the same as that of 30 paths. Comparing fitted circles in Figure 5 and Figure 3, the local energy discrete path level set method is superior to the only discrete path level set method. In Figure 5, it is perfect match on the subspaces of bottle caps seals and surfaces by the circle spaces.

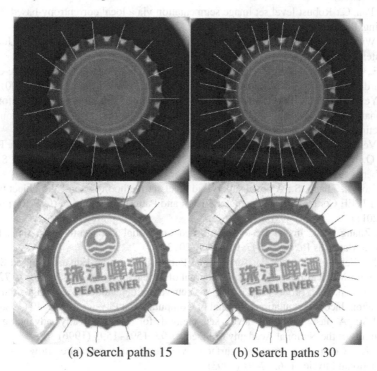

(a) Search paths 15   (b) Search paths 30

Fig. 5. Local energy discrete paths level set method

4   Conclusion

In this paper, the subspace match method based on the level set is studied, and the discrete paths level set method is proposed to raising the evolution speed of the level set. On this basis, the local energy discrete paths level set method is developed, which overcomes the disturbance raised by the uneven illumination and gripping device image. Experiments on practical images of bottle caps on the produce lines show that the proposed method is the most effective in searching the defect of caps. This method is robust for various species of crown caps. This precision subspace matching is the good foundation for further detecting defect of caps accurately.

References

Prisacariu, V.A., Reid, I.: Nonlinear shape manifolds as shape priors in level set segmentation and tracking. In: 2011 IEEE Conference on Computer Vision and Pattern Recognition (CVPR), pp. 2185–2192 (2011)

Zhang, K., Zhang, L., Song, H., Zhou, W.: Active contours with selective local or global segmentation: A new formulation and level set method. Image and Vision Computing 28, 668–676 (2010)

Wang, L., Pan, C.: Robust level set image segmentation via a local correntropy-based K-means clustering. Pattern Recognition 47, 1917–1925 (2014)

Kass, M., Witkin, A., Terzopoulos, D.: Snakes: Active contour models. International Journal of Computer Vision 1, 321–331 (1988)

Chan, T.F., Vese, L.A.: Image segmentation using level sets and the piecewise-constant Mumford-Shah model. Tech. Rep. 0014, Computational Applied Math Group (2000)

Tsai, A., Yezzi Jr., A., Willsky, A.S.: Curve evolution implementation of the Mumford-Shah functional for image segmentation, denoising, interpolation, and magnification. IEEE Transactions on Image Processing 10, 1169–1186 (2001)

Chan, T., Vese, L.: An active contour model without edges. In: Nielsen, M., Johansen, P., Fogh Olsen, O., Weickert, J. (eds.) Scale-Space 1999. LNCS, vol. 1682, pp. 141–151. Springer, Heidelberg (1999)

Sun, X., Yao, H., Zhang, S.: A novel supervised level set method for non-rigid object tracking. In: 2011 IEEE Conference on Computer Vision and Pattern Recognition (CVPR), pp. 3393–3400 (2011)

Jiang, X., Zhang, R.J.: Image Restoration Based on Partial Differential Equations (PDEs). Advanced Materials Research 647, 912–917 (2013)

Khanian, M., Feizi, A., Davari, A.: An Optimal Partial Differential Equations-based Stopping Criterion for Medical Image Denoising. Journal of Medical Signals and Sensors 4, 72 (2014)

Osher, S., Sethian, J.A.: Fronts propagating with curvature-dependent speed: Algorithms based on Hamilton-Jacobi formulations. Journal of Computational Physics 79, 12–49 (1988)

Sethian, J.A.: A fast marching level set method for monotonically advancing fronts. Proceedings of the National Academy of Sciences 93, 1591–1595 (1996)

Chopp, D.L.: Computing minimal surfaces via level set curvature flow. Journal of Computational Physics 106, 77–91 (1993)

Adalsteinsson, D.: A fast level set method for propagating interfaces. Citeseer (1994)

Mumford, D., Shah, J.: Optimal approximations by piecewise smooth functions and associated variational problems. Communications on Pure and Applied Mathematics 42, 577–685 (1989)

Chan, T.F., Vese, L.A.: Active contours without edges. IEEE Transactions on Image Processing 10, 266–277 (2001)

Research on Visual Environment Evaluation System of Subway Station Space

Fengqun Guo and Hui Xiao

College of Electronic and Information Engineering, Tongji University, Shanghai 201804

Abstract. Based on the energy crisis, LED with its energy-saving and environmental friendly is gradually used to the subway station space lighting. But now, there are little materials about the visual environment evaluation for semiconductor lighting, so that the use of LED lighting lacks theoretical basis and data support. So, in order to promote the LED lighting in subway station space, it's very important to evaluate the visual environment. Therefore, the core of this paper was to build a theoretical model to evaluate the visual environment of subway station space using Particle Swarm Optimization. Firstly, chose 16 evaluation indexes which were fit for the subway station visual environment evaluation and got the initial judgment matrix through pair wise comparison, after that, established the non-linear consistency correction model. Finally, used Particle Swarm Optimization to calculate the judgment matrix with better consistency and the corresponding index weight, and constructed the theoretical model.

Keywords: LED lighting, visual environment, evaluation system, particle swarm optimization.

1 Introduction

With fluorescent technology maturing gradually, fluorescent lamps are being widely used in subway station space lighting, but related research is mainly concentrated on the application of lighting technology. As the development of semiconductor technology and the demand for energy conservation, LED used in subway station space lighting becomes the hot topic in lighting area. At now, Shenzhen metro line 2 is the first metro line with LED for lighting directly in our country. However, there is less research on the visual environment with semiconductor lighting, and leading to the lack of theoretical basis and data supports for LED lighting in the subway station space. To promote the application of LED lighting in subway station space, this paper attempted to build a visual environment comprehensive evaluation system on the basis of subjective experience, to build the hierarchical organization of subway station space visual environmental assessment [1].

The core of this theoretical model included evaluation index and index weight. In this paper, chose 16 evaluation indexes which were fit for the subway station visual

M. Fei et al. (Eds.): LSMS/ICSEE 2014, Part II, CCIS 462, pp. 169–179, 2014.
© Springer-Verlag Berlin Heidelberg 2014

environment evaluation and got the initial judgment matrix through pair wise comparison, after that, established the non-linear consistency correction model. And then, used Particle Swarm Optimization to calculate the judgment matrix with better consistency and the corresponding index weight, and ultimately built a relationship model between objective evaluation and subjective feelings. Meanwhile, the theoretical model was applied to evaluate the visual environment of subway station space lighting to get the best subjective evaluation value.

2   Subway Station Space Visual Environment Evaluation Indexes

Subway station space visual environment evaluation indexes include two categories: the functionality and energy efficiency indicators, functional indicators are divided into light environment and space environment. Light environmental factors, include the level of illumination, illumination uniformity, brightness level, brightness distribution, glare index, color temperature, color rendering index, visual induction, recognition, and so on. Spatial environment factors include lamps appearance, three-dimensional, environmental coordination, light level, artistic lighting and so on. Energy-saving factors include control strategy, control means, lighting power density value (LPD), and energy saving lamps, new energy utilization.

This paper used Delphi method to filter out 16 indicators to build the index system: illumination level(D1), illumination uniformity(D2), brightness level(D3), brightness distribution(D4), glare(D5), color temperature(D6), color rendering(D7), visual induction(D8), recognition(D9), three-dimensional(D10), environmental coordination (D11), artistic lighting(D12), and control strategy(D13), control means(D14), lighting power density(D15), energy-saving lamps(D16), and established the hierarchical structure of visual environment evaluation.

3   Calculating the Index Weight Sector

The reasonable index weight is the prerequisite for the proper evaluation, after choosing sixteen indexes, calculating the index weight reasonably is the most important part of visual environment evaluation system. The accuracy of index weight is more related to the rightness and scientific of the final evaluation results. Based on this, this paper used AHP to construct hierarchical structure, created the initial judgment matrix, and used particle swarm optimization to calculate index weight so that established the city subway station space lighting visual environment evaluation system model [2]. The calculation process is as follows:

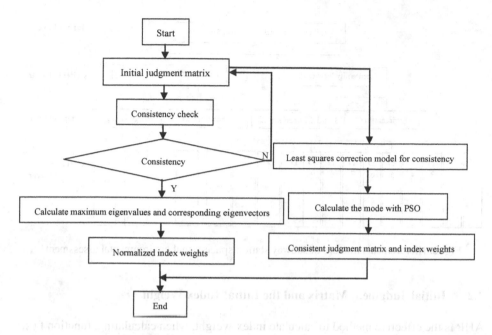

Fig. 1. Index weight calculating process

3.1   Hierarchical Structure of Subway Station Space Visual Environmental Evaluation

The key of AHP is to establish hierarchical relationships of indexes, to decompose a complex decision problem into a number of interrelated levels, the uppermost layer named target, then the layer to reflect the target characteristics named criteria, followed by an index layer and sub-index layer to reflect criteria characteristics, and the last layer is the solution layer composed of evaluated objects. Through the analysis of indexes, built the hierarchical structure of subway station space lighting visual environmental evaluation hierarchical structure [3], as Figure 2:

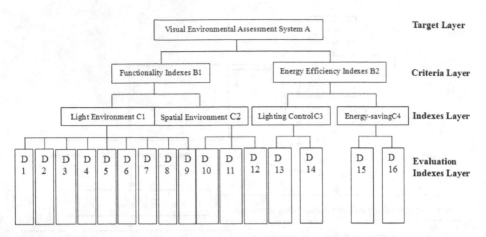

Fig. 2. Hierarchical structure of subway station space visual environmental assessment

3.2　Initial Judgment Matrix and the Initial Index Weight

AHP is the effective method to calculate index weight, when calculating, function f (x, y) indicates the importance between x and y. If f (x, y)> 1, indicates x is more important than y, if f (x, y) <1, y is described more important than x, if and only if f (x, y) = 1, indicating x and y is equally important, and f (y, x) = 1 / f (x, y).

For the subway station space visual environment evaluation, taking $U = \{u_1, u_2, ..., u_n\}$ as the index collection, by pairwise comparison, got the results matrix $A^0 = (a_{ij})_{n \times n}$, in it, $a_{ij} = f(u_i, u_j)$, where A^0 is called the initial judgment matrix.

$$A^0 = \begin{bmatrix} 1 & 3 & 1 & 1/3 & 3 & 3 & 4 & 1/2 & 1/2 & 5 & 7 & 9 & 5 & 5 & 1 & 5 \\ 1/3 & 1 & 1/3 & 1/5 & 2 & 2 & 3 & 1/3 & 1/4 & 4 & 6 & 8 & 4 & 4 & 1/2 & 4 \\ 1 & 3 & 1 & 1/3 & 3 & 3 & 4 & 1/2 & 1/2 & 5 & 7 & 9 & 5 & 5 & 1 & 5 \\ 3 & 5 & 3 & 1 & 7 & 8 & 9 & 5 & 7 & 9 & 9 & 9 & 6 & 6 & 5 & 8 \\ 1/3 & 1/2 & 1/3 & 1/7 & 1 & 3 & 2 & 1/5 & 1/5 & 3 & 5 & 7 & 2 & 2 & 1/7 & 1/3 \\ 1/3 & 1/2 & 1/3 & 1/8 & 1/3 & 1 & 1 & 1/3 & 1/3 & 3 & 5 & 7 & 2 & 2 & 1/3 & 1/2 \\ 1/4 & 1/3 & 1/4 & 1/9 & 1/2 & 1 & 1 & 1/3 & 1/3 & 3 & 5 & 7 & 2 & 2 & 1/3 & 1/2 \\ 2 & 3 & 2 & 1/5 & 5 & 3 & 3 & 1 & 1 & 5 & 7 & 9 & 1 & 1 & 1/3 & 1 \\ 2 & 4 & 2 & 1/7 & 5 & 3 & 3 & 1 & 1 & 5 & 7 & 9 & 1 & 1 & 1/3 & 1 \\ 1/5 & 1/4 & 1/5 & 1/9 & 1/3 & 1/3 & 1/3 & 1/5 & 1/5 & 1 & 1 & 3 & 1/5 & 1/5 & 1/7 & 1/5 \\ 1/7 & 1/6 & 1/7 & 1/9 & 1/5 & 1/5 & 1/5 & 1/7 & 1/7 & 1 & 1 & 3 & 1/5 & 1/5 & 1/7 & 1/5 \\ 1/9 & 1/8 & 1/9 & 1/9 & 1/7 & 1/7 & 1/7 & 1/9 & 1/9 & 1/3 & 1/3 & 1 & 1/7 & 1/7 & 1/9 & 1/8 \\ 1/5 & 1/4 & 1/5 & 1/6 & 1/2 & 1/2 & 1/2 & 1 & 1 & 5 & 5 & 7 & 1 & 1 & 1/3 & 1 \\ 1/5 & 1/4 & 1/5 & 1/6 & 1/2 & 1/2 & 1/2 & 1 & 1 & 5 & 5 & 7 & 1 & 1 & 1/3 & 1 \\ 1 & 2 & 1 & 1/5 & 7 & 3 & 3 & 3 & 3 & 7 & 7 & 9 & 3 & 3 & 1 & 3 \\ 1/5 & 1/4 & 1/5 & 1/8 & 3 & 2 & 2 & 1 & 1 & 5 & 5 & 8 & 1 & 1 & 1/3 & 1 \end{bmatrix}$$

3.3   Building the Least Squares Consistency Correction Model

AHP is a subjective weighting method, the subjective choice and preferences play a very important role in this decision-making problem and have great unreliability. When solving problem with AHP, it requires decisions to be consistent, only through consistency test, the results of weight will provide a valuable reference for the actual decision-making problems. But currently, the judgment matrix is often difficult to meet consistency, leading to judgment matrix inconsistent with the actual thinking and relative weight distortion, which is the current decision-making problem to be solved.

Because the subway station space visual environment comprehensive evaluation system contains many indexes, the order of initial judgment matrix is big so that it is difficult to check the consistency of initial judgment matrix and can not make accurate judgment. Therefore, for this 16-step initial judgment matrix, to test the consistency and calculate reasonable weight vector, this paper tried to build a proper consistency correction model and optimized with an intelligent algorithms, through this way, got the judgment matrix with better consistency and the corresponding weight vector quickly. After that, gave the result to expert to know whether the expert could accept the judgment matrix revised and index weight to assess the practical problem. If opinion is yes, the adjusted judgment matrix is acceptable and can do the further calculating and evaluation.

$$\min Y = \sum_{i=1}^{n} \sum_{j=1}^{n} \left[\lambda_1 (x_{ij} - a_{ij})^2 + \lambda_2 (x_{ij} - \omega_i / \omega_j)^2 \right]$$

$$s.t. \quad \omega_i > 0; \sum_{i=1}^{n} \omega_i = 1;$$

$$\lambda_1 + \lambda_2 = 1, \lambda_1, \lambda_2 \geq 0;$$

$$x_{ij} = 1 / x_{ji};$$

$$x_{ij} \in \left[(1-\theta)a_{ij}, (1+\theta)a_{ij} \right];$$

$$0 < \theta < 1, i, j = 1, 2, ..., n.$$

(1)

Smaller the objective function Y is, smaller the magnitude of adjustment in the case of good consistency, and higher quality of the modified matrix. Among them, λ_1, λ_2 is the weighting vector, their values can be decided based on the practical problem, one is the degree of expert judgment matrix followed; one is the degree of consistency index requirements. In this paper, the main target is to improve the consistency, so the degree of consistency is higher than adjustment degree of judgment matrix [4]. θ is the constraint index for every factor adjustment in initial judgment matrix, smaller is better. a_{ij} is the factor of initial judgment matrix, x_{ij}, w_i are the revised matrix and corresponding index weight.

3.4   Solving the Least Squares Consistency Correction Model

After extensive research, the paper proposed particle swarm optimization (PSO) to solve the least square model established by AHP to obtain judgment matrix with better consistency, adjusted the index weight from both subjective and objective aspects, and improved the accuracy of the index weight. The basic principle is: There are M particles in D-dimensional space, the particle motion space is the solution space, the function to be optimized is the fitness of the particle, particle position vector indicates variables of optimizing problems in the solution space, and the motion process of particles is the solution search process. When searching, according to the advantage of individual history and the most advantage of all the particles within the group's history, updates flight speed and position [5].

[6][7] shows the principle of PSO: among M particles, the position of i particle is $x_i = \left(x_{i1}, x_{i2}, \ldots, x_{iD} \right)$, flight speed is $v_i = \left(v_{i1}, v_{i2}, \ldots, v_{iD} \right)$, the best history position is the local optimal location $P_i = \left(P_{i1}, P_{i2}, \ldots, P_{iD} \right)$, that is P_{best}, the best history position of all particles is G_{best}. The speed and position updating formula are as follows:

$$v_i^{k+1} = \omega^* \bullet v_i^k + c_1 \bullet rand_1 \bullet (P_i^k - x_i^k) + c_2 \bullet rand_2 \bullet (G^k - x_i^k);$$
$$x_i^{k+1} = x_i^k + v_i^{k+1}; \qquad i = 1, 2, \ldots, M \, . \tag{2}$$

In it, k represents iteration number, rand1 and rand2 are rand numbers among [0, 1], c1 and c2 are acceleration weights, ω^* is inertia weight. The solving process of the least squares consistency correction model is as follows:

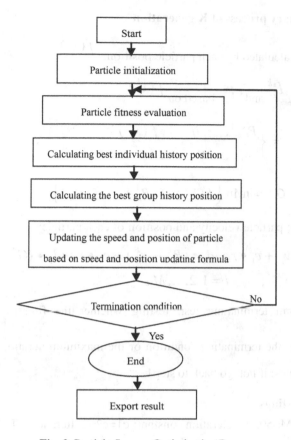

Fig. 3. Particle Swarm Optimization Process

The solving process with Matlab is as below:

(1) Initializing

First, set an initial particle population: the particle swarm contains M = 50 particles, each particle have D = 136 dimensional vectors, all these 50 particles have been given an initial position and velocity:

$$X^0 = (x_1^0, x_2^0, ..., x_i^0, ..., x_M^0), x_i^0 = \left(x_{i1}^0, x_{i2}^0, ..., x_{iD}^0 \right), x_i^0 \in \left[x_{lb}, x_{up} \right] ; \quad (3)$$

$$V^0 = (v_1^0, v_2^0, ..., v_i^0, ..., v_M^0), v_i^0 = \left(v_{i1}^0, v_{i2}^0, ..., v_{iD}^0 \right), v_i^0 \in \left[-v_{max}, v_{max} \right] . \quad (4)$$

The local optimal location Pbest and the global optimal location Gbest is:

$$P_i^0 = x_i^0; \qquad G^0 = \min \left\{ f(P_1^0), ..., f(P_i^0), ..., f(P_M^0) \right\} . \quad (5)$$

(2) The evolutionary process of K generation

Step1: fitness calculated by each particle position: $f(x_i^k)$;

Step2: updating P_i^k and G^k based on $f(x_i^k)$;

$$P_i^k = \begin{cases} P_i^{k-1}, & if & f(x_i^k) \geq f(P_i^{k-1}) \\ x_i^k, & if & f(x_i^k) < f(P_i^{k-1}) \end{cases} ; \qquad (6)$$

$$G^k = \min\left\{ f(P_1^k), f(P_2^k), ..., f(P_M^k) \right\} 。 \qquad (7)$$

Step3: updating particle velocity and position of each particle:

$$v_i^{k+1} = \omega^* \bullet v_i^k + c_1 \bullet rand_1 \bullet (P_i^k - x_i^k) + c_2 \bullet rand_2 \bullet (G^k - x_i^k);$$
$$x_i^{k+1} = x_i^k + v_i^{k+1}; \qquad i = 1, 2, ..., M 。 \qquad (8)$$

Step4: algorithm terminating test: testing whether the fitness related to the updated G^k arrives the terminating condition or the maximum iteration, if match, the algorithm terminates; if not, go back to step1.

(3) Parameter settings

Particle scale: M=50; Acceleration constant: c1=c2=2; Iterations: T=1000.

Inertia weight: $\omega^* \in [0.8, 1.2]$, updated with the particle velocity and position updating.

Weighting factor: $\lambda_1 \in [0.1, 0.4]$, $\lambda_2 \in [0.6, 0.9]$, updated with the particle velocity and position updating.

Constraints: $\theta \in [0.1, 0.3]$, updated with the particle velocity and position updating.

(4) Realization of PSO algorithm for solving the least squares consistency correction model with MATLAB:

① **Algorithm Initialization**

```
v(:,:,i)=rand(n+1,n);      % initializing particle velocity
pbp(:,:,i)=zeros(n+1,n);   %  initializing  individual
optimal location
gbp=zeros(n+1,n);          % initializing global optimal
location
pbf=inf*ones(1,N);         %  initializing  individual
optimal fitness
gbf=inf;                   % initializing global optimal
fitness
```

② **Iteration updating**

```
v(:,:,i)=w(iter)*v(:,:,i)+c1*rand(1)*(pbp(:,:,i)-partic
le(:,:,i))+c2*rand(1)*(gbp-particle(:,:,i));
% updating  velocity
particle(:,:,i)=particle(:,:,i)+v(:,:,i);     % updating
location
```

③ **Results**

Run the program above for 100 times using MATLAB and obtained optimal results:

Parameters   corresponding   to   the   optimal   weight:
$\omega^* =1.0959$, $\lambda_1 =0.3219$, $\lambda_2 =0.6781$, $\theta=0.2479$

The index weight vector W and judgment matrix with better consistency are:

W =[0.0901   0.0729   0.1015   0.1184   0.0620   0.0881   0.0219   0.0743
0.1024   0.0199   0.0139   0.0124   0.0454   0.0503   0.0755   0.0511].

$$A = \begin{bmatrix} 1.0000 & 3.7546 & 1.2938 & 0.6153 & 3.9484 & 4.0837 & 4.6059 & 0.8770 & 0.8656 & 6.3956 & 7.9125 & 7.0804 & 5.2415 & 2.7360 & 0.9681 & 4.7196 \\ 0.2663 & 1.0000 & 0.4604 & 0.2051 & 2.4751 & 2.0203 & 2.6421 & 0.6505 & 0.5377 & 5.3838 & 4.1874 & 8.8592 & 2.5011 & 4.7027 & 0.8945 & 3.0621 \\ 0.7729 & 2.1719 & 1.0000 & 0.8178 & 3.2751 & 3.0136 & 4.8943 & 0.6192 & 0.5409 & 6.1263 & 5.9787 & 10.2223 & 4.9411 & 6.4967 & 1.1709 & 2.6859 \\ 1.6253 & 4.8751 & 1.2229 & 1.0000 & 8.2009 & 7.1862 & 8.4860 & 5.4709 & 6.1582 & 8.5504 & 7.2102 & 8.1443 & 5.3956 & 5.3448 & 5.1981 & 6.7086 \\ 0.2533 & 0.4040 & 0.3053 & 0.1219 & 1.0000 & 2.9746 & 2.3533 & 0.8178 & 0.4460 & 2.7866 & 4.5073 & 7.9453 & 1.7674 & 1.4836 & 0.4745 & 0.7858 \\ 0.2449 & 0.4950 & 0.3318 & 0.1392 & 0.3362 & 1.0000 & 1.9491 & 0.5555 & 0.8258 & 4.2157 & 4.4808 & 8.0263 & 2.4458 & 1.8148 & 0.3586 & 0.4013 \\ 0.2171 & 0.3785 & 0.2043 & 0.1178 & 0.4249 & 0.5131 & 1.0000 & 0.8762 & 0.5265 & 2.4495 & 4.5098 & 6.4763 & 2.7428 & 2.5728 & 0.2279 & 0.3984 \\ 1.1402 & 1.5372 & 1.6149 & 0.1828 & 1.2228 & 1.8000 & 1.1413 & 1.0000 & 1.4129 & 4.5270 & 6.0832 & 7.4503 & 0.9881 & 1.5295 & 0.4834 & 1.2822 \\ 1.1552 & 1.8597 & 1.8488 & 0.1624 & 2.2420 & 1.2109 & 1.8994 & 0.7077 & 1.0000 & 5.0884 & 7.3295 & 7.8836 & 1.2976 & 0.8414 & 0.3237 & 1.3586 \\ 0.1564 & 0.1857 & 0.1632 & 0.1170 & 0.3589 & 0.2372 & 0.4083 & 0.2209 & 0.1965 & 1.0000 & 0.7788 & 2.6962 & 0.2372 & 0.2697 & 0.5625 & 0.3075 \\ 0.1264 & 0.2388 & 0.1673 & 0.1387 & 0.2219 & 0.2232 & 0.2217 & 0.1644 & 0.1364 & 1.2841 & 1.0000 & 2.4934 & 0.1924 & 0.3379 & 0.2524 & 0.2675 \\ 0.1412 & 0.1129 & 0.0978 & 0.1228 & 0.1259 & 0.1246 & 0.1544 & 0.1342 & 0.1268 & 0.3709 & 0.4011 & 1.0000 & 0.3593 & 0.1884 & 0.1774 & 0.4551 \\ 0.1908 & 0.3998 & 0.2024 & 0.1853 & 0.5658 & 0.4089 & 0.3646 & 1.0121 & 0.7707 & 4.2156 & 5.1980 & 2.7834 & 1.0000 & 1.3251 & 0.3242 & 1.3149 \\ 0.3655 & 0.2126 & 0.1539 & 0.1871 & 0.6740 & 0.5510 & 0.3887 & 0.6538 & 1.1885 & 3.7077 & 2.9591 & 5.3083 & 0.7547 & 1.0000 & 0.5164 & 1.2545 \\ 1.0329 & 1.1179 & 0.8541 & 0.1924 & 2.1074 & 2.7889 & 4.3877 & 2.0687 & 3.0890 & 1.7777 & 3.9616 & 5.6371 & 3.0842 & 1.9366 & 1.0000 & 2.1600 \\ 0.2119 & 0.3266 & 0.3723 & 0.1491 & 1.2726 & 2.4918 & 2.5103 & 0.7799 & 0.7360 & 3.2525 & 3.7378 & 2.1972 & 0.7605 & 0.7972 & 0.4630 & 1.0000 \end{bmatrix}$$

4   Establishing the Subway Station Space Visual Environment Evaluation System

Because the expert could accept the revised judgment matrix and corresponding index weight vector, used the judgment matrix corrected as the final judgment matrix and the weight vector as the evaluation index weight to build a subway station space visual environment comprehensive evaluation model. The model is in the following table:

Table 1. Visual environmental assessment system of subway station space

	Criterion Layer B	Index Layer C	Evaluation Index Layer D	Index Weight
Subway Station Space Visual Environment Evaluation System A	Functional Index B1	Light Environment C1	Illumination Level D1	0.0901
			Illumination Uniformity D2	0.0729
			Brightness Level D3	0.1015
			Brightness Distribution D4	0.1184
			Glare D5	0.0620
			Color Temperature D6	0.0881
			Color Rendering D7	0.0219
			Visual Induction D8	0.0743
			Recognition D9	0.1024
		Space Environment C2	Three-dimensional D10	0.0199
			Environmental Coordination D11	0.0139
			Artistic Lighting D12	0.0124
	Energy-saving Index B2	Lighting Control C3	Control Strategy D13	0.0454
			Control Means D14	0.0503
		Lighting Energy-saving C4	Lighting Power Density D15	0.0755
			Energy-saving Lamps D16	0.0511

5   Conclusions and Prospects

This paper used Particle Swarm Optimization to build a visual environment comprehensive evaluation model to evaluate the visual environment created by LED and fluorescent lighting in subway station space. Firstly, used AHP to create the hierarchical structure and further established the initial judgment matrix. Secondly, built the least squares consistency correction model to get the revised judgment with good consistency and the corresponding index weight, and then built a theoretical model. Finally, applied this theoretical model to evaluate the visual environment of subway station space in Shenzhen Metro Line 2 and Line 3 and got the optimal visual environment.

References

1. Zhang,Y.: Research on Lighting Environmental Quality Assessment and Technology System of Hotel. Chongqing University (2005)
2. Du, Y.: Multi-attribute Problem Weights Solving and Its Application in the Emergency Logistics Based on PSO and AHP. University of Science and Technology of China (2011)
3. Liu, S., Liu, Z.: Research on Fuzzy Comprehensive Evaluation Model for Subway Safety and Its Application. Railway Engineering Society (2011)
4. Ding, B., Du, Y.: Least Squares Model for Multi-attribute Decision Making Problems Weights Solving Based on PSO and AHP. System Engineering (2010)
5. Yin, D.: Research on Aero Engine Model Solution Algorithm and Parameter Estimation in Performance Seeking Control. University of National Defense Technology (2011)
6. Kennedy, J.: The particle warm: Social adaptation of knowledge. In: Proceedings of 1997 IEEE International Conference on Evolutionary Computation, pp. 303–308 (1997)
7. Lu, W.Z., Fan, H.Y., Lo, S.M.: Application of evolutionary neural network method in predicting pollutant levels in downtown area of Hong Kong. Neurocomputing, 387–400 (2003)

Study on Pattern Recognition of Hand Motion Modes Based on Wavelet Packet and SVM

Fuxin Liang, Chuanjiang Li[*], Yunling Gao, Chongming Zhang, and Jiajia Chen

College of Information, Mechanical and Electrical Engineering,
Shanghai Normal University, Shanghai, China
licj@shnu.edu.cn

Abstract. For pattern recognition-based myoelectric prosthetic hand control, high accuracy of multiple discriminated hand motions is presented in related literature. But in practical applications of myoelectric control, considering cost and simple installation, fewer sensors are expected to be used. A method of pattern recognition based on the wavelet packet decomposition and support vector machine (SVM) is proposed in this paper. Firstly, energy spectrum as feature vectors of the surface electromyography (sEMG) signal is acquired by wavelet packet transform. Then, SVM is used for pattern recognition of hand motion modes. Four channels of sEMG signals obtained from sensors placed on different positions of forearm are used to experiment of hand motion recognition. And different combinations of 2 or 3 signals are tried to recognize hand motion modes. The results show that recognition rate of proposed method can get 92.5% while using 4 sEMG signals to recognize 8 different hand motions, which is 2.5% higher than using traditional method. And when using 3 sEMG signals from specific positions, it can reaches as high as 90%. When using 2sEMG signals only 6 motions can be discriminated with more than 90% recognition rate. Thus, the proposed method can meet the demands of sEMG prosthetic hand control and has high practical value.

Keywords: Surface electromyography signal, Wavelet packet decomposition, Support vector machine.

1 Introduction

Surface electromyography signal is a kind of one-dimensional time-series signal, which is guided through electrode from the surface of muscles and recorded activity of the neuromuscular system [1], it accurately reflects muscles' activity status and functional status in a non-injury status in real-time [2].

Sampling multi-channel sEMG signals on the skin surface is a safe, non-invasive measurement for the movement control of artificial prostheses. With the development of detection technology, signal processing methods and computer technology, how to use sEMG for prosthetic control has attracted more concerns, especially in the fields of rehabilitation medicine and intelligent robot [3] [4]. Feature extraction is an

---

[*] Corresponding author - An associate professor in Shanghai Normal University, Shanghai, China. His current research interest is Biomedical Signal Processing and Intelligent Control.

M. Fei et al. (Eds.): LSMS/ICSEE 2014, Part II, CCIS 462, pp. 180–188, 2014.
© Springer-Verlag Berlin Heidelberg 2014

important process of pattern recognition. the selection of feature vectors is the key to enhance identification ability of recognition systems. Traditional sEMG action recognition methods usually extract time domain or frequency domain statistical characteristics as feature vectors such as the integration of absolute value, the number of zero-crossings, variance, power spectrum analysis, cepstrum analysis, etc. ,or model sEMG time series and extract AR model coefficients as feature vectors. Integral absolute mean, variance, auto-regressive (AR) model coefficients, linear cepstrum coefficients and the adaptive spectrum parameters are used as feature vectors to achieve sEMG pattern recognition in literature [5] [6]. Based on the frequency domain characteristics of sEMG, a method is proposed in literature using power spectrum ratio to indicate feature information on limb movements [7]. These methods only analyze data in the time domain or frequency domain, and treat non stationary sEMG signals as stationary signals or piecewise stationary signals. In literature [8], a new wavelet packet based feature extraction method is proposed, which applies wavelet packet transform on sEMG signals from 6 hand motion modes of forearm muscles and the maximum eigenvalue of the covariance matrix of the resulting signals was used as an effective feature.

Wavelet packet transform can be used to analyze non-stationary signals and is able to extract more reliable signal characteristics [9]. A method of hand motion recognition based on wavelet packet transform and SVM is proposed in this paper. The results show that the proposed method improves the recognition rate of sEMG hand motion modes.

Since pattern recognition based on 4 or more sEMG signals has been a hot spot, the existing literature does not refer to the influence of sensors' number and position. From practical point of view, using sensors as little as possible is wanted in order to install easily and reduce cost. So, we focus on using the minimum number of sensors and the best placement for actual application.

2   Overall Idea

The proposed method of pattern recognition of sEMG signal is based on the wavelet packet decomposition (WPD) and SVM. The flow chart of the algorithm is showed in Fig. 1.

Four surface electromyography electrodes are used to acquire measurement data of hand motion. Four stage wavelet packet transform is applied to the signals from the forearm muscles which carries the information on eight kinds of hand motion modes. Wavelet packet coefficients energy spectrum of 1-4 sub-band are extracted respectively.

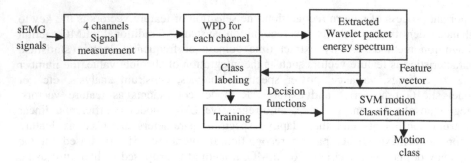

Fig. 1. Blockdiagram for sEMG pattern recognition

3   FeatureExtraction

3.1   Preprocessing of Surface EMG

The steady state sEMG signal is used for hand motion modes recognition. Moving average method is used to deal with instantaneous energy of surface EMG signal sequence, and active segment is detected by comparing with the threshold. The algorithm steps are as follows:

(1)   Calculating the mean square of four surface EMG with the following formula :

$$sEMG_{ave}(i) = \frac{1}{4}\sum_1^4 [sEMG(i)]^2 \tag{1}$$

where, $sEMG_{ave}$ is the mean square of four sEMG signals which reflect instantaneous energy of the signal, $sEMG(i)$ is the sample data of four sEMG signals.

(2)   Using the activity window for $sEMG_{ave}$ processing, instantaneous energy of the signal is processed by itemized moving average with window width $P = 64$:

$$sEMG_{MAV}(i) = \frac{1}{P} \sum_{j=i}^{i+P-1} sEMG_{ave}(j) \qquad i \geq P \tag{2}$$

(3)   Setting an appropriate threshold TH. Each $sEMG_{MAV}(i)$ value will be set to zero if it is less than TH, otherwise it will be unchanged. The choice of the threshold value TH is based on the experimental results in specific application.

$$sEMG_{rec}(i) = \begin{cases} 0 & sEMG \leq TH \\ sEMG_{MAV}(i) & sEMG > TH \end{cases} \tag{3}$$

(4)   According to the data after the threshold processing, very small non-zero data segment should be treated as noise and should be removed. For each data segment, the first data point of 64 consecutive non-zero points should be selected as the start point of EMG activity, and the first data point of 64 consecutive zero points should be the end point of the active segment.

3.2   Wavelet Packet Decomposition(WPD)

In the wavelet multi-resolution analysis, scale function $\varphi(t)$ and wavelet function $\psi(t)$ satisfy the two-scale equation:

$$\varphi(t) = \sqrt{2}\sum_k h(k)\psi(2t-k) \tag{4}$$

$$\psi(t) = \sqrt{2}\sum_k g(k)\varphi(2t-k) \tag{5}$$

With $\mu_0(t) = \varphi(t)$, $\mu_1 = \psi(t)$, the two-scale equation rewrites as the following recursive form:

$$\mu_{2n}(t) = \sqrt{2}\sum_k h(k)\mu_n(2t-k) \tag{6}$$

$$\mu_{2n+1}(t) = \sqrt{2}\sum_k g(k)\varphi(2t-k) \tag{7}$$

Thus it can be defined as an orthogonal wavelet packet of orthogonal scaling functions [10].

Wavelet packet space comes from $\varphi(t)$ stretching by the translation system, each of which consists of rectangular spatial composition. Operator forms for the multi-resolution analysis are as follows:

$$H[S_k](j) = \sum_k S_k h_{k-2j} \tag{8}$$

$$G[S_k](j) = \sum_k S_k g_{k-2j} \tag{9}$$

Let signal $f(t) \in U_j^m$, namely $f(t) = \sum_k S_k^j \mu_n(2^{-j}-k)$, then

$$f(t) = 2^{-\frac{1}{2}}\sum_i H[S_k^j](i)\mu_{2n}(2^{-j-1}t-i) + \\ 2^{-\frac{1}{2}}\sum_i G[S_k^j](i)\mu_{2n+1}(2^{-j-1}t-i) \tag{10}$$

This is the wavelet packet decomposition formula, whose principle is that the signals in the wavelet packet decomposition can be decomposed into two parts: one is the signals transformed throughthe Hfunction,the other is the signals transformed through the Gfunction. The decomposing process is showed in Fig.2.

High-frequency part of the signals of wavelet packet sequence is re-decomposition. It is better and more subtle than wavelet decomposition of localized functions [11].

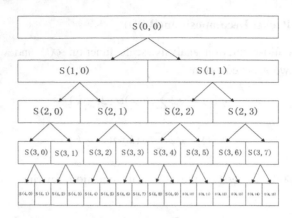

Fig. 2. Sketch map of wavelet packet decomposition

3.3   Wavelet Packet Transform Energy Spectrum

In the wavelet transform, 2-norm of the original signal $f(x)$ in $L^2(R)$ is defined as

$$\| f \|_2^2 = \int_R |f(x)|^2 dx \tag{11}$$

Thus, in the wavelet transform,2-norm of the signals is equivalent to the original signal energy in the time domain. If the basic wavelet $\psi(t)$ is a wavelet allowed, then there exists

$$\int_R da \int_R db \left| W_\psi f(a,b)/a \right|^2 = \| f \|_2^2 \qquad f \in L^2(R) \tag{12}$$

It means an equivalence relation exists between the wavelet energy and the energy of the original signal,so that energy band can be extracted from the result of the wavelet packet decomposition to represent the energy distribution of the original signal.

Leteach sub-band signal sequences of the result of wavelet packet transform expressed as $\{S_{i,j,k} \mid k = 1,2,\cdots,L\}$,where i is decomposition level; j is decomposed band number; $j = 1, 2, \cdots, 2^i - 1$; L is the sample length sequence for each band of the wavelet packet. The total energy of the whole band is set to 1. With each sub-band of the energy normalized, thenormalized energy of each sub-band signal is

$$E_{i,j} = \sum_{k=1}^{L} \left| S_{i,j,k} \right|^2 / \sum_{j=0}^{2^i-1} \sum_{k=1}^{L} \left| S_{i,j,k} \right|^2 \tag{13}$$

4   Experimental Result and Discuss

Firstly, subjects' extensor pollicis brevis, extensor indicisproprius, extensor digitorumcommunis and extensor digitiquintiproprius iscleanedwithalcohol.Surface electromyography test system, Trigno wireless surface electromyography collection instrument produced byDELSYS company of USAwas used to collect four-channel surface EMG. Then the signal would be amplified to input into the acquisition card for samplecollection. For each hand motion, participants repeated 40 times at a pre-defined sequence. Each hand motion startsfrom a relaxed state to the target state, and targetstateshould be maintained 3 seconds. Then subjects returned to the relaxed state, and after 3 seconds repeat the hand motion. Having 40 groups of data for each of eight hand motion modes, we get totally 320 groups of data for each subject. 160 groups of randomly selected data areused to train the classifier, and the other 160 groups of data areused for prediction. Three subjects participate in this experiment, one woman and two men.

The eight kinds of basic hand motion modes are showed in Fig.3.

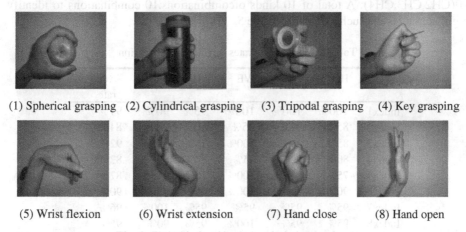

(1) Spherical grasping   (2) Cylindrical grasping   (3) Tripodal grasping   (4) Key grasping

(5) Wrist flexion   (6) Wrist extension   (7) Hand close   (8) Hand open

Fig. 3. Eight kinds of basic hand motion modes

Table 1. Recognition rates for hand motion modes

Hand motionmodes	Traditional method Recognition rate	WPD and SVM method Recognition rate
Spherical grasping	85%	87.5%
Cylindrical grasping	80%	85%
Tripodal grasping	85%	87.5%
Key grasping	90%	92.5%
Wrist flexion	95%	97.5%
Wrist extension	97.5%	95%
Hand close	92.5%	97.5%
Hand open	95%	97.5%
Average rate	90%	92.5%

Performance of proposed method and traditional method is compared in Table 1. In traditional method, MAV(Mean Absolute Average), SD(Standard Deviation),

MPF(mean power frequency) and MF(median frequency) BP Neural Network are used as classifier. The recognition rate of proposed method is averages 92.5%, which is 2.5% higher than the traditional methods. For the eight hand modes, the first four grasp modes are difficult to be recognized than the last four wrist motions. Especially, Cylindrical grasping is easy to be mistake for Spherical grasping or Tripodal grasping. The proposed method in this paper has much better performance than traditional methods in the first four grasp modes recognition.

In order to find a minimum numberand the best position of sensors, this paper adopts different combination of channel 2 and channel 3 sensors for pattern recognition research.The combination of dual channel has six kinds: Index1(CH1, CH2),Index2(CH1,CH3),Index3(CH1,CH4),Index4(CH2,CH3),Index5(CH2,CH4),Index6(CH3,CH4); The combination of the three channels has four kinds:Index7(CH1,CH2,CH3),Index8(CH1,CH2,CH4),Index9(CH1,CH3,CH4),Index 10(CH2,CH3,CH4); A total of 10 kinds ofcombinations.10 combinations to identify eight action results such as Table 2-Table 5.

Table 2. Recognition rates forfive kinds of action

	HO	HC	WE	WF	NIP	Average rate
Index1	100%	100%	100%	95%	90%	97%
Index2	85%	95%	75%	60%	90%	81%
Index3	85%	100%	100%	75%	100%	92%
Index4	80%	90%	75%	80%	85%	82%
Index5	75%	95%	100%	70%	95%	87%
Index6	90%	100%	100%	65%	95%	90%
Index7	95%	95%	95%	95%	100%	96%
Index8	95%	95%	100%	95%	90%	95%
Index9	90%	100%	100%	80%	90%	92%
Index10	100%	95%	100%	80%	100%	95%

Table 3. Recognition rates forsix kinds of action

	HO	HC	WE	WF	CYL	NIP	Average rate
Index1	100%	100%	100%	100%	80%	90%	95%
Index2	85%	95%	75%	60%	70%	90%	79.17%
Index3	90%	100%	100%	80%	70%	100%	90%
Index4	80%	90%	60%	80%	80%	85%	79.17%
Index5	70%	95%	100%	75%	80%	100%	86.67%
Index6	90%	100%	100%	65%	80%	100%	89.17%
Index7	100%	95%	95%	95%	100%	100%	97.5%
Index8	95%	95%	100%	90%	95%	100%	95.83%
Index9	90%	100%	100%	90%	80%	100%	91.67%
Index10	100%	95%	100%	90%	100%	100%	97.5%

Table 4. Recognition rates forseven kinds of action

	HO	HC	WE	WF	CYL	TRI	NIP	Average rate
Index1	95%	90%	70%	95%	65%	90%	70%	82.14%
Index2	80%	90%	60%	55%	90%	50%	90%	73.57%
Index3	75%	100%	90%	80%	100%	55%	100%	85.71%
Index4	80%	95%	65%	85%	80%	95%	75%	82.14%
Index5	85%	95%	100%	70%	85%	85%	80%	85.71%
Index6	75%	95%	90%	60%	85%	35%	100%	77.14%
Index7	100%	95%	100%	95%	80%	100%	85%	93.57%
Index8	95%	95%	100%	95%	80%	80%	95%	91.43%
Index9	80%	100%	100%	75%	95%	45%	100%	74.38%
Index10	95%	95%	100%	80%	80%	90%	85%	89.29%

Table 5. Recognition rates for eight kinds of action

	HO	HC	WE	WF	CYL	SPH	TRI	NIP	Average rate
Index1	90%	90%	65%	85%	50%	100%	95%	70%	80.63%
Index2	85%	90%	65%	55%	90%	60%	40%	90%	71.89%
Index3	75%	100%	90%	80%	95%	75%	45%	95%	81.88%
Index4	80%	95%	50%	70%	75%	55%	95%	80%	75%
Index5	75%	95%	95%	75%	65%	80%	85%	80%	81.25%
Index6	75%	100%	90%	70%	90%	65%	45%	95%	78.75%
Index7	100%	95%	100%	95%	75%	65%	90%	90%	88.5%
Index8	95%	95%	100%	90%	75%	95%	70%	100%	90%
Index9	100%	95%	100%	95%	75%	65%	90%	90%	88.5%
Index10	100%	95%	100%	80%	75%	60%	90%	85%	85.63%

It can be seen from these tables, the average recognition rate of the two sensors to identify eight kinds of action is low, especially in 3 different grasping actions, some mistakes happened ,and the average recognition rate of using three sensors for eight kinds of action is higher than 85%, which can be used in the actual control system, the effect of using three channels（CH1,CH2, CHz4）is the best one .

the results reflect that the effect of using double channels（CH1,CH2）is the best one.The average recognition rate of using double channel （CH1,CH2）sEMG signalto recognise5 actionsis higher than 95%,which is satisfied the requirements of the electrical control system.The average recognition rate ofidentifying six actionsis higher than 90%, the average recognition rate ofidentifyingseven actionsis higher than 80%.

5   Conclusions and Future Work

In this paper, surface EMG for eight kinds of human forearm hand motion modes are collected. Firstly,to cope with the non-stationary signal property of the sEMG, features are extracted by wavelet packetdecomposition.Then, support vector machine (SVM) is used for pattern recognition of eight kinds of hand motion modes. Experiments results show that:

Four channels of sEMG signals obtained from sensors placed on different position of forearm are used to experiment of hand motion recognition. And different combinations of 2 or 3 signals are tried to recognize hand motion modes. The result shows that recognition rate of proposed method can get 92.5% while using 4 sEMG signals to recognize 8 different hand motions, which is 2.5% higher than using traditional methods. And when using 3 sEMG signals from specified positions, it can reach as high as   90%. When using 2 sEMG signals only 6 motions can be discriminated with more than 90% recognition rate. Thus, the proposed method can meet the demands of EMG prosthetic hand control and has high practical value.

In future work, we will compare the experimental results performed on forearm amputees with those on healthy people for algorithm optimization purpose. Then a lightweight version of the optimized algorithm will be realized on embedded systems, and will ultimately be used in prostheses control system.

Acknowledgements. This work is supported by the Program of Shanghai Normal University (DZL126, SK201318). We thank professorMinruiFei, Shanghai University, and professor Huosheng Hu, for use of experimental equipment and their guidance.

References

1. Oskoei, M.A., Hu, H.: Myoelectric Control Systems - A survey. Journal of Biomedical Signal Processing & Control, 275–294 (2007)
2. Englehart, K., Hudgins, B.: A robust, real-time control scheme for multifunction myoelectric control. Journal of Biomedical Engineering 50(4), 848–854 (2003)
3. Hudgins, B., Parker, P., Scott, R.N.: A new strategy for multifunction myoelectric control. IEEE Transactions on Biomedical Engineering, 82–94 (1993)
4. Luo, Z.Z., Zhao, P.F.: The application of nonlinear PCA in surface EMG electrical signal feature extraction. Journal of Sensing Technology 20(10), 2164–2168 (2007)
5. Park, S.H., Lee, S.P.: EMG Pattern Recognition Based on Artificial Intelligence Techniques. IEEE Transactions on Biomedical Engineering 6(4), 263–271 (1998)
6. Li, F., Zhang, Y., Gao, K.N.: Pattern Recognition of Finger Motion's EMG Signal based on Improved BP Neural Networks. In: IEEE International Conference on Computer Science and Network Technology (ICCSNT), pp. 1266–1269 (2011)
7. Luo, Z.Z., Zhao, P.F.: The application of nonlinear PCA in surface EMG electrical signal feature extraction. Journal of Sensing Technology 20(10), 2164–2168 (2007)
8. Shi, J., Zhou, M.J., Zhu, Z.Z., Fu.: Electromyographic signal feature extraction based on wavelet packet transform. Journal of Microcomputer Information 7(2), 224–230 (2010)
9. Liu, J., Liu, Z., Xiong, Y.: Underwater target recognition based on wavelet packet energy spectrum and SVM. Journal of Wuhan University of Technology 36(2), 361–365 (2012)
10. Li, J., Zhao, L., Ren, S.Y., et al.: Feature extraction algorithm research of surface EMG electrical signal based on virtual instrument. Journal of Machine Tool & Hydraulics 39(3), 41–43 (2011)
11. Gao, Y.Y., Gao, Q.S., Meng, M.: Knee prosthesis gait recognition method based on multi-source information fusion. Journal of Scientific Instrument 31(12), 2682–2688 (2010)

Image Segmentation Using Multiphase Curve Evolution Based on Level Set

Li Liu[1,3], Xiaowei Tu[1], Wenju Zhou[2,3], Minrui Fei[1],
Aolei Yang[1,*], and Jun Yue[3]

[1] Shanghai Key Laboratory of Power Station Automation Technology,
School of Mechatronic Engineering and Automation,
Shanghai University, 200072, Shanghai, China
liulildu@163.com, tuxiaowei@comac.cc, mrfei@staff.shu.edu.cn,
aolei.yang@gmail.com
[2] School of Computer Science and Electronic Engineering,
University of Essex, CO4 3SQ, Colchester, UK
wzhoua@essex.ac.uk
[3] School of Information Science and Electrical Engineering,
Ludong University, 264025, Yantai, China
yuejuncn@sina.com

Abstract. A novel multiphase curve evolution based on level set (MCELS) is presented, which is used for image segmentation. The MCELS method introduces N level set functions partition 2^N sub-regions, which reduces the computational complexity. The double curve function is developed on the modified penalty function during the evolution. The experimental objects employ tablet packaging images. From the simulation results, the MCELS method can be used to partition multiple gray regions images for the noise, uneven gray scale, and intensity inhomogeneities. Comparing with recent researches based on level set methods, the characteristics of MCELS for image segmentation are superior robustness for noise, less run time and preferable computational efficiency.

Keywords: Image segmentation, multiphase curve evolution, level set, tablet packaging image.

1 Introduction

The multiphase regional segmentation is a kind of fundamental problem which widely applies in image processing and computer vision. In reality, due to the

* This work was supported by the Science and Technology Commission of Shanghai Municipality under "Yangfan Program" (14YF1408600), the Shanghai Municipal Commission Of Economy and Informatization under Shanghai Industry-University-Research Collaboration Grant (CXY-2013-71), Natural Science Foundation of Shandong Province (ZR2012FM008), The Project of Shandong Province Science and Technology Development Program (2013GNC11012) , National Natural Science Foundation of China (61100115). Corresponding author: Aolei Yang, aolei.yang@gmail.com

M. Fei et al. (Eds.): LSMS/ICSEE 2014, Part II, CCIS 462, pp. 189–198, 2014.
© Springer-Verlag Berlin Heidelberg 2014

overlaps, some images are intensity inhomogeneity. Such as, the magnetic resonance image (MRI), vehicle detection, complex medical image, tablet packaging image, and so on.

The traditional Chan-Vese (C-V) method [1] is generally applied to two phases image segmentation, which is divided into the target and background regions. It is difficult to achieve the desired segmentation effect, when the target and background information are similar. At present, the effective method is used multiphase level set evolution function for multiple target segmentation. [2] presented a framework for using multiple level set segmentation probability. The bias correction method adopts the maximum-likelihood (ML) and the expectation maximization (EM) [3]. But it is expensive to the initialization variables. Li et al. [4] proposed a weighted K-means variational level set (WKVLS) approaching to bias correction, which is an approximation process in bias field and restore the true signals. In [5], Li et al. introduced one order variational level set method based on the heterogeneous characteristics of the brain MRI. On the basis, Zhu et al. [6] proposed a segmentation region for the entire image, whose disadvantage is greatly increasing the computational complexity. In addition, Gao et al. [7] presented a novel unified level set multiphase image segmentation framework, which developed a new weighted distance function (WDF). The unified level set framework requires the establishment of a unified tensor, which expresses the multiphase level set function evolution. The model used the Gaussian distribution for the strength of the local mean and variance statistics. In machine vision field, Zhou et al. [8, 9] put theory well apply to practical, which the image segmentation is applied to the cap detection.

This paper investigates a new method of multiphase curve evolution based on level set (MCELS). This method employs the Gaussian mixture model (GMM) and the modified double curves penalty functions. The objective is the entire image region is changed and using level set functions to partition sub-regions.

The paper is organized as follows. The section two introduces the traditional region-based level set method. The section three proposes the MCELS method for images segmentation. The simulation studies for tablet packaging images are brought forth in the section four. The rest summarizes the current and future work.

2   Preliminary Theory of Multiphase Level Set Function

Mansouri et al. [10] proposed the multi level set segmentation model, which used $N - 1$ level set functions to partition N class inhomogeneity regional. Every level set function expresses one region, the remaining region is the forth. The characteristic function [11] of each region is defined by equation (1).

$$R_1 : H_1 = (1 - H(\phi_0)) H(\phi_1), H(\phi_0) \equiv 0,$$
$$R_2 : H_2 = (1 - H(\phi_0)) (1 - H(\phi_1)) H(\phi_2),$$
$$R_3 : H_3 = (1 - H(\phi_0)) (1 - H(\phi_1)) (1 - H(\phi_2)) H(\phi_3),$$
$$R_4 : H_4 = (1 - H(\phi_0)) (1 - H(\phi_1)) (1 - H(\phi_2)) (1 - H(\phi_3)) H(\phi_4), . \quad (1)$$
$$H(\phi_4) \equiv 1$$
$$R_i : H_i = \prod_{j=0}^{i-1} [1 - H(\phi_j)] H(\phi_i), \quad H_i = \begin{cases} 1, (x,y) \in R_i \\ 0, (x,y) \notin R_i \end{cases}$$

The C-V model [1] presents the curve evolution contains region data terms and regularization terms. A given image $I : \Omega \subset R^d$ includes that the image field Ω, the image dimension d and the image pixel vector x. The energy function of C-V model is expressed as (2). According to function (2), the regularization term is introduced, in order to make curve evolution sufficiently smooth and short. It is the sum of all the energy within the level set curve evolution. The multi level set method energy function has some parameters are set. $u = \{u_{R_i}, i = 1, \cdots, N\}$ represents the average of pixel on the level set curve, λ_i, $i = 1, 2, \cdots, N$ is the weight coefficient on each energy terms, and μ is the weight coefficient of the regularization term. Particularly, λ_i and μ are positive constant, respectively.

$$
\begin{aligned}
E(\Gamma_i(t), u) &= E^D(\Gamma_i(t), u) + E^R(\Gamma_i(t)) \\
&= \lambda_1 \int_{R_1} (I(x) - u_{R_1})^2 dx + \lambda_2 \int_{R_2} (I(x) - u_{R_2})^2 dx + \cdots + \\
&\quad \lambda_k \int_{R_k} (I(x) - u_{R_k})^2 dx + \cdots + \lambda_{N-1} \int_{R_{N-1}} (I(x) - u_{R_{N-1}})^2 dx + \\
&\quad \lambda_N \int_{R_N} (I(x) - u_{R_N})^2 dx + \mu \sum_{i=1}^{N-1} \int_{\Gamma_i} ds \\
&= \lambda_i \sum_{i=1}^{N} \int_{R_i} (I(x) - u_{R_i})^2 dx + \mu \sum_{i=1}^{N-1} \int_{\Gamma_i} ds
\end{aligned}
\quad (2)
$$

3 Multiphase Curve Evolution Based on Level Set

The computational complexity is augmented, because of using $N - 1$ level set functions achieve to N classes in heterogeneous region segmentation. In order to overcome the weakness of the multi level set segmentation model, this paper investigates the multiphase curve evolution which using N level set functions for 2^N regions. As shown as Figure 1, it is assumed that original image I is divided into four no-overlapping regions by two level set functions ϕ_1 and ϕ_2.

3.1 Energy Fitting Function

The C-V model based on Mumford-Shah [12] model, is the piecewise invariable-ness multiphase image segmentation (P-S). Refer to [13], the proposed MCELS method corresponding with likelihood function (LF), the energy fitting function is defined as

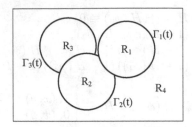

Fig. 1. Two level set functions express four regions

$$E\left(\theta\right) = \sum_{i=1}^{2^N} \int_\Omega \int_{R_i} \kappa_r\left(x,y\right)\left(\log\left(\sqrt{2\pi\sigma_i^2}\right) + \left(I\left(y\right) - U\left(x\right)V_i\left(x\right)\right)^2/\left(2\sigma_i^2\right)\right)dydx,$$

$$\sigma_i = 1/\sqrt{2\pi}$$

$$E\left(\theta\right) = \sum_{i=1}^{2^N} \int_\Omega \int_{R_i} \kappa_r\left(x,y\right)\pi(I\left(y\right) - U\left(x\right)V_i\left(x\right))^2 dydx \tag{3}$$

The energy fitting function is similar to the WKVLS method [4], if is denote as Gaussian kernel function [14].

3.2   Update Regularization Term

To avoided the appearance of small regions, the energy regularization term is described as

$$E^R\left(\phi_i\right) = E^L\left(\phi_i\right) + E^P\left(\phi_i\right) . \tag{4}$$

Length Smooth Term. MCELS method increase the length smooth regularization term to the evolution curve is enough smooth and avoid the small region appearance as far as possible. It is presented as follow

$$E^L\left(\phi_i\right) = \mu\sum_{i=1}^{N}\int_\Omega |\nabla H\left(\phi_i\left(y\right)\right)|dy = \mu\sum_{i=1}^{N}\int_\Omega \delta\left(\phi_i\right)|\nabla\phi_i\left(y\right)|\,dy . \tag{5}$$

Wherein, μ is a positive constant, $H\left(\phi_i\right)$ represents Heavide function and $\delta\left(\phi_i\right)$ represents Dirac Delta function, respectively, which are equal to

$$H_\varepsilon\left(\phi_i\right) = \tfrac{1}{2}\left[1 + \tfrac{2}{\pi}\arctan\left(\tfrac{\phi_i}{\varepsilon}\right)\right]$$
$$\delta_\varepsilon\left(\phi_i\right) = H_\varepsilon{}'\left(\phi_i\right) = \tfrac{\varepsilon}{\pi}\tfrac{1}{\varepsilon^2+\phi_i{}^2} \tag{6}$$

Penalty Function. Refer to [15], the level set cluster $\{\Gamma_i\}_{i=1}^{N}$ needs to be structured the signed distance function (SDF). To avoid re-initializing SDF after each iteration, the energy penalty term is developed. The energy penalty term can be expressed as follow

$$E^P\left(\phi_i\right) = \frac{\nu}{2}\sum_{i=1}^{N}\int_\Omega \left(|\nabla\phi_i\left(y\right)| - 1\right)^2 dy . \tag{7}$$

In order to raise the computing efficiency, MCELS method uses a double curve evolution method. Li et al. in [16] proposed double-well distance regularized level set evolution (DRLSE). According to the energy penalty function (7), a kind of potential function term $\eta(|\nabla\phi|)$ is developed, which sets the function $\eta : [0, \infty) \to \Re^2$ and is defined as follow

$$\eta(s) = \begin{cases} \frac{1}{(2\pi)^2}(1 - \cos(2\pi s)), & if \ s \le 1 \\ \frac{1}{2}(s-1)^2, & if \ s > 1 \end{cases}$$

$$\eta'(s) = \begin{cases} \frac{1}{2\pi}\sin(2\pi s), & if \ s \le 1 \\ s - 1, & if \ s > 1 \end{cases} \qquad (8)$$

When $s = 0$ and $s = 1$, $\eta(s)$ attends the minimum. $l(|\nabla\phi|) = \frac{\eta'(|\nabla\phi|)}{|\nabla\phi|} < 1$ contains three cases, which are $|\nabla\phi| > 1$, $\frac{1}{2} < |\nabla\phi| < 1$ and $|\nabla\phi| < \frac{1}{2}$.

3.3   MCELS Curve Evolution

Depend on the above analysis the fitting energy of MCELS model contains three parts, such as the regional fitting energy, the length smooth regularization term and the penalty regularization term. Thus, the entire energy fitting function is expressed as

$$E^{MCELS}(\phi) = \sum_{i=1}^{2^N} \int_\Omega g_i(y)\Phi_i dy + \mu \sum_{i=1}^{N} \int_\Omega \delta(\phi_i)|\nabla\phi_i(y)| dy +$$

$$\frac{\nu}{2} \sum_{i=1}^{N} \int_\Omega (|\nabla\phi_i(y)| - 1)^2 dy,$$

$$g_i(y) = \int_{R_i} \kappa_r(x,y)\left[\log(\sqrt{2\pi}\sigma_i) + (I(y) - U(x)V_i(x))^2 / (2\sigma_i^2)\right] dx, \quad (9)$$

$$\begin{cases} \Phi_1 = H(\phi_1)H(\phi_2) \\ \Phi_2 = H(\phi_1)(1 - H(\phi_2)) \\ \Phi_3 = (1 - H(\phi_1))H(\phi_2) \\ \Phi_4 = (1 - H(\phi_1))(1 - H(\phi_2)) \end{cases}$$

Obviously, the energy function is determined by region parameter $g = \{g_i, i = 1, 2\}$ and the GMM standard deviation $\sigma = \{\sigma_i, i = 1, 2, 3, 4\}$. $U(x)V_i(x)$ is the mean of region gray, $U(x)$ is the bias field, which is ensured by the normalized convolution [17], and σ are defined as follow

$$\begin{cases} V_1 = \frac{\int(\kappa_r * U)I(y)H(\phi_1)H(\phi_2)dy}{\int(\kappa_r * U^2)H(\phi_1)H(\phi_2)dy} \\ V_2 = \frac{\int(\kappa_r * U)I(y)H(\phi_1)(1-H(\phi_2))dy}{\int(\kappa_r * U^2)H(\phi_1)(1-H(\phi_2))dy} \\ V_3 = \frac{\int(\kappa_r * U)I(y)(1-H(\phi_1))H(\phi_2)dy}{\int(\kappa_r * U^2)(1-H(\phi_1))H(\phi_2)dy} \\ V_4 = \frac{\int(\kappa_r * U)I(y)(1-H(\phi_1))(1-H(\phi_2))dy}{\int(\kappa_r * U^2)(1-H(\phi_1))(1-H(\phi_2))dy} \end{cases}$$

$$U = \frac{\sum_{i=1}^{2^N} \kappa_r * (I(y)\Phi_i) \cdot \frac{V_i}{\sigma_i^2}}{\sum_{i=1}^{2^N} \kappa_r * (\Phi_i) \cdot \frac{V_i}{\sigma_i^2}}, \qquad (10)$$

$$\sigma_i = \sqrt{\frac{\int\int \kappa_r(y,x) * (I(y) - U(x)V_i)^2 \Phi_i(y)dydx}{\int\int \kappa_r(y,x) * \Phi_i(y)dydx}}$$

According to the definition of equation (9) and (10), it is universal for various modalities images, and de-noising effect is significant. In term of the potential penalty function $\eta\left(|\nabla\phi|\right)$ and equation (7), the gradient descent flow of potential penalty term is expressed as follow

$$
\begin{aligned}
\frac{\partial E^P}{\partial\phi} &= \frac{\partial}{\partial\phi}\sum_{i=1}^{N}\int_{\Omega}\left(|\nabla\phi_i\left(y\right)|-1\right)^2 dy \\
&= div\left(l\left(\nabla\phi_1\right)\nabla\phi_1\right)+div\left(l\left(\nabla\phi_2\right)\nabla\phi_2\right)
\end{aligned} \tag{11}
$$

The curvature of evolution curve is expressed as $div\left(\frac{|\nabla\phi|}{\nabla\phi}\right)$. Similarly, the gradient descent flow of length regularization term for the MCELS method is determined as

$$
\begin{aligned}
\frac{\partial E^L}{\partial\phi} &= \frac{\partial}{\partial\phi}\sum_{i=1}^{N}\int_{\Omega}\delta_\varepsilon\left(\phi_i\right)\left(|\nabla\phi_i|\right)dxdy \\
&= \delta_\varepsilon\left(\phi_1\right)div\left(\frac{\nabla\phi_1}{|\nabla\phi_1|}\right)+\delta_\varepsilon\left(\phi_2\right)div\left(\frac{\nabla\phi_2}{|\nabla\phi_2|}\right)
\end{aligned} \tag{12}
$$

First of all, the gradient descent flow of the region control term $E\left(\phi\right)$ can be expressed as follow

$$
\begin{aligned}
\frac{\partial E(\phi)}{\partial\phi_1} &= \frac{\partial}{\partial\phi_1}\int_{\Omega}g_1 H\left(\phi_1\right)H\left(\phi_2\right)dy+\frac{\partial}{\partial\phi_1}\int_{\Omega}g_2 H\left(\phi_1\right)\left(1-H\left(\phi_2\right)\right)dy+ \\
&\quad \frac{\partial}{\partial\phi_1}\int_{\Omega}g_3\left(1-H\left(\phi_1\right)\right)H\left(\phi_2\right)dy+\frac{\partial}{\partial\phi_1}\int_{\Omega}g_4\left(1-H\left(\phi_1\right)\right)\left(1-H\left(\phi_2\right)\right)dy \\
&= \left[\left(g_1-g_2-g_3+g_4\right)H\left(\phi_2\right)+g_2-g_4\right]\delta_\varepsilon\left(\phi_1\right)\quad, \\
\frac{\partial E(\phi)}{\partial\phi_2} &= \frac{\partial}{\partial\phi_2}\int_{\Omega}g_1 H\left(\phi_1\right)H\left(\phi_2\right)dy+\frac{\partial}{\partial\phi_2}\int_{\Omega}g_2 H\left(\phi_1\right)\left(1-H\left(\phi_2\right)\right)dy+ \\
&\quad \frac{\partial}{\partial\phi_2}\int_{\Omega}g_3\left(1-H\left(\phi_1\right)\right)H\left(\phi_2\right)dy+\frac{\partial}{\partial\phi_2}\int_{\Omega}g_4\left(1-H\left(\phi_1\right)\right)\left(1-H\left(\phi_2\right)\right)dy \\
&= \left[\left(g_1-g_2-g_3+g_4\right)H\left(\phi_1\right)+g_3-g_4\right]\delta_\varepsilon\left(\phi_2\right)
\end{aligned} \tag{13}
$$

Next, the gradient descent flow of the regularization term can be evolved by

$$
\begin{aligned}
\frac{\partial E^R}{\partial\phi} &= \frac{\partial E^L}{\partial\phi_1}+\frac{\partial E^L}{\partial\phi_2}+\frac{\partial E^P}{\partial\phi_1}+\frac{\partial E^P}{\partial\phi_2} \\
&= div\left(\frac{\nabla\phi_1}{|\nabla\phi_1|}\right)\delta_\varepsilon\left(\phi_1\right)+div\left(\frac{\nabla\phi_2}{|\nabla\phi_2|}\right)\delta_\varepsilon\left(\phi_2\right)+\cdot \\
&\quad div\left(l\left(\nabla\phi_1\right)\nabla\phi_1\right)+div\left(l\left(\nabla\phi_2\right)\nabla\phi_2\right)
\end{aligned} \tag{14}
$$

Therefore, according to (13) and (14), the gradient descent flow corresponding to the minimizing energy function is represented

$$
\begin{cases}
\begin{aligned}
\frac{\partial\phi_1}{\partial t} &= -\left[\left(g_1-g_2-g_3+g_4\right)H\left(\phi_2\right)+g_2-g_4\right]\delta_\varepsilon\left(\phi_1\right)- \\
&\quad div\left(\frac{\nabla\phi_1}{|\nabla\phi_1|}\right)\delta_\varepsilon\left(\phi_1\right)+div\left(l\left(\nabla\phi_1\right)\nabla\phi_1\right) \\
\frac{\partial\phi_2}{\partial t} &= -\left[\left(g_1-g_2-g_3+g_4\right)H\left(\phi_1\right)+g_3-g_4\right]\delta_\varepsilon\left(\phi_2\right)- \\
&\quad div\left(\frac{\nabla\phi_2}{|\nabla\phi_2|}\right)\delta_\varepsilon\left(\phi_2\right)+div\left(l\left(\nabla\phi_2\right)\nabla\phi_2\right)
\end{aligned}
\end{cases} \tag{15}
$$

Setting the spacing of discrete grid mesh is h, and time step is Δt. After n periods of time, the discrete multi-phase level set function cluster $\{\phi_i\}_{i=1}^{N}$ is expressed as $\{\phi_i\left(ah,bh,n\Delta t\right)\}_{i=1}^{N}$. The forward differential discrete level set function cluster can be expressed as

$$
\left(\phi_{i,ab}^{n+1}-\phi_{i,ab}^{n}\right)\Big/\Delta t = L\left(\phi_{i,ab}^{n}\right)\ . \tag{16}
$$

Depend on equation (14), the level set evolution equation can be updated by the following form

$$
\begin{aligned}
\phi_{i,ab}^{n+1} &= \phi_{i,ab}^n + \Delta t \cdot \Big\{ -\delta_\varepsilon \left(\phi_{i,ab}^n \right) \\
&\quad \left[(g_1 - g_2 - g_3 + g_4) \cdot H\left(\phi_j\right) + g_2 - g_4 + k_i \right] + \\
&\quad \nu \cdot \left(\phi_{i,a+1b}^n + \phi_{i,a-1b}^n + \phi_{i,ab+1}^n + \phi_{i,ab-1}^n - 4\phi_{i,ab}^n + k_i \right) \Big\}, \\
k_i &= div \left(\frac{\nabla \phi_i}{|\nabla \phi_i|} \right) = \frac{\phi_{i,xx}\phi_{i,y}^2 - 2\phi_{i,x}\phi_{i,y}\phi_{i,xy} + \phi_{i,yy}\phi_{i,x}^2}{\left(\phi_{i,x}^2 + \phi_{i,y}^2 \right)^{3/2}}, \\
j &= 2^{i-1} + (-1)^{3-i}
\end{aligned}
\tag{17}
$$

3.4 Algorithm Steps

The algorithm steps of the proposed multiphase curve evolution based on level set are summarized as the following.

Step 1: Initializing parameters. Δt represents the time step, h is grid spacing, and ε represents the regularization parameter of Heavide function.

Step 2: Initializing level set curves. To be SDF the evolution cluster is $\{\phi_i\}_{i=1}^N$.

Step 3: Updating the evolution curve. According to equation (17) it is suitable to update the evolution curve during the evolution.

Step 4: Determining the termination conditions. If condition is satisfied, output the segmentation result. Otherwise, updating the initial level set function $\phi^{n+1} = \phi^n$ as the next iteration, and go to Step 3.

4 Experimental Results

The implementation section compares the experimental results with RSF, 3-Phase LSE and MCELS. In order to validate the availability of MCELS method, the experiments introduce the tablet packaging images. The simulation environment uses MATLAB 2010, the computer configuration is Windows XP operating system, Intel 2.2GHz CPU, and 3G capacity memory.

4.1 Experimental Evaluation for Tablet Packaging Images

Experiment one. The original image is one single-color and multi-tablets packaging image, which size is 220*138 pixels. Table 1 shows the image segmentation effect using the initial contour, finial contour and final level set function.

Experiment two. The original image is one non-tablets packaging image, which size is 220*136 pixels. Table 2 shows the image segmentation effect using each level set method for the initial contour, finial contour and final level set function.

Table 1. Method verification for experiment one

Original image	Method	Initial contour	Finial contour	Final level set function
	RSF			
	3-Phase LSE			
	MCELS			

Table 2. Method verification for experiment two

Original image	Method	Initial contour	Finial contour	Final level set function
	RSF			
	3-Phase LSE			
	MCELS			

Table 3. The parameter comparison for each level set method

	Parameter	RSF	3-Phase LSE	MCELS
	Δt	0.1	0.1	0.45
	σ	3.0	4.0	4.5
	ε	1.0	1.0	1.0
	Iterations number	200	Outer iterations=100, Inner iterations=10	40
Exp 1	Total time	19.3125	23.4688	2.3438
Exp 2	Total time	16.5156	22.2813	2.3281

4.2 Simulation Analysis

In experiment one and two, setting some initial contour parameters and the experimental result value are listed by Table 3. The simulation analysis is summarized as follow.

The RSF [18] method, whose initializing contour is a rectangular, is able to divide contour such as tablet, but the segmentation effect is non-significant.

The 3-Phase LSE [19] method uses $N - 1$ level set functions partition N inhomogeneity regions. The level set functions are randomly initialized, the texture also is divided. So the image segmentation result does not prioritize.

The MCELS method uses N level set functions expression 2^N inhomogeneity regions, which randomly initialized contour. Adopting the double curves decrease the iterations and run time. The tablet contour, production date and other decorative parts can be well segmented.

5 Conclusion and Future Work

This paper presents a new method for image segmentation, which is multiphase curve evolution based on level set (MCELS). The method defines N level set functions to partition 2^N sub-regions. The profile curve is not sensitive to the initial conditions. The double curves evolution improves the arithmetic speed and decreases the iterations. The experimental verification uses tablet packaging images. The simulation results demonstrate that MCELS can able to divide the intensity in-homogeneities into multiple gray regions. The MCELS method is superior robustness for noise images, and the texture segmentation is obvious. The future work will be improved the accuracy of image segmentation.

Acknowledgments. The authors would like to thank the Science and Technology Commission of Shanghai Municipality, and the editors and anonymous reviewers for their valuable comments and helpful suggestions, which greatly improved the papers quality.

References

1. Chan, T.F., Vese, L.A.: Active Contours without Edges. IEEE Transactions on Image Processing 10, 266–277 (2001)
2. Do, M.N., Vetterli, M.: The Contourlet Transform: An Efficient Directional Multiresolution Image Representation. IEEE Transactions on Image Processing 14, 2091–2106 (2005)
3. Vovk, U., Pernus, F., Likar, B.: A Review of Methods for Correction of Intensity Inhomogeneity in MRI. IEEE Transactions on Medical Imaging 26, 405–421 (2007)
4. Li, C.-M., Huang, R., Ding, Z., Gatenby, C., Metaxas, D., Gore, J.: A Variational Level Set Approach to Segmentation and Bias Correction of Images with Intensity Inhomogeneity. In: Metaxas, D., Axel, L., Fichtinger, G., Székely, G. (eds.) MICCAI 2008, Part II. LNCS, vol. 5242, pp. 1083–1091. Springer, Heidelberg (2008)
5. Wang, L., Chen, Y., Pan, X., Hong, X., Xia, D.: Level Set Segmentation of Brain Magnetic Resonance Images based on Local Gaussian Distribution Fitting Energy. Journal of Neuroscience Methods 188, 316–325 (2010)
6. Lei, Z., Jing, Y.: Fast Multi-Object Image Segmentation Algorithm based on CV Model. Journal of Multimedia 6, 99–106 (2011)

7. Gao, X., Wang, B., Tao, D., Li, X.: A Unified Tensor Level Set Method for Image Segmentation. In: Lin, W., Tao, D., Kacprzyk, J., Li, Z., Izquierdo, E., Wang, H. (eds.) Multimedia Analysis, Processing and Communications. SCI, vol. 346, pp. 217–238. Springer, Heidelberg (2011)

8. Zhou, W., Fei, M., Li, K., Wang, H., Bai, H.: Accurate Image Capturing Control of Bottle Caps based on Iterative Learning Control and Kalman Filtering. Transactions of the Institute of Measurement and Control, 0142331213507077 (2013)

9. Zhou, W., Fei, M., Zhou, H., Li, K.: A Sparse Representation based Fast Detection Method for Surface Defect Detection of Bottle Caps. Neurocomputing 123, 406–414 (2014)

10. Mansouri, A.-R., Mitiche, A., Vzquez, C.: Multiregion Competition: A Level Set Extension of Region Competition to Multiple Region Image Partitioning. Computer Vision and Image Understanding 101, 137–150 (2006)

11. Pan, Z., Li, H., Wei, W., Guo, Z., Zhang, C.: A Variational Level Set Method of Multiphase Segmentation for 3D Images. Chinese Journal of Computers 32, 2464–2474 (2009)

12. Vese, L.A., Chan, T.F.: A Multiphase Level Set Framework for Image Segmentation Using the Mumford and Shah Model. International Journal of Computer Vision 50, 271–293 (2002)

13. Zhang, K., Zhang, L., Zhang, S.: A Variational Multiphase Level Set Approach to Simultaneous Segmentation and Bias Correction. In: 17th IEEE International Conference on Image Processing (ICIP 2010), pp. 4105–4108 (2010)

14. Cheng, Y.: Mean Shift, Mode Seeking, and Clustering. IEEE Transactions on Pattern Analysis and Machine Intelligence 17, 790–799 (1995)

15. Li, C., Xu, C., Gui, C., Fox, M.D.: Level Set Evolution without Re-initialization: A New Variational Formulation. In: IEEE Computer Society Conference on Computer Vision and Pattern Recognition (CVPR 2005), pp. 430–436 (2005)

16. Li, C., Xu, C., Gui, C., Fox, M.D.: Distance Regularized Level Set Evolution and Its Application to Image Segmentation. IEEE Transactions on Image Processing 19, 3243–3254 (2010)

17. Knutsson, H., Westin, C.-F.: Normalized and Differential Convolution. In: IEEE Computer Society Conference on Computer Vision and Pattern Recognition (CVPR 1993), pp. 515–523 (1993)

18. Li, C., Kao, C.-Y., Gore, J.C., Ding, Z.: Minimization of Region-scalable Fitting Energy for Image Segmentation. IEEE Transactions on Image Processing 17, 1940–1949 (2008)

19. Li, C., Huang, R., Ding, Z., Gatenby, J., Metaxas, D.N., Gore, J.C.: A Level Set Method for Image Segmentation in the Presence of Intensity Inhomogeneities with Application to MRI. IEEE Transactions on Image Processing 20, 2007–2016 (2011)

Experimental Platform Design and Implementation for Plate Structure Shape Perception and Reconstruction Algorithm

Mingdong Li[1], Hesheng Zhang[1,2], Kaining Liu[1], and Xiaojin Zhu[1,*]

[1] School of Mechatronic Engineering and Automation,
Shanghai University, Shanghai, 200072, P.R. China
[2] Shanghai Institute of Aerospace Electronic Technology,
China Aerospace Science and Technology Corporation, Shanghai 201109, China
mgzhuxj@shu.edu.cn

Abstract. For the problem of experimental verification for plate structure shape perception and reconstruction algorithm, an experimental verification platform was designed and constructed consisting of experiment base station, excitation system, measurement system and relative software, to the fitting algorithm based on plane curve as reference algorithm and the static error analysis and dynamic error analysis for the effect of reconstruction was conducted precisely. The results showed that the experimental platform was with good real-time capacity and high accuracy to meet the needs for the verification and data analysis of algorithms.

Keywords: Experimental platform, Shape perception and reconstruction algorithm, Surface fitting, Plate structure.

1 Introduction

For the problem of plate structure shape perception and visual reconstruction, there are two type of the existing methods: based on vision[1,2]and based on non-vision, among them the basic idea of based on non-vision approach is to obtain the strain information from the surface of the structure, then convert the strain to curvature or displacement and conduct geometric recursive and superimpose displacement field, make use of computer graphics technology to achieve the shape perception and visual reconstruction. As the method based on non-visual owning the advantages of less data collection and good real-time capacity, is the current research focus. Xiaojin Zhu decomposed the space surface into an array of plane curves, using curve fitting algorithm based on one-dimensional curvature information to achieve the shape perception and reconstruction of the flexible plate structure[3]; Hongtao Wang divided the space surface into surface patches array, using the internal relationship of

---

* Corresponding author - This research is supported by the National Nature Science Foundation of China (No.51175319), and key program of Shanghai Municipal Education Commission (No.13ZZ075).

M. Fei et al. (Eds.): LSMS/ICSEE 2014, Part II, CCIS 462, pp. 199–209, 2014.

two orthogonal directions curvature to establish surface patches equations and obtain the equation recursively for each surface patch using genetic algorithms to achieve shape perception and reconstruction of the flexible plate structure[4,5]; Rapp used modal displacement matrix by linear combination the modal information to achieve the estimation of plate structural displacement field[6,7].

Among these types of shape perception and reconstruction method, the method based on the plane curve array was matured and easy to implement, and had been verified experimentally[8]; the method based on surface patches just obtained a simulation test and still need further study and comparative analysis of the experiment; the method based on modal displacement matrix only achieved an estimation of structural displacement field, and the accuracy of static reconfiguration was not given, which also needed further experimental analysis. To verify the capacity of static and dynamic reconstruction as the above described algorithms, a measurement verification experimental platform with high-precision and good real-time capability is a must.

In this article, for the experimental verification of the method of plate structure shape perception and reconstruction, an experimental verification platform was designed and constructed consisting of experiment base station, excitation system, measurement system and related software. Considering the fitting reconstruction based on plane curve array as the reference algorithm, the platform was verified from both static and dynamic fitting accuracy, and the results showed that the platform with good real-time capacity and high accuracy to meet the needs for the verification and data analysis of algorithms.

2   Method of Shape Perception and Reconstruction Based on Non-vision

2.1   Fitting Algorithm Based on Plane Curve Array

The idea of the fitting algorithm based on plane curve array is dividing the plate structure into plane curve array, using the fitting algorithm based on unidirectional curvature to achieve reconstruction of the entire surface, the recursive process of curvature shown in Fig.1:

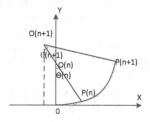

Fig. 1. The fitting schematic diagram of a recursive algorithm

In Fig.1, the tangential direction of the starting point is considered as x axis, $P(n)$ represents the curvature point of n and $O(n)$ is the center of the arc corresponding to the curvature point of n. Assuming that r_n is the corresponding radius of curvature, θ_n is the angle between the line curvature $P(n)O(n)$ and vertical direction. The process of center recursion and displacement recursion are showed as formula (1) and formula (2):

$$\begin{cases} O(n+1).x = O(n).x + [r_n - r_{n+1}] * sin\theta_n \\ O(n+1).y = O(n).y + [r_n - r_{n+1}] * cos\theta_n \end{cases} \tag{1}$$

$$\begin{cases} P(n+1).x = O(n+1).x + r_{n+1} * sin\theta_{n+1} \\ P(n+1).y = O(n+1).y - r_{n+1} * cos\theta_{n+1} \end{cases} \tag{2}$$

Through formula (1), we can obtain the center coordinates of the next displacement point, and can calculate the corresponding coordinates of the displacement point by formula (2) and the result of formula (1). The each of curve coordinates of plane curve array can be coordinates by the above described method, and we obtain the dense set of points in the plane surface to achieve the reconstruction of the plate structure. This method calculates the coordinates of the points directly and is easy to implement in computer.

2.2   Fitting Algorithm Based on Surface Patches Array

The idea of the fitting algorithm based on surface patches array is dividing the plate structure into surface patches array, using orthogonal curvature information to achieve reconstruction of the entire surface. Taking one surface patch from the array as research object, the mutually perpendicular normal curvature information from four corners are obtained, they are $K_{i,j}^{n1}, K_{i,j}^{n2}$ ($i, j = 1, 2, 3, 4$).

Assuming that the curved surface as formula (3):

$$f_{xy}(x, y) = a_1 + a_2 x + a_3 y + a_4 xy + a_5 x^2 + a_6 y^2 \tag{3}$$

here $a_1, a_2, a_3, a_4, a_5, a_6$ are undetermined coefficient, and the first basic amount E, F, G and the second basic amount L, M, N of quadric equation are expressed as in formula (4):

$$\begin{cases} E = 1 + (a_2 + a_4 y + 2a_5 x)^2 \\ F = (a_2 + a_4 y + 2a_5 x)(a_3 + a_4 x + 2a_6 y) \\ G = 1 + (a_3 + a_4 x + 2a_6 y)^2 \\ L = 2a_5 / \sqrt{(a_2 + a_4 y + 2a_5 x)^2 + (a_3 + a_4 x + 2a_6 y)^2 + 1} \\ M = a_4 / \sqrt{(a_2 + a_4 y + 2a_5 x)^2 + (a_3 + a_4 x + 2a_6 y)^2 + 1} \\ N = 2a_6 / \sqrt{(a_2 + a_1 y + 2a_5 x)^2 + (a_3 + a_4 x + 2a_6 y)^2 + 1} \end{cases} \tag{4}$$

Assuming that the function g_{xy} expressed in formula (5):

$$g_{xy} = \frac{EN - 2FM + GL}{FG - F^2} - \left(K_{i,j}^{n1} + K_{i,j}^{n2} \right) \tag{5}$$

According to the constraint equations of four control points and two boundary constraint equations from surface patch, formula (6) can be obtained:

$$\begin{cases} g_1(a_1, a_2, a_3, a_4, a_5, a_6) = 0 \\ g_2(a_1, a_2, a_3, a_4, a_5, a_6) = 0 \\ g_3(a_1, a_2, a_3, a_4, a_5, a_6) = 0 \\ g_4(a_1, a_2, a_3, a_4, a_5, a_6) = 0 \\ f_5(a_1, a_2, a_3, a_4, a_5, a_6) = 0 \\ f_6(a_1, a_2, a_3, a_4, a_5, a_6) = 0 \end{cases} \tag{6}$$

Through formula (6), the surface equation of patch can be solved, and the fitting equation of each surface patch and coordinates of control points are obtained by analogy to achieve the fitting reconstruction process of surface patch.

2.3   Method of Shape Reconstruction Based on Modal Superposition Method

The idea of the fitting algorithm based on modal superposition method is analyzing each degree vibrational mode of plate structure, using the strain displacement transformation matrix to convert the strain information to displacement information, linear superposition each degree mode to achieve reconstruction of the curve surface.

Assuming that displacement $a(t)$ is the linear combination of each degree displacement mode ϕ_i ($i = 1, 2, \cdots, n$), so the transformation formula (7) is expressed:

$$a(t)_{M \times 1} = \sum_{i=1}^{n} \phi_i x_i = \Phi X(t) \tag{7}$$

Where Φ is the modal matrix $M \times n$, M is the number of nodes selected from finite element model, n is the modal degree.

Assuming that strain $\varepsilon(t)$ is the linear combination of each degree strain mode, and is expressed as formula (8):

$$\varepsilon(t)_{N\times 1} = \sum_{i=1}^{n} \varphi_i x_i = \Psi X(t) \tag{8}$$

Where Ψ is the modal matrix $N \times n$, N is the required nodes of the finite element model, n is the modal degree. We define formula (9):

$$DST_{M\times N} = \Phi \left(\Psi^T \Psi \right)^{-1} \Psi^T \tag{9}$$

Through formula (7), formula (8) and formula (9), we obtain formula (10):

$$a(t)_{M\times 1} = DST_{M\times N} \varepsilon(t)_{N\times 1} \tag{10}$$

Where $DST_{M\times N}$ is the transformative matrix of strain and displacement, while $N \ll M$, we can measure the strain of fewer nodes, using the displacement transformation matrix to get more points displacement and superimposing each degree vibration modes to achieve reconstruction of curve surface.

3   Design and Construction of the Experimental Platform

3.1   Selection of the Experimental Base Station

As a carrying device, experimental base station as a carrying device not only to be able to suppress the generated vibration of the exciter and avoid resonance with the bearer, but also take the advantages of easy installation, fix and movement, thus considering many aspects, the optics experimental test platform GZ103PTB was chosen as the experimental base station of the platform. Its natural frequency is 0.88-1.0Hz, far less than the measured frequency of vibration and the amplitude is less than 1um, while the surface is an array of M6 threaded bore and convenient to install various types of experimental apparatus and fixed components.

3.2   Excitation System

Excitation system is composed of function signal generator, power amplifier and exciter. The working process is shown in Fig.2:

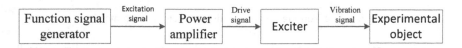

Fig. 2. The working process of excitation system

In Fig.2, the function signal generator can produce sine, square, ramp and white noise signal as the signal source into the power amplifier; the power amplifier amplifies the excitation and drives the exciter action; the exciter applies the corresponding vibration signal to the experimental object, leading the object vibrating according to preset vibration signal. Here the model of the function signal generator is SFG-2110, the model of power amplifier is YE5872 and the model of the exciter is JZK-10.

3.3   Measurement System

Measurement system comprises two parts, the measurement of strain information and the measurement of displacement information. The measurement of strain information is composed by FBG strain sensors, fiber grating network analyzer and data acquisition software, its working process is showed in Fig.3:

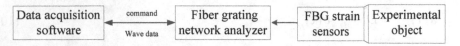

Fig. 3. The working process of strain measurement

In Fig.4, data acquisition software sends command of acquisition to the fiber grating network analyzers; the fiber grating network analyzers acquires the strain data (wavelength data) of FBG according to the set frequency after receiving command and sends the data to the data acquisition software.

The model of fiber grating network analyzers is FONA-2008C, it owns 9 acquisition channels, each channels can collect 60 grating points; the frequency of acquisition is 200Hz; the wavelength range of acquisition is 1532-1568 nm and the resolution is 1pm.

The measurement of displacement information comprises laser displacement sensor, three-dimensional slider, control box and control software. Its working process is showed in Fig.4:

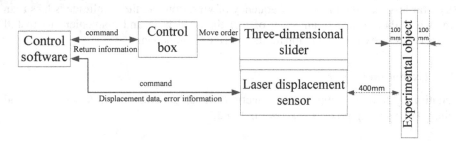

Fig. 4. The working process of displacement measurement

In Fig.5, control software sends move command to control box; the microcontroller in control box operates the three-dimensional slider move according to the set direction and speed; the laser displacement sensor installed on the three-dimensional slider sends the displacement information to the control software after the three-dimensional slider moving to the set point.

The model of laser displacement sensor is LK-G400, it use 650nm red laser as measurement signal, and the measuring range of ± 100mm while reference distance is 400mm, and the measurement accuracy is 0.01um, in addition, the smallest movement unit of three-dimensional slider is 2.5um, therefore the precision of displacement measurement system can reach for 2.5um.

3.4   Software Platform

The software platform comprises client and server, the detailed function is showed in Fig.5.

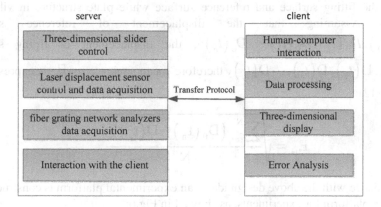

Fig. 5. The function of software system

In Fig.6, server can operate the movement of three-dimensional slider, control and receive data acquisition of the laser displacement sensor, control and receive data acquisition of the fiber grating network analyzers and interact with the client; the client provides the service of human-computer interaction, data processing, three-dimensional display and error analysis; server and client interact in accordance with established protocol.

To examine the effect of fitting algorithm, the fitting result needs for quantitative analysis, therefore error analysis is one of the key elements of the experimental platform design. Error analysis is composed by static deformation error analysis and dynamic deformation error analysis. Static deformation error analysis is that the point-to-point mean square error and extreme error of the fitting surface and reference surface while plate structure in fixed displacement. Mean square error represents the comprehensive situation of fitting error and can examine the accuracy of fitting algorithm; control software sends move command to control box; extreme error represents the maximum error of fitting algorithm and is also an important reference to examine the effect of fitting algorithm. Since the coordinates of the fitting surface is corresponding to the coordinates of the reference, therefore the distance can be calculated, meanwhile the greater the distance, the greater the fitting error. Assuming that the displacement of reference surface is $D_0(1), D_0(2), D_0(3), \cdots D_0(n)$, the displacement of fitting surface is $D(1), D(2), D(3), \cdots D(n)$, therefore mean square error E_a and extreme error E_m are expressed as formula (11):

$$\begin{cases} E_a = \sqrt{\dfrac{\sum_{n=1}^{N}(D_0(n)-D(n))^2}{N}} \\ E_m = Max\big((D_0(1)-D(1)),(D_0(2)-D(2)),\cdots(D_0(n)-D(n))\big) \end{cases} \tag{11}$$

Dynamic deformation error analysis is that the point-to-point dynamic mean square error of the fitting surface and reference surface while plate structure in vibration process. Assuming that the displacement of reference surface is $D_0(t_1), D_0(t_2), D_0(t_3), \cdots D_0(t_n)$, the displacement of fitting surface is $D(t_1), D(t_2), D(t_3), \cdots D(t_n)$, therefore mean square error E_{ta} is expressed as formula (12):

$$E_{ta} = \sqrt{\dfrac{\sum_{t=t_1}^{N}(D_0(t_n)-D(t_n))^2}{N}} \tag{12}$$

Accordance with the above design ideas, an experimental platform is constructed as building a platform for experiments, as showed in Fig.6:

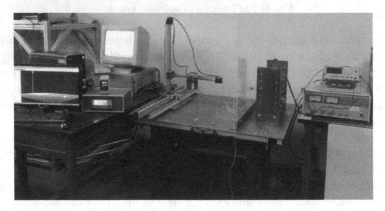

Fig. 6. Experimental platform for plate structure shape reconstruction algorithm

4   Experimental Result

According to the characteristic of the devices, Bragg grating is selected to be the experimental object, and the fitting algorithm based on plane curve array is chosen as the reference algorithm. In the experiment, plate structure is secured to one end of the size of 800mm * 800mm, showed in Fig.7:

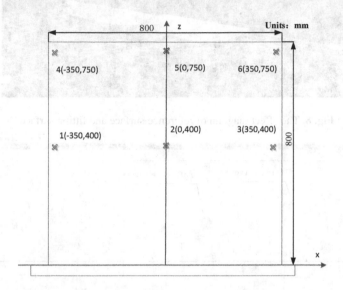

Fig. 7. Experimental verification platform for plate structure shape reconstruction algorithm

In Fig.8, 6 points were selected to do error analysis, and not only did accurate analysis of static deformation error, but also acquired analyzed real-time dynamic deformation experimental data, and the results are showed in Table 1 and Table 2:

Table 1. Result of static error analysis(units:mm)

Point	1	2	3	4	5	6	Max
E_m	1.120	1.124	1.116	1.142	1.157	1.156	1.157
E_a	1,112	1.103	1.105	1.123	1.126	1.129	1.129

In Table 1, for 10 experiments of plate structure, the maximum of extreme error is 1.157mm while the maximum of mean square error is 1.129mm, far higher than the accuracy of data in literature [9]. The effect diagram of reference surface and fitting surface is showed in Fig.8.

Table 2. Result of dynamic error analysis(units:mm)

Point	1	2	3	4	5	6	Max
E_{ta}	1.121	1.211	1.102	1.098	1.129	1.131	1.211

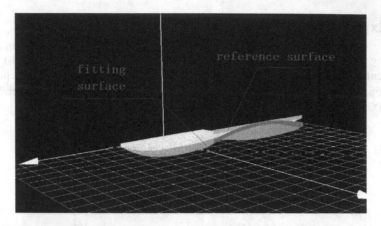

Fig. 8. The effect diagram of reference surface and fitting surface

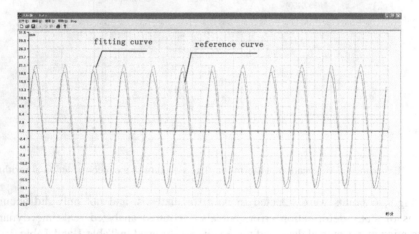

Fig. 9. Vibration graph of reference curve and fitting curve

In Table 2, for 6 experiments of plate structure, the maximum of dynamic mean square error is 1.211mm, far higher than the accuracy of data in literature [9]. These data suggest that this platform is with good real-time capacity and high accuracy to meet the needs for the verification and data analysis of a variety of algorithms. The vibration graph of reference curve and fitting curve is showed in Fig.9.

5   Conclusion

For the problem of experimental verification for plate structure shape perception and reconstruction algorithm, an experimental verification platform was designed and constructed consisting of experiment base station, excitation system, measurement systems and relative software, considering the fitting algorithm based on plane curve as reference algorithm, and the platform not only analyzes accurately for static deformation error, but also acquires and analyzes the real-time experimental data of dynamic deformation. Results suggest that the experimental platform can conduct experimental verification and data analysis of a variety of algorithms with good real-time capacity and high measurement accuracy.

Acknowledgment. This research is supported by the National Nature Science Foundation of China (No.51175319), and key program of Shanghai Municipal Education Commission (No.13ZZ075), Shanghai University "11th Five-Year Plan" 211 Construction Project and Shanghai Key Laboratory of Power Station Automation Technology.

References

1. Zhang, Q.B., Zhao, J.: Determination of mechanical properties and full-field strain measurements of rock material under dynamic loads. International Journal of Rock Mechanics and Mining Sciences 60, 423–439 (2013)
2. Lin, X.L., Qian, Q.J.: Automatic damage detection system for satellite solar array. Chinese Journal of Optics and Applied Optics 3(2), 140–145 (2010)
3. Zhu, X.J., Zhang, H.S., Xie, C.N.: Analysis of Curve Surface Fitting Algorithm Based on Curvatures for Space Sailboard Structure. Journal of System Simulation 19(11), 2496–2499 (2007)
4. Wang, H.T., Sun, X.H.: Smooth B-Spline Surface Reconstruction from Triangular Mesh of Airplane Configuration Parts. Journal of Nanjing University of Aeronautics & Astronautics 39(3), 323–328 (2007)
5. Lo, S.H.: Automatic merging of tetrahedral meshes. International Journal for Numerical Methods in Engineering 93(11), 1191–1215 (2013)
6. Stephan, R., Kang, L.H., et al.: Displacement field estimation for a two-dimensional structure using fiber Bragg grating sensors. Smart Materials and Structures 18, 2566(12p.) (2009)
7. Glondu, L., Marchal, M., Dumont, G.: Real-time simulation of brittle fracture using modal analysis. IEEE Transactions on Visualization and Computer Graphics 19(2), 201–209 (2013)
8. Fan, H.C., Qian, J.W.: Sensor network design for flexible surface shape measurement. Optics and Precision Engineering 16(6), 1087–1092 (2008)
9. Di, H.T.: Three dimensional reconstruction of curved shape based on curvature fiber optic sensor. Optics and Precision Engineering 18(5) (2010)

Improved Mean-Value Coordinates Algorithm for Image Fusion

Cai Fu[1], Yuying Shao[2], Li Deng[1,*], and Gen Lu[1]

[1] School of Mechatronics Engineering and Automation, Shanghai University,
Shanghai, 200072, China
[2] State Grid Shanghai Municipal Electric Power Company,
Shanghai, 200025, China
dengli@shu.edu.cn

Abstract. Image fusion is an advanced image processing, in which mean-value coordinates (MVC) algorithm based on Poisson image is a fast and effective algorithm. However, the algorithm may have unsatisfactory results if the source image and target image have many variations of color on the image boundary and image details. To solve the problem, this paper proposes two optimization methods, preserving color based on geodesic distance and matching details with modified detail layer. To verify the feasibility of the methods, the improved MVC results are compared with the original MVC results by experiments. The comparison results show that the improved approach can achieve better performance in image fusion.

Keywords: image fusion, mean-value coordinates, improved mean-value coordinates, image color, image detail.

1   Introduction

Image fusion is a useful image editing operation. It means to integrate and synthesize the information of source image and target image to generate a single image with high quality and accurate description. In these years, many image fusion algorithms are developed to produce a seamless 2D image more rapidly and improve the fusion performance more favorably.

The gradient-domain techniques applied in the image fusion can accomplish the editing task efficiently. The algorithm based on this technique was initially proposed by Fattal et al. in 2002 [1]. Perez et al. proposed a algorithm based on Poisson equation, which is a most useful tool in gradient-domain techniques in 2003 [2]. However, the Poisson fusion is inefficient because it needs to take a lot of time out of solving Poisson equation. Farbman et al. propose a coordinate-based approach -- mean-value coordinates(MVC) fusion, with which the fusion tasks can be fast and straight forward to implement [3].

---

[*] Corresponding author.

M. Fei et al. (Eds.): LSMS/ICSEE 2014, Part II, CCIS 462, pp. 210–218, 2014.

The implementation of MVC needs the boundary vertexes' weights and the boundary interpolation, with which, the source image can be modified. Patching the modified source image into target image, the result image is obtained. Like Poisson fusion, the MVC result image is totally dependent on the boundary information of the image. So the MVC fusion cannot work well when the error along the boundary is obvious, especially the existed color deviation may cause color distortion, which results in unrealistic image. In addition, if the source image and the target image have big difference in details, the result may not be harmonious.

Petrovic et al. present a multi-resolution image fusion based on the gradient Laplacian pyramid. This fusion approach transfers visual information from input image into result image without loss of information or distortion accurately. However, the noise may be blended in because of the operation for quantization and threshold setting in the processing procedure of Laplacian transform coefficient [4-5].

Guo et al. present an approach to solve the color distortion by a image inpainting approach with user's marks. The approach needs user's brief marks, with which the result image can be modified to the correct color [6]. Sunkavalli et al. propose an approach to solve the problem of image harmonization in 2010 [7]. They present a framework that can explicitly matches the visual appearance of image before fusing. They use the multi-scale technique to transfer the appearance of source image to that of target image.

In this paper, a new approach based on the MVC is introduced. Two preprocess approaches are combined with MVC and solve the problem of color and detail. Only a little user's mark is needed in this fusion operation. After modifying the details of source image with image inpainted and correcting color with geodesic distance to correct, a harmonious and realistic result image can be obtained.

2   Algorithm

2.1   Mean-Value Coordinates Fusion

The algorithm is an approach of mean-value interpolation. Consider a closed polygonal boundary curve, the mean-value coordinate of each inside pixel is given by

$$\lambda_i(x) = \frac{w_i}{\sum_{j=0}^{m-1} w_j} \tag{1}$$

Where

$$w_i = \frac{\tan(\alpha_{i-1}/2) + \tan(\alpha_i/2)}{\|p_i - x\|} \tag{2}$$

The angle α_{i-1}, α_i and the boundary p_i is shown in Fig. 1, once computed, the coordinates can be used to interpolate for the value of inside pixels.

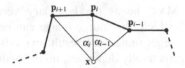

Fig. 1. Angle definitions

$$r(x) = \sum_{i=0}^{m-1} \lambda_i(x)(f^* - g)(p_i) \qquad (3)$$

$r(x)$ is the modified function, the result may be got by $f = g + r$.

To make MVC fusion fast, two optimizations can be used.

1. Generating a Delaunay adaptive mesh in fusion area, only the vertices of the mesh need to use mean-value interpolation. The value at each other pixels is obtained by linear interpolation.

2. Sampling the boundary hierarchically to make different inside vertices use different vertices at the boundary. The number of sampled boundary vertices is inversely proportional to the distance between the pixels and boundary vertices [8].

2.2   Modified Algorithm

This section explains the detail how to solve the problem of color distortion and details mismatch. The image editing process is shown in Fig. 2. As is shown in Fig. 2,   the workflow is consist of three main procedures, such as layer decomposition, detail layer operation and color layer operation. Since the two optimization approaches are based on different layers, the layer decomposition is necessary. Detail layer operation and color layer operation are the main approaches to implement the modification.

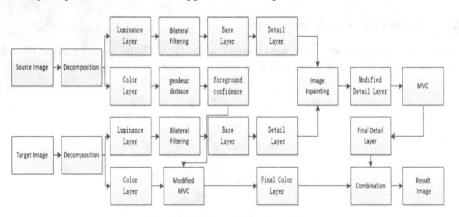

Fig. 2. Workflow of the algorithm

2.2.1   Layer Decomposition

To modify the problem of details and color respectively, layer decomposition needs to be implemented firstly. After decomposing layer, the luminance layer can be acquired, from this layer the detail layer can be separated. Specifically, it is obtained by converting to the CIELAB color space firstly, which includes luminance layer L and two color layer a, b, and the detail layer D is given by

$$D = L - B \tag{4}$$

Where L is the luminance layer and B is the base layer. The base layer containing the framework of image is got by filtering the luminance layer. The approach is a bilateral filter based on the adaptive mesh, which is accomplished by filtering the mesh using local neighborhoods. The approach is often used in 3D image denoising [9] [10]. Combining with the adaptive mesh that generated in MVC preprocessing, each vertex v is filtered to

$$\tilde{v} = \frac{\sum_{q \in N(v)} w_c w_s I(q)}{\sum_{q \in N(v)} w_c w_s} \tag{5}$$

Where $w_c w_s$ is weight, N(v) is the adjacent neighborhood vertexes. Like bilateral filter, this approach also preserves the edge of image [11].

As shown in formula (4), the detail layer D used in modifying detail harmony and color layer a, b used in color preservation (or correction) can be finally got.

2.2.2   Operation on Detail layer

Since the MVC algorithm implements the fusion tasks without taking consider in the details, especially which of the target image may get lost seriously and result in an unharmonious fusion image. In other words, the details of fusion area may not match the other areas in the target image. Furthermore, some unwanted details may blended in the result image [12].

Criminisi et al. present an approach to implement image inpainting, which combines the advantage of the texture synthesis with image inpainting [13]. The approach of inpainting the interested area by using the texture patch contains the pixels outside the area. The detail layer of the source image is modified by extracting the detail patch outside the fusion area in the target image, the modified detail layer can be obtained by

$$D_b = \gamma_s D_s + \gamma_t D'_t \tag{6}$$

Where D_s is the detail layer from source image, and D'_t is the modified layer for source image. γ_s、γ_t are respectively the weight for these layers. The weight affects the details of source image and target image used in the result image. If the details of both image have little difference, γ_t may be set to 0 with the γ_s set to 1. The default value of γ_s and γ_t is set to 0.1 and $1 - \gamma_s$. Note that γ_t cannot be set to a small value, this may cause the loss of details.

Finally, composing the modified detail layer and base layer, a new luminance layer of source image can be generated, after which, using MVC to fuse the luminance layer.

2.2.3 Operation on Color Layer

Since the MVC is totally dependent on the boundary information, it may cause more or less color distortion unless the color of boundary of both images are identical. If the color between two images has large difference, big color error may generate in the result image, especially the foreground area in it. This is because the foreground distortion is more sensitive than the background distortion. To preserve color of the foreground image, the foreground confidence is defined to control the value of color, with which, not only the foreground color can be kept, but also smooth transition can also be implemented.

In fact, the foreground confidence is considered as the weight for the variations of MVC process. So the value of foreground confidence is inversely proportional to the distance between the pixel and the foreground and also inversely proportional to final color variations. To make the foreground color transit to background color naturally, the foreground confidence for the areas with same feature should be smooth. So the foreground confidence $B(x)$ is given as smooth interpolation of foreground distance function $D(x)$

$$B(x)=3\left(\frac{D(x)-D_{\min}}{D_{\max}-D_{\min}}\right)^2-2\left(\frac{D(x)-D_{\min}}{D_{\max}-D_{\min}}\right)^3 \tag{7}$$

Where $D_{\min}$ is the minimum distance and $D_{\max}$ is the maximum distance. The distance is not got by calculating Euclidean distance, which only shows the relations of pixels in coordinates space. The geodesic distance is used to replace the Euclidean distance [14-16], which combines the color space with the coordinates space, for each pixel x, the distance $D(x)$ can be given by

$$D(x)=\min d(f,x) \tag{8}$$

Where f is the marked pixels, geodesic distance $d(f,x)$ is obtained by

$$d(x_1,x_2)=\min_{S_{x_1,x_2}}\sum|p(x)-p(y)| \tag{9}$$

Where S_{x_1,x_2} is a set of path between the pixel x_1 and pixel x_2, $|p(x)-p(y)|$ is the value between the adjacent pixel. The formula (8) is shown that the geodesic distance is to search the minimum distance between two pixels. The formula (9) is shown that the foreground distance is to search the minimum distance between the given pixel x and all the marked pixels. The foreground confidence $B(x)$ can be finally got by formula (8-9). With the foreground confidence viewed as the weight of color variations, the modified MVC algorithm is defined as

$$f(x) = g(x) + B(x)r(x) \tag{10}$$

The modified color layer a and b is obtained by using formula (10). At last, after combining the two color layers with the luminance layer given in 2.2.2, the result image with preserved color and harmonious details can be obtained.

3   Experimental Results and Analysis

In this paper, we modify the MVC algorithm to achieve a more harmonious and realistic image only with a little user's mark and test the feasibility of this approach by OPENCV on VS2010. Firstly the original MVC algorithm is tested. If the color difference between the source image and target image is small, the result image is good, as shown in the Fig. 3(3).

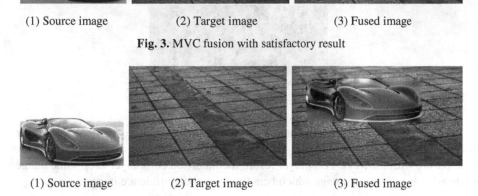

(1) Source image   (2) Target image   (3) Fused image

Fig. 3. MVC fusion with satisfactory result

(1) Source image   (2) Target image   (3) Fused image

Fig. 4. MVC fusion with unsatisfactory result

However, if source image is fused as in Fig. 4(1) and (2), the result may be unsatisfactory, just as shown in the Fig. 4(3), because of the big difference between the source image and target image, the color distortion become obvious, the car as the foreground of the image is yellowish overall and some background details doesn't match the target image. The improved algorithm is used to solve the problem as described above. On request, firstly the source image need to be marked as shown in Fig.5 and then the program can be run. With detail layer inpainting process and MVC process, the composited luminance layer can be obtained as shown in the Fig. 6(3). Since the detail feature of target image is not obvious and rich, so the composited luminance layer does not have large differences with the MVC luminance layer result.

Fig. 5. Marked source image

Red line: foreground mark   Blue line: background mark

(1) Source luminance layer image   (2) MVC luminance layer image (3) Modified luminance layer image

Fig. 6. Comparison in luminance layer image

With the marked pixels, the confidence distribution (or weight distribution) image can be easily obtain as shown in Fig. 7. The color of a pixel is blacker; the confidence for the pixel is smaller. Notably, the confidence distribution image is totally dependent on the marked lines. So, different marked lines have different confidence.

Fig. 7. Foreground confidence image

According to the weight information in the confidence image, the final color layer image can be obtained with the modified MVC as shown in right of Fig. 8-9. Comparing with the MVC color layer result, the modified color layer keep some image regions' color of source image which benefits from the confidence image.

(1) Layer image   (2) MVC color layer image   (3) Modified color layer image

Fig. 8. Comparison in a color layer image

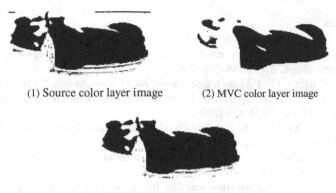

(1) Source color layer image   (2) MVC color layer image

(3) Modified color layer image

Fig. 9. Comparison in b color layer image

Compositing all the layer above and patching to the target image, a more harmonious and realistic image can be finally got as shown in the right of Fig. 10. For comparison, the original MVC approach is tested and the image is obtained in the left of Fig. 10. It is shown that the car as the foreground image is more similar to the car of source image (color preserved), although the detail optimization is not obvious, this is shown that the source image matches the target image better. Notably, the difference between source image and target image is inevitable existence.

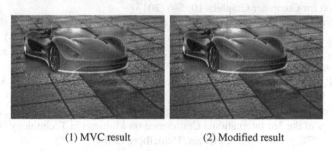

(1) MVC result   (2) Modified result

Fig. 10. Comparison between MVC result and modified result

4   Conclusion

In this paper, an improved mean-value coordinates algorithm is presented for image fusion. After decomposing the image to a luminance layer and color layers, the texture patch is used to modify the source image detail, with which the fusion area is made to match the target image for getting a more harmonious image. Afterwards, the geodesic distance is applied to estimate the distance between the pixels and the marked foreground pixels, with which the foreground image color can be preserved. Both approaches above are based on MVC, in other words, they are only the preprocessing procedure. The approaches only need a little user mark, and can easily get more satisfactory results.

References

1. Fattal, R., Lischinski, D., Werman, M.: Gradient domain high dynamic range compression. ACM Trans. Graph. 21(3), 249–256 (2002)
2. Perez, P., Gangnet, M.: Poisson image editing. ACM Trans. Graph. 22(3), 313–318 (2003)
3. Farbman, Z., Hoffer, G., Lipman, Y., Lischinski, D.: Coordinates for instant image cloning. ACM Trans. Graph. 28(3), 1–9 (2009)
4. Petrovic, V.S., Xydeas, C.S.: Gradient-based mutliresolution image fusion. IEEE Trans. on Signal Processing 13(2), 228–237 (2004)
5. Burt, P.J., Adelson, E.: The Laplacian pyramid as a compact image code. IEEE Communication Society 31(4), 532–540 (1983)
6. Guo, D., Sim, T.: Color me right-seamless image compositing. In: Jiang, X., Petkov, N. (eds.) CAIP 2009. LNCS, vol. 5702, pp. 444–451. Springer, Heidelberg (2009)
7. Sunkavalli, K., Micah, K.J., Matusik, W., Pfister, H.: Multi-scale image harmonization. ACM Trans. Graph. 29(4), 125–134 (2010)
8. Floater, M.S.: Mean value coordinates. Comput. Aided Geom. 20(1), 19–27 (2003)
9. Fleishman, S., Drori, I., Cohen-Or, D.: Bilateral Mesh Denoising. ACM Trans. Graph. 22(3), 950–953 (2003)
10. Zheng, Y.Y., Fu, H.B., Au, O.K.-C., Tai, C.-I.: Bilateral Normal Filtering for Mesh Denoising. IEEE Transactions on Image Processing 17(10), 1521–1530 (2011)
11. Tomasi, C., Manduchi, R.: Bilateral filtering for gray and color images. In: Proceedings of the Sixth International Conference on Computer Vision, pp. 1–5. IEEE, Bombay (1998)
12. Hin, Y., Hideo, Y.: Poisson Image Analogy: Texture-Aware Seamless Cloning. European Association for Computer Graphics 10, 5–6 (2013)
13. Criminisi, A., Perez, P., Toyama, K.: Region filling and object removal by exemplar-based Inpainting. IEEE Transactions on Image Processing 13(9), 1200–1212 (2004)
14. Criminisi, A., Sharp, T., Rother, C., Perez, P.: Geodesic image and video editing. ACM Trans. Graph. 29(5), 1–15 (2010)
15. Brian, L.P., Bryan, M., Scott, C.: Geodesic graph cut for interactive image segmentation. CVPR 13(118), 3161–3168 (2010)
16. Wu, H., Xu, D.: Color preserved image compositing. In: Farag, A.A., Yang, J., Jiao, F. (eds.) Proceedings of the 3rd International Conference on Multimedia Technology (ICMT 2013). LNEE, vol. 278, pp. 325–333. Springer, Heidelberg (2014)

Study of Face Recognition Technology Based on STASM and Its Application in Video Retrieval

Chunlei Liu, Kun Chen, and Yongjin Xu

Department of Automation, College of Mechatronics Engineering and Automation,
Shanghai University; Shanghai Key Laboratory of Power Station Automation Technology,
200072 Shanghai, China
liuchunlei602A@126.com, chenkun10086@163.com,
xuyj@mail.shu.edu.cn

Abstract. The gradual perfection of video retrieval technology has a positive effect in maintaining public order. However, with the improving complexity of monitoring environment, the increase of related video data requires further improvement to the efficiency of video retrieval technology. Video retrieval technology aiming at processing massive video data is needed urgently and it has become hot research subject in multimedia retrieval area. In this paper, the application of face recognition technology in video retrieval is discussed. To improve the retrieval efficiency, STASM algorithm based on OpenCV software platform is designed. The research involves the acquisition of video image frame data, face recognition and detection. Experimental results demonstrate the effectiveness and efficiency of the algorithms.

Keywords: face detection, face recognition, STASM, video retrieval.

1 Introduction

The development of video retrieval technology provides technical support to maintain public order. Among the massive related video data, however, how to find the surveillance information about a specific person is a problem need to be solved urgently. Traditional video retrieval technology has the drawbacks such as strong subjectivity, slow speed and high error rate. Such drawbacks restrict the efficiency of video retrieval, so its application in emergency situation would be limited. To solve the problem, video retrieval technology based on content has become the hot research subject in multimedia research recent years.

Currently, the related references about the application of face recognition technology in video retrieval are not so many. Everingham proposed face clustering method; Arandjelovic and Sivic put forward the video face recognition method on the front face for video retrieval, etc.

Due to the complexity of video data, traditional video retrieval technology has low efficiency and high difficulty, and cannot make intelligent retrieval on video content.

M. Fei et al. (Eds.): LSMS/ICSEE 2014, Part II, CCIS 462, pp. 219–227, 2014.

Existing systems commonly used linear sequence storage strategy, so it is essential to acquire the related video time information in advance. Otherwise, a wide range of video playback mode is the only choice for video retrieval. To determine whether a specific face image occurs in a period of surveillance video, traditional method requires retrieving all the data in order. Thus, an automatic retrieval algorithm is particularly important. The paper presents STASM retrieval algorithm, which can get more accurate retrieval results [1].

2   The Retrieval System Frame

The entire frame retrieval system is presented in Figure 1, including four parts [2][3]:

1. The establish of a STASM video face detection model in special conditions;

2. To detect the face region as a template, two-way track to the frame region of which was not detected and collect missed faces;

3. The face region can be divided into a sequence of packet: Group 1, Group 2, ..., and Group i. The order of grouped Eigen Face remains unchanged;

4. To recognize human faces, if group i is similar to the target face, Gabor wavelet transform + pattern matching is used to identify retrieval.

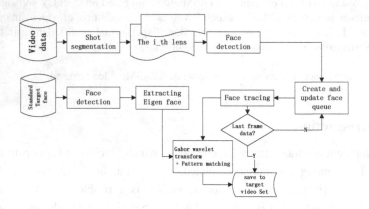

Fig. 1. Structure diagram of the retrieval system

3   The Algorithm Introduction

3.1   Stasm Face Detection Algorithm

Stasm is a C++ software package to locate facial landmarks in human face. Input a face image and the positions of the landmarks returns. Stasm is designed to work on front views of approximately upright faces with neutral expressions. Poor effect may be achieved on faces with complex expressions. The Histogram Array Transform (HAT) descriptors, which is similar to SIFT descriptors, is used by Stasm for template matching.

Active Shape Model (ASM) algorithm is a global statistical shape model constraint local texture matching results. Active Appearance Model (AAM) shape and texture combined statistics (apparent) model parameters are optimized so that the best match with the input model. The effective combination of the two algorithms can be very precise to mark the human face facial features points. ASM is a point distribution model (Point Distribution Model, PDM) algorithm. In PDM, shapes similar to the object, such as a human face, hand, heart, lungs and other geometric shapes may be a shape vector, which is formed by a series of several feature points (landmarks) [1].

3.2 Face Recognition Algorithm

Wavelet Transform and Image Matching Based on Gabor of Shape

The algorithm uses an improved Harris corner detection to extract the first corner, gets the coordinate of corner, filters the two-dimensional image with the reference image, gets the wavelet coefficients, regards it as the characterization, and then introduces the two similar factors to match. A large number of different images through experiments that the algorithm to select the appropriate parameters, while using the case of the longest common subsequence metric factor of the same name can be successfully extracted more points, and can achieve a higher matching rate [4].

Face Recognition Algorithm Based on a Particular Subspace

All people use a face subspace in "feature face" which builds a face subspace for each human face. It not only can describe the diversity between different individual faces better, but also discard the adverse noise and class diversity which is harm to face recognition. Thus it has a better distinguish ability compared to the traditional one. Moreover, we put forward a technology based on a single sample that generates multiple training samples for face recognition that has the single training sample, therefore, the method that needs many training samples can be used to face recognition for the single training sample [4].

4   Facial Feature Extraction and Recognition

ATSM/AAM resulting model can cover a good variety of facial geometry and face texture subspace, and can be a good part of the region of non-face. The experienced knowledge of the parameters included in the model to extract the target has a good guide, and thus having robustness to blocking, degradation and other issues.

In the practical application, STASM includes training and search: STASM Training; Establish the shape of the mode.

4.1   Collected Training Samples

If STASM training is needed in the critical areas of human face, n sample images containing personal facial area should be collected. It is noted that people's facial region should be included in the collected images, while the normalization issue can be ignored.

4.2   Manual Records or k Key Feature Points Marked with a Small Program

As shown in Fig.2 and Fig.3, for any training image, the coordinate information of a plurality of critical feature points should be recorded and stored in a text file.

Fig. 2. STASM Single Eigen Face landmark

Fig. 3. STASM Multiple Eigen Face landmark

4.3   Construction of the Training Set Shape Vector

One shape vector (1) consists of k key feature points marked in a graph:

$$a_i = (x^i_1, y^i_1, x^i_2, y^i_2, \ldots \ldots x^i_k, y^i_k), i = 1, 2, 3 \ldots \ldots n \tag{1}$$

(x^j_i, y^j_i) is the coordinate of the $j-$th feature point in the number $i-$th training sample. Define n as the number of training samples, thus n shape vectors are achieved [5].

4.4   Shape Normalization

The purpose of this step is to make the face shape calibrated manually be normalized or aligned. With this method, the non-shape interference, which is caused by external factors, such as different angles of the image, distance and posture transformation, can be eliminated. A more efficient point distribution model is achieved. Mainly through translation, rotation, scaling transformations, a point distribution model is conducted aligned without changing the point distribution model. STASM uses Procrustes to make point distribution model aligned, the steps are described as follow:

(1) Align all face models in training set to the first person to face model
(2) Calculate the average face model $\overline{\alpha}$
(3) Align all face models in training set to face model $\overline{\alpha}$
(4) Repeat (2) (3) until convergence

4.5   The Shape of the Alignment Vector After PCA Process

(1) Calculate the average shape vector:

$$\overline{a} = \frac{1}{n} \sum_{i=1}^{n} (a_i) \tag{2}$$

(2) Calculate covariance matrix:

$$S = \frac{1}{n} \sum_{i=1}^{n} (a_i - \overline{a})^T (a_i - \overline{a}) \tag{3}$$

(3)Calculate the Eigen values of covariance matrix S and sorted in order from largest to smallest.

This will get$\lambda_1, \lambda_2, \ldots \ldots \lambda_q$, ($\lambda_i > 0$). Select t eigenvectors $P = (p_1, p_2, \ldots, p_t)$ so that its corresponding Eigen values satisfy [6]:

$$\frac{\sum_{i=1}^{t} \lambda_i}{\sum_{i=1}^{q} \lambda_s} > f_c V_T \tag{4}$$

f_c is a feature vector by a scaling factor to determine the number, the value is generally 95%, but V_T is the sum of all the Eigen values. Namely:

$$V_T = \sum \lambda_i \tag{5}$$

For any shape of a vector for the training can be expressed as:

$$a_i \approx \overline{a} + P_s b_s \tag{6}$$

b_s is a vector include t parameters.

$$b_s(i) = P^T(a_i - \overline{a}) \tag{7}$$

4.6   Final Construction of Local Feature Point for Each Feature

In the both sides of $i-$th feature point in the $j-$th training image, m pixels are selected in a direction perpendicular to the point of change after the two feature points to form a vector of $2m+1$ length. The local texture g_{ij} is calculated by derivation of pixel gray value contained in the vector. For $i-$th feature point in other training samples images, n local texture $g_{i1}, g_{i2}, \ldots, g_{in}$ of $i-$th feature point is made by the same operation and then be averaged:

$$\overline{g}_i = \frac{1}{n} \sum_{j=1}^{n} g_{ij} \tag{8}$$

And the variance:

$$S_i = \frac{1}{n} \sum_{j=1}^{n} (g_{ij} - \overline{g}_i)^T \cdot (g_{ij} - \overline{g}_i) \tag{9}$$

Then obtain the local feature of the $i-$th feature point and do same operation to all the other feature points and local features of each feature points can be obtained. The similarity between new feature point g of a feature point and its trained local feature can be represented by Mahalanobis distance:

$$f_{sim} = (g - \overline{g}_i) S_i^{-1} (g - \overline{g}_i)^T \tag{10}$$

5   The Application of Stasm in Video Retrieval

Continuance in time is a very important characteristic of video image, and the uncertainty of the face information is generated. The biggest difference between the face recognition algorithm based on static images and face recognition algorithm is the use of time information when tracking and recognizing face image. At present such algorithms can be approximately divided into two categories:

The first method is Tracking-then-Identification. Human face is first detected and facial feature is traced over time. When a frame image meeting the certain standards (size, position) is captured, face recognition algorithm based on static image is adopted. Tracking and identification methods are performed independently, while time information is only used in the tracking stage. That is to say, time information is still based on static image.

The other method is Track - and – Identification. With such method, face tracking and recognition is carried out simultaneously, and time information is both used in the tracking stage and recognition stage [7].Video retrieval test system is presented in Figure .4:

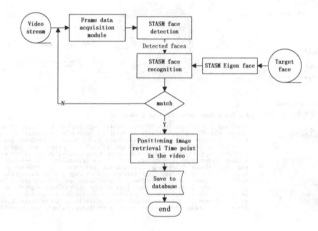

Fig. 4. The frame of STASM test system

The main library functions are described as below:

We read the data from the video frame image using AVI CvCapture class cvCreateFileCapture() and cvQueryFrame() to achieve access to the data of the image frame. Using stasm_search_single() method to detect human faces and obtain STASM facial feature and features dot matrix. Using cvMahalonobis() to identify with the normalization of two Eigen Face's position vector. Finally cvWriteFrame() function to retrieve all relevant written data frames to the destination video files , to achieve the purpose of retrieving .

Table 1. Verification data and results analysis

Test video set	Total frame	Correctly detection	Uncorrected detection	The ratio of retrieved face	Retrieval accuracy
Single face	803	645	61	80.30%	90.54%
Multiple face	624	802	123	78.32%	84.66 %
Mix face	687	545	68	79.33 %	87.52%

As can be seen from Table1, if the video contains more than one face image, the retrieval accuracy and efficiency will decrease to some extent. Miss of some facial expression, rotation of face image, complexity of image background and inaccuracy of training set are the main reasons. Experimental results show that the retrieval time for compressed video data is shorter. To improve the retrieval efficiency, video data is required to be compressed in advance. The WinForm of experiment is presented in Figure.5.

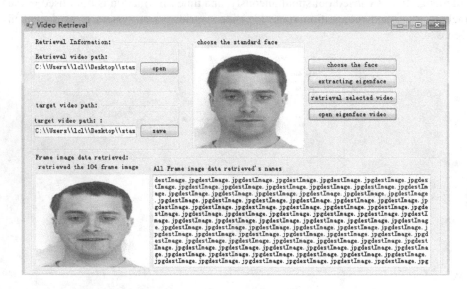

Fig. 5. The windows of STASM Eigen Face system

The main point of this paper is to achieve information retrieve in surveillance video by face recognition technology. The retrieve result shows its practicality in some aspects of this field. When applied to video retrieval, it avoids the drawbacks of traditional retrieve method based on massive manual playback, which improves work efficiency extremely and had a good positive prospect.

6   Conclusion

This paper designs and implements an example of video retrieval experiment based on the technology of face recognition. Video retrieval experiment includes face recognition and target face retrieving in the specific video. To some degree, the results show that this method can meet user's demand for video retrieval and achieve the desired purpose. However, undetected error and false retrieval caused by the rotation of the face and complex expressions is inevitable, which is what we need to solve in the future. A more efficient, more robust feature extraction algorithm need to be designed.

Acknowledgments. This work is supported by Department of Automation, College of Mechatronics Engineering and Automation, Shanghai University and Shanghai Key Laboratory of Power Station Automation Technology. Thanks my teacher Professor Yong jin Xu.

References

1. Tan, H.C., Zhang, Y.J.: Computing Eigenface from Edge Images for Face Recognition Based on Hausdorff Distance. In: Proceedings of the 4th International Conference on Image and Graphics, Chengdu, China: [s. n.] (2007)
2. Barkan, O., Weill, J., Wolf, L., Aronowitz, H.: Fast high dimensional vector multiplication face recognition. In: ICCV (2013)
3. Simonyan, K., Parkhi, O.M., Vedaldi, A., Zisserman, A.: Fisher vector faces in the wild. In: BMVC (2013)
4. Li, H., Hua, G., Lin, Z., Brandt, J., Yang, J.: Probabilistic elastic matching for pose variant face verification. In: CVPR (2013)
5. Cootes, T.F., Edwards, G., Taylor, C.J.: Comparing Active Shape Models with Active Appearance Models. In: Proceedings of British Machine Vision Conference (1999)
6. Kong, J., Han, C.: Content-based video retrieval system research. In: IEEE Conference on Computer Science and Information Technology (2010)
7. Smeulders, Worring, M., Santini, S.: Content-based image retrieval at the end of the early years. EEE Trans. Pattern Anal. Machine Intel. (2000)

Robust Blurred Palmprint Recognition
via the Fast Vese-Osher Model

Danfeng Hong[1,*], Wanquan Liu[2], Jian Su[3], Zhenkuan Pan[1], and Xin Wu[1]

[1] Qingdao University, Qingdao, 266071 China
[2] Curtin University, Perth, WA 6102 Australia
[3]The University of Electronic Science and Technology of China, Chengdu, 611731 China
{sj890718,hongdanfeng}@gmail.com, w.liu@curtin.edu.au

Abstract. In this paper, we propose a new palmprint recognition system by using the fast Vese-Osher decomposition model to process the blurred palmprint images. First, a Gaussian defocus degradation model (GDDM) is proposed to extract the structure layer and texture layer of blurred palmprint images by using the fast Vese-Osher decomposition model, and the structure layer is proved to be more stable and robust than texture layer for palmprint recognition. Second, a novel algorithm based on weighted robustness with histogram of oriented gradient (WRHOG) is proposed to extract robust features from the structure layer of blurred palmprint images, which can address the problem of translation and rotation to a large extent. Finally, the normalized correlation coefficient (NCC) is used to measure the similarity of palmprint features for the new recognition system. Extensive experiments on the PolyU palmprint database and the blurred PolyU palmprint database validate the effectiveness of the proposed recognition system.

Keywords: Biometrics, robustness, stable feature extraction, the fast VO decomposition model, weighted robustness with histogram of oriented gradient.

1   Introduction

Biometrics techniques are aimed to verify the identity of a living person effectively with physiological or behavioral characteristics. As one of these techniques, palmprint recognition has attracted much attention due to its various advantages, including high recognition accuracy, low-cost of hardware, easy availability, etc. In the past decades, palmprint recognition as an emerging technology in the field of biometric identification has achieved significant progress [1]. According to the representation methods of palmprint features, the current approaches can be roughly classified into three categories: principal line extraction [2], subspace learning [3-4], and texture coding [5-7].Texture coding methods have high recognition accuracy, which is one

---

[*] This work was supported in part by the Natural Science Foundation of China under Grant No. 61170106 and a project of Shandong Province Higher Educational Science and Technology Program No.J14LN39.

M. Fei et al. (Eds.): LSMS/ICSEE 2014, Part II, CCIS 462, pp. 228–238, 2014.

type of the most effective recognition methods currently with an assumption that palmprint images must be clean. Usually, the clean palmprint images can only be obtained by using contact devices. Due to this strict limitation, the non-contact palmprint image acquisition and recognition gradually become the mainstream of research [8-9]. However, the Non-contact system also has some inherent defects. For example, it is easy to produce blurred images due to defocus because the palm sometimes lies outside the realm of depth of focus, and this would decrease the system performance. The image blur is a common problem in biometrics, and it has not attracted significant attention in palmprint recognition systems. Some typical works include [2, 3, 9]. All these approaches are just used the low frequency features directly and we believe there is plenty of room for improvement.

In this paper, we propose an effective robust and fast blurred palmprint recognition method, which not only can achieve high recognition accuracy but also meet real-time requirement for a larger database. We first introduce the theory of blurry images with the Gaussian defocus degradation model (GDDM) and extract the structure layer of blurred palmprint images by utilizing the fast VO decomposition model. To further extract the stable features from the structure layer, we choose the histogram of oriented gradient (HOG) that is a desirable descriptor on the direction characteristics. In order to make this feature more robust for the translation and rotation, a fractal weight is added to the improved HOG for further improving the validity of characteristics. The proposed method is named as fast VO-WRHOG. Finally, we use the normalized correlation coefficient (NCC) to measure the similarity, and select the palmprint category for classification.

The remainder of this paper is organized as follows: In Section II, the image blur theory and fast VO model are introduced. In Section III, the proposed fast VO-WRHOG is presented including extraction of stable features from blurred palmprint images and description of the feature matching method. In Section IV, we first introduce the PolyU palmprint database briefly and then we report a series of experimental results. Finally we give some conclusions in Section V.

2   Image Burr Theory and Fast VO Model

2.1   The Image Blur Theory

In [9], a blur Image can be considered to be equivalent to a clean image in convolution with a degradation function in the spatial domain, which can be shown as follow

$$d(i, j) = f(i, j) * h(i, j) + n(i, j) \tag{1}$$

where (i, j) is the position of image, $f(i, j)$ is a clean image, $h(i, j)$ is the degradation function, $n(i, j)$ is additive noise, $d(i, j)$ is blurred image, '*' is an operator of convolution. In [10], Wang et al. listed several common degradation functions, and the Gaussian defocus degradation model (GDDM) is one of the most effective models for simulating the image blurriness. The GDDM can be expressed by using the following degradation function.

$$h(i, j) = \frac{1}{2\pi\sigma} e^{-\left(\frac{x^2+y^2}{2\sigma^2}\right)} \tag{2}$$

where σ is the sampling width of the filter, which controls the degree of image degradation.

2.2   The Fast VO Decomposition Model

2.2.1   The VO Decomposition Model

Meyer pointed out in [11] that an image can be divided into the structure layer and texture layer by using an image decomposition model, which is given by

$$f = u + v \tag{3}$$

where f is the original image, u is the structure layer of image, v is the texture layer of image. Based on this theory, Meyer [11] presented the concept of G space, which used the L2-norm of total variation model (TV) to describe oscillating component of an image. Consequently, many numerical computation methods were proposed for TV-G model based on Mayer's idea and validate that G space is effective for describing the oscillating component of an image. Among these methods, Vese and Osher [12] established the VO decomposition model.

In fact, the solution procedure of VO model is very complicated, which is evitable to make the obtained results inaccurate with too much time consumption. Therefore, we need a fast algorithm for simplifying the solution procedure, and fortunately the Split Bregman algorithm [13] can address the problem effectively and accelerate the convergence speed. In detail, we need to solve the following optimization problem

$$\inf_{u,g_1,g_2,w} \left\{ E(u,g_1,g_2,w) = \int_\Omega |w| \, dxdy + \lambda \int_\Omega \left(f-u-\nabla\cdot\vec{g}\right)^2 dxdy + \mu \int_\Omega \left(\sqrt{g_1^2+g_2^2}\right) dxdy + \theta \int_\Omega \left(w-\nabla u-b^{n+1}\right) dxdy \right\} \tag{4}$$

$$b^{n+1} = b^n + \nabla u^n - w^n \tag{5}$$

where θ is a penalty parameter. For such purpose, we obtained the corresponding Euler-Lagrange equation as follows.

$$\begin{cases} u = f - \partial_x g_1 - \partial_y g_2 + \dfrac{\theta}{2\lambda}\left(\nabla\cdot(\nabla u) + \nabla\cdot b^{n+1} - \nabla\cdot w\right) \\[2mm] w^{n+1} = \left(\nabla u + b^{n+1}\right) - \dfrac{1}{\theta}\dfrac{w^{n+1}}{|w^{n+1}|} \\[2mm] \mu\dfrac{g_1}{\sqrt{g_1^2+g_2^2}} = 2\lambda\left[\dfrac{\partial(u-f)}{\partial x} + \partial_{xx}^2 g_1 + \partial_{xy}^2 g_2\right] \\[2mm] \mu\dfrac{g_2}{\sqrt{g_1^2+g_2^2}} = 2\lambda\left[\dfrac{\partial(u-f)}{\partial y} + \partial_{xy}^2 g_1 + \partial_{yy}^2 g_2\right] \end{cases} \tag{6}$$

where w^{n+1} can be solved by using the generalized soft threshold formula as follows

$$w^{n+1} = Max\left(\left|\nabla u + b^{n+1}\right| - \frac{1}{\theta}, 0\right) \frac{\nabla u + b^{n+1}}{\left|\nabla u + b^{n+1}\right|} \quad, \quad 0\frac{0}{0} = 0 \tag{7}$$

Hence, we repeatedly use explicit iterative method to deal with Eq.(6) and Eq.(7) for obtaining u and v until the equation to achieve convergence.

3   The Weighted Robustness with Histogram of Oriented Gradient

3.1   Histogram of Oriented Gradient

Dalal et al. [14] proposed the histogram of oriented gradient (HOG). In order to obtain the HOG of a palmprint image, we first derive an orientation map of palmprint image by utilizing the gradient operator, which is defined as

$$f_x = I * W, f_y = I * W^{\mathrm{T}},$$

$$Mag(i, j) = \sqrt{f_x^2(i, j) + f_y^2(i, j)}, Ang(i, j) = \tan^{-1}\left(\frac{f_y(i, j)}{f_x(i, j)}\right).$$

where I stands for original image with the size of $M \times M$, '*' is the operator of convolution, $W=[-1,0,1]$ is a mask of convolution, $Mag(i, j)$ and $Ang(i, j)$ are gradient magnitude and angle of $I(i, j)$, $-\pi/2 < Ang(i, j) < \pi/2$. Here, Ang is considered as the orientation map of I. We first transform $Ang(i, j)$ from $(-\pi/2, \pi/2)$ to $(0, 2\pi)$, which is given as follows

$$Ang_m(i, j) = \begin{cases} Ang(i, j) & 0 < Ang(i, j) < \pi/2 \\ Ang(i, j) + 2\pi & -\pi/2 < Ang(i, j) \le 0 \end{cases} \tag{8}$$

$$Ang_f(i, j) = \begin{cases} Ang_m(i, j) & f_x(i, j) \ge 0 \\ Ang_m(i, j) + \pi & f_x(i, j) < 0 \& Ang_m(i, j) < \pi/2 \\ Ang_m(i, j) - \pi & f_x(i, j) < 0 \& Ang_m(i, j) > 3\pi/2 \end{cases} \tag{9}$$

Ang_f is the updated orientation map via Eq. (8) and Eq. (9). Then, HOG is obtained as follows

$$F_k = F_k + Mag(i, j) \quad if \quad (k-1) \times (2\pi/N) < Ang_f(i, j) < k \times (2\pi/N) \tag{10}$$

where $k = 1, 2, \ldots\ldots N$, F_k represents the value corresponding to each bin of HOG. Therefore, the feature of HOG is shown as

$$HOG = \left(F_1, F_2, \cdots\cdots, F_N \right) \tag{11}$$

HOG is usually normalized as $HOG = HOG \bigg/ \sum_{k=1}^{N} F_k$.

3.2   Robust Histogram of Oriented Gradient

As illustrated in Fig 1, P is a pixel point as the center of a circle with radius r and P_n is one of the sample points in this circle, where the sample point quantity is 8 and r is 2. Then, a local x—y coordinate system can be established by P and P_n for each sample point; Therein, as shown in the Fig 1, $\overline{P_n^4 P_n^2}$ and $\overline{P_n^3 P_n^1}$ are defined as the positive y axis and x axis, respectively. Obviously, we can obtain a rotation invariant gradient operator (RIGO) as follows.

$$RIGO_x(P) = \frac{1}{8} \sum_{n=1}^{8} \left(I\left(P_n^1 \right) - I\left(P_n^3 \right) \right), RIGO_y(P) = \frac{1}{8} \sum_{n=1}^{8} \left(I\left(P_n^2 \right) - I\left(P_n^4 \right) \right).$$

where P_n^i, $i = 1, 2, 3, 4$ are P_n's surrounding points along the x axis and y axis and $I\left(P_n^i \right)$ is the gray value at P_n^i. By observing and analyzing the RIGO, we can see that the RIGO is not only rotation invariant but also robust for the noise and illumination because of using the average gradient.

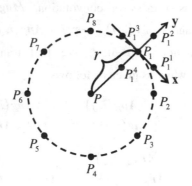

Fig. 1. Sketch of rotation for invariant gradient operator

Therefore, the rotation invariant histogram of oriented gradient (RIHOG) is defined as

$$RIHOG = \left(FN_1, FN_2, \cdots\cdots, FN_N \right) \tag{12}$$

where $\left(FN_1, FN_2, \ldots\ldots, FN_N \right)$ stands for the updated $\left(F_1, F_2, \ldots\ldots, F_N \right)$ by utilizing the RIGO.

On the other hand, we divide the palmprint images into non-overlapped blocks of $s \times s$ pixels to reduce the interference of translation and improve the distinguishability of extracted features, and then extract RIHOG from each block. Finally, the robust histogram of oriented gradient (RHOG) was obtained by putting the RIHOG of each block together. Therefore, the palmprint image is divided into $\dfrac{M \times M}{s \times s}$ blocks, the RHOG is then defined as follows.

$$RHOG = \left(RIHOG_{block1}, RIHOG_{block2}, \ldots\ldots, RIHOG_{block\frac{M\times M}{s\times s}} \right) \qquad (13)$$

3.3   Fractal Dimensions

Fractal dimension is usually used to measure the regularity of the object's surface. In this paper, we combine fractal dimension with RHOG in this paper to obtain a superior descriptor for the palmprint images. There are several methods for solving the fractal dimension. Among these methods, differential box counting (DBC) method [15] is one of the most effective methods, which is described as follows.

An image of size $M \times M$ is divided into non-overlapped grids of $s \times s$ pixels, where s stands for the current scale of image. Considering the image as a three-dimensional space with coordinates (x, y, z), where (x, y) stands for a point in the plane coordinate system, and z is corresponding to the gray value at the position of (x, y). Now we fill the grid by using boxes with the size of $s \times s \times s$. If the minimum and the maximum gray value of each grid locate into the h-th box and the l-th box, respectively, the total number of boxes in the grid is $n_r(x) = l - h + 1$ with $r = s/M$, the total number of boxes in the image is $N_r = \sum\limits_{x=1}^{X} n_r(x)$ with $X = (M \times M)/(s \times s)$. Then the fractal dimension D is obtained by the following equation

$$D = \frac{\log N_r}{\log\left(\dfrac{1}{r}\right)} \qquad (14)$$

Then the final feature of WRHOG is defined as follows.

$$WRHOG = \left(D_1 \times RIHOG_{block1}, D_2 \times RIHOG_{block2}, \ldots\ldots, D_{\frac{M\times M}{s\times s}} \times RIHOG_{block\frac{M\times M}{s\times s}} \right) \quad (15)$$

where $\left(D_1, D_2, \ldots\ldots, D_{\frac{M\times M}{s\times s}} \right)$ is a vector of block fractal dimension corresponding to each block of image. Now we can present the framework of the proposed VO-WRHOG method in Fig 2.

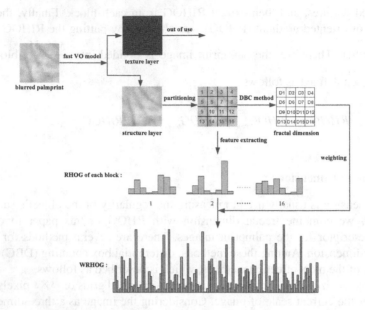

Fig. 2. Outline of the proposed fast VO-WRHOG method

3.4   Feature Matching

Based on above proposed framework, we need criteria to measure the similarity of two palmprint images. We use the Normalized correlation coefficient (NCC) [16] as the matching score to scale the comparability between palmprint features. Suppose that $A = (a_1, a_2, \ldots, a_n)$ and $B = (b_1, b_2, \ldots, b_n)$ are two vectors of WRHOG, and their NCC is defined as follows.

$$\text{NCC} = \frac{\left| \sum_{i=1}^{n} (a_i - \mu_A)(b_i - \mu_B) \right|}{l \times \sigma_A \times \sigma_B} \tag{16}$$

where $\mu_A (\mu_B)$ is the mean of $A(B)$, $\sigma_A (\sigma_B)$ is the standard deviation of $A(B)$, l is the length of A or B. The value of NCC is between -1 and 1. If NCC is close to 1, which implies that the palmprint images are largely resulting from the same one; otherwise, it is more likely to be different from each other.

4   Experiments

4.1   PolyU Palmprint Database

The PolyU palmprint database [5] includes 7752 palmprint images, which were captured from 386 different palms. Samples from each of these palms were collected in two separate sessions. The average time interval between two sessions was two months, whilst 10 samples were captured in the first session and the second session, respectively.

The proposed fast VO-WRHOG method is implemented by using MATLAB2010a on a desktop with the CPU (2.90GHZ), and 2GB random access memory. With the purpose of verifying effectiveness of the proposed method, we use two different palmprint databases to test our method; one is PolyU palmprint database, the other is the blurred PolyU palmprint database which is obtained by using GDDM with random scale of blurring (the range of σ is from 0 to 10, as shown Eq. (2)) for PolyU palmprint database, we carry out the same pattern to compute the recognition accuracy that each palmprint image is matched with all the other palmprint images, the calculation pattern and results coincide with [16]. False rejection rate (FRR) and false acceptance rate (FAR) are used as two evaluation standards, and they are defined as follows:

$$FRR = \frac{NFR}{NEA} \times 100\% \qquad (17)$$

$$FAR = \frac{NFA}{NIA} \times 100\% \qquad (18)$$

where NEA (Number of Enrollee Attempts) and NIA (Number of Impostor Attempts) stand for the true matching number and false matching number, respectively; NFR (Number of False Rejections) and NFA (Number of False Acceptance) are the number of false rejection and false acceptance.

4.2   Experiment Results

In the proposed fast VO-WRHOG method, the block size ($s \times s$) and divided orientation number (N) for the proposed WRHOG are two important factors. Here, the blocks with different sizes (4×4, 8×8, 16×16, 32×32) and different orientation numbers (N=4,6,8,10,12) are used to perform some experiments in order to obtain the optimal parameters for the blurred palmprint database. We found that when the block size is 16×16 and the orientation number is 12; the EER achieves the lowest value (0.1324%).

We list the results in table 1 and 2 for results in comparison with other features and approaches.

Table 1. EER values for different features

Method	EER(%)
HOG	1.7068
fast VO-HOG	1.3146
RHOG	1.1979
fast VO-RHOG	0.7532
WRHOG	0.6222
fast VO-WRHOG	0.1324

Table 2. Comparisons of EERs among reported methods and the fast VO-WRHOG method

Method	EER(%)	
	PolyU palmprint database	Blurred PolyU palmprint database
2DPCA [2]	5.2653	6.5943
LST [9]	2.7208	2.4676
DCT-BEPL [3]	1.9249	1.6173
PalmCode [5]	0.9810	5.2653
FusionCode [6]	0.8156	3.5215
Competitive Code[8]	0.4684	2.0037
RLOC [7]	0.1685	1.1149
fast VO-WRHOG	0.1421	0.1324

From these results in table 1 and 2, we can obtain some conclusions. The EER of fast VO-WRHOG method can achieve 0.1324% on the blurred database, which is much lower than the rest of methods. In other words, the proposed fast VO-WRHOG method can achieve the desirable recognition result for blurred palmprint recognition.

In addition, we should highlight the time-consuming for the proposed method because we use the fast algorithm in the paper. So the time-consuming T for handling each palm from collection to feature matching can be computed as follows

$$T = T_{ia} + T_{ra} + T_{fe} + T_{fm} \qquad (19)$$

where T_{ia}, T_{ra}, T_{fe} and T_{fm} stand for image acquisition time, ROI acquisition time, feature extraction time and feature matching time, respectively. The table 3 shows the time-consuming for each step between VO-WRHOG and fast VO-WRHOG. According to the table 3, we can calculate the total time-cost of the proposed method for each palm via Eq. (19) (approximately 1s), which is quick enough to meet the real-time requirement. It should be noted that although the time-cost of the feature extraction for each palm has subtle change between VO-WRHOG and fast VO-WRHOG, the gap for the time-cost is very obvious in a large-scale database. For example, there are 10000 palmprint images in certain database, the total time-consuming for the feature extraction is a large gap between VO-WRHOG and fast VO-WRHOG (1198s VS 345s). Hence, the fast VO-WRHOG method can greatly reduce the time-cost in the aspects of feature extraction. In other words, the proposed method can cost down by shortening the time-cost.

Table 3. Time-consuming for each step

Step	Time(ms)	
	VO-WRHOG	fast VO-WRHOG
Image Acquisition[5]	<1000	<1000
ROI Acquisition[5]	138	138
Feature Extraction	**119.8**	**34.5**
Feature Matching	0.059	0.059

5   Conclusions

In this paper, we have proposed the fast VO-WRHOG method, which can not only solve the problem of blurred palmprint recognition, but also address the common problem in the palmprint recognition, such as translation and rotation. The structure layer of the blurred image, which is obtained by using the fast VO model, is considered as the stable information through theoretical analysis. Then, the WRHOG is designed to extract the robust features from the structure layer. In comparison with the previous high-performance palmprint recognition methods, the proposed fast VO-WRHOG not only can obtain a more stable recognition result on the different palmprint databases, but also it can achieve a desirable EER. Also, we add the fast algorithm to the VO model, making the recognition speed so fast that the proposed method can extend to the large database.

References

1. Kong, A., Zhang, D., Kamel, M.: A survey of palmprint recognition. Pattern Recognition 42, 1408–1418 (2009)
2. Lin, S., Yuan, W.Q., Wu, W., Fang, T.: Blurred palmprint recognition based on DCT and block energy of principal line. Journal of Optoelectronics Laser 23, 2200–2206 (2012)
3. Sang, H.F., Liu, F.: Defocused palmprint recognition using 2DPCA. In: IEEE International Conference on Artificial Intelligence and Computational Intelligence, pp. 611–615. IEEE Press, Shanghai (2009)
4. Xue, M.L., Liu, W.Q., Liu, X.D.: A novel weighted fuzzy LDA for face recognition using the genetic algorithm. Neural Computing and Applications 22, 1531–1541 (2013)
5. Zhang, D., Kong, W.K., You, J., Wong, M.: Online palmprint identification. IEEE Transactions on Pattern Analysis and Machine Intelligence 25, 1041–1050 (2003)
6. Kong, W.K., Zhang, D., Kamel, M.: Palmprint identification using feature-level fusion. Pattern Recognition 39, 478–487 (2006)
7. Jia, W., Huang, D.S., Zhang, D.: Palmprint verification based on robust line orientation code. Pattern Recognition 41, 1521–1530 (2008)
8. Kanhangad, V., Kumar, A., Zhang, D.: A unified framework for contactless hand verification. IEEE Transactions on Information Forensics and Security 6, 1014–1027 (2011)
9. Lin, S., Yuan, W.Q.: Blurred palmprint recognition under defocus status. Optics and Precision Engineering 21, 734–741 (2013)
10. Wang, J.Z., Wang, Y.Y.: The degradation function of defocused images. In: Asia-Pacific Conference on Information Theory, pp. 64–67. Scientific Research Publishing, Xi'an (2010)
11. Meyer, Y.: Oscillating patterns in image processing and nonlinear evolution equations. American mathematical society academic publishers (2001)
12. Vese, L., Osher, S.: Modeling textures with total variation minimization and oscillating patterns in image processing. Journal of Scientific Computing 19, 553–572 (2003)
13. Goldstein, T., Osher, S.: The split bregman method for L^1 regularized problems. UCLA CAM Report, 08-29 (2008)

14. Dalal, N., Triggs, B.: Histograms of oriented gradients for human detection. In: IEEE Computer Society Conference on Computer Vision and Pattern Recognition (CVPR), pp. 886–893. IEEE Press, San Diego (2005)
15. Hong, D.F., Pan, Z.K., Wu, X.: Improved differential box counting with multi-scale and multi-direction: A new palmprint recognition method. Optik-International Journal of Light Electron Optics 125, 4154–4160 (2014)
16. Hong, D.F., Su, J., Hong, Q.G., Pan, Z.K., Wang, G.D.: Blurred palmprint recognition based on stable-feature extraction using a Vese–Osher decomposition model. PLoS ONE 9(7), e101866 (2014)

Affinity Propagation Clustering
with Incomplete Data

Cheng Lu[1,2], Shiji Song[1,*], and Cheng Wu[1]

[1] Department of Automation, Tsinghua University, Beijing 100084, China
[2] Army Aviation Institute, Beijing 101123, China
lu-10@mails.tsinghua.edu.cn,
shijis@mail.tsinghua.edu.cn,
wuc@tsinghua.edu.cn

Abstract. Incomplete data are often encountered in data sets for clustering problems, and inappropriate treatment of incomplete data will significantly degrade the clustering performances. The Affinity Propagation (AP) algorithm is an effective algorithm for clustering analysis, but it is not directly applicable to the case of incomplete data. In view of the prevalence of missing data and the uncertainty of missing attributes, we put forward improved AP clustering for solving incomplete data problems. Three strategies(WDS, PDS and IPDS) are given, which involve modified versions of the AP algorithm. Clustering performances at different missing rates are discussed, and all approaches are tested on several UCI data sets with randomly missing data.

Keywords: Incomplete data, AP algorithm, Missing rate.

1 Introduction

Cluster learning is an important research in machine learning. Recently, different types of clustering models and algorithms have been developed([1],[2]). When complex data become the subject of data sources, how to find the hidden class structures has become an important research both in analysis and engineering systems. Affinity Propagation(AP) is a relatively new clustering algorithm introduced by Frey and Dueck (2007)([3]), which can handle large data sets in a relatively short period to get satisfactory results. AP algorithm is superior to other similar algorithms in terms of processing speed and clustering performances. Unlike most prototype-based clustering algorithms, AP does not require the pre-specified number of clusters and initial cluster centers, which attracts the attention of many scholars([4],[5],[6]).

With developments of sensors and database technology, the ability to obtain information and data is growing. However, in practice, many scenarios result in incomplete data due to various reasons, such as bad sensors, mechanical failures to collect data, illegible images due to low pixels and noises, unanswered

---

* Corresponding author.

M. Fei et al. (Eds.): LSMS/ICSEE 2014, Part II, CCIS 462, pp. 239–248, 2014.

questions in surveys, etc. When we apply clustering methods to these data to explore more information, we are often faced with the problem of incomplete data, which makes the traditional clustering models inapplicable([7],[8]).

Missing data can be classified into three categories[9]: missing completely at random (MCAR), missing at random (MAR) and not missing at random (NMAR). The first two cases also called ignorable missing mechanisms are more realistic models than the last one. The approaches to deal with incomplete data include listwise deletion(LD), imputation, model-based method and direct analysis. There is a close relationship between these methods on their implementation. LD ignores those samples with missing values, which will lose a lot of sample information. Imputation and model-based methods usually assume that the data are missing at random, then substitute the missing value with an appropriate estimate in order to construct a complete data set. However, it takes a long time to do the imputation, and these techniques are also prone to cause larger estimation errors as dimensionality and incompleteness increase. EM algorithm[10] is a commonly used iterative algorithm based on maximum likelihood estimation in missing data analysis. When there is a large number of clusters variables that the current statistical imputation method is limited in the application due to the robustness and implementation difficulties etc.

The current research on missing data problems in machine learning mainly focuses on model-based methods and direct analysis. Direct analysis can improve performance through improved clustering models. Despite the lack of clustering data everywhere, there is no available principle method to the clustering of variables with missing data. Neither statistical methods nor machine learning methods for dealing with missing data can not meet the current actual needs. Various methods for handling missing data still need to be further optimized. There is few research on direct modeling method without the prior imputation for missing data.

The existing research on improved methods for unsupervised clustering models mainly concentrates on the fuzzy C-means clustering (FCM) algorithm[11]. In 1998, imputation and discarding/ignoring were proposed to handle missing values in FCM[12]. In 2001, Hathaway and Bezdek[13] proposed four strategies to continue the FCM clustering of incomplete data and proved the convergence of the algorithms. In addition, Hathaway and Bezdek (2002)[14] used triangle inequality-based approximation schemes(NERFCM) to cluster incomplete relational data. Li et al.[15] put forward a FCM algorithm based on nearest-neighbor intervals to solve incomplete data. Zhang et al.[16] introduced the kernel method into the standard FCM algorithm.

FCM algorithms are sensitive to the initial centers, which makes the clustering results with great uncertainty. Especially when some data are missing, the selection of the initial cluster centers becomes more important. To address this issue, we consider the AP algorithm, which does not require initial cluster centers and the number of clusters. AP does not require a vector space structure and the exemplars are chosen among the observed data samples and not computed as hypothetical averages.

The remainder of this paper is organized as follows. Section 2 presents a description of AP algorithm. Three strategies for solving AP clustering of incomplete data sets are given in section 3. We present experimental results for UCI data sets in section 4. Finally, some remarks are given in section 5.

2 AP Clustering Algorithm

AP algorithm and k-means algorithm have the same objective function, but there are some differences in the principle of the algorithm.

AP is a clustering algorithm based on the nearest neighbor information. With similarity matrix of the data as input, all the samples are regarded as potential clustering centers in the initial stage of the algorithm, while each sample point is considered as a node in the network. Attraction information transmit along the node connection recursively until the optimal set of class representative points is found(representative point must be the actual point of the data set, called the exemplar), so as to maximize the sum of the similarity that all the data points to the nearest representative point. Among them, the attraction information is the degree that the data point is suitable to be selected as the class representatives of other data points.

The mathematical model of AP algorithm:

Let the data set $X = \{x_1, x_2 \ldots, x_N\}, \forall x_i \in \mathbb{R}$. There are some relatively close clusterings in the feature space. Each data point only corresponds to a cluster, and $x_{C(i)}$ $(1 < C < N)$ represents the exemplar for given x_i. Clustering error function is defined as follows

$$J(C) = \sum_{i=1}^{N} d^2(x_i, x_{C(i)}). \tag{1}$$

The goal of AP is to find the optimal exemplar set by minimizing the clustering error function

$$C^* = argmin[J(C)]. \tag{2}$$

Firstly, AP simultaneously considers all data points as potential exemplars, then establishes the attractiveness information between each sample point and other sample points, i.e., the similarity between any two sample points.

The similarity can be set according to the particular research questions, mainly including similarity coefficient function and distance function. In traditional clustering problems, similarity is usually set as the negative of Euclidean distance squared

$$s(i,j) = -d^2(x_i, x_j) = -\|x_i - x_j\|_2^2, i \neq j, \tag{3}$$

where $s(i, j)$ is stored in a similarity matrix S, representing the suitability that the data point x_i is the exemplar of the point x_j. The bias parameter $s(i, i)$ is set for each data point, which is greater, the more possible that the corresponding point is selected as the exemplar. Algorithms usually assume the same possibility that all sample points are selected to be the exemplar, that is to set all the $s(i, i)$ for the same value P. Under normal circumstances, set P as the similarity mean of the similarity matrix (median$(s(i, j))$, $i \neq j$).

In order to select the appropriate clustering center, AP is constantly searching for two different information: responsibility and availability. when meet the termination conditions, algorithm ends.

As a common and effective clustering algorithm, AP clustering algorithm is applicable to the case of complete data similar to the traditional clustering model. To date, no AP algorithm with incomplete data has been available. We propose three strategies for solving the problem by changing the similarity in the next section.

3   AP Clustering Algorithm with Incomplete Data

In view of the prevalence of missing data and the uncertainty of missing attributes, we select the exemplars using information transmission mechanism of AP, and the strategies for doing AP clustering with incomplete data sets are as follows

1 Whole Data Strategy(WDS)

If the proportion of incomplete data is small, then it may be useful to simply delete all incomplete data and apply AP to the remaining complete data. We will refer to this as whole data strategy(WDS), which is divided into WDS1 and WDS2 depending on whether the properties of the original data set are considered. WDS1 deletes throughly the samples containing the missing data, a new complete data set is formed by the remaining samples (non-missing data), which is the input of AP. WDS2 considers all the samples whether attribute values are missing, using AP clustering for the complete data set as WDS1, samples containing missing data are automatically classified into one of the clusters, and they are classified into cluster 1 in this paper.

For example, there are two samples $x_i = \{1, ?, 3, 4, ?\}$ and $x_j = \{3, 4, ?, 6, 7\}$, $1 \leqslant i < j \leqslant N$, then $X1 = \{x_1, \ldots, x_{i-1}, x_{i+1}, \ldots, x_{j-1}, x_{j+1}, \ldots, x_N\}$. WDS-AP directly computes the similarity matrix of $X1$ by equation 3 as the input of AP, in which the number of samples is N-2 and N respectively in WDS1 and in WDS2. That is, x_i and x_j are discarded in WDS1-AP, and classified into cluster in WDS2-AP.

WDS1 is the idealized case, it will not directly provide cluster membership information for all samples. WDS2 is the more realistic case, in which information of all samples considered.

2 Partial Data Strategy(PDS)

The second approach to AP with incomplete data is based on the partial distance, which is called partial data strategy(PDS). It consists of calculating partial (squared Euclidean) distances using all available (i.e., nonmissing) feature values, and then scaling this quantity by the reciprocal of the proportion of components used. That is, we will only calculate the data on the dimension of observation data, and a similarity matrix of AP algorithm is formed using the distance between each point and exemplar. For example, there are two samples $x_1 = \{1, ?, 3, 4, ?\}$ and $x_2 = \{3, 4, ?, 6, 7\}$, then the distance between two points is as follows

$$D_{12} = \|x_1 - x_2\|_2^2 = \frac{5}{5-3} \left[(1-3)^2 + (4-6)^2 \right]. \tag{4}$$

When x_a and x_b are incomplete, the distance $\|x_a - x_b\|_2^2$ cannot be directly obtained, it can be handled as follows

$$\|x_a - x_b\|_2 = \sqrt{\sum_{j=1}^{M} d_j(x_{aj}, x_{bj})^2 \times \frac{M}{\omega}}, \tag{5}$$

where

$$d_j(x_{aj}, x_{bj}) = \begin{cases} 0, & (1 - m_{aj})(1 - m_{bj}) = 0, \\ d_N(x_{aj}, x_{bj}), & others \end{cases} \tag{6}$$

$$d_N(x_{aj}, x_{bj}) = |x_{aj} - x_{bj}|, \tag{7}$$

$$m_{ij} = \begin{cases} 1, & x_{ij} \text{ is missing} \\ 0, & x_{ij} \text{ is not missing,} \end{cases} \tag{8}$$

where $1 \le j \le M, 1 \le i \le N$. $d_j(x_{aj}, x_{bj})$ represents the distance on the jth attribute between the two samples. ω is the feature dimension that the two samples are both not missing, M is the dimensions of all feature. m_{ij} is indicator function to explain whether the variable is missing.

As the input of AP, similarity matrix of X can be calculated by equation 5. Then the two informations update alternately, which are both zero in the initial stage, and the update process of which as follows

$$r(i, j) \leftarrow s(i, j) - \max[a(i, j') + s(i, j')], \tag{9}$$

$$a(i, j) \leftarrow \begin{cases} \min_{i \ne j} \left\{ 0, r(j, j) + \sum_{i' \ne i, i' \ne j} \max[0, r(i', j)] \right\} & i \ne j \\ \sum_{i' \ne j} \max[0, r(i', j)] & i = j. \end{cases} \tag{10}$$

To avoid the numerical oscillation, the damping factor λ is introduced as follows

$$\begin{cases} R_i = (1 - \lambda)R_i + \lambda R_{i-1} \\ A_i = (1 - \lambda)A_i + \lambda A_{i-1}. \end{cases} \tag{11}$$

3 Inproved Partial Data Strategy(IPDS)

In most of the cases standardized Euclidean distance $\|x_a - x_b\|_2^2 = \dfrac{\sum\limits_{j=1}^{M} d_j(x_{aj}, x_{bj})^2}{\delta}$ gives better performance than the Euclidean distance. The calculation of standard deviation(δ) is not easy when some data are missing. So we introduce the Weighted Euclidean distance based on the range of feature values.

The range of the variable on each dimension is considered on the basis of the second approach, then a similarity matrix is formed. We reformulate Equation 7 as Equation 12 by introducing a weight in which the maximum and minimum are considered.

$$d_N(x_{aj}, x_{bj}) = \frac{|x_{aj} - x_{bj}|}{\max(x_j) - \min(x_j)}, \tag{12}$$

where, $max(x_j)$ and $min(x_j)$ are the maximum and minimum of the observation data when there is missing data. Other parameters are the sane as strategy 2. When x_a and x_b are incomplete, the distance $\|x_i - x_j\|_2^2$ can be handled by formula 5, where $d_N(x_{aj}, x_{bj})$ is calculated by Equation 12. The similar weights obtained by above calculation, then AP clustering algorithm can be adopted to solve the clustering problem with incomplete data.

Preference P is a very important parameter for AP especially when the data is incomplete. It determines the clustering number, it also exercises an crucial influence over AP convergence rate. Through a lot of studies and experiments, it shows that better experiment result can be gained when P is in the range: median(s)$*2^{-5}$ $\sim$ median(s)$*2^5$, where median(s)=(median($s(i,j)$), $i \neq j$). In the iterative, we restrict P in the interval [median(s)$*2^5$, median(s)$*2^{-5}$].

4   Simulation Analysis

In order to test the proposed clustering algorithm, we compare the proposed WDS1, WDS2, PDS and APDS of AP using artificially generated incomplete data sets. The scheme for artificially generating an incomplete data set X is to randomly select a specified percentage of components and designate them as missing. The random selection of missing attribute values is constrained so that(Hathaway and Bezdek 2001)[13]

(1)each original feature vector x_k retains at least one component;

(2)each feature has at least one value present in the incomplete data set X.

At least one dimensional data exists for each vector data, and at least one or more data exist for each dimension. That is, the data in each row are not empty, each column of data cannot be null.

In the following experiments, we tested the performance of proposed algorithm on commonly used UCI data sets: Iris, Seeds and WDBC, which are taken from the UCI machine repository [17], and often used as standard databases to test the performance of clustering algorithms.

According to the characteristics of the Iris as the data source, the Iris data contains 150 four-dimensional attribute vectors, which include petal length, petal width, sepal length and sepal width. The Iris data set contains three kinds of plant(Setosa, Versicolor and Virginica), each containing 50 vectors.

The Seeds data set comprised kernels belonging to three different varieties of wheat: Kama, Rosa and Canadian, 70 elements each. The Seeds data contains 210 7-dimensional attribute vectors, which include area, perimeter, compactness, length of kernel, width of kernel, asymmetry coefficient and length of kernel groove.

The Wisconsin Diagnostic Breast Cancer (WDBC) data set comprises 569 samples, and for each of the samples, there are 30 attributes. The individuals are divided into two groups(malignant and benign).

To test the clustering performance, the clustering results of WDS1-AP, WDS2-AP, PDS-AP and IPDS-AP are compared. For the three data sets, damping factor $\lambda = 0.7$, decreasing step of preferences $pstep = 0.01$, max iteration time $nrun = 2000$, convergence condition $nconv = 100$. Fowlkes-Mallows index[18] is used to measure the clustering performance based on external criteria. The larger the FM value is, the better the clustering performance is.

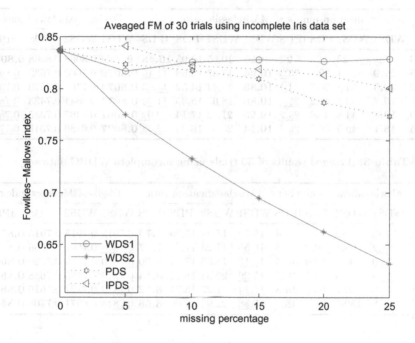

Fig. 1. Aveaged FM of 30 trials using incomplete Iris data set

Because missing data was randomly selected, different tests lead to different results. To eliminate the significant variation in the results from trial to trial, Tables 1, 2, and 3 present the averages obtained over 30 trials on incomplete Iris, Seeds and WDBC data sets. The same incomplete data set is used in each trial for each of the three approaches, so that the results can be correctly compared. As shown in Figure1, it demonstrates the clustering performances(aveaged Fowlkes-Mallows index) for WDS1, WDS2, PDS and IPDS using incomplete Iris data set. WDS1 is always the best performer except for the 5% cases of incomplete Iris data set, where IPDS gives suboptimal solutions.

Table 1. Aveaged results of 30 trials using incomplete Iris data set

%miss	Misclassification number				%Misclassification ratio				FowlkesCMallows index			
	WDS1	WDS2	PDS	IPDS	WDS1	WDS2	PDS	IPDS	WDS1	WDS2	PDS	IPDS
0	14	14	14	14	9.33	9.33	9.33	9.33	0.8365	0.8365	0.8365	**0.8370**
5	15.5	19.9	16.0	**14.1**	11.89	12.67	11.33	**9.33**	0.8165	0.7752	0.8232	**0.8407**
10	**14.0**	23.8	16.9	16.3	**10.33**	15.89	11.24	10.87	**0.8258**	0.7324	0.8176	0.8216
15	**13.0**	27.3	18.1	16.7	**10.10**	18.22	12.04	11.11	**0.8278**	0.6948	0.8097	0.8191
20	**12.5**	30.9	21.5	17.6	**10.25**	20.64	14.33	11.71	**0.8266**	0.6625	0.7865	0.8129
25	**11.9**	34.4	22.0	19.0	**10.16**	22.93	14.67	12.67	**0.8288**	0.6320	0.7736	0.8004

Table 2. Aveaged results of 30 trials using incomplete Seeds data set

%miss	Misclassification number				%Misclassification ratio				FowlkesCMallows index			
	WDS1	WDS2	PDS	IPDS	WDS1	WDS2	PDS	IPDS	WDS1	WDS2	PDS	IPDS
0	23	23	23	**22**	10.95	10.95	10.95	**10.48**	0.8068	0.8068	0.8068	**0.8095**
5	**21.9**	28.1	30.8	23.3	**10.94**	13.37	14.68	11.10	**0.8070**	0.7660	0.7632	0.8003
10	**20.5**	34.5	30.5	24.4	**10.83**	16.44	14.52	11.63	**0.8074**	0.7164	0.7576	0.7924
15	**19.7**	39.1	32.2	25.0	**10.86**	18.63	15.35	11.90	**0.8074**	0.6844	0.7433	0.7888
20	**18.8**	44.8	36.0	25.2	**10.95**	21.32	17.14	12.02	**0.8061**	0.6467	0.7385	0.7862
25	**18.1**	49.5	38.5	26.3	**10.14**	23.57	18.32	12.52	**0.8027**	0.6186	0.7315	0.7791

Table 3. Aveaged results of 30 trials using incomplete WDBC data set

%miss	Misclassification number				%Misclassification ratio				FowlkesCMallows index			
	WDS1	WDS2	PDS	IPDS	WDS1	WDS2	PDS	IPDS	WDS1	WDS2	PDS	IPDS
0	83	83	83	44	15.59	15.59	15.59	**7.73**	0.7915	0.7915	0.7915	**0.8707**
5	84.2	98.1	99.0	**44.8**	15.55	17.23	17.40	**7.87**	0.7828	0.7601	0.7656	**0.8689**
10	78.0	106.0	99.9	**45.5**	15.13	18.63	17.55	**8.00**	0.7870	0.7370	0.7720	**0.8671**
15	74.9	115.6	103.3	**46.3**	15.29	20.32	18.15	**8.14**	0.7848	0.7185	0.7648	**0.8653**
20	70.6	120.7	95.3	**46.8**	15.14	21.22	16.74	**8.22**	0.7870	0.7127	0.7616	**0.8643**
25	68.8	130.5	97.3	**48.8**	15.58	22.94	17.09	**8.58**	0.7843	0.6971	0.7708	**0.8596**

In the three tables, the optimal solutions in each row are highlighted in bold, and the suboptimal solutions are underlined.

From Tables1, 2, and 3, WDS and PDS reduce to regular AP for 0% missing data. The results of IPDS are better on Seeds and WDBC data sets. For other cases, different methods for handling missing attributes in AP lead to different clustering results. With the growth rate of missing data, the misclassification number and ratio of WDS2, PDS and IPDS are increased, FM is decreased.

In terms of misclassification ratio, WDS1 is always the best performer for incomplete Iris and Seeds data sets, where IPDS gives suboptimal solutions. IPDS is always the best performer for incomplete WDBC data set, where WDS1 gives suboptimal solutions. Although the results of WDS1 are mainly the optimal solutions, which is at the expense of all the missing data as a precondition, it is unreasonable because of discarding all the samples containing missing data. The samples of the data set generally should be considered. The results of WDS are the worst if all samples of the data set are considered seeing the results of WDS1 and WDS2. The above experimental results imply that IPDS is the best among the three strategies for solving AP clustering of incomplete data.

Conclusion

In this paper, we considered three strategies for solving AP clustering of incomplete data sets. An important implication of our numerical experiments is that although the simple approach of deleting incomplete data (WDS-AP) works well if the information contained in the samples with missing data can be ignored, other approaches are generally superior if a larger proportion of data is missing. IPDS provides the highest accurate in the case of maximally incomplete data sets.

The three strategies are simple and easily implemented methods which directly deal with incomplete data set using AP algorithm. If we are rich in resources and do not consider the samples containing missing data, WDS1-AP can be selected. If the samples containing missing data can not be ignored, IPDS AP can be selected to cluster the incomplete data sets. In the future, our work will focus on the selection of P and the damping factor λ with theoretical basis, and the improvement on the similarity measurement of AP when the missing percentage is large, which will be helpful to solve clustering problems with various missing percentages.

Acknowledgements. The paper is supported by Research Fund for the Doctoral Program of Higher Education (No. 20120002110035, No.20130002130010) and Natural Science Foundation of China (No. 61273233).

References

1. Jain, A.K., Murty, M.N., Flynn, P.J.: Data clustering: A review. ACM Computing Surveys (CSUR) 31(3), 264–323 (1999)
2. Aggarwal, C.C., Reddy, C.K.: Data Clustering: Algorithms and Applications. CRC Press (2013)
3. Frey, B.J., Dueck, D.: Clustering by passing messages between data points. Science 315(5814), 972–976 (2007)
4. Shang, F., Jiao, L.C., Shi, J., et al.: Fast affinity propagation clustering: A multilevel approach. Pattern Recognition 45(1), 474–486 (2012)
5. Wang, K., Zhang, J., Li, D., et al.: Adaptive affinity propagation clustering. arXiv preprint arXiv, 0805–1096 (2008)
6. Xiao, Y., Yu, J.: Semi-supervised clustering based on affinity propagation algorithm. Journal of Software 19(11), 2803–2813 (2008)
7. Wagstaff, K.: Clustering with missing values: No imputation required. Springer, Heidelberg (2004)

8. García Laencina, P.J., Sancho-Gómez, J.L., Figueiras-Vidal, A.R.: Pattern classification with missing data: A review. Neural Computing Applications 19(2), 263–282 (2010)
9. Little, R.J.A., Rubin, D.B.: Statistical Analysis with Missing Data, 2nd edn. Wiley, New York (2002)
10. Dempster, A.P., Laird, N.M., Rubin, D.B.: Maximum likelihood from incomplete data via the EM algorithm. Journal of the Royal Statistical Society 39(1), 1–38 (1977)
11. Bezdek, J.C.: Pattern Recognition with Fuzzy Objective Function Algorithms. Plenum Press, New York (1981)
12. Miyamoto, S., Takata, O., Umayahara, K.: Handling missing values in fuzzy c-means. In: Proc. of the Third Asian Fuzzy Systems Symposium, pp. 139–142 (1998)
13. Hathaway, R.J., Bezdek, J.C.: Fuzzy c-means clustering of incomplete data. IEEE Transactions on Systems Man and Cybernetics Part B-Cybernetics 31(5), 735–744 (2001)
14. Hathaway, R.J., Bezdek, J.C.: Clustering incomplete relational data using the non-Euclidean relational fuzzy c-means algorithm. Pattern Recognition Letters 23(1-3), 151–160 (2002)
15. Li, D., Gu, H., Zhang, L.Y.: A fuzzy c-means clustering algorithm based on nearest-neighbor intervals for incomplete data. Expert Systems with Applications 37(10), 6942–6947 (2010)
16. Zhang, D.Q., Chen, S.C.: Clustering incomplete data using kernel-based fuzzy c-means algorithm. Neural Processing Letters 18(3), 155–162 (2003)
17. UCI machine learning repository, http://archive.ics.uci.edu/ml/datasets.html
18. Halkidi, M., Batistakis, Y., Vazirgiannis, M.: Cluster Validity Methods: Part I. In: Proceedings of the ACM SIGMOD International Conference on Management of Data, vol. 31(2), pp. 40–45 (2002)

An Improved Algorithm for Camera Calibration Technology Research

Gen Lu[1], Yuying Shao[2], Li Deng[1,*], and Minrui Fei[1]

[1] School of Mechatronics Engineering and Automation, Shanghai University,
Shanghai 200072, China
[2] State Grid Shanghai Municipal Electric Power Company, Shanghai 200025, China
dengli@shu.edu.cn

Abstract. Computer vision technology has wide application value in daily life and industrial production. Camera calibration is the base of computer vision technology, which is the key and necessary step to get three-dimensional spatial information from a two-dimensional image. In the paper, the geometric parameter of camera is considered as the research object. Firstly, the relationship model of camera calibration is established and used to unify the world coordinate system, the camera coordinate system and the image coordinate system. It takes the image pixel point as the optimization goal, then a differential evolution combined with particle swarm optimization algorithm is proposed to calibrate camera. Experimental simulation results show that the improved algorithm has good optimization ability and used for camera calibration has validity and reliability.

Keywords: camera calibration, intrinsic parameter, external parameter, differential evolution, particle swarm algorithm.

1   Introduction

Camera calibration is the base of computer vision technology, which is the key and necessary step to get three-dimensional spatial information from a two-dimensional image [1]. Currently, the theory of camera calibration is already very mature and there are many calibration methods proposed. The basic methods of camera calibration can be divided into the traditional camera calibration methods and the camera self-calibration methods [2-3]. The traditional calibration method has a high calibration precision, but need specific calibration reference substance. The self-calibration method does not rely on calibration reference substance, but the calibration results are relatively unstable. Tsai [4] proposed a most common two-step calibration method, which can effectively obtain the most of camera parameters. Zhang [5] proposed a camera calibration method based the planar template, which is flexible and very simple. Ma [6] proposed a self-calibration method based on active vision. The calibration

---

* Corresponding author.

M. Fei et al. (Eds.): LSMS/ICSEE 2014, Part II, CCIS 462, pp. 249–256, 2014.

method is simple and can get linear solution, but which used inflexibly and has a high cost. With the developing of the intelligent algorithms, many algorithms are applied to camera calibration. Deep et al. [7] presented a camera calibration method based on particle swarm optimization and can solve the camera parameters very well. Tian et al. [8] presented a camera calibration method based on BP neural network, which raises the camera calibration precision and robustness. But the limitations of these intelligent algorithms are unable to find the optimal solution and also increase the complexity of the calibration. So some simple stable and efficient algorithms are looked for applying to camera calibration, which has the very important research significance.

This paper is devoted to use particle swarm optimization and differential evolution algorithm for camera calibration technology research. Particle Swarm Optimization (PSO) [9] is proposed by Kennedy and Eberhart, which is a kind of global random search algorithm based on swarm intelligence. It has less adjustable parameters and has better ability of global optimization. Differential Evolution (DE) [10] is proposed by Stron and Price, which is a kind of global random search algorithm based on real parameter optimization problem. It has fast convergence speed and good robustness. In view of the characteristics of the two algorithms and combining to avoid the defect of particle swarm algorithm which is easy to fall into local optimum, then which can be used for camera calibration technology research.

2   Camera Calibration Model

The camera imaging model is the basis of camera calibration. When determined the imaging model, the camera internal and external parameters and solving methods can be determined [11]. Assuming a three-dimensional point of the world coordinate is $(x_w, y_w, z_w)^T$, a homogeneous coordinate of the camera coordinate system is $(x_c, y_c, z_c)^T$, two-dimensional image pixel coordinate is $(u, v)^T$, the camera imaging relationship as follows.

(I) The transformation between the world coordinate system and the camera coordinate system

$$\begin{bmatrix} x_c \\ y_c \\ z_c \\ 1 \end{bmatrix} = \begin{bmatrix} R_{3\times3} & T_{3\times1} \\ 0 & 1 \end{bmatrix} \begin{bmatrix} x_w \\ y_w \\ z_w \\ 1 \end{bmatrix} = M_{R,T} \begin{bmatrix} x_w \\ y_w \\ z_w \\ 1 \end{bmatrix} \tag{1}$$

According to the relationship of rotation matrix R and translation matrix T, the coordinate transformation between the two coordinate systems can be realized.

(II) The transformation between the camera coordinate system and the image coordinate system

The ideal perspective projection transformation under pinhole model is established as follows:

$$\begin{bmatrix} x \\ y \\ 1 \end{bmatrix} = \frac{1}{z_c} \begin{bmatrix} f & 0 & 0 & 0 \\ 0 & f & 0 & 0 \\ 0 & 0 & 1 & 0 \end{bmatrix} \begin{bmatrix} x_c \\ y_c \\ z_c \\ 1 \end{bmatrix} \tag{2}$$

$(x, y)^T$ is the homogeneous coordinate of the image physical coordinate system. The transformation between the image physical coordinate system and pixel coordinate system is expressed as:

$$\begin{bmatrix} u \\ v \\ 1 \end{bmatrix} = \begin{bmatrix} 1/d_x & 0 & u_0 \\ 0 & 1/d_y & v_0 \\ 0 & 0 & 1 \end{bmatrix} \begin{bmatrix} x \\ y \\ 1 \end{bmatrix} \tag{3}$$

u_0 and v_0 are the intersection coordinate between the optical axis center and image plane. From (2) and (3), the transformation between the camera coordinate system and the image pixel coordinate system is expressed as:

$$\begin{bmatrix} u \\ v \\ 1 \end{bmatrix} = \frac{1}{z_c} \begin{bmatrix} 1/d_x & 0 & u_0 \\ 0 & 1/d_y & v_0 \\ 0 & 0 & 1 \end{bmatrix} \begin{bmatrix} f & 0 & 0 & 0 \\ 0 & f & 0 & 0 \\ 0 & 0 & 1 & 0 \end{bmatrix} \begin{bmatrix} x_c \\ y_c \\ z_c \\ 1 \end{bmatrix} = \frac{1}{z_c} \begin{bmatrix} f/d_x & 0 & u_0 & 0 \\ 0 & f/d_y & v_0 & 0 \\ 0 & 0 & 1 & 0 \end{bmatrix} \begin{bmatrix} x_c \\ y_c \\ z_c \\ 1 \end{bmatrix} \tag{4}$$

(Ⅲ) The transformation between the world coordinate system and the image coordinate system

According to the transformation relationship between the above coordinate systems, from (1) and (4), the final camera imaging model is expressed as:

$$z_c \begin{bmatrix} u \\ v \\ 1 \end{bmatrix} = \begin{bmatrix} f_x & 0 & c_x & 0 \\ 0 & f_y & c_y & 0 \\ 0 & 0 & 1 & 0 \end{bmatrix} \begin{bmatrix} R_{3\times3} & T_{3\times1} \\ 0 & 1 \end{bmatrix} \begin{bmatrix} x_w \\ y_w \\ z_w \\ 1 \end{bmatrix} = AM_{R,T} \begin{bmatrix} x_w \\ y_w \\ z_w \\ 1 \end{bmatrix} \tag{5}$$

Where $f_x = f/d_x$ and $f_y = f/d_y$ are respectively the camera image plane scale factor of horizontal axis x and vertical axis y. c_x and c_y are respectively the offset of horizontal direction and vertical direction between the camera center and optical axis. Calibrated camera internal parameter is mainly solving these four parameters (f_x, f_y, c_x, c_y). A is the camera internal parameter array, $M_{R,T}$ is the camera external parameter array. They represent the basic relationship between the two-dimensional image coordinate and three-dimensional world coordinate, which can determine the camera calibration model.

3   Algorithm Design and Application

3.1   Differential Evolution Particle Swarm Optimization

DE and PSO algorithm are based on the evolution of population, because of its own superiority, which has been successfully applied in many optimization problems [12-13]. Because there are some shortcomings of PSO, this paper proposes a kind of algorithm based on differential evolution particle swarm optimization (DEPSO). The mutation, crossover and selection of DE are introduced into the PSO. The mutation and crossover operation are adopted in each iteration, which can maintain the diversity of population particles and select the optimal particle of each iteration to the next iteration.

It can improve the convergence of the algorithm and prevent the particles into premature convergence [14].

Assuming population size is N and each individual has a D-dimensional vector, the location target vector and speed test vector are respectively represented as $X_i = (x_{i1}, x_{i2}, \cdots, x_{iD})^T$ and $V_i = (v_{i1}, v_{i2}, \cdots, v_{iD})^T$, $i = 1, 2, \cdots, N$. The initial population is $S = \{X_1, X_2, \cdots, X_N\}$. The PSO renewal equations of speed and position are expressed as follows:

$$v_{id}^{k+1} = wv_{id}^k + c_1 rand\,()(pbest - x_{id}^k) + c_2 rand\,()(gbest - x_{id}^k) \tag{6}$$

$$x_{id}^{k+1} = x_{id}^k + v_{id}^{k+1} \tag{7}$$

The each target vector individual of G generation is represented as $X_{i,G}$. In the paper, the mutation operator of DE is often used as shown below:

$$v_{i,G} = x_{r1,G} + F(x_{r2,G} - x_{r3,G}) \tag{8}$$

From (8), the mutation mechanism is introduced into the particle swarm iteration computation, which generates mutation after each iteration and prevents particle swarm premature into local optimum.

In order to increase the diversity of population, the crossover operation is introduced. The test vector V_i and the target vector X_i are permeated to achieve the purpose of improving the population global search ability. Assuming the cross vector is represented as $U_i = (u_{i1}, u_{i2}, \cdots, u_{iD})^T$. The crossover operator is expressed as:

$$u_{j,i,G} = \begin{cases} v_{j,i,G}, & (rand_j \leq CR)\ or\ (j = j_{rand}) \\ x_{j,i,G}, & else \end{cases} \tag{9}$$

Where $j = 1, 2, \cdots, N$, $j_{rand} \in [1, N]$. CR is a cross control parameter and general value is between [0, 1]. If the value is greater, which generates the probability of crossover is greater and the diversity of population is better.

According to the above mutation and crossover, the candidate individual $U_{i,G}$ is evaluated by fitness function and decided whether to select the new generation individual. The select operator is expressed as:

$$x_{i,G+1} = \begin{cases} u_{i,G}, & f(u_{i,G}) \leq f(x_{i,G}) \\ x_{i,G}, & else \end{cases} \tag{10}$$

It is assume that the optimal value of objective function is the minimum value which is the fitness evaluation standard for select operation. The optimal individual is chose in each iteration and able to achieve optimization purpose.

3.2   Algorithm Application

According to the analysis of the camera model, the camera parameters are used as the optimization goal. The DEPSO algorithm is applied to solve the camera calibration. Then on the basis of the solved parameters, the image two-dimensional coordinates can be obtained. The application process of particle swarm optimization based on differential evolution is as follows:

Step1 Population initialization: it randomly generates the position and speed of N particles within the allowed scope, and sets the upper and lower limit of particle velocity.

Step2 Selecting fitness function: the objective function is considered the distance of obtained pixels and actual pixels, which is used as the fitness evaluation standard and calculated to get the individual extremum and global extremum of the initialization population. Fitness function is expressed as:

$$f = \min \sum_{i=1}^{n} \sqrt{(u_i - x_i)^2 + (v_i - y_i)^2} \tag{11}$$

Step3 Renewing population: according to Eq.(6) and (7), the speed and position of each particle are renewed.

Step4 Selective renewal: using the selection strategy of differential evolution, the fitness of the renewed particle is compared with the fitness of particle before renewing, and choosing high fitness particle to update location.

Step5 Crossover operation: increasing diversity of the population and ensuring that the excellent individuals have a high fitness.

Step6 Mutation operation: according to comparing fitness value, the low fitness individuals are generated mutation with greater probability, which is beneficial to produce excellent model and guarantee the existence of superior individuals. Thus a new generation of excellent population is forming.

Step7 Renewing the population extremum: according to the fitness value of a new generation population, renewing the individual extremum and global extremum of population.

Step8 Determining whether the termination condition is satisfied: if it reaches the maximum number of iteration, then the end of the loop and output results, otherwise go to step3 to continue iteration.

4   Experimental Analysis

In order to verify the application performance of the proposed algorithm, taking a company visual identification project as application background. The calibrating camera for experiment is the ARTCAM-150PIII CCD camera of ARTRAY Company, Japanese Seiko lens TAMRON 53513, the effective pixel is 1392×1040. To ensure the fairness of contrast, the experiment is carried out on Windows XP system platform, clocked at 2.67GHz, RAM is 2.00GB and the development environment of Matlab. Unified setting particle population size N is 30, the maximum number of iteration is 1000 and 100 times continuous optimization.

The calibration image used for experiment is the classic black and white chessboard whose size is 8×10, namely 80 corners. According to the camera internal and external parameters, solving all the geometric parameters will need at least two perspective images. In order to ensure the convergence and accuracy of the camera parameters, the camera calibration chooses 10 calibration board images of different perspective in the paper and the corners are considered as calibration points. This paper proposed the algorithm is used for camera calibration and compared with the planar pattern calibration method of Zhang Zhengyou which is the most widely used method at present. In view of the different visual angles, the camera external parameters are uncertain. So only the camera internal parameter is calibrated in the paper, calibration results are shown in Table 1.

Table 1. Camera parameters calibration results

Parameters	DEPSO	DE	PSO	Zhang
f_x (pixels)	3580.468994	3585.466215	3588.167245	3590.147217
f_y (pixels)	3620.209961	3624.256194	3627.341156	3628.451172
c_x (pixels)	597.373413	600.623577	600.551365	602.444153
c_y (pixels)	249.571625	251.239451	251.276542	252.145920

As shown in Table 1, it can be seen that the experimental simulation results are close to the calibration result of Zhang Zhengyou and the relative error is small. It is show that the algorithms used for camera calibration is feasible. According to the model transformation relationship, using the calibrated camera parameters and the three-dimensional space coordinates (x_w, y_w, z_w) to solve the corresponding two-dimensional image coordinates (u, v), and which are compared with the actual image coordinates $(\tilde{u}, \tilde{v})$ obtained by the image processing. Then it can verify the validity and accuracy of the calibration method results. Experiment is basis on the visual identification project and 10 groups of the actual measured data are randomly selected to contrast, the verification results are shown in Table 2.

Table 2. Validity verification results

x_w (mm)	y_w (mm)	z_w (mm)	$\tilde{u}$ (pixels)	u (pixels)	$\lvert\tilde{u}-u\rvert$ (pixels)	$\tilde{v}$ (pixels)	v (pixels)	$\lvert\tilde{v}-v\rvert$ (pixels)
-10.5	420	126	563.5468	563.4729	0.0739	484.4662	484.5679	0.1016
-31.5	420	147	368.1478	368.1244	0.0234	276.8552	276.6524	0.2027
10.5	420	147	773.9859	774.0655	0.0795	274.4552	274.4674	0.0122
-17.5	420	133	504.3714	504.4749	0.1034	412.2836	412.3035	0.0198
17.5	420	140	842.4265	842.5655	0.1389	342.3060	342.1068	0.1992
-3.5	420	119	640.1724	640.2547	0.0822	547.5030	547.5612	0.0581
-31.5	420	112	370.5156	370.3410	0.1746	616.4668	616.4727	0.0059
24.5	420	98	912.9185	912.7060	0.2124	748.3865	748.3856	0.0009
10.5	420	133	774.6569	774.7635	0.1066	410.9766	410.8161	0.1605
3.5	420	105	708.0079	708.2405	0.2325	682.8162	682.8112	0.0050

From Table 2 contrasting analysis results show that the derived image coordinates have a very small difference comparing with the actual image coordinates. Then it is show that this calibration method has good reliability and stability. In terms of the objective function f_{obj} as the evaluation object, the object mean of all images pixel coordinates are calculated and compared with the other calibration methods. Comparison results are shown in Table 3.

Table 3. Comparison results of the object mean

Object mean	DEPSO	DE	PSO	Zhang
f_{obj} (pixels)	0.167186	0.261793	0.305274	0.327475

The comparison results in Table 3 show that the object mean which is solved by the algorithms in the paper is less than the Zhang Zhengyou method result. Differential evolution combined with the particle swarm optimization can effectively avoid falling into local optimum. It is show that the application of the DEPSO algorithm has a better optimization effect and calibration results have a higher precision. Comparing the above experiment results show that using the differential evolution particle swarm optimization has good feasibility and reliability for camera calibration.

5   Conclusions

Camera calibration has a very important position in the computer vision and is the basis of developing the other aspect research. Then what methods can be used to calibrate camera more accurate, which has become the current center problem of the camera calibration research. Based on actual project as the background in this paper, the relationship model of camera calibration is established and the differential evolution particle swarm optimization is used to calibrate camera parameters. The simulation results show that the algorithm is feasible and effective, which has a simple operation and good optimization ability. Currently, there are many methods proposed for camera calibration, the application of intelligent learning algorithm is a very good research direction.

References

1. Xu, Y., Guo, D.X., Zheng, T.X., Cheng, A.Y.: Research on camera calibration methods of the machine vision. In: 2011 Second International Conference on Mechanic Automation and Control Engineering, pp. 5150–5153 (2011)
2. Sun, J., Gu, H.B.: Research of linear camera calibration based on planar pattern. World Academy of Science, Engineering and Technology 60, 627–631 (2011)

3. Song, L.M., Wu, W.F., Guo, J.R., Li, X.H.: Survey on camera calibration technique. In: Proceedings of 2013 5th International Conference on Intelligent Human-Machine Systems and Cybernetics, vol. 2, pp. 389–392 (2013)

4. Tsai, R.Y.: An Efficient and Accurate Camera Calibration Technique for 3D Machine Vision. In: Proceedings of IEEE Conference on Computer Vision and Pattern Recognition, pp. 364–374 (1986)

5. Zhang, Z.Y.: A flexible new technique for camera calibration. IEEE Transactions on Pattern Analysis and Machine Intelligence 22(11), 1330–1334 (2000)

6. Ma, S.D.: A self-calibration technique for active vision systems. IEEE Transactions on Robotics and Automation 12(1), 114–120 (1996)

7. Deep, K., Arya, M., Thakur, M., Raman, B.: Stereo camera calibration using particle swarm optimization. Applied Artificial Intelligence 27(7), 618–634 (2013)

8. Tian, Z., Xiong, J.L., Zhang, Q.: Camera calibration with neural networks. Applied Mechanics and Mechanical Engineering 29-32, 2762–2767 (2010)

9. Kennedy, J., Eberhart, R.: Particle swarm optimization. In: Proceedings of the 4th IEEE International Conference on Neural Networks, pp. 1942–1948. IEEE Service Center, Piscataway (1995)

10. Storn, R., Price, K.: Differential evolution - A simple and efficient heuristic for global optimization over continuous spaces. Journal of Global Optimization 11(4), 341–359 (1997)

11. Wang, Q., Fu, L., Liu, Z.Z.: Review on camera calibration. In: 2010 Chinese Control and Decision Conference, pp. 3354–3358 (2010)

12. Zhang, J.D., Lu, J.G., Li, H.L., Xie, M.: Particle swarm optimization algorithm for non-linear camera calibration. International Journal of Innovative Computing and Applications 4(2), 92–99 (2012)

13. De la Fraga, L.G.: Self-calibration from planes using differential evolution. In: Bayro-Corrochano, E., Eklundh, J.-O. (eds.) CIARP 2009. LNCS, vol. 5856, pp. 724–731. Springer, Heidelberg (2009)

14. Li, G.Y., Liu, M.G.: The summary of differential evolution algorithm and its improvements. In: 2010 3rd International Conference on Advanced Computer Theory and Engineering, vol. 3, pp. 153–156 (2010)

Gearbox Fault Diagnosis Method Based on SVM Trained by Improved SFLA

Lu Ma, Guochu Chen[*], and Haiqun Wang

Electric Engineering School, Shanghai DianJi University, Shanghai, 200240 China
chengc@sdju.edu.cn

Abstract. A method of fault diagnosis based on support vector machine trained by the improved shuffled frog leaping algorithm (ISFLA-SVM) is proposed to promote the classification accuracy of the wind turbine gearbox fault diagnosis. Because the parameter selection for penalty factor and kernel function in support vector machine (SVM) have a great impact on the classification accuracy, we may use the improved shuffled frog leaping algorithm to select excellent SVM parameters, use the optimized parameters to train machine. Then three groups of data in UCI are used for performance evaluation. Finally ISFLA-SVM model will be applied to the wind turbine gearbox fault diagnosis. The result of the diagnosis indicates that the common fault of wind turbine gearbox can be exactly identified by this method.

Keywords: gearbox, shuffled frog leaping algorithm, support vector machine, fault diagnosis, optimization, accuracy.

1 Introduction

Gearbox is an indispensable key component of wind turbine, its main function is passed the dynamic which is generated by wind wheel under the action of wind to the generator and make the generator get corresponding speed [1]. For the wind turbine installed in a high tower, once the gear box failure, the maintenance cost will be high. According to Spanish EHN company data statistics, gearbox is one of the highest failure rate components for the wind turbine. Therefore, strengthening the on-line monitoring and fault diagnosis for gearbox plays a decisive role in reducing repair costs and improving the recovery efficiency of wind turbine.

Support vector machine (SVM) based on the statistical learning theory can not only maximizing discover the hidden classification knowledge in the data, but also solved some problems such as small sample learning, high dimension in machine learning. From the generalized point of view, SVM is more suitable for wind turbine gearbox fault diagnosis. At present, the parameter selections of SVM are still no specific rules to follow. In recent years, researchers are constantly use new intelligent optimization algorithm to SVM parameters optimization. Literature [2] proposed the improved

---

[*] Corresponding author.

M. Fei et al. (Eds.): LSMS/ICSEE 2014, Part II, CCIS 462, pp. 257–263, 2014.

PSO which was applied to the SVM parameter optimization for fault diagnosis of transformer. Literature [3] proposed the simulated annealing algorithm which was applied to SVM parameter optimization for mid-long term load forecasting .The more optimal parameters are acquired to use these methods, but they are also easy to fall into local optimization and the iterations are large and they are time-consuming.

Shuffled frog leaping algorithm (SFLA) was emerging in 2003, there is no literature in gearbox fault diagnosis to be investigated. In view of the fault problems of the gearbox, a method of fault diagnosis based on SVM trained by improved shuffled frog leaping algorithm (ISFLA-SVM) is proposed. The improved SFLA (ISFLA) algorithm can adjust the balance between global and local search capabilities suitably and find the optimal values of SVM parameters. Firstly this paper puts forward ISFLA, and then use three groups of data in UCI for performance evaluation, finally gearbox fault diagnosis model will be applied to the wind turbine gearbox fault diagnosis and come to a conclusion.

2   The Improved SFLA and Its Performance Analysis

2.1   Shuffled Frog Leaping Algorithm (SFLA)

An initial population of F frogs is created randomly for a d dimensional problem. A frog i is represented by d variables and the i frog indicated as $X_i = (x_{i1}, x_{i2}, ..., x_{id})$. Frogs are sorted in descending order based on their fitness values, and write down the global best individual X_g, then the entire population is divided into m meme groups. Within each meme groups, frogs with the best and the worst fitness are identified as X_b and X_w. To improve the worst solution, perform the local search.

For the basic rules of SFLA known the update formula is:

$$S = r * (X_b - X_w) \tag{1}$$

$$X_w' = X_w + S, \quad ||S|| \le S_{max} \tag{2}$$

Where, r is a random number between 0 &1, S_{max} is the maximum step size, S is the step of worst frog.

2.2   Improved SFLA (ISFLA)

(1) Rand function can't guide the frog forward to the optimum direction. Through verification, in the early evolution, to some extent increase the local search ability of SFLA, can expand the search scope, in the middle and later of evolution, reduce the local search can escape from local optimal solution. According to the simulation experiment, the literature [4] shows that the logarithmic function is more suitable for the mobile factor, so this article will take the logarithm function as adaptive mobile factor.

$$\theta_t = log_N t, t = 1, 2, ..., N \tag{3}$$

In the formula, θ_t on behalf of adaptive mobile factor, N on behalf of the times of each meme groups searched.

(2) SFLA is easy to fall into local optimal solution. In this article, join the mutation operation to the SFLA global search process, carrying out random mutation for F frogs. This can greatly increase the diversity of population, prevent SFLA into local optimum.

Get a number between$(0,1)$randomly, if greater than 0.5, variate the frog, procedure is as follows:

$$h = \text{ceil}(2 * \text{rand});$$

$$\text{if } h == 1 \quad p(i,h) = (20 - 1) * \text{rand} + 1(i = 1,2,...,F)$$

$$\text{if } h == 2 \quad \text{The new initialization of } p(i,h)$$

If less than 0.5, the frog will not change.

(3) The bigger parameter c of SVM will lead to over learning state. We join the threshold limit to the frog individual update. If both the iterative optimal value minus the global optimal value is less than a certain threshold (δ) and the value which is on behalf of the c less than the value which is on behalf of global optimal frog c , assuming the frog first dimensional x_{i1} on behalf of c. If formula (11) (12) satisfied, the global optimal value and the global optimal solution replaced by that iterative optimal value; Else, not for the update operation.

$$fitness(i) - X_g\_fitness < \delta \tag{4}$$

$$p(i,1) < X_g(i,1) \tag{5}$$

3   The SVM Model Based on ISFLA

3.1   Using ISFLA Choose the Best c and g Parameters for SVM

The fitness function defined in this paper is the classification accuracy achieved by the class test that SVM deals with the test data. The optimization procedure is as follows:

The first step: to determine the fitness function in the sense of CV.

The second step: the initialization of meme groups$m$, frog number of each group v, local search N, global iteration number G_{max}, and the relevant parameters of SVM.

The third step: calculate the fitness value.

The fourth step: divide the frog individual into m meme groups.

The fifth step: for each meme group, perform local search N times.

The sixth step: when the local search is completed, fitness scaling.

The seventh step: To judge whether meet the termination conditions or not, if meet, output the global optimal solution and the individual of optimal solution; otherwise, return to the third step and recount.

3.2   Simulation Experiment

The experiment uses Transformer Faults, Breast Cancer, Wine, three data sets of the UCI database (Table. 1) to do the simulation test. Transformer Faults data set for 5 class classification problems, containing 33 samples, each sample has 3 properties; Breast Cancer data set for 2 class classification problems, there are 569 samples, each sample has 10 attributes. The Wine data set for 3 class classification problems, there are 178 samples, each sample has 13 attributes.

Table 1. UCI standard data set

Data sets	Training sample	Test sample
Transformer Faults	23	10
Breast Cancer	300	269
Wine	89	89

ISFLA parameter setting: $v = 10, m = 10, N = 5, G_{max} = 20$. The IPSO[5] parameter setting: c1=1.5, c2=1.7, maxgen=200, sizepop=20.

In order to verify the effectiveness of improved shuffled frog leaping algorithm process and prevent random optimization results, compare ISFLA with SFLA, and algorithm of each case is independently operated 30 times. From the Fig. 1 we can see that the optimization effect of ISFLA is obviously better than the SFLA. ISFLA both for Breast Cancer data sets of multiple Transformer Faults data and for less data set all achieved good optimization results. The optimization effect is stable. So it proves the validity of ISFLA.

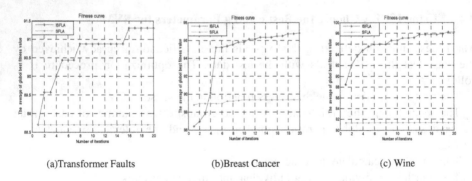

(a)Transformer Faults   (b)Breast Cancer   (c) Wine

Fig. 1. Average iterative curve of classification accuracy

In order to verify the classification performance of ISFLA-SVM, compare ISFLA-SVM with SFLA-SVM, IPSO-SVM (Improved PSO, IPSO). What can be seen from table.2 is that ISFLA-SVM and IPSO-SVM have the same classification accuracy for Transformer Faults which is higher than that of SFLA-SVM. For the two classification problems of Breast Cancer data, the classification accuracy of ISFLA-SVM is higher than IPSO-SVM and SFLA -SVM classification accuracy. For three classification problem of Wine data, ISFLA-SVM only misclassifies a set of data. It can be concluded that ISFLA-SVM can achieve good results in classification problems in general and reflects the superiority of its performance.

Table 2. Classification accuracy

Data set \ Methods	SFLA-SVM	IPSO-SVM	ISFLA-SVM
Transformer Faults	60% (6/10)	90% (9/10)	90% (9/10)
Breast Cancer	94.05%(253/269)	96.28% (259/269)	97.026% (261/269)
Wine	75.28% (67/89)	97.75% (87/89)	98.87% (88/89)

4   Application in Fault Diagnosis of Wind Turbine Gearbox

4.1   The Experimental Design

A simulation experiment was carried out on the gear fault fan transmission system through the experimental platform of Shanghai Dianji University wind engineering research center. Gearbox fault diagnosis model is established by the characteristics of the gearbox fault, as shown in Fig. 2. The vibration signal is collected by the Fourier transform, and the normal gear vibration spectrum as a reference, obtain the 45 groups of gearbox diagnosis samples (Listed 16 samples of data in Table.3), T'_{30} as the amplitude of each wave in the spectrum, T'_{31} as corresponding frequency, and T'_{32} as the frequency difference is near the crest, and T'_{33} as crest edge band frequency difference.

In this section, respectively using the '①', '②', '③', '④', '⑤' representing the normal, pitting, broken teeth, wear , broken teeth and wear five kinds of modes, each mode has 9 groups data. In order to ensure the reliability of the classification results from each fault type, random selected 6 samples as training data, 3 groups of samples as the diagnostic data.

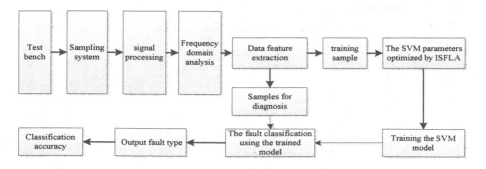

Fig. 2. The gearbox fault diagnosis model

Table 3. The samples of gearbox for diagnosis

Sequence number	T'_{30}	T'_{31}	T'_{32}	T'_{33}	Failure state	Sequence number	T'_{30}	T'_{31}	T'_{32}	T'_{33}	Failure state
1	0.360	2383.125	10.000	24.375	①	9	23.227	2335.625	6.250	24.375	③
2	0.377	2407.500	10.000	25.000	①	10	23.836	2360.000	6.875	24.375	③
3	0.667	2456.875	9.375	24.375	①	11	24.972	2382.500	10.625	25.000	④
4	27.738	2134.375	5.625	24.375	②	12	29.048	2308.750	10.625	25.000	④
5	28.841	2158.750	5.625	24.375	②	13	27.463	2333.750	10.625	24.375	④
6	23.439	2208.125	5.625	24.375	②	14	30.041	2363.125	0	-24.375	⑤
7	29.511	2286.250	6.875	25.000	③	15	22.981	2412.500	0	-24.375	⑤
8	26.005	2311.250	6.375	24.375	③	16	19.488	2436.875	0	-25.000	⑤

4.2 Experimental Results and Analysis

Train the ISFLA-SVM classifier by using training samples, the parameter of kernel function g=3.2224, penalty coefficient c=10.4005. Fault diagnosis to 3 groups samples for diagnosis which randomly extracted from each failure type by using the identified parameters. It can be seen from the results (Fig.3), the actual and predicted values were consistent, and the correct diagnostic rate was 100%.

In order to validate ISFLA-SVM on classification performance of gearbox fault diagnosis, repeat the above steps for 6 times and compare the correct rate of fault diagnosis with SFLA-SVM and IPSO-SVM. It is seen from Table.4, the method proposed in this paper in gearbox fault diagnosis accurate rate is higher than IPSO-SVM and SFLA-SVM. The average classification time of the method in this paper is not only lower than the IPSO-SVM average classification time, but also largely lower than the SFLA-SVM classification time. So it verified the speed superiority of ISFLA-SVM in fault diagnosis of gearbox. Therefore, ISFLA-SVM in wind turbine gearbox fault diagnosis has good stability, speed superiority and classification accuracy.

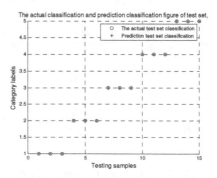

Fig. 3. The test samples of gear box fault classification

Table 4. Comparison of different gearbox fault diagnosis methods

Fault diagnosis methods	SFLA-SVM	IPSO-SVM	ISFLA-SVM
The average diagnostic accuracy	88.89%	96.67%	98.89%
The average classification time/s	114.75	25.86	22.02

5   Conclusions

An efficient approach with ISFLA-SVM has been proposed in this paper for solving the wind turbine gearbox fault diagnosis problem. The accuracy and stability of SVM classification is improved by ISFLA. The simulation results show that the method is suitable for wind turbine gearbox fault classification and has achieved fast classification, good stability, high classification accuracy, and provides a new approach for fault diagnosis of wind turbine gearbox.

Acknowledgments. This work was financially supported by scientific research innovation projects of Shanghai municipal education commission (Grant No.13YZ140) and the key disciplines of Shanghai Municipal Education Commission of China (Grant No. J51901).

References

1. Gong, J.: Wind engineering technical manuals. China Machine Press, Beijing (2004) (in Chinese)
2. Jia, R., Zhang, Y., Hong, G.: Parameter optimization of least squares support vector machine based on improved particle swarm optimization in fault diagnosis of transformer. Power System Protection and Control 38(17), 121–124 (2010) (in Chinese)
3. Li, J., Liu, J., Wang, J.: Mid-long term load forecasting based on simulated annealing and SVM algorithm. Proceedings of the CSEE 31(16), 63–66 (2011) (in Chinese)
4. Ma, P.: Research on shuffled frog leaping algorithm, pp. 23–26. Xidian University (2013) (in Chinese)
5. Chinese BBS about MATLAB. Thirty case analysis about MATLAB neural network, pp. 122–131. Beihang University Press (2010) (in Chinese)

Discrete Chaotic Synchronization and Its Application in Image Encryption[*]

Hua Wang, Jiu-Peng Wu, Xiao-Shu Sheng, and Peng Zan

School of Mechatronic Engineering and Automation,
Shanghai Key Laboratory of Power Station Automation Technology,
Shanghai University, Shanghai, 200072, China
running_lunzi@sina.com, zanpeng@shu.edu.cn

Abstract. Based on the synchronization principle of three-dimensional discrete Henon chaos, this paper presents a new image encryption algorithm. A set of keys are generated by chaotic iteration from the sending part. Then the image can be encrypted. Via the same keys generated by chaotic synchronization, the receiving part can get decrypted image by inverse transformation. From the view of cryptology, high dimensional chaos in this paper is more complex than the general chaos, it is difficult to predict; and this algorithm can not only substitute the pixel values, but also scramble the pixel locations at the same time, but in most cases some algorithms can only encrypt the image in a way. As a supplement to the encryption process, this algorithm introduces the technology of chaotic synchronization and bits scrambling, increases the difficulty of cracking, and enhances the algorithm security. Simulation results illustrate the effectiveness of the proposed method.

Keywords: Discrete Systems, Henon chaos, Chaotic Synchronization, Bits Scrambling, Image Encryption.

1 Introduction

Because of the characteristic of the pseudo randomness and sensitivity to initial values, chaotic map is suitable for image encryption. Even if the difference is very small between the parameters of two chaotic systems, their trajectories will be exponential divergent [1]. However, there is no breakthrough on the research of chaotic synchronization over a long time. Until 1990, Pecora and Carroll realized synchronization by driving the Lyapunov exponent of the error equation negative [2]. This finding opens up a new way to apply the theory of chaos to practical affairs.

In 2006, Meng etc. [3] proposed an improved synchronization method for Logistic mapping via M type nonlinear feedback control. With the deeper research of chaotic synchronization in various fields, it shows great potential applications in secure communication. In 2013, Xu etc. [4] proposed an image encryption method via

---

[*] This research is supported by the National Nature Science Foundation of China (11202121, 61104006, and 31370998).

M. Fei et al. (Eds.): LSMS/ICSEE 2014, Part II, CCIS 462, pp. 264–272, 2014.

different structure chaotic synchronization. In 2013, Liu etc.[5]studied the hybrid synchronization of unified chaotic system in image encryption. Generally speaking, image encryption is a hot topic in recent years. But how to use chaotic synchronization in image encryption is a new area and the research in this direction is relatively rare.

Since the actual hardware system can only process discrete digital signal, so we should discrete a continuous signal in realization. In another way, this paper firstly studied the synchronization of discrete chaotic system. We can use the discrete sequence to design the encryption algorithm directly. This can save the necessary sampling time of continuous system. So it can improve the speed of image encryption [6].

This paper presents a new synchronization method based on discrete Henon chaos. On the sending part, we sort the chaotic sequence generated by the driving chaos firstly. This can help to get the address conversion codes [7]; Thanks to the irregular sort, bit positions can be scrambled [8]. So the image can be encrypted. On the receiving part, the response chaotic system is used to get the synchronized address codes. Then the image is decrypted by inverse transformation.

2   System Description

Definition 1: A continuous, strictly increasing function $W:[0, \infty) \to [0, \infty)$ with $W(0) = 0$ and $W(u) > 0$ if $u > 0$ is called a wedge function. (We denote wedge functions in the sequel by W or W_i, where i is an integer.)

Consider the following cascade discrete-time systems:

$$\begin{cases} x(k+1) = f(x(k)) + h(x(k), y(k)) \\ y(k+1) = g(y(k)) \end{cases} \qquad (1)$$

Where $x \in R^n$, $y \in R^m$, f and g are assumed to be smooth mappings and satisfying the following conditions:

$$f(0) = 0; \quad h(x; 0) = 0; \quad g(0) = 0.$$

Lemma 1 [9]: If the zero solutions of system $x(k+1) = f(x(k))$ and $y(k+1) = g(y(k))$ are globally asymptotically stable and every solution to (1) is bounded, then the zero solution $o = (0, 0)$ of (1) is globally asymptotically stable.

Theorem 1: Let $V(x(n))$ is a positive function satisfying

(1) $W_1(\|x(n)\|) \leq V(x(n)) \leq W_2(\|x(n)\|)$

(2) $\Delta V_{(1)}(x(n)) \leq -W_3(\|x(n)\|) + M$ for some constant $M > 0$

Where $W_1(u)$, $W_3(u) \to \infty$ as $u \to \infty$, then the solutions of (1) are bounded.

Proof: because M is a constant, then we can choose a constant B_1 satisfying $B_1 > W_3^{-1}(M)$. Now we want to prove that

$$V(x(n)) \leq W_2(B_1) + M, \quad \text{for all } n \geq 0. \qquad (2)$$

We will use the reduction to absurdity method in mathematic to prove this fact.
Suppose there is an $n_1 > 0$ such that

$$V(x(n)) \leq W_2(B_1) + M \text{ , for } 0 < n < n1. \tag{3}$$

but

$$V(x(n_1 + 1)) > W_2(B_1) + M. \tag{4}$$

We discuss two possibilities (A) and (B):

(A): $V(x(n_1)) \leq W_2(B_1)$ and $V(x(n_1) + 1) > W_2(B_1) + M$.

Then let $L = W_1^{-1}(W_2(B_1) + M)$. From assumption (ii) in theorem 2

$$\Delta V_{(1)}(x(n)) \leq -W_3(\| x(n) \|) + M.$$

We can get

$$V(x(n_1 + 1)) \leq V(x(n_1)) - W_3(\|x(n)\|) + M \tag{5}$$
$$\leq W_2(B_1) - W_3(\|x(n)\|) + M$$
$$\leq W_2(B_1) + M$$

This is a contradiction to (4).

(B): $V(x(n_1)) > W_2(B_1)$ and $V(x(n_1) + 1) > W_2(B_1) + M$.

Then, from assumption (ii) in theorem 2

$$\Delta V_{(1)}(x(n)) \leq -W_3(\|x(n)\|) + M$$

We can get

$$V(x(n1 + 1)) \leq V(x(n1)) - W_3(\|x(n)\|) + M. \tag{6}$$

From condition (B) $V(x(n1)) > W_2(B_1)$ we know this implies that $\|x(n_1)\| > B_1$.

Then,

$$W_3(\|x(n_1)\|) > W_3(B_1) > M, \text{ i.e. } M - W_3(\|x(n_1)\|) < 0.$$

We arrive at

$$V(x(n_1 + 1)) \leq V(x(n_1)) - W_3(\|x(n)\|) + M < V(x(n_1)) \tag{7}$$

Combined with (3) we know that

$$V(x(n_1 + 1)) \leq V(x(n_1)) < W_2(B_1) + M. \tag{8}$$

Again, this is a contradiction to (4). Therefore, (2) holds and thus,

$$\|x(n)\| \leq W_1^{-1}(W_2(B_1) + M) \equiv B_2, \text{ for all } n > 0 \tag{9}$$

Here completes the proof.

3   Chaotic Synchronization

In general, two dynamic systems in synchronization are called the master system and the slave system respectively. A well designed controller will make the trajectory of the slave system track the trajectory of the drive system, that is, the two systems will be synchronous. In this section, we will research on the synchronization of two chaotic systems. This is a base of image encryption.

Consider the following master chaotic system [10]

$$\begin{cases} x_1(k+1) = a - x_2^2(k) - bx_3(k) \\ \quad x_2(k+1) = x_1(k) \\ \quad x_3(k+1) = x_2(k) \end{cases} \tag{10}$$

The slave system is

$$\begin{cases} y_1(k+1) = a - y_2^2(k) - by_3(k) \\ \quad y_2(k+1) = y_1(k) + u_1(k) \\ \quad y_3(k+1) = y_2(k) + u_2(k) \end{cases} \tag{11}$$

Denote the response errors as $e_i(k) = y_i(k) - x_i(k)$ $(i = 1, 2, 3)$. Subtracting Equation (10) from Equation (11) yields the error system as follows

$$\begin{cases} e_1(k+1) = -(2x_2(k) + e_2(k))e_2(k) - be_3(k) \\ \quad e_2(k+1) = e_1(k) + u_1(k) \\ \quad e_3(k+1) = e_2(k) + u_2(k) \end{cases} \tag{12}$$

Design the linear controller as $u_1(k) = -e_1(k) + \dfrac{1}{2}e_2(k)$, $u_2(k) = \dfrac{1}{2}e_1(k)$.

The error system (12) can be expressed as the following cascade system:

$$\begin{cases} e_1(k+1) = -(2x_2(k) + e_2(k))e_2(k) - be_3(k) \\ \quad e_3(k+1) = e_2(k) + \dfrac{1}{2}e_1(k) \end{cases} \tag{13}$$

And

$$e_2(k+1) = \frac{1}{2}e_2(k). \tag{14}$$

Now it will prove that system (12) can be globally asymptotically stabilized by the linear controller. We will prove this in two steps based on Lemma1.

Step 1: It is obvious that the solution of (14) is

$$e_2(k) = (\frac{1}{2})^k e_2(0).$$

It is globally asymptotically stable and its solutions are bounded by $|e_2(k)| < |e_2(0)|$. Take $e_2(k) = 0$ and we can get

$$\begin{cases} e_1(k+1) = -be_3(k) \\ e_3(k+1) = \dfrac{1}{2}e_1(k) \end{cases} \qquad (15)$$

Consider the following positive function

$$V_1(e_1(k), e_3(k)) = e_1^2(k) + e_3^2(k) \qquad (16)$$

Calculate the difference of $V_1(e_1(k), e_3(k))$ with respect to (15) and we can get

$$\begin{aligned} \Delta V_1(e_1(k), e_3(k)) &= e_1^2(k+1) + e_3^2(k+1) - e_1^2(k) - e_3^2(k) \\ &= b^2 e_3^2(k) + \frac{1}{4}e_1^2(k) - e_1^2(k) - e_3^2(k) \qquad (17) \\ &= -\frac{3}{4}e_1^2(k) - (1-b^2)e_3^2(k) \end{aligned}$$

So we can get $\Delta V_1(e_1(k), e_3(k))$ is negative with respect to 0<b<1. From Lyapunov theorem we know that system (15) is globally asymptotically stable.

Step 2: In this step, we will prove that the solutions of the error system (12) are bounded.

In Step 1, we have shown that the solution of the subsystem $e_2(k+1) = \dfrac{1}{2}e_2(k)$ is bounded by $|e_2(k)| < |e_2(0)|$.

Now consider the following positive function

$$V_2(e_1(k), e_3(k)) = e_1^2(k) + e_3^2(k). \qquad (18)$$

Calculate the difference of $V_2(e_1(k), e_3(k))$ with respect to the first and the third equation of (12) and we can get

$$\begin{aligned} \Delta V_2(e_1(k), e_3(k)) &= e_1^2(k+1) + e_3^2(k+1) - e_1^2(k) - e_3^2(k) \\ &= (-(2x_2(k) + e_2(k))e_2(k) - be_3(k))^2 + (e_2(k) + \frac{1}{2}e_1(k))^2 - e_1^2(k) - e_3^2(k) \\ &= -\frac{3}{4}e_1^2(k) - (1-b^2)e_3^2(k) + ((2x_2(k) + e_2(k))e_2(k))^2 \\ &\quad + 2b(2x_2(k) + e_2(k))e_2(k)e_3(k) + e_2(k)e_1(k) \\ &\leq -(\frac{3}{4} - \frac{1}{2}\rho_2)e_1^2(k) - (1 - b^2 - \rho_1)e_3^2(k) + \frac{1}{2\rho_2}e_2(k))^2 \\ &\quad + [e_2(k)(2x_2(k) + e_2(k))]^2 + \frac{1}{\rho_1}[be_2(k)(2x_2(k) + e_2(k))]^2 \end{aligned} \qquad (19)$$

Where ρ_1 and ρ_2 are positive constants that can be chosen freely. Because the master system is chaotic, its states $x_i(k)$, i = 1, 2, 3 are bounded by some constant $\beta_1 > 0$. Taking in account that $e_2(k)$ is bounded by $|e_2(k)| < |e_2(0)|$, so there exists a constant $\beta_2 > 0$ such that

$$\left|\frac{1}{2\rho_2}(e_2(k))^2 + [e_2(k)(2x_2(k) + e_2(k))]^2 + \frac{1}{\rho_1}[be_2(k)(2x_2(k) + e_2(k))]^2\right| < \beta_2$$

So we can get

$$\Delta V_2(e_1(k), e_3(k)) < -(\frac{3}{4} - \frac{1}{2}\rho_2)e_1^2(k) - (1 - b^2 - \rho_1)e_3^2(k) + \beta_2 \quad (20)$$

From Theorem 1 we know that the solutions of the error system (12) are bounded. Here completes the proof of Step 2.

From lemma 1 we know the error system (12) is globally asymptotically stable, so the master system (10) and the slave system (11) realize chaos synchronization via the linear controller $u_1(k) = -e_1(k) + \frac{1}{2}e_2(k), u_2(k) = \frac{1}{2}e_1(k)$.

4   Image Encryption Design

4.1   Secure Communication

On the sending part, the signal which needs to be transmitted is coupled with the chaotic signal. Synthesized signal will be sent out through the public channel. On the receiving part, there is a chaotic system which is similar with the sending part. The controller drives the local chaotic system to be synchronized with the chaotic system in the sending part. Then the reconstructed signal can be decoupled from the receiving composite signal. Thereby, the most primitive information is obtained [11]. Principle is shown in Fig.1.

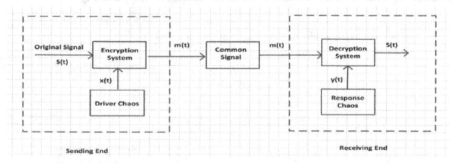

Fig. 1. Secure communication

4.2   Image Encryption

This algorithm is based on the discrete chaotic synchronization method. On the sending part, we sort the chaotic sequence generated by the driving chaos firstly. In the generation process of keys, we consider and remove the front chaotic sequences that have not been synchronized. The appropriate sequences of synchronized are selected as the keys; and the keys are also known as the address conversion codes. Thanks to the irregular sort, image can be encrypted. For example, this algorithm uses the size of M×N RGB image, and gives the initial values X_1, X_2, X_3. Encryption steps are as follows:

Step 1: According to the driving equation (10), after several iterations, take the current values as the initial values of key.

Step 2: Make M iterations and N×8 iterations to generate two dimensional chaotic sequences.

Step 3: According to ascending order, it sorts the chaotic sequences to get the address conversion codes P_{3xm} and Q_{3xnx8}.

Step 4: According to P_{3xm}, make line conversion on RGB components of the image.

Step 5: The RGB components of the image data are converted into 8-bits.

Step 6: According to Q_{3xnx8}, make column conversion on RGB components of the image.

After these steps, the image can be encrypted.

Decryption algorithm is the reverse process of encryption algorithm. According to the response equation (11), in the case of different initial values, it can achieve chaotic synchronization after several iterations. Then we can get the same address conversion codes, which can help to decrypt the image.

4.3   Image Encryption

4.3.1   Synchronous Simulation

In the simulation process, the parameters are chosen as $a = 1.76, b = 0.1$ the initial values of the driving system are $(x_1(0), x_2(0), x_3(0)) = (0.1, 1.5, 2.1)$ the initial values of the response system are $(y_1(0), y_2(0), y_3(0)) = (1, -10, 5)$. According to the above algorithm, the synchronic simulation results are shown in Fig.2.

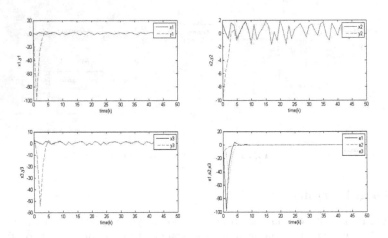

Fig. 2. Synchronic simulation

From this, we can see it only takes 5 seconds to realize the chaotic synchronization because of the simple controller design. But generally speaking it needs to spend more time in other algorithms [3-5].

4.3.2 Encryption Simulation

In the encryption process, this paper selects 128×128 lean image (Fig.3 Original image). The image is encrypted by the driving chaotic system (Fig.3 Encrypted image). Then according to the synchronization of chaotic response, the image can be decrypted (Fig.3 Decrypted image).

Original image | Encrypted image   Decrypted image

Fig. 3. Image processing

Compared with the general chaos, just like Logistic chaos, three-dimensional chaotic system is more complex, and the space of keys is lager, and the chaotic sequence is more erratic and unpredictable[12]. Because of chaotic synchronization, the keys are more convenient to manage compared with the single secret key system. At the same time, the RGB components of the image data are converted into 8-bits.This step is very important. When bit positions are scrambled, each pixel is simultaneously scrambled and spread. It can not only modify the positions of the pixel, but also modify the values of the pixel.

Experimental results show that the effect of encryption, decryption is good. However, the simulation results of some other algorithms are shown in Fig.4. This algorithm meets the design requirements.

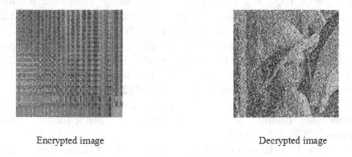

Encrypted image   Decrypted image

Fig.4. The simulation results of some other algorithms

5   Conclusion

In the information age, communication security is particularly important. Image as a common medium of life, it carries a lot of, or important, or privacy information. How to prevent the image leak becomes the original intention of many researchers. In this paper, a new improved chaotic image encryption algorithm is proposed. And compared with the one dimensional chaotic systems, three dimensional chaotic systems is more complex. Meanwhile, with the help of synchronization principle, both sides of the sending part and the receiving part use asymmetric secret key system to facilitate the secret key management, and the controller design is simpler. In the encryption process, through sorting and scrambling bit positions, we give full play to the randomness of chaotic sequence. Finally, the successful implementation of the image encryption and the decryption in practice exhibits a strong application prospect.

Acknowledgments. This research is supported by the National Nature Science Foundation of China (No.51175319), and key program of Shanghai Municipal Education Commission (No.13ZZ075), Shanghai University "11th Five-Year Plan" 211 Construction Project and Shanghai Key Laboratory of Power Station Automation Technology.

References

1. Ma, W.H., Gai, R.D.: Research on synchronization using parametric adaptive control. Computer Engineering and Design 29(19) (2008)
2. Pecora, L.M., Carroll, T.L.: Synchronization in chaotic systems. Physical Review Letters 64(8), 821–824 (1990)
3. Meng, J., Song, W.: Study on synchronization of Logistic mapping using M type nonlinear feedback control. Journal of Circuits and Systems 11(4), 82–84 (2006)
4. Xu, J.Y., Bu, J.R., Huang, F.: Study on digital image encryption method based on different structure chaos synchronization system. Mechanical Science and Technology for Aerospace Engineering 32(006), 899–903 (2013)
5. Liu, H.J., Zhu, Z.L., Yu, H.: Application of hybrid synchronization of the unified chaotic system in image encryption. Journal of Chinese Computer Systems 34(007), 1680–1684 (2013)
6. Zang, H.Y., Min, L.Q., Wu, C.X., Zhao, G.: An image encryption scheme based on generalized synchronization theorem for discrete chaos system. Journal of University of Science and Technology Beijing 29(1), 96–101 (2007)
7. Lu, D.X., Liao, X.F., Han, J., Li, M.: Picture scrambling algorithm based on Logistic mapping and sort transformation. Computer Technology and Development 17(12), 27–30 (2007)
8. Tuo, C.Y., Qin, Z., Li, Q.: Color image encryption algorithm based on 2D Logistic map and bit rearrange. Computer Science 40(8), 300–302 (2013)
9. Iggidr, A., Bensoubaya, M.: New results on the stability of discrete-time systems and applications to control problems. Journal of Mathematical Analysis and Applications 219(2), 392–414 (1998)
10. Baier, G., Klein, M.: Maximum hyperchaos in generalized Hénon maps. Physics Letters A 151(6), 281–284 (1990)
11. Zhang, H.M., Fan, J.L.: Research of secure communication based on chaotic synchronization. In: Proceedings of 2011 Asia-Pacific Youth Conference on Communication (APYCC 2011), vol. 1 (2011)
12. Nian, F.Z., Wang, X.Y., Li, M.: LSB encryption based on high-dimensional chaotic system. Computer Engineering and Design 31(4), 709–712 (2010)

Structural Shape Reconstruction through Modal Approach Using Strain Gages

Li Li, Wuqian Li, Pingan Ding, Xiaojin Zhu, and Wei Sun*

Department of Automation, College of Mechatronics Engineering and Automation,
Shanghai University, Shanghai Key Laboratory of Power Station Automation Technology,
Shanghai, China
elec_soft@shu.edu.cn

Abstract. It is significant of the strain detection and shape reconstruction of flexible structures for guaranteeing the safe and reliable operation of large-scale and precision equipment, such as spacecraft, space station, satellite, et al. This paper presents a structural shape reconstruction method based on the modal approach using the real-time sensing strain data. Firstly, the displacement-strain transformation relationship is derived using a modal approach and the finite element method is used as a numerical analysis method. From the implementation point of view, the united simulation is employed by the software of MATLAB and ANSYS. Moreover, a plate flexible structure is regarded as the research object and the simulation of shape reconstruction is investigated. The simulation results show that the structural shape reconstruction method based on the modal approach is a novel strategy. Furthermore, the experiments are performed with strain gages and the reconstructed displacements are compared to the measured displacement data from laser displacement sensor. The error analysis is demonstrated and it is indicated that the modal approach based structural shape reconstruction method is an effective approach for structural displacement monitoring of flexible structures.

Keywords: shape reconstruction, modal approach, finite element analysis, strain gages.

1 Introduction

The dynamic performance of requirements is commanded attention in modern flexible structures such as high large precision spacecraft wings [1] and space reflectors [2]. Especially in civil engineering, the structural deformation should be limited in a range to guarantee the safety and serviceability of the buildings [3]. Therefore, the active monitoring of the structural displacement has been becoming a hot topic and it is also a challenging issue. In practice, it is difficult to the direct measurement of displacement during operation.

For active monitoring of structural deformation, there are some indirect detection methods such as measuring acceleration data [4] and strain data [2, 3, 5]. Thus, the whole displacement field can be estimated through the measured data by a certain

M. Fei et al. (Eds.): LSMS/ICSEE 2014, Part II, CCIS 462, pp. 273–281, 2014.

transformation approach. Based on the relationship between strain and displacement, it is convenient of shape reconstruction through the strain data. There are some researches in different fields, such as mechanical engineering [1, 5] and civil engineering [3, 4]. However, the transformation from measured strain data to whole displacement filed is a very important issue. There are two main kinds of methods, one is based on the geometry curvature information and another one is built from the modal observation method. In the first kind of method, the geometry curvature acts as a bridge between structural strain and deformation, such as shape reconstruction based on B-spline fitting [5] and curve fitting using polynomial function [6]. Since some approximate treatments are needed to reconstruct the shape from curvature information, the error is also induced especially for the low precision algorithm. In addition, a certain number of sensors are required to improve the reconstruction accuracy. In order to solve these problems, a modal approach based structural shape reconstruction method [7] was employed. Then, the displacement field of a two-dimensional structure was estimated using the modal approach [8, 9], in which the fiber Bragg grating sensors were used to detect the discrete strain data. In this method, the numerical error of the transformation from strain to displacement is small and acceptable [7]. Moreover, the number of strain sensors is related to the interesting mode shapes, thus, it can provide enough accuracy for shape reconstruction with few numbers of strain sensors in static or lower frequency system.

Therefore, in this study, the modal approach based structural shape reconstruction method is employed for real-time monitoring structural deformation. In addition, the finite element method is implemented to solve the structural mode shapes effectively. In the experiment, strain gages are used in order to measure strains due to the guaranteed precision, easy operation and low cost.

The structure of this paper is as follows. We first derive the relationship between strain and displacement based on the structural mode shapes in Section 2. Then the implementation of structural shape reconstruction is described using the united simulation of MATLAB and ANSYS in Section 3. The simulation process is demonstrated in Section 4. The experiments are presented in Section 5 to illustrate the modal approach based structural shape reconstruction method. Finally, conclusions are made in Section 6.

2   Displacement-Strain Transformation Relationship

In this study, the structural displacement field is estimated using a modal approach, in which the global displacement field is reconstructed using several locally measured strains through the displacement-strain transformation relationship.

Generally, the displacement of a structure can be expressed by an infinite number of displacement mode shapes. Practically, the whole displacement {D} is calculated through the displacement mode shapes $[\phi_d]$ multiplying the modal coordinates {q}, i.e.,

$$\{D\}_{N\times 1} = [\phi_d]_{N\times n} \cdot \{q\}_{n\times 1}, \tag{1}$$

where N is the number of displacements, n is the number of modes used.

Similarly, the strain $\{S\}$ can be transformed through the strain mode shapes $[\phi_s]$ and modal coordinates $\{q\}$, which is expressed as [7],

$$\{S\}_{M\times 1} = [\phi_s]_{M\times n} \cdot \{q\}_{n\times 1}, \tag{2}$$

where M is the number of measured strain data.

In Equation (2), the modal coordinates can be approximated by way of least squares expressed as,

$$\{\hat{q}\}_{n\times 1} = \left([\phi_s]_{M\times n}^T \cdot [\phi_s]_{M\times n}\right)^{-1} \cdot [\phi_s]_{M\times n}^T \cdot \{S\}_{M\times 1}. \tag{3}$$

Equation (3) is back submitted to Equation (1), thus, the estimated displacement $\{\hat{D}\}$ can be obtained by the measured strain, and is given as,

$$\{\hat{D}\}_{N\times 1} = [\phi_d]_{N\times n} \cdot \left([\phi_s]_{M\times n}^T \cdot [\phi_s]_{M\times n}\right)^{-1} \cdot [\phi_s]_{M\times n}^T \cdot \{S\}_{M\times 1}. \tag{4}$$

In a given system, the estimated displacement $\{\hat{D}\}$ can be provided by the measured strains $\{S\}$ through the displacement-strain transformation relationship. For convenience, this transformation matrix is called DST matrix as expressed in the following way,

$$[DST]_{N\times M} = [\phi_d]_{N\times n} \cdot \left([\phi_s]_{M\times n}^T \cdot [\phi_s]_{M\times n}\right)^{-1} \cdot [\phi_s]_{M\times n}^T. \tag{5}$$

However, it is noted that the maximum rank of the [DST] matrix is equal to M. In this way, the adequate accuracy is expected of estimating the displacement, and the enough strain data is needed. In the experiment, the strain sensors are should be guaranteed. Therefore, M is greater than or equal ton, i.e.,

$$M \geq n. \tag{6}$$

3   Implementation Combining MATLAB and ANSYS

To implement the transformation from measured strain data to the displacement, the finite element method is employed in the presented modal approach to calculate the displacement mode shapes and strain mode shapes. The numerical calculation is programed by the software of MATLAB and ANSYS.

Firstly, the platform is built in the MATLAB due to its powerfully computational performance. It is including the definition of structure, boundary condition, material properties, load case, images displaying. However, the analysis of finite element method calls the commercial software of ANSYS using the system function of MATLAB. It is a good choice for using the ANSYS as a tool of finite element analysis due to its powerful analysis capability. The function can be described in the following way,

```
system('D:\Program\ANSYSInc\v140\ansys\bin\intel\ANSYS1
40.exe -b -p ansys -product -feature - i input file -o out
file.bat'),
```

where b denotes batch mode, p specifies product such as ANSYS/Mechanical, i is the input file and o is the output file.

Secondly, the finite element analysis is carried out by the ANSYS Parametric Design Language (APDL) which is a scripting language to execute the common tasks automatically. Therefore, it can be used in united simulation and optimization problem related to the finite element analysis. In this study, the APDL is employed and the main operation includes the input data (Read), analysis process, and the output command (Write). For more details, the reader can refer to the User Guide such as the ADPL Guide of ANSYS [10] and other related literatures [11].

Thirdly, the result of analysis is transformed from ANSYS to MATLAB, which includes the output operation from ANSYS and the input process of MATLAB. It is needed to pay attention to the data format and file transfer between the two softwares.

4   Simulation

4.1   Simulation Modal

In this study, the investigation is a cantilever plate with dimensions of $0.8m \times 0.4m \times 0.0015m$. The material of this plate is epoxy resin with Young's modulus as $2.0 \times 10^{10} N/m^2$, Poisson ration as 0.16 and the density as $1730 Kg/m^3$. The left side is fixed and the uniform force is applied in the vertical direction of the structural thickness. The finite element model is established in ANSYS and the element type of SHELL181 is used with the thickness of $1.5 \times 10^{-3} m$. In the finite element analysis model, the nodes of 3321 are employed to discrete the structure of cantilever plate.

4.2   The Optimization of Sensor Placements

It can be seen from Equation (5) that the DST matrix is composed of $[\phi_d]_{N \times n}$ and $[\phi_s]_{M \times n}$. To calculate the displacement and strain transformation matrix, modal analysis for the finite element model is performed. The displacement mode shapes can be obtained directly by using ANSYS. But the strain mode shapes matrix is dependent on sensor placement, sensor orientation, the number of sensors, which means the estimated accuracy is closely related to strain measurement nodes. Therefore, the optimization of strain measurement nodes is very important to the displacement reconstruction of structures.

The objective of optimization is to minimize the estimated error. According to Rapp et al [9] and Li et al [12], the condition number CN of the DST matrix can be used as an indicator for the optimization of strain measurement nodes. Generally, the smaller the condition number, the better the estimated results. It is defined as,

$$CN = \|DST\| \cdot \|DST^{-1}\|, \tag{7}$$

where $\|DST\|$ is the norm of DST matrix.

In this project, 32 points are defined to install strain sensors, and two sensors are orthogonal at each point. Fig. 1 shows the two distribution schemes of sensors. Fig. 1

(a) is array distribution of sensors based on prior knowledge and the property of strain field. Moreover, the condition number CN of the transformation matrix for this distribution is relatively small. Meanwhile, the optimized distribution of strain sensors is achieved by a simulated annealing algorithm as shown in Fig.1 (b).

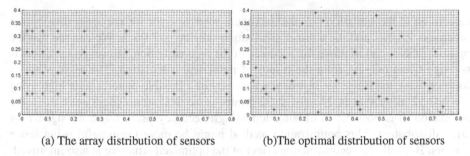

(a) The array distribution of sensors   (b)The optimal distribution of sensors

Fig. 1. The distribution schemes of strain sensors

4.3   Simulation Results

Since the simple static deformation is investigated in the simulation, it is enough to estimate the displacement using the first 12 mode shapes. The finite element model has been established in Section 4.1. When the uniform load of 0.8N is applied to the free end of the structure, the reconstruction results are shown in Fig. 2 with the two distribution schemes of sensors as displayed in Fig. 1.

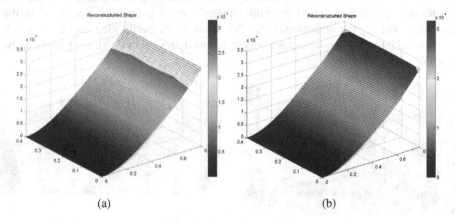

(a)   (b)

Fig. 2. Simulation results

In Fig. 2, picture (a) is the shape reconstruction result for the array distribution scheme and its maximum displacement is $3.1466 \times 10^{-4}m$. Picture (b) is the shape reconstruction result for the optimal distribution scheme and the maximum displacement is $3.1507 \times 10^{-4}m$. In order to contrast the accuracy between the two schemes, the relative maximum Error is defined as Equation (8). $D_{e(i)}$ represents the estimated displacement at every specific point and $D_{a(i)}$ represents the real displacement in ANSYS at the same point.

$$Error = max\left(\frac{D_{e(i)} - D_{a(i)}}{D_{a(i)}}\right) \times 100\%. \tag{8}$$

According to the Equation (8), the maximum relative error for the array distribution scheme is 2.14%, and the optimal distribution scheme is 2.05%. We can see that the optimal scheme can improve the accuracy of the reconstruction. However, the effect is not significant judging from the results of numerical calculation. The reasons are as follows. First, the shape reconstruction accuracy of a flexible plate is quite high using the first 12 mode shapes. Moreover, 32 groups of vertical strain gages are used in the simulation. It is possible to obtain enough high accuracy using 64 strain signals for array distribution. This optimization method might be more meaningful using fewer sensors. But still, the reconstruction effect of the optimized structure is also improved.

5   Experiment

5.1   Experimental Platform

This experimental platform is shown in Fig. 3 (a), which includes epoxy resin plate, strain gage set, strain test system DH3817, signal test and analysis software system DHDAS, etc. In this experiment, 64 strain gages are needed to detect the strain data, while one DH3817 only has eight channels and its price is quite expensive. Therefore, the monitoring platform based on multi-channel switch controller is adopted in the experiment. During the operation, 8 parallel strain gages are divided into a set, which is controlled by one channel using 8 electronic switches, then 64 strain gages can be detected by one DH3817.

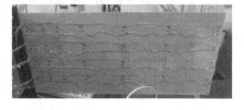

(a) Experimental platform   (b)The installation of strain gages

Fig. 3. Experimental platform with array distribution of sensors based on multi-channel switch controller

5.2   The Installation of Strain Gages

Resistance strain gage, short for strain gage, used to detect strain data, is made up of sensitive gate. It can convert the change of strain into the change of resistance value. Dong [13] presented the relationship between strain and resistance value as

$$\frac{dR}{R} = K\varepsilon ,$$ (9)

where K is the sensitivity coefficient of strain gage.

It can be seen that there is a linear relation between the strain and the relative variation of resistance in Equation (9). Thus, the strain-resistance transformation can be obtained and strain gages of BX120-4AA are used in our experiment.

For the experimental maneuverability, the scheme of array distribution of sensors is employed in this experiment. Since this experiment is mainly used to verify the effectiveness of displacement reconstruction based on modal approach, the array distribution scheme is enough to meet the accuracy requirement of the study. In order to detect the strain information of one point in the horizontal direction and vertical direction at the same time, two orthogonal strain gages are bounded on the both sides at the same placement of the plate. According to the knowledge of elastic mechanics, the strain information of two sides of a plate is symmetrical, thus, the strain values of the two sides at same point are the same. The installation of strain gages are shown in Fig. 3 (b).

5.3   Experiment Results

Pure bending deformation of the epoxy resin plate is evaluated in this experiment. The pure bending deformation experiments include 0.1m pure bending deformation and 0.15m pure bending deformation defined Case 1 and Case 2 respectively. It is noticed that the 0.1m pure bending deformation means that the maximum displacement of the cantilever plate is 0.1m with the pure bending load case. And 3 points are selected at the right side defined Point 1, Point 2 and Point 3, which are at the highest point of the right side, the middle point of the right side and the lowest point of the right side respectively. The displacement calculated by using the modal method is called estimated displacement and the displacement measured by laser displacement sensor is called measured displacement, which are used as the reference. In order to compare the estimated displacements $D_{estimated}$ to the measured displacements $D_{measured}$, the Relative error is defined as,

$$\text{Relative error} = \frac{D_{estimated} - D_{measured}}{D_{measured}} \times 100\% .$$ (10)

The experimental results are listed in the Table 1. For Case 1, the maximum error between the estimated displacement and the measured displacement is 7.63%, and the maximum error for Case 2 is 4.23%. However, the errors for the displacement estimation using strain gages are acceptable. The errors might be caused by the follow reasons. Firstly, although the thickness of the strain gage is quite small, the effect of the strain gage thickness is needed to be evaluated. Secondly, the strain coefficient of strain gages will change a little after solidification of the glue used in the experiment. Thirdly, the mechanical properties of the structure are influenced due to the bounding of the strain gages to the surface of the plate. For the estimation algorithm based on modal approach, the finite element method has a lot of matrix operations, and the deformation of some matrix will inevitably introduce errors. Therefore, reasonable mesh and appropriate element type are very important to the displacement estimation. Moreover, since the mechanical properties between the actual plate and the simulation plate are not exactly the same, the DST matrix calculated by ANSYS is not very accurate for the actual plate. We also can see that the accuracy of Case 2 is higher than the Case 1. This is can be explained as follows, the strain information is easy to measure under the case of larger deformation. In addition, for the same case, it is more accuracy for laser displacement sensor to measure the displacement.

Table 1. The experiment results

Test case	Test point	Coordinate (mm)			Measured coordinate (mm)				Estimated displacement (mm)	Relative error
		X	Y	Z	X	Y	Z	displacement		
	Point 1	800	-180	0	780	-180	99.85	101.83	94.15	-7.54%
Case 1	Point 2	800	0	0	780	0	99.85	101.83	94.06	-7.63%
	Point 3	800	180	0	780	180	99.85	101.83	94.20	-7.49%
	Point 1	800	-180	0	780	-180	147.5	148.85	142.55	-4.23%
Case 2	Point 2	800	0	0	780	0	147.5	148.85	142.72	-4.12%
	Point 3	800	180	0	780	180	147.5	148.85	142.60	-4.20%

6   Conclusion

In this study, the structural shape reconstruction method based on the modal approach is investigated with a flexible plate structure. The displacement and strain transformation relationship is built to reconstruct the whole displacement field through the measured strain data. In addition, the finite element analysis is employed for calculating the displacement mode shapes and strain mode shapes. From the implementation view of point, the united simulation is used of MATLAB and ANSYS.

In the simulation process, a sensor location scheme of an array distribution is performed as a comparison case, and the maximum error of displacement is 2.14%. However, the optimal distribution of sensors has a better performance, and the maximum error of displacement is 2.05%. These results show that it is a great way to reconstruct structural shape with a modal approach. Meanwhile, the advantage of this optimization scheme is less obvious and there are two possible reasons. Firstly, the array distribution of sensors is based on the characters of strain field and the strain information can be detected effectively with less error. Secondly, this optimization scheme reveals a great performance with using fewer sensors. Therefore, the optimization scheme of sensor location is still a challenge research issue for future study.

To verify the proposed shape reconstruction method, the experiments are performed. The two load cases of pure bending are applied and the maximum error is 7.63%. The error is considerably and the error might be caused by the experiment operation, the accuracy of measurement using the strain gages, bounding skills of strain gages, and reconstruction algorithm itself. However, the shape reconstruction method is an effective approach and it is technically feasible in the practical applications. To achieve more accurate strain measurements, the Fiber Bragg Grating (FBG) sensors can be adopted to detect the strain information in future research.

References

1. Priou, A.: Exploratory Team on Smart Structures and MEMS of NATO, Panel 3, AC 243 RSG. In: Proceedings of Smart Materials and Structures, San Diego, CA, pp. 12.1–12.6 (1996)
2. Albert, B., George, C.K., David, G.B., Alan, D.K., Tsai, T.E., Friebele, E.J.: Strain-displacement Mapping for a Precision Truss using Bragg-gratings and Identified ModeShapes. In: Proceedings of SPIE 2717, Smart Structures and Materials, San Diego, CA, pp. 341–349 (1996)
3. Xia, Y., Zhang, P., Ni, Y.Q., Zhu, H.P.: Deformation Monitoring of a Super-tall Structure using Real-time Strain Data. Engineering Structures 69, 29–38 (2014)
4. Chan, W.S., Xu, Y.L., Ding, X.L., Dai, W.J.: An Integrated GPS – accelerometer Data Processing Technique for Structural Deformation Monitoring. J. Geodesy 80, 705–719 (2006)
5. Zhu, X.J., Jiang, L.N., Sun, B., Zhang, H.S., Yi, J.C.: Shape Reconstruction of FBG Intelligent Flexible StructureBased on B-spline Fitting. Optics and Precision Engineering 19, 1627–1634 (2011)
6. Payo, I., Feliu, V., Cortazar, O.D.: Fibre Bragg Grating (FBG) Sensor System for Highly Flexible Single-link Robots. Sensors and Actuators A: Physical 150, 24–39 (2009)
7. Foss, G.C., Haugse, E.D.: Using Modal Test Results to Develop Strain to Displacement Transformation. In: Proceedings of the 13th International Modal Analysis Conference, Nashville, Tennessee, pp. 112–118 (1995)
8. Kang, L.H., Kim, D.K., Han, J.H.: Estimation of Dynamic Structural Displacements using Fiber Bragg Grating Strains Sensors. Journal of Sound and Vibration 305, 534–542 (2007)
9. Rapp, S., Kang, L.H., Han, J.H., Mueller, U.C., Baier, H.: Displacement Field Estimation for a Two-dimensional Structure using Fiber Bragg Grating Sensors. Smart Materials and Structures 18, 025006–025017 (2009)
10. ANSYS Parametric Design Language Guide, http://www.ansys.com
11. Song, H.W., Liu, H.: Optimum Structural Design Based on MATLAB and ANSYS. Journal of Dalian Nationalities University 13, 284–287 (2011)
12. Li, C.J., Ulsoy, A.G.: High-precision measurement of tool-tip displacement using strain gauges in precision flexible line boring. Mechanical Systems and Signal Processing 13, 531–546 (1999)
13. Dong, W.: On sticking skills for resistance strain chip. Shanxi Architecture 37, 46–48 (2011)

A Fast Colorization Algorithm for Infrared Video

Mengchi He, Xiaojing Gu, and Xingsheng Gu

Key Laboratory of Advanced Control and Optimization for Chemical Processes,
Ministry of Education, East China University of Science and Technology,
Shanghai 200237, China
xsgu@ecust.edu.cn

Abstract. Color night-vision technology increases the representation ability of monochrome night-vision imagery by adding color to it, making observers' understanding easier. Usually the color night-vision methods require the infrared and the low-light-level images at the same time, which hinders their application in the environment where totally without light or covered by heavy rain and thick fogs. To expand the application area of color night-vision technology, we propose a quickly colorization method based only on single band infrared video, which can provide all weather condition working. This method only requires a few pixels to be manually set with chrome values, and then the entire frame as well as the following frame sequence is automatically colorized. Experiments show that the colorization results are satisfactory and the algorithm is running fast.

Keywords: infrared thermal image, color night vision, color fusion, fast algorithm.

1 Introduction

Night-vision sensors increase the observing ability of human in night time, helping us to see the scenery around when there is no light. But standard night-vision images are monochrome and human's observing perception is much sensitive to color images comparing to monochrome ones. So the recent developed color night-vision technique highly improves observer's performance on scene understanding and reaction time by displaying the monochrome night-vision image in colors. It is useful in security, antiriot, military and detection fields.

The key problem of colorization is to find features in the monochrome image that can help us to get its chromatic value, this is more difficult in night-vision area because night-vision images are lack of textures and have low contrast, current night-vision technologies can mainly divided into two categories: colorize by single band night-vision image or colorize by pseudo-color image fused by night-vision images with different bands.

To give the pixels in a monochrome image with chromatic values, Welsh[1] invents a method that use a chromatic image as reference to transfer its color to the target image. This method finds every pixel's matching pixel in the reference image and

M. Fei et al. (Eds.): LSMS/ICSEE 2014, Part II, CCIS 462, pp. 282–292, 2014.

transfers the color between them, although the result of this method is close to the reality[2-6], it can't used for night-vision images because their features are not strong enough to find the correct matching. To make up this defect, Toet[7,8] use images in different bands to get more features, he first fuses a low-light-level image and an infrared image into a pseudo-color image, then use Reinhard's[9] method to transfer a reference image's statistical properties to it and build a look-up table to colorize other images automatically. Since multi-band features can provide more information than single-band features, this method can colorize night-vision image correctly. To get more realistic results, Toet uses a natural light image with identical scene to build the table. This method uses image fusion to get useful features in the night, there are many other similar methods in color night-vision area[10-22]. Waxman[23] and Toet[24] also make some real-time equipments to colorize night-vision images.

Although the methods above can colorize night-vision images quickly and have desirable results, their demand of the equipment is high, since they need multi-band images as input. Low-light-level equipments will lose their efficacy when there is totally without visible lights or under cloudy and fog weathers. To decrease the requirements of the equipments, some scholars propose colorization methods based on single-band infrared imagery. Gu[25] proposes a kernel based method to train a model that can predict the color of the image, Haman[26] finds the best reference image from the database to colorize a single band infrared image based on its texture information. These methods all use machine learning to train the colorization models; they are usually computational expensive and time consuming.

In this paper, we propose a method to colorize the single-band infrared image/video based on color seeds technique [27-35]. This method first sets a few color seeds on one frame manually and then propagate the colors to the whole image/video sequence automatically. Some accelerating strategy is employed along to ensure the implement speed. Compared to other methods, our method can provide a coloring infrared video display in desirable time.

2   Fast Colorization for Infrared Video

The procedure of the colorization process is as follows. After we enhancing the luminance contrast, we first assign the chromatic values for a set of pixels (seeds) in a key frame and give each pixel chromatic values by blending the colors of all seeds according to their grayscale distances to this pixel. Then we extract the next frame of the video, if the contexts in these two frames change a lot, we color this frame by redefine some seeds in it, otherwise the algorithm transfer the color from the previous frame to it by finding matching pixels based on some features suitable for infrared videos.

Before the colorization operation, we first change the color space of the image from RGB to YCbCr, where Y is the luminance channel; Cb and Cr are the deviation of blue and red. We use this color space because its three channels are independent and we don't want to change the grayscale of a pixel when we change its color. Then we set some seeds in the image, the color of the whole image will decided by these seeds.

2.1   Key Frame Colorization

The colorization method we use is based on an assumption: the colors of adjacent pixels with similar grayscales are similar. This assumption is always tenable in real images. Before we fuse the color of each pixel, we need to know its distance relation to all seeds, the seed closer to it is more important in fusing. The distance relation here is the sum of the grayscale changing on the shortest path between them which is called grayscale distance.

So for each seed, we need to compute its grayscale distance to every pixel in the image. We consider a monochrome image as a graph; each pixel represents a node and links to four adjacent pixels, the length of the path is their grayscale difference, so we can use Dijkstra[15] algorithm to compute the distance from a seed to all other pixels and thus we can obtain a pixel's grayscale distance to all seeds after we run the algorithm to all seeds. Then we can fuse its color according to these distances and the color of seeds. The fusing strategy of a pixel p is shown in Equation (1):

$$C(p) = \frac{\sum_{s \in seeds} W(D(s.p)) * C(s)}{\sum_{s \in seeds} W(D(s.p))} \tag{1}$$

Where $W(D(s.p))$ is the weight of seed s, it should be big when the grayscale distance between them is small, in our experiment we use $D(s.p)^{-6}$ so the weight is ∞ when the grayscale distance is zero and zero when the distance is ∞, the size of the index is not important, it just has little impact on the result. Since the grayscale distances between a seed and two adjacent same grayscale pixels are same, their color obtained by Equation (1) is also the same, this conforms to the assumption we complied with.

2.2   Optimization for Key Frame Colorization

According to the analysis above, the time spent by the algorithm are mainly used for computing the distance between the seeds and other pixels. For each seed we need to iterate n times to calculate its distance to all pixels if there are n pixels in the image, and we need to find the pixel with smallest distance from all pixels in each iteration, so the time complexity of a seed is $O(n^2)$ and it will be very slow when we process large images. We can use priority queue to store the data during the iteration to speed up, the priority of a pixel is its grayscale distance from the seed and the priority is high when the distance is small.

Ordinary priority queue uses binary tree structure, the element in it stores according to its priority, we take the first element each time and push the new element into the position corresponding to its priority. The time complexity of the pushing operation is $O(\log(n))$, so the overall time complexity is $O(n\log(n))$.

To make the running speed faster, we refer to the untidy priority queue [36] and make improvements on its storage format for our purpose. The time complexities of push and pop operation of the untidy priority queue are both $O(1)$, so the time complexity of the image is $O(n)$, the running speed is significantly improved. Untidy priority queue cuts the entire queue into many small queues, each queue contains a specific range of priority and the ranges between adjacent queues are linked together. The elements in each queue all have same priorities. When we get a new element, we decide which queue it belongs to and push it to the end of this queue and we pop the first element in the first queue when we need the element with the highest priority. When the first queue is empty, we update its priority range and make it to be the last queue.

This method distributes the priority range of each queue flexibly, uses least queues to finish the job. Since the of each queue's priority range is small, the sequence of the elements in untidy priority queue and ordinary priority queue are similar and the error in the result usually can't detected by naked eyes. But the size of each queue's range is difficult to choose, we use Δ to denote the priority range of each queue, if Δ is too small, the total range of the queue will also be small, when we go to the edge of an object, the grayscale differences of these pixels may be larger than this range, so we can but ignore these pixels. If Δ is too big, more pixels with different priorities will be pushed into same queue and this will increase the error of the result.

So we change the structure of the untidy priority queue here, maintain small result error and large priority range at same time. We initialize many small queues and fix the priority ranges of them, push the pixel to the end of the corresponding queue and pop first pixel in the first queue which is not empty. In this way we need to set the max range of the queue, we consider that there is no relation between two pixels if the grayscale distance between them is larger than a threshold, since the range of the grayscale is 0~255, 255 is the biggest difference between two pixels, for convenience we think that two pixels will have no relation if their grayscale difference is more than 300, this means that when we are dealing with a seed pixel, we ignore the pixel which grayscale distance from it is more than 300. We set 30 priority queues for each contains the priority range of 10; put the pixel which grayscale distance belongs to 0~9 into the first queue; 10~19 into the second queue and pop the first pixel in the first queue which is not empty when we need the pixel with smallest grayscale distance, this method avoids the problem of choosing the range for each queue, we can finish the work when all the queues are empty, thus the time complexity reduces from $O(n^2)$ to $O(n)$ and the result is satisfactory.

2.3 Video Sequence Colorization

After we colorize the key frame of the video, we use it as the reference image to make the following frames into chromatic.

The existing colorization methods based on reference images are usually finding each pixel's corresponding pixel in the reference image by its feature. The original method to transfer a monochrome image into chromatic is setting colors for each grayscale level manually, this is the simplest method that uses a pixel's feature to find its corresponding color and the feature using here is the grayscale, but since a grayscale level may corresponds to several colors in an image, we can't find the accurate color by this method. Other scholar's methods are all adding other features based on this foundation, such as Welsh's texture feature and Toet's grayscale of two images.

Welsh's method pays more attention on the grayscale when the texture is indistinct; this will cause errors when we are handling an infrared image because the texture information is very weak and the temperature of different objects might be similar. So these methods just reduce the probability that a single feature corresponds to more than one color and will not find exact color for every pixel when the image is complex or the information in reference image is incomplete.

$$\Delta(p_1 \cdot p_2) = W_t \times (T(p_1) - T(p_2))^2 + W_g \times (G(p_1) - G(p_2))^2 + W_l \times (P(p_1) - P(p_2))^2 \qquad (2)$$

To further reduce the probability that the feature of a pixel corresponds to multiple colors, we add the position of a pixel as its feature here, since the movement of an object in two successive frames is small, the change of its position will be very small and its feature will be similar in adjacent frames. So a pixel will find its matching pixel only in a small area in the previous frame first and find farther pixels if there is no similar pixel nearby, this method reduces the searching area of a pixel and thereby reduce the probability that a feature corresponding to more than one chromatic values. The features of a pixel we use is shown in Fig.1, we use grayscale level, position, texture and the texture information here are nine Laws' masks of a pixel. The difference between two pixels is shown in Equation (2), where W_t, W_g and W_l are the weights of texture, grayscale and position. Different videos can use different weights; the weight of grayscale should be bigger if the grayscales of different objects are very different in the image; the weight of texture should be bigger if the texture is distinct and we can increase the weight of position if the movements of the objects in the video are small. We find the matching pixel of every pixels in the monochrome image from the reference image, set color of the pixel with minimize result of Equation (2) to the pixel we are colorizing.

Grayscale/Temperature

Nine Laws' masks

Ordinate

Fig. 1. Features we used for a pixel

2.4   Optimization for Video Sequence Colorization

If we compare the pixel we want to color to all the pixels in the reference image, the time spend for colorizing a frame will be very huge, so similar to Welsh's method, we set 100 pixels symmetrically in the reference image to be the reference pixels and find the matching pixel between them, the number of the reference pixels can be increased if the content of the video is complex.

Although this method can significantly improve the colorization speed, the time spent for colorizing a frame is still about 3 seconds and this can't meet the practical requirements. So we need other methods to increase the speed because the colorization result will be bad if we decrease the number of reference pixels. Since movements of the objects in a video are usually continuous and not very huge, we can consider that a pixel's color doesn't change if its grayscale and texture are identical in two successive frames. So we first compare the grayscale and texture of a pixel with the pixel at same position in the previous frame when we colorize it, considering the noise of the video, we think a pixel's color doesn't change if the grayscale and texture difference is less than a threshold and transfer its color directly. So we save the time of comparing to reference pixels for most pixels and make the colorization process faster. The time for coloring a single frame using this method is about 200~500 ms, it has great improvement compared with previous.

The colorization results of the following frames inherit the previous frames well and the color of the output video is satisfactory, yet some pixel will have error matching if we don't use position as a feature, the result is shown in Fig.10.

3   Experimental Results

We use images and videos collected by Flir's infrared thermal imager to test our method.

Fig. 2. Comparison of two colorization methods

Table 1. Processing time of two methods

Ordinary	54.22s	78.01s	77.23s	64.55s	86.16s	56.5s	59.3s
Untidy priority queue	0.79s	0.76s	0.82s	0.76s	1.23s	0.74s	0.83s

Fig.2 is the comparison of colorization results whether using untidy priority queue or not. Left side are the monochrome infrared images with seeds, middle are the results without untidy priority queue and right side are the results using it. We can see that the output images by these two methods are similar; this confirms that the error is small when we use untidy priority queue. Table.1 is the time spent by these two methods; we can find out that untidy priority queue makes the colorization process faster and almost with no error.

Table 2. Time spent by different methods

Image size	Ordinary method	Untidy priority queue	Our method
160*100	452ms	32ms	32ms
200*200	2699ms	62ms	62ms
361*233	11793ms	234ms	109ms
320*240	9906ms	176ms	125ms
400*262	18283ms	390ms	156ms

Table.2 is the time spent by these two methods for a single seed in images with different size, we can find out that the cost of untidy priority queue increases linearly with the size of the image and the cost of the original method increases exponentially, so the time is unacceptable when the image is huge. The modified untidy priority queue used in our method can also save more time by add the threshold to ignore the pixels far away from the seed, so we needn't to iterate all pixels when the image is huge. But if there are only few seeds, some pixels will have no color when we use this method; we can use the mean of the pixels with similar grayscale at this time.

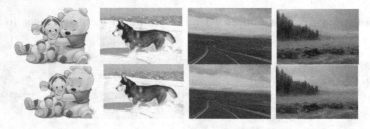

Fig. 3. Colorization result of normal pictures

Fig.3 is the colorization result of normal images, set different seeds for an image can receive different results, as shown in Fig.4, so this method can also used to color monochrome photos, movies or cartoons.

Fig. 4. Result of setting different seed pixels

Fig.5 is the colorization results of a night-vision video weather use position as a pixel's feature or not. The result images are selected from 50 following frames. Fig.5(a) is the result that only uses grayscale and texture as the feature of a pixel. We can see that this method pays more attention to grayscale when the texture information is weak and transfer the colors from another part of the image. Fig.5(b) is the result of our method that add position as one feature, it's effect is better.

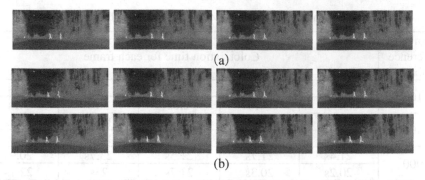

(a)

(b)

Fig. 5. Video colorization result by two methods

Fig.6 is the result that using our colorization method to other videos, the result is desirable when we use it on infrared videos, but there are some error matching when the grayscales and the textures are similar in adjacent objects. The reason is that we take only 100 pixels to be the reference pixel and this makes it easy to match the nearby pixel with small grayscale and texture difference. To reduce this error we can set more reference pixels, but this will slow down the color speed. It takes about 2.5~3s to color a frame which size is 320*240 when there are 100 reference pixels, 10s when there are 200 reference pixels and more than 20s when there are 900 reference pixels, so we usually choose reference pixels less than 200 and 100 is the most effective number. The operation times of different size of reference pixels are shown in Table.3; the size of the video is 320*240 and we find every pixel's matching pixel from the reference pixels. Table.4 is the result that using a threshold to decide whether a pixel changes in adjacent frames to color the video, since this method transfers the color for the pixel which grayscale and texture changes small directly and the feature for matching is position, grayscale and texture, so the result is same to that finding the matching pixel and better when there are few reference pixels, because the distances between reference pixels are large this time, when the object doesn't move a pixel may have difference with the nearest reference pixel and can't find its accurate color, but use the color at the same position in previous frame will not have this problem.

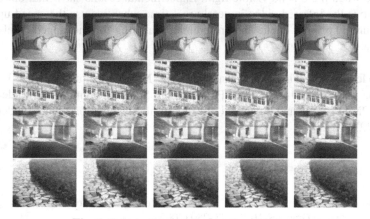

Fig. 6. Colorization results for other videos

Table 3. Operation time for different size of reference pixel for a single frame

Numberof reference pixels	Colorization time for each frame				
100	2.8s	2.6s	2.6s	2.9s	2.7s
	3.5s	3.1s	2.6s	3.6s	3.2s
400	10.6s	9.9s	10.3	10.2s	11.5s
	10.6s	9.9s	10s	10.8s	10.4s
900	21.4s	21.7s	21.5s	23.7s	20.9s
	20.2s	20.3s	21.7s	23s	22.1s

Table 4. Time using comparison for whether using threshold or not

Colorization method	Colorization time for each frame				
Find the matching pixel for each pixel	2.8s	2.6s	2.6s	2.9s	2.7s
	3.5s	3.1s	2.6s	3.6s	3.2s
Find unchanged pixels first	764ms	484ms	328ms	281ms	500ms
	437ms	484ms	468ms	655ms	592ms

4 Conclusion

This article proposes a fast colorization algorithm that reconstructs the color distribution of infrared video. We first assign the chromatic values for a set of pixels (seeds) to colorize a key frame image and then use this image as reference to colorize the following frames in the video sequence. This algorithm can provide the infrared video with a satisfied colorization result and its implementation is very fast. Compared to other color night-vision methods, we use only single channel infrared videos as input and avoid the training procedure; this makes our method very easy to achieve and available when there is no visible light. Experimental results show that our algorithm can colorize the infrared video in nearly real time when we use the speed-up strategy such as untidy priority queue and preferential matching for unchanged pixels in video sequence.

Acknowledgments. This paper is supported by National Natural Science Foundation of China (No.61205017), China Postdoctoral Science Foundation funded project (No.2012M511058), and Shanghai Postdoctoral Sustentation Fund funded project (No.12R21412500).

References

1. Welsh, T., Ashikhmin, M., Mueller, K.: Transferring color to grayscale images. ACM Transactions on Graphics 21(3), 277–280 (2002)

2. Hertzmann, A., Jacobs, C.E., Oliver, N.: Image analogies. In: Proceedings of the 28th Annual Conference on Computer Graphics and Interactive Techniques, pp. 327–340 (2001)
3. Gauge, C., Sasi, S.: Automated Colorization of Grayscale Images Using Texture Descriptors. ACEEE Int. J. on Information Technology 01(01) (2011)
4. Irony, R., Cohen-Or, D., Lischinski, D.: Colorization by Example. In: Eurographics Symposium on Rendering/Eurographics Workshop on Rendering Techniques - EGSR, pp. 201–210 (2005)
5. Kumar, S., Singh, D.: Colorization of Gray Image in Lαβ Color Space Using Texture Mapping and Luminance Mapping. International Journal of Computational Engineering Research 2(5) (2008)
6. Uruma, K., Konishi, K., Takahashi, T., Furukawa, T.: An image colorization algorithm using sparse optimization. In: IEEE International Conference on Acoustics, Speech and Signal Processing (ICASSP), May 26-31, pp. 1588–1592 (2013)
7. Hogervorst, M.A., Toet, A.: Fast natural color mapping for night-time imagery. Information Fusion 11(2), 69–77 (2010)
8. Hogervorst, M.A., Toet, A.: Progress in color night vision. Optical Engineering 51(1) (January 2012)
9. Reinhard, E., Ashikhmin, M., Gooch, B.: Color transfer between images. IEEE Computer Graphics and Applications, 34–40 (September/October 2001)
10. Li, G., Wang, K.: Applying daytime colors to nighttime imagery with an efficient color transfer method. In: Proc. SPIE 6559, Enhanced and Synthetic Vision 2007, 65590L (2007)
11. Tsagaris, V., Anastassopoulos, V.: Fusion of visible and infrared imagery for night color vision. Displays 26(4-5), 191–196 (2005)
12. Hossny, M., Nahavandi, S., Creighton, D.: Color map-based image fusion. In: Proc. IEEE Int. Conf. Ind. Info. 2008 (INDIN 2008), pp. 52–56. IEEE Press, Los Alamitos (2008)
13. Kong, W., Lei, Y., Ni, X.: Fusion technique for grey-scale visible light and infrared images based on non-subsampled contourlet transform and intensity-hue-saturation transform. IET Sig. Proc. 5(1), 75–80 (2011)
14. Sun, F., Li, S., Yang, B.: A new color image fusion method for visible and infrared images. In: Proc. IEEE Int. Conf. on Robotics Biomim., pp. 2043–2048. IEEE Press, Los Alamos (2007)
15. Zaverietal, T.: An optimized region-based color transfer method for night vision application. In: Proc. 3rd IEEE Int. Conf. Sig. Imag. Process. (ICSIP 2010), pp. 96–101. IEEE Press, Los Alamitos (2010)
16. Christinal, J.J., Jebaseeli, T.J.: A Novel Color Image Fusion for Multi Sensor Night Vision Images. International Journal of Computer Applications Technology and Research 2(2), 155–159 (2013)
17. Si, T., Zhang, J.: A pseudo-color Fusion Algorithm of Night Vision Image Based on Environment-adaptive Color Transfer. In: 2013 8th International Conference on Computer Science & Education (ICCSE), pp. 411–415 (2013)
18. Zheng, Y.: An overview of night vision colorization techniques using multispectral images From color fusion to color mapping. In: 2012 International Conference on Audio, Language and Image Processing (ICALIP), pp. 134–143 (2012)
19. Qian, X., Han, L., Wang, Y., Wang, B.: Color contrast enhancement for color night vision based on color mapping. Infrared Physics & Technology 57, 36–41 (2013)
20. Yang, S., Liu, W., Deng, C., Zhang, X.: Color Fusion Method for Low-Light-Level and Infrared Images in Night Vision. In: 2012 5th International Congress on Image and Signal Processing (CISP), pp. 534–537 (2012)

21. Wang, Y., Wu, Y., Shi, X., Ye, Y.: The Color Fusion of Infrared and Visual Images Based on NSCT. In: 2013 Seventh International Conference on Image and Graphics (ICIG), pp. 597–602 (2013)
22. Lee, K., Kriesel, J., Gat, N.: Night Vision Camera Fusion with Natural Colors Using a Spectral/Texture Based Material Identification Algorithm. In: Meeting of the Military Sensing Symposia (MSS) on Passive Sensors (2010)
23. Waxman, A.M., Mario, A., Fay, D.A., Ireland, D.B., Racamato, J.P., Ross, W.D., Carrick, J.E., Gove, A.N., Seibert, M.C., Savoye, E.D.: Solid-State Color Night Vision_fusion of low light visible and thermal infrared imagery. Lincoln-Laboratory-Journal 11(1), 41–60 (1998)
24. Hogervorst, M.A., Jansen, C., Toet, A., Bijl, P., Bakker, P., Hiddema, A.C., Vlie, S.F.: Colour-the-INSight combining a direct view rifle sight with fused intensified and thermal imagery. In: SPIE Proceedings, vol. 8407-24 (2012)
25. Gu, X., Sun, S., Fang, J., Zhuo, P.: Kernel based color estimation for night vision imagery. Optics Communications 285(7), 1697–1703 (2012)
26. Hamam, T., Dordek, Y., Cohen, D.: Single-Band Infrared Texture-Based Image Colorization. In: IEEE 27th Convention of Electrical and Electronics Engineers in Israel, pp. 1–5 (2012)
27. Yatziv, L., Sapiro, G.: Fast image and video colorization using chrominance blending. IEEE Transactions on Image Processing 15(5), 1120–1129 (2006)
28. Levin, A., Lischinski, D., Weiss, Y.: Colorization using optimization. ACM Transactions on Graphics (TOG) - Proceedings of ACM SIGGRAPH 23(3), 689–694 (2004)
29. Zhang, Z., Cui, H., Lu, H.: A Colorization Method Based on Fuzzy Clustering and Distance Transformation. In: 2nd International Congress on Image and Signal Processing, CISP 2009, pp. 17–19 (2009)
30. Horiuchi, T.: Colorization algorithm using probabilistic relaxation. Image and Vision Computing 22(3), 197–202 (2004)
31. Kawulok, M., Kawulok, J., Smolka, B.: Image colorization using discriminative textural features. In: The 12th IAPR Conference on Machine Vision Applications (June 13-15, 2011)
32. Luan, Q., Wen, F., Cohen, D.: Natural Image Colorization. In: EGSR 2007 Proceedings of the 18th Eurographics Conference on Rendering Techniques, pp. 309–320 (2007)
33. Kalia, A.: Coloring of Grayscale Images using Prioritized Source Propagation method. In: Science and Information Conference (SAI), October 7-9, pp. 455–458 (2013)
34. Pang, J., Au, O.C., Tang, K., Guo, Y.: Image colorization using sparse representation. In: 2013 IEEE International Conference on Acoustics, Speech and Signal Processing (ICASSP), May 26-31, pp. 1578–1582 (2013)
35. Yu, C., Sharma, G., Aly, H.: Computational Efficiency Improvements for Image colorization. In: SPIE Proceedings, vol. 9020 (2014)
36. Yatziv, L., Bartesaghi, A., Sapiro, G.: Implementation of the Fast Marching Algorithm. Journal of Computational Physics 212(2), 393–399 (2006)

A New Networked Identification Approach for a Class of Hammerstein Systems

Weihua Deng[1], Kang Li[2], Jing Deng[2], Enyu Jiang[1,*], and Qiang Zhu[3]

[1] College of Electrical Engineering,
Shanghai University of Electric Power, Shanghai, 200090, China
dwh197859@126.com
[2] School of Electronics, Electrical Engineering and Computer Science,
Queen's University Belfast, United Kingdom
[3] Shanghai Automation Instrumentation Co., Ltd.
dwh197859@126.com

Abstract. This paper investigates the problem of identification of a class of Hammerstein systems over a wireless network. An iterative identification method is implemented over a physical IEEE 802.11b wireless channel. Every time the identified model is used into next identification process to produce the estimated values of the plant outputs for compensating the influence of network delays. Finally the identified model can be optimized through the multiple iterations. The effectiveness of the proposed approach is demonstrated by numerical examples.

1 Introduction

In last decades Hammerstein system identification have been studied in many literatures (Boutayeb et al., 1996; Bai et al., 2010; Mao et al. , 1966). But the identification by use of networks is not yet concerned. This work is meaningful because the local computing sources are often limited. So we can implement complex identification algorithms by using rich hard resources over networks. However this will bring another problem, namely, the influence of networks such as network-induced delay and packet loss. The model-based approach (Montestruque et al., 2008; Polushin et al., 2008) is often used to solve these problems. But the model was given in advance and it is not practical. Based on the above reasons, this paper designs a networked identification approach, which not only finishes identification over a wireless network but also compensates network delays.

This paper is based on the previous work reported in (Bai et al., 2010; Deng et al., 2013, 2014) and it proves the iterative algorithm from another point of view. Main contributions of the present paper are concluded as follows

(a) The idea of switching between input-output method and subspace one is used to analyze problem. Thus they can provide more suitable platforms to apply iterative algorithm into Hammerstein system identification.

---

* Corresponding author.

M. Fei et al. (Eds.): LSMS/ICSEE 2014, Part II, CCIS 462, pp. 293–299, 2014.

(b) The idea of making full use of remote computers which have the strong computation ability and the huge storage space through a wireless network is adopted to implement the identifier. Thus it is possible to finish complex iterative algorithms and to provide enough space to store the great quantities of data used for system identification.

(c) The idea of using the identified model to estimate the delayed plant outputs is explored to design identification strategy above the wireless network. Thus it can compensate the influence of the wireless network on system identification.

2   Problem Formulation

The input-output equation of a general Hammerstein system is given as

$$
\begin{aligned}
y(k) &= d_1 y(k-1) + d_2 y(k-2) + \cdots + d_n y(k-n) \\
&\quad + b_1 f(k-1) + b_2 f(k-2) + \cdots + b_n f(k-n)
\end{aligned}
\tag{1}
$$

where $f(k) = f(u(k)) = a_1 g_1(u(k)) + \cdots + a_l g_l(u(k))$. The above system is transformed into the equivalent state space description as

$$
\begin{aligned}
x(k+1) &= Ax(k) + Bf(u(k)) \\
y(k) &= Cx(k)
\end{aligned}
\tag{2}
$$

where

$$
A = \begin{bmatrix}
0 & 1 & 0 & \cdots & 0 \\
0 & 0 & 1 & \cdots & 0 \\
\vdots & \vdots & \vdots & & \vdots \\
0 & 0 & 0 & \cdots & 1 \\
d_n & d_{n-1} & d_{n-2} & \cdots & d_1
\end{bmatrix}, B = \begin{bmatrix}
0 \\
0 \\
\vdots \\
0 \\
1
\end{bmatrix}
$$

$$
C = \begin{bmatrix} b_n & b_{n-1} & \cdots & b_2 & b_1 \end{bmatrix}
$$

We will use (1) and (2) in the system identification and model-based compensation respectively. The purpose of transformation is to adopt the methods which are more suitable for solving problems. It is obvious that equation (1) is more convenient to implement iterative algorithm (Bai et al., 2010) in system identification and equation (2) is much easier to establish the model-based approach (Montestruque et al., 2008; Polushin et al., 2008). And the transformation between (1) and (2) is also easier in theory and practical operation.

Then we give the description of Hammerstein system model. Let $\tilde{k} = 0, 1, \cdots, N$, N represents iteration number. The $\tilde{k}$th identified input-output model and state space one are described as

$$
\begin{aligned}
y_{\tilde{k}}(k) &= d_{\tilde{k}1} y_{\tilde{k}}(k-1) + d_{\tilde{k}2} y_{\tilde{k}}(k-2) + \cdots + d_{\tilde{k}n} y_{\tilde{k}}(k-n) \\
&\quad + b_{\tilde{k}1} f_{\tilde{k}}(k-1) + b_{\tilde{k}2} f_{\tilde{k}}(k-2) + \cdots + b_{\tilde{k}n} f_{\tilde{k}}(k-n) \\
f_{\tilde{k}}(k) &= f_{\tilde{k}}(\hat{u}_{\tilde{k}}(k)) = a_{\tilde{k}1} g_1(\hat{u}_{\tilde{k}}(k)) + \cdots + a_{\tilde{k}l} g_l(\hat{u}_{\tilde{k}}(k))
\end{aligned}
\tag{3}
$$

and

$$x_{\tilde{k}}(k+1) = A_{\tilde{k}}x_{\tilde{k}}(k) + Bf_{\tilde{k}}(\hat{u}_{\tilde{k}}(k))$$
$$y_{\tilde{k}}(k) = C_{y_{\tilde{k}}(k)}x_{\tilde{k}}(k) \tag{4}$$

where

$$A_{\tilde{k}} = \begin{bmatrix} 0 & 1 & 0 & \cdots & 0 \\ 0 & 0 & 1 & \cdots & 0 \\ \vdots & \vdots & \vdots & & \vdots \\ 0 & 0 & 0 & \cdots & 1 \\ d_{\tilde{k}n} & d_{\tilde{k}n-1} & d_{\tilde{k}n-2} & \cdots & d_{\tilde{k}1} \end{bmatrix}, B = \begin{bmatrix} 0 \\ 0 \\ \vdots \\ 0 \\ 1 \end{bmatrix}$$
$$C_{\tilde{k}} = \begin{bmatrix} b_{\tilde{k}n} & b_{\tilde{k}n-1} & \cdots & b_{\tilde{k}2} & b_{\tilde{k}1} \end{bmatrix}$$

We assume that there is some errors between the $\tilde{k}$th identified model and real system and they are described as

$$A = A_{\tilde{k}} + \Delta A_{\tilde{k}}, C = C_{\tilde{k}} + \Delta C_{\tilde{k}}$$
$$f(u_{\tilde{k}}(k)) = f_{\tilde{k}}(u_{\tilde{k}}(k)) + \Delta f_{\tilde{k}}(u_{\tilde{k}}(k))$$
$$\Delta A_{\tilde{k}} = \begin{bmatrix} 0 & 1 & 0 & \cdots & 0 \\ 0 & 0 & 1 & \cdots & 0 \\ \vdots & \vdots & \vdots & & \vdots \\ 0 & 0 & 0 & \cdots & 1 \\ \Delta d_{\tilde{k}n} & \Delta d_{\tilde{k}n-1} & \Delta d_{\tilde{k}n-2} & \cdots & \Delta d_{\tilde{k}1} \end{bmatrix} \tag{5}$$
$$\Delta C_{\tilde{k}} = \begin{bmatrix} \Delta b_{\tilde{k}n} & \Delta b_{\tilde{k}n-1} & \cdots & \Delta b_{\tilde{k}2} & \Delta b_{\tilde{k}1} \end{bmatrix}$$
$$\Delta f_{\tilde{k}}(u_{\tilde{k}}(k)) = \Delta a_{\tilde{k}1}g_1(u(k)) + \cdots + \Delta a_{\tilde{k}l}g_l(u(k))$$

And for $\forall k > 0$ these errors are bounded as

$$\left|\Delta d_{\tilde{k}i}\right| \le \bar{d}_i, i = 1, \cdots, n$$
$$\left|\Delta b_{\tilde{k}i}\right| \le \bar{b}_i, i = 1, \cdots, n$$
$$\left|\Delta f_{\tilde{k}}(u_{\tilde{k}}(k))\right| \le f_{\tilde{k}}, \forall k > 0$$

3   Wireless Network Delays

The delays in a wireless network are random. So it is reasonable to model the delays using suitable statistical models. In this paper an inverse Gaussian distribution which is first used into describing wireless network delays by (Deng et al., 2013, 2014) will be used to characterize the wireless network delays. The random delay $\tau(k)$ is inverse Gaussian-distributed with mean μ and shape parameter λ, we write $\tau(k) \sim IG(\mu, \lambda)$. Its cumulative density function is given by

$$F(\tau(k)) = \Phi(\sqrt{\frac{\lambda}{\tau(k)}}(\frac{\tau(k)}{\mu} - 1)) + \exp(\frac{2\lambda}{\tau(k)})\Phi(-\sqrt{\frac{\lambda}{\tau(k)}}(\frac{\tau(k)}{\mu} + 1)) \tag{6}$$

We need to make the following change for the effective network controller design. Firstly solve the probability values using cumulative density function $F(\tau(k))$:

$$\Pr(\tau(k) \le \tau^*) = p$$
$$\Pr(\tau(k) > \tau^*) = 1 - p \tag{7}$$

where τ^* depicts the maximum delay time which the identifier can wait for. Then let one indicator function be

$$\delta(k) = \begin{cases} 1, & \tau(k) \le \tau^* \\ 0, & \tau(k) > \tau^* \end{cases} \tag{8}$$

where $\delta(k) = 1$ represents real output arrive before τ^*, otherwise it means this data lost. And it is easy to get

$$\begin{aligned} \Pr(\delta(k) = 1) &= p \\ \Pr(\delta(k) = 0) &= 1 - p \end{aligned} \tag{9}$$

4   Networked Iterative Identification

4.1   The Identification Algorithm

The purpose of identification is to identify the unknown parameters

$$d = [\,d_n\ d_{n-1}\ \cdots\ d_1\,]^T, b = [\,b_n\ b_{n-1}\ \cdots\ b_1\,]^T, a = [\,a_1\ a_2\ \cdots\ a_l\,]^T$$

based on the input-output measurements $\{u(k), y(k)\}_{k=1}^{N}$ over wireless networks. And at the $\tilde{k}$th identification process let

$$J_{1N}(b_{\tilde{k}}, d_{\tilde{k}}) = \frac{1}{N} \sum_{k=1}^{N} (y_{\tilde{k}}(k) - y(k))^2$$

$$= \frac{1}{N} \sum_{k=1}^{N} (\sum_{i=1}^{n} d_{\tilde{k}i} y(k-i) + \sum_{j=1}^{n} b_{\tilde{k}j} \sum_{i=1}^{l} a_{\tilde{k}i} g_i(u(k-j))] - y(k))^2$$

$$J_{0N} = \frac{1}{N} \sum_{k=1}^{N} (y_{0\tilde{k}}(k) - y(k))^2$$

$$= \frac{1}{N} \sum_{k=1}^{N} (d_{\tilde{k}}^T \phi(k) + \sum_{i=1}^{l} \bar{G}_i(k) a_{\tilde{k}i} b_{\tilde{k}} + c_{\tilde{k}} - y(k))^2$$

and

$$J_{1N\tilde{k}}(b_{\tilde{k}}, d_{\tilde{k}}) = \frac{1}{N} \sum_{k=1}^{N} (y_{1\tilde{k}}(k) - (\delta(k)y(k) + (1 - \delta(k))y_{1\tilde{k}-1}(k)))^2$$

$$= \frac{1}{N} \sum_{k=1}^{N} (\sum_{i=1}^{n} d_{\tilde{k}i} y(k-i) + \sum_{i=1}^{n} b_{\tilde{k}i} f_{\tilde{k}}(k-i) - (\delta(k)y(k) + (1 - \delta(k))y_{1\tilde{k}-1}(k)))^2$$

$$\begin{aligned} J_{0N\tilde{k}} = \frac{1}{N} \sum_{k=1}^{N} (y_{0\tilde{k}}(k) - (\delta(k)y(k) + (1 - \delta(k))y_{0\tilde{k}-1}(k)))^2 \\ = \frac{1}{N} \sum_{k=1}^{N} (d_{\tilde{k}}^T \phi(k) + \sum_{i=1}^{l} \bar{G}_i(k) a_{\tilde{k}i} b_{\tilde{k}} + c_{\tilde{k}} \\ -\delta(k) d^T \phi(k) - (1 - \delta(k)) d_{\tilde{k}-1}^T \phi(k) \end{aligned}$$

$$-\delta(k) \sum_{i=1}^{l} \bar{G}_i(k) a_i b - (1 - \delta(k)) \sum_{i=1}^{l} \bar{G}_i(k) a_{\tilde{k}-1i} b_{\tilde{k}-1} - \delta(k) c - (1 - \delta(k)) c_{\tilde{k}-1}))^2$$

where

$$y_{0\tilde{k}}(k) = d_{\tilde{k}}^T \phi(k) + \sum_{i=1}^{l} \bar{G}_i(k) a_{\tilde{k}i} b_{\tilde{k}} + c_{\tilde{k}}$$

is obtained from

$$y(k) = \sum_{i=1}^{n} d_{\tilde{k}i} y(k-i) + \sum_{j=1}^{n} b_{\tilde{k}j} \underbrace{\sum_{i=1}^{l} a_{\tilde{k}i}[g_i(u(k-j)) - \frac{1}{N_{\tilde{k}}} \sum_{m=1}^{N_{\tilde{k}}} g_i(u(m-j))]}_{\bar{g}_i(u(k-j))}$$

$$+ \underbrace{\sum_{j=1}^{n} b_{\tilde{k}j} \sum_{i=1}^{l} a_{\tilde{k}i} \frac{1}{N_{\tilde{k}}} \sum_{m=1}^{N_{\tilde{k}}} g_i(u(m-j))}_{c_{\tilde{k}}}$$

$$= d_{\tilde{k}}^T \underbrace{\begin{pmatrix} y(k-1) \\ \vdots \\ y(k-n) \end{pmatrix}}_{\phi(k)} + \sum_{i=1}^{l} \underbrace{\big(\bar{g}_i(u(k-1)) \cdots \bar{g}_i(u(k-n)) \big)}_{\bar{G}_i(k)} a_{\tilde{k}i} b_{\tilde{k}} + c_{\tilde{k}}$$

So we will give the following algorithm.

- the initial stage: given the initial estimates $b_{\tilde{k}}(0) = [1 \quad 0 \cdots \quad 0]$ and arbitrary $d_{\tilde{k}}(0)$ and $c_{\tilde{k}}(0)$
- at iteration stage $k \geq 1$, if $\tilde{k} = 0$: find, for the fixed $b_{\tilde{k}}(k-1), d_{\tilde{k}}(k-1)$ and $c_{\tilde{k}}(k-1)$,

$$\tilde{a}_{\tilde{k}} = \arg\min J_0(\bar{a}_{\tilde{k}}, b_{\tilde{k}}(k-1), d_{\tilde{k}}(\bar{k}-1), c_{\tilde{k}}(k-1))$$

else find, for the fixed $b_{\tilde{k}}(k-1), d_{\tilde{k}}(k-1)$ and $c_{\tilde{k}}(k-1)$,

$$\tilde{a}_{\tilde{k}} = \arg\min J_{0N\tilde{k}}(\bar{a}_{\tilde{k}}, b_{\tilde{k}}(k-1), d_{\tilde{k}}(k-1), c_{\tilde{k}}(k-1))$$

Normalize $a_{\tilde{k}}(k) = \pm \frac{\bar{a}_{\tilde{k}}}{\|\tilde{a}_{\tilde{k}}(k)\|}$ so that $\|a_{\tilde{k}}\| = 1$ and the first nonzero element is positive. And substitute $f_{\tilde{k}}(k) = \sum_{j=1}^{l} a_{\tilde{k}} g_j(u(k))$ and if $\tilde{k} = 0$ find, for the fixed $a_{\tilde{k}}$,

$$(b_{\tilde{k}}(k), d_{\tilde{k}}(k)) = \arg\min J_1(a_{\tilde{k}}(k), \bar{b}_{\tilde{k}}, \bar{d}_{\tilde{k}})$$

else for the fixed $a_{\tilde{k}}$,

$$(b_{\tilde{k}}(k), d_{\tilde{k}}(k)) = \arg\min J_{1N\tilde{k}}(a_{\tilde{k}}(k), \bar{b}_{\tilde{k}}, \bar{d}_{\tilde{k}})$$

Finally, set

$$c_{\tilde{k}}(k) = \sum_{j=1}^{n} b_{\tilde{k}j}(k) \sum_{i=1}^{l} a_{\tilde{k}i} \big(\frac{1}{N_{\tilde{k}}} \sum_{m=1}^{l} g_i(u(m-j)) \big)$$

. Replace k by $k+1$, the process is repeat.

In the above algorithm, the parameter $\delta(k)$ is very important, which reflects the the influences of wireless network delays on identification process. In fact, in here the delays has been transformed into dropouts problem.

4.2   The Proof of Algorithm Convergence

Definition 1: If $\left\| \widehat{\beta} - \beta \right\| \leq \varepsilon$, then $\widehat{\beta}$ is convergent in sense of norm.

Proof: After a series of deduction we can get

$$\left\| b_{\tilde{k}} - b \right\| \leq \left\| b_{\tilde{k}} \right\| + \|b\|$$

$$\leq (1 + q^{\tilde{k}} \frac{\left| \sum\limits_{j=1}^{l} a_{0j} \right|}{\left| \sum\limits_{j=1}^{l} a_{\tilde{k}j} \right|} + (1 - q^{\tilde{k}}) \frac{\left| \sum\limits_{j=1}^{l} a_{j} \right|}{\left| \sum\limits_{j=1}^{l} a_{\tilde{k}j} \right|}) \|b\| + q^{\tilde{k}} \frac{\left| \sum\limits_{j=1}^{l} a_{0j} \right|}{\left| \sum\limits_{j=1}^{l} a_{\tilde{k}j} \right|} (\sum\limits_{i=1}^{n} \varepsilon_{bi}^{2})^{1/2}$$

$$\left\| \bar{a}_{\tilde{k}} - \bar{a} \right\| = \left\| b_{\tilde{k}}(1) a_{\tilde{k}} - b(1) a \right\| \leq \left| b_{\tilde{k}}(1) \right| + |b(1)|$$
$$\leq 2\, |b(1)| + (1 - p)^{k} \varepsilon_{b1}$$

$$\left\| d_{\tilde{k}} - d \right\| \leq \left\| d_{\tilde{k}} \right\| + \|d\|$$
$$\leq 2\, \|d\| + q^{k} \varepsilon_{d0}$$

If $N \to \infty$, we can get their limitation respectively as

$$\lim_{\tilde{k} \to \infty} \left[(1 + q^{\tilde{k}} \frac{\left| \sum\limits_{j=1}^{l} a_{0j} \right|}{\left| \sum\limits_{j=1}^{l} a_{\tilde{k}j} \right|} + (1 - q^{\tilde{k}}) \frac{\left| \sum\limits_{j=1}^{l} a_{j} \right|}{\left| \sum\limits_{j=1}^{l} a_{\tilde{k}j} \right|}) \|b\| + q^{\tilde{k}} \frac{\left| \sum\limits_{j=1}^{l} a_{0j} \right|}{\left| \sum\limits_{j=1}^{l} a_{\tilde{k}j} \right|} (\sum\limits_{i=1}^{n} \varepsilon_{bi}^{2})^{1/2} \right]$$

$$= (1 + \frac{\left| \sum\limits_{j=1}^{l} a_{j} \right|}{\left| \sum\limits_{j=1}^{l} a_{\tilde{k}j} \right|}) \|b\|$$

$$\lim_{\tilde{k} \to \infty} \left[2\, |b(1)| + q^{\tilde{k}} \varepsilon_{b1} \right] = 2\, |b(1)|$$

$$\lim_{\tilde{k} \to \infty} (2\, \|d\| + q^{\tilde{k}} \varepsilon_{d0}) = 2\, \|d\|$$

According to definition 1 the result is obtained.

5   The Result of Simulation

The Hammerstein system parameters are assumed as

$$d = \begin{bmatrix} 0.3\ 0.2\ -0.3 \end{bmatrix}$$
$$b = \begin{bmatrix} 2\ 1\ 1 \end{bmatrix}$$
$$a = 1$$

If let $\tilde{k} = 100$, we can get the relation of bounds of $\left\| \bar{a}_{\tilde{k}} - \bar{a} \right\|, \left\| d_{\tilde{k}} - d \right\|, \left\| b_{\tilde{k}} - b \right\|$ and themselves, see Fig 2, Fig3 and Fig 4, where $\bar{a}_{\tilde{k}} = b_{\tilde{k}}(1) a_{\tilde{k}}$.

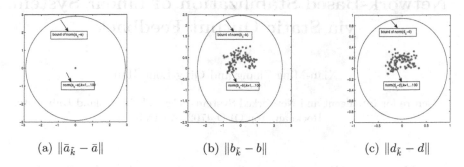

(a) $\|\bar{a}_{\bar{k}} - \bar{a}\|$ (b) $\|\bar{b}_{\bar{k}} - b\|$ (c) $\|d_{\bar{k}} - d\|$

Fig. 1. The convergence of the identification parameters in norm senses

6 Conclusions

A new networked identification algorithm is proposed. This approach is different with the ones in (Bai et al., 2010; Deng et al., 2013, 2014) because its convergence is proved in the sense of norm. And numerical simulation examples have confirmed the efficacy of the proposed approaches.

Acknowledgment. This work was supported by Shanghai Green Energy Grid Connected Technology Engineering Research Center under Grant 13DZ2251900, The Young Teacher Training Program and Industry-Study-Research Cooperation Project from Shanghai Education Commission under Grant ZZsdl13008, CXYsdl14012.

References

1. Bai, E.W., Li, K.: Convergence of the iterative algorithm for a general Hammerstein system identification. Automatica 46, 1891–1896 (2010)
2. Boutayeb, M., Rafaralahy, H., Darouach, M.: A robust and recursive identification method for Hammerstein model. In: IFAC World Congress, San Francisco, pp. 447–452 (1996)
3. Deng, W.H., Li, K., Irwin, G.W., Fei, M.R.: Identification and control of Hammerstein systems via wireless networks. International Journal of Systems Science 44, 1613–1625 (2013)
4. Deng, W.H., Li, K., Fei, M.R.: Identification and output tracking control of Hammerstein systems through wireless networks. Transactions of the Institute of Measurement and Control 36, 3–13 (2014)
5. Mao, K.Z., Billings, S.A.: An iterative method for the identification of nonlinear systems using a Hammerstein model. IEEE Transactions on Automatic Control 11, 546–550 (1966)
6. Montestruque, L.A., Antsaklis, P.J.: On the model-based control of networked systems. Automatica 39, 1837–1843 (2008)
7. Polush, I.G., Liu, P.X., Lung, C.H.: On the model-based approach to nonlinear networked control systems. Automatica 44, 2409–2414 (2008)

Network-Based Stabilization of Linear Systems via Static Output Feedback

Xian-Ming Zhang and Qing-Long Han

Centre for Intelligent and Networked Systems, Central Queensland University,
Rockhampton QLD 4702, Australia

Abstract. This paper is concerned with the problem of static output feedback control for networked systems with a logic zero-order-hold (ZOH). First, the networked closed-loop system is modeled as a discrete-time linear system with a time-varying delay, whose upper bound can be regarded as not only the admissible maximum delays induced by the network but also the admissible maximum number of packet dropouts between any two consecutive updating instants of ZOH. Then, in order to obtain a larger upper bound of the time-varying delay, a generalized finite-sum inequality is introduced, based on which, a less conservative stability condition is derived by incorporating with convex combination technique. By using the cone complementary linearization approach, the desired output feedback controllers can be designed by solving a nonlinear minimization problem subject to a set of linear matrix inequalities. Finally, some examples are given to show the effectiveness of the proposed method.

1 Introduction

During the past decades, networked control systems (NCSs) have been attracting much interesting of researchers in the field of control. In an NCS, the data transmission between a physical plant and a controller is completed through a digit network with constraint quality of service (QoS). On the one hand, the introduction of networks has some advantages, such as lower costs, easier installation and maintenance and higher reliability, and thus NCSs have been applied to a wide range of systems, see, e.g. [1,2]. On the other hand, the introduction of networks also brings some disadvantages due to the constraint QoS. With the limitation of network bandwidth, network-induced delays and data dropouts are inevitable during signal transmission. These unfavorable factors possibly cause system performance degradation or even instability of an NCS. Therefore, it is quite significant to investigate the effects of network-induced delays and data dropouts on an NCS. Much effort has been made on this issue, and one can refer to [2–8] and references therein.

Recalling some results reported in the literature, it is found that network-induced delays and data dropouts are usually handled separately or simultaneously to model NCSs as different systems. In [9] the network-induced delays from sensor to controller and from controller to actuator are modeled as two Markov

M. Fei et al. (Eds.): LSMS/ICSEE 2014, Part II, CCIS 462, pp. 300–309, 2014.

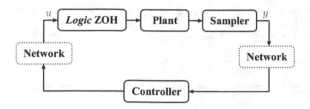

Fig. 1. A networked control system with logic ZOH

chains and the closed-loop NCS is then modeled as a jump linear system with two modes. The effects of network-induced delays on the NCS are studied. As for an NCS with packet dropouts, different models are proposed. For example, an NCS with packet dropouts is modeled as a stochastic system with a Bernoulli distributed white sequence in [10], a discrete-time switched system in [11] and an asynchronous dynamic system in [12]. Consequently, a number of results have been derived on how the packet dropouts affect the stability of an NCS. In [13], network-induced delays and packet dropouts are handled simultaneously to model the closed-loop NCS as a linear system with a time-varying delay. This approach has gained considerable attention of researchers. However, the timing mechanism scheduled in [13] cannot express the network-induced delays or the number of packet dropouts explicitly. As a result, the effects of network-induced delays or packet dropouts on an NCS is implicit. Moreover, data packet disorder phenomena are not considered in [13], which means that one cannot ensure the newest data packet to drive the physical plant at each updating instant of the zero-order-hold (ZOH).

In [14], Xiong and Lam proposed a general framework to deal with the network-induced delays and data dropouts simultaneously. By introducing a logic ZOH to store the newest control information, an NCS subject to both network-induced delays and data dropouts can be modeled as a linear system with a time-varying delay in the discrete-time domain. Since both network-induced delays and data dropouts can be derived from the logic ZOH, this approach is more convenient than that in [13] to be used to investigate the effects of network-induced delays and data dropouts on the NCS. Employing this model, Xiong and Lam discussed the problem of state feedback stabilization, but the obtained results are somewhat conservative. Moreover, when the state of the physical plant is unmeasurable, these results are inapplicable.

This paper deals with the problem of output feedback stabilization of an NCS with a logic ZOH in the discrete-time domain. Similar to [14], the closed-loop NCS is modeled as a linear discrete-time system with a time-varying delay. Then, a generalized finite sum inequality is introduced to formulate some stability criterion for the closed-loop NCS, which includes the one in [14] as a special case. Finally, by employing the cone complementary linearization approach, suitable output feedback controllers can be designed based on the proposed stability criterion, whose effectiveness is confirmed by some numerical examples.

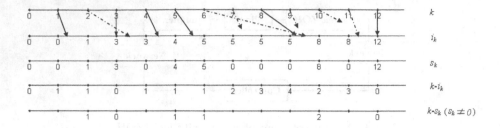

Fig. 2. An example of packet lost and used (dagged line: lost; real line: used)

2   System Description

Consider an NCS shown in Fig. 1, where the plant is a continuous-time system. Assume that both the sampler and the zero-order-hold (ZOH) are clock-driven. The output signals of the plant are sampled by the sampler at $t_k = kT_s$, $k = 1, 2, \cdots$, where T_s is the sampling period. At each sampling instant t_k, the output signal $y(t_k)$ and its time stamp are encapsulated into a packet and sent to controller through a network. Once this packet arrives at the controller, the controller immediately calculates and generates a new control signal, which together with its time stamp is encapsulated again into a control packet. The control packet is finally transmitted by the network to the ZOH. The ZOH is configured to be a logic ZOH, which is synchronized with the sampler in the sense that the ZOH adjusts its output only at the sampling instants. This logic ZOH compares the time stamps of the arrived control packets with the time stamp currently stored in the ZOH such that the newest control packet can be chosen to drive the physical plant. Concisely, the logic ZOH is described as follows.

Logic ZOH: Given $u(0)$, let $i_0 = 0$, $k = 0$ and $s_q = 0$ for all $q \in \{1, 2, 3, \cdots\}$.

1. At sampling instant t_k, ZOH changes its control input packet to $u(t) = u(i_k T_s)$ for $t_k \le t < t_{k+1}$. Let $i_{k+1} = i_k$;
2. During $t_k < t \le t_{k+1}$, if a packet $u(jT_s)$ arrives and $j > i_{k+1}$, then ZOH stores $u(jT_s)$ and lets $i_{k+1} = j$ and $s_{k+1} = j$;
3. Repeat Step 2) until t reaches the next sampling instant t_{k+1}. Let $k = k+1$ and go to Step 1).

The above description of logic ZOH is slightly different from that in [14]. Here, we introduce the sequence $\{s_1, s_2, \cdots, s_q, \cdots\}$ to report the updating instants of the ZOH. $s_q \ne 0$ means that at the sampling instant $k = q$, ZOH updates its output with the newest control input packet whose time stamp is s_q, while $s_q = 0$ means that at the sampling instant $k = q$, ZOH does not update and still uses the previous input packet to drive the physical plant. For example, in Fig. 2, $s_5 = 4 \ne 0$ means that at the sampling instant $k = 5$ the ZOH updates its output with the control input packet whose time stamp is 4 to drive the physical plant. In the following, we call the packet whose time stamp is s_q ($s_q \ne 0$) an *effective data packet*.

Define $d(k) = k - i_k$ and suppose $0 \leq d(k) \leq d_{\max}$, where $d_{\max}$ is constant. Then the network-induced delay of an *effective data packet* and the number of data dropouts between two consecutive updating instants of ZOH can be expressed explicitly.

- Denote by N_{s_q} the network-induced delay of the effective data packet whose time stamp is s_q. Then we have

$$N_{s_q} = q - s_q, \quad \text{if } s_q \neq 0, \quad \forall q \in \mathbb{N} \tag{1}$$

For example, in Fig. 2, $N_{s_2} = 1, N_{s_3} = 0$ and $N_{s_{10}} = 2$. Moreover, notice that $i_q = s_q$ when $s_q \neq 0$. In consequence, $N_{s_q} \leq d_{\max}$ ($q \in \mathbb{N}$).
- One can seen that packet dropouts occur in the case of $d(k+1) < d(k)$. When $d(k+1) < d(k)$, denote by $N_{dropout}^{k+1}$ the number of packet dropouts between the current updating instant $k+1$ and the previous updating instant k of the ZOH. Then we have

$$N_{dropout}^{k+1} = d(k) - d(k+1), \quad \text{if } d(k+1) < d(k) \; \forall k \in \mathbb{N} \tag{2}$$

For example, in Fig. 2, $N_{dropout}^3 = 1, N_{dropout}^{10} = 2$ and $N_{dropout}^{12} = 3$. Moreover, it is clear that $N_{dropout}^{k+1} \leq d_{\max}$.

Similar to [14], the physical plant in Fig. 1, together with the sampler and the logic ZOH, can be descretized as a linear discrete-time system with input delays:

$$\begin{cases} x(k+1) = Ax(k) + Bu(k - d(k)) \\ y(k) = Cx(k) \\ x(0) = x_0 \end{cases} \tag{3}$$

where $x(k) \in \mathbb{R}^n, y(k) \in \mathbb{R}^p$ and $u(k) \in \mathbb{R}^r$ are the state, measured output and control input, respectively; $A \in \mathbb{R}^{n \times n}, B \in \mathbb{R}^{n \times r}$ and $C \in \mathbb{R}^{p \times n}$ are constant real matrices and x_0 is the initial condition. $d(k)$ is a time-varying delay. Without loss of generality, suppose that

$$0 \leq d_1 \leq d(k) \leq d_2 = d_{\max} \tag{4}$$

where d_1 is not necessarily equal to zero.

In this paper, we aim to design a static output feedback controller of form

$$u(k) = Ky(k) \tag{5}$$

where K is the controller gain to be determined, such that the resulting closed-loop system

$$x(k+1) = Ax(k) + BKCx(k - d(k)) \tag{6}$$

is asymptotically stable. The initial condition sequence of the system (6) is supplemented as $\{\phi(k) | k = -d_2, -d_2+1, \cdots, 0\}$ with $\phi(0) = x_0$.

Remark 1. *It is clear that with a logic ZOH, the closed-loop NCS is modeled as a linear system with a time-varying delay, where network-induced delays and packet dropouts are taken into account. Let $N_{delay}^{\max} := \max_{q \in \mathbb{N}}\{N_{s_q}\}$ represent the upper bound of network-induced delays, and let $N_{dropout}^{\max} := \max_{k \in \mathbb{N}}\{N_{dropout}^{k+1}\}$ stand for the maximum number of packet dropouts. Then one has*

$$N_{delay}^{\max} \leq d_{\max}, \qquad N_{dropout}^{\max} \leq d_{\max} \tag{7}$$

which means that upper bound $d_{\max}$ of $d(k)$ reflects the endurability of networks on network-induced delays and packet dropouts. It is worth pointing out that, however, similar model is proposed in [13, 15] and [16], but the relationship among $N_{delay}^{\max}$, $N_{dropout}^{\max}$ and $d_{\max}$ is given by

$$N_{delay}^{\max} + N_{dropout}^{\max} \leq d_{\max} \tag{8}$$

which indicates that $d_{\max}$ reflects the endurability of networks on the sum of network-induced delays and packet dropouts.

To end this section, two lemmas are introduced for the stability analysis in the next section.

Lemma 1. *[17, 18] For any constant matrix $R \in \mathbb{R}^{m \times m}$ with $R = R^T > 0$, integers r_1 and r_2 with $r_2 > r_1 > 0$, vector function $w : \{r_1, r_1 + 1, \cdots, r_2\} \to \mathbb{R}^m$, the following inequality holds*

$$\sum_{j=r_1}^{r_2-1} w^T(j)Rw(j) \geq \frac{1}{r_2-r_1}\left(\sum_{j=r_1}^{r_2-1} w^T(j)\right) R \left(\sum_{j=r_1}^{r_2-1} w(j)\right).$$

Lemma 2. *(A generalized finite-sum inequality [19]). For any constant matrix $R \in \mathbb{R}^{m \times m}$ with $R = R^T > 0$, constant integers or time-varying integer-valued functions r_1 and r_2 with $r_2 > r_1 > 0$, vector function $w : \{r_1, r_1 + 1, \cdots, r_2\} \to \mathbb{R}^m$, if there exist $E \in \mathbb{R}^{m \times q}$ and $\psi \in \mathbb{R}^{q \times 1}$ such that $\sum_{j=r_1}^{r_2-1} w(j) = E\psi$, then for any matrix $M \in \mathbb{R}^{m \times q}$, the following inequality holds*

$$\sum_{j=r_1}^{r_2-1} w^T(j)Rw(j) \geq \psi^T \Upsilon \psi \tag{9}$$

where $\Upsilon := E^T M + M^T E - (r_2 - r_1)M^T R^{-1} M$.

3   Main Results

In this section, we first present a stability criterion for the closed-loop system (6). Then a cone complementary approach is employed to design suitable static output feedback controllers.

For simplicity of presentation, denote

$$\xi(k) := \mathrm{col}\{x(k), x(k - d(k)), x(k - d_1), x(k - d_2)\}$$

and

$$e_1 = [I\ 0\ 0\ 0], \ e_2 = [0\ I\ 0\ 0], \ e_3 = [0\ 0\ I\ 0], \ e_4 = [0\ 0\ 0\ I]$$

where I is an $n \times n$ real identity matrix. Thus

$$x(k+1) = \mathscr{C}_1 \xi(k), \quad \eta(k) = \mathscr{C}_2 \xi(k)$$

where $\eta(k) := x(k+1) - x(k)$ and

$$\mathscr{C}_1 = A e_1 + B K C e_2, \quad \mathscr{C}_2 = (A - I) e_1 + B K C e_2 \tag{10}$$

The Lyapunov functional candidate is chosen as

$$V(k) = V_1(k) + V_2(k) + V_3(k) + V_4(k) \tag{11}$$

where

$$V_1(k) := x^T(k) P x(k)$$

$$V_2(k) := \sum_{j=k-d_1}^{k-1} x^T(j) S_1 x(j) + \sum_{j=k-d_2}^{k-d_1-1} x^T(j) S_2 x(j)$$

$$V_3(k) := \sum_{j=-d_2+1}^{-d_1+1} \sum_{i=k+j-1}^{k-1} x^T(i) S_3 x(i)$$

$$V_4(k) := d_1 \sum_{j=-d_1}^{-1} \sum_{i=k+j}^{k-1} \eta^T(i) R_1 \eta(i) + \sum_{j=-d_2}^{-d_1-1} \sum_{i=k+j}^{k-1} \eta^T(i) R_2 \eta(i)$$

We now state the following result.

Proposition 1. *For given scalars d_1 and d_2 satisfying $d_2 \geq d_1 > 0$, the system (6) is asymptotically stable if there exist $n \times n$ real matrices $P > 0, S_1 > 0, S_2 > 0, S_3 > 0, R_1 > 0, R_2 > 0$ and $n \times 4n$ real matrices M_1 and M_2 and $r \times p$ real matrix K such that*

$$\begin{bmatrix} \Omega_1 + \Omega_2 + \Omega_3 & d_{12} M_1^T \\ d_{12} M_1 & -d_{12} R_2 \end{bmatrix} < 0, \quad \begin{bmatrix} \Omega_1 + \Omega_2 + \Omega_3 & d_{12} M_2^T \\ d_{12} M_2 & -d_{12} R_2 \end{bmatrix} < 0 \tag{12}$$

where $d_{12} = d_2 - d_1$, and

$$\Omega_1 := \mathscr{C}_1^T P \mathscr{C}_1 + e_1^T [S_1 + (d_{12} + 1) S_3 - P] e_1$$
$$- e_2^T S_3 e_2 + e_3^T (S_2 - S_1) e_3 - e_4^T S_2 e_4 \tag{13}$$

$$\Omega_2 := \mathscr{C}_2^T (d_1^2 R_1 + d_{12} R_2) \mathscr{C}_2 \tag{14}$$

$$\Omega_3 := -(e_1 - e_3)^T R_1 (e_1 - e_3) + (e_4 - e_2)^T M_1$$
$$+ M_1^T (e_4 - e_2) + (e_2 - e_3)^T M_2 + M_2^T (e_2 - e_3) \tag{15}$$

Proof. The proof is omitted due to page limitation. □

In case of $d_1 = 0$, modifying the Lyapunov functional in (11) slightly and following the same line of the proof of Proposition 1, then we have the following result without proof.

Corollary 1. *For a given scalar d_2, the system (6) is asymptotically stable if there exist $n \times n$ real matrices $P > 0, S_1 > 0, S_2 > 0, Z > 0$ and $n \times 3n$ real matrices M_1 and M_2 and $r \times p$ real matrix K such that*

$$\begin{bmatrix} \Omega_{01} + \Omega_{02} & d_2 M_1^T \\ d_2 M_1 & -d_2 Z \end{bmatrix} < 0, \quad \begin{bmatrix} \Omega_{01} + \Omega_{02} & d_2 M_2^T \\ d_2 M_2 & -d_2 Z \end{bmatrix} < 0 \tag{16}$$

where

$$\Omega_{01} := \mathscr{C}_{01}^T P \mathscr{C}_{01} - \tilde{e}_2^T S_2 \tilde{e}_2 - \tilde{e}_3^T S_1 \tilde{e}_3 + \tilde{e}_1^T [S_1 + (d_2 + 1) S_2 - P] \tilde{e}_1$$

$$\Omega_{02} := d_2 \mathscr{C}_{02}^T Z \mathscr{C}_{02} + (\tilde{e}_3 - \tilde{e}_2)^T M_1 + M_1^T (\tilde{e}_3 - \tilde{e}_2) + (\tilde{e}_2 - \tilde{e}_1)^T M_2 + M_2^T (\tilde{e}_2 - \tilde{e}_1)$$

with $\tilde{e}_1 = [I \ \ 0 \ \ 0], \tilde{e}_2 = [0 \ \ I \ \ 0], \tilde{e}_3 = [0 \ \ 0 \ \ I], \mathscr{C}_{01} = A\tilde{e}_1 + BFC\tilde{e}_2$ and $\mathscr{C}_{02} = (A - I)\tilde{e}_1 + BFC\tilde{e}_2$.

Corollary 1 includes Theorem 1 in [14] as a special case. In fact, let $S_1 \to 0^+$, $S_2 \to 0^+$, $M_1 = 0$ and $M_2 = [T_1^T \ \ T_2^T \ \ 0]$, then the matrix inequalities in (16) with $C = I$ reduces to matrix inequality (4) in [14] after some simple algebraic manipulation. Therefore, Corollary 1 is less conservative than Theorem 1 in [14].

In the following, we aim to design suitable output feedback controllers of form (5) based on Proposition 1. Notice that matrix inequalities in (12) are nonlinear due to nonlinear terms such as $\mathscr{C}_1^T P \mathscr{C}_1$ and $\mathscr{C}_2^T (d_1^2 R_1 + d_{12} R_2) \mathscr{C}_2$. The cone complementary linearization (CCL) proposed in [20] can be used to convert the non-convex feasible problem into nonlinear minimization problem subject to linear matrix inequalities. In doing so, we rewrite Proposition 1 in an equivalent form as follows.

Proposition 2. *For given scalars d_1 and d_2 satisfying $d_2 \geq d_1 > 0$, the system (6) is asymptotically stable if there exist $n \times n$ real matrices $P > 0, L > 0, Z > 0, S_1 > 0, S_2 > 0, S_3 > 0, R_1 > 0, R_2 > 0$ and $n \times 4n$ real matrices M_1 and M_2 such that*

$$\begin{bmatrix} \tilde{\Omega}_1 + \Omega_3 & d_{12} M_1^T & \mathscr{C}_1^T & \mathscr{C}_2^T \\ \star & -d_{12} R_2 & 0 & 0 \\ \star & \star & -L & 0 \\ \star & \star & \star & -Z \end{bmatrix} < 0, \quad \begin{bmatrix} \tilde{\Omega}_1 + \Omega_3 & d_{12} M_2^T & \mathscr{C}_1^T & \mathscr{C}_2^T \\ \star & -d_{12} R_2 & 0 & 0 \\ \star & \star & -L & 0 \\ \star & \star & \star & -Z \end{bmatrix} < 0 \tag{17}$$

$$PL = I, \quad Z(d_1^2 R_1 + d_{12} R_2) = I \tag{18}$$

where Ω_3 is defined in (15) and

$$\tilde{\Omega}_1 := e_1^T [S_1 + (d_{12} + 1) S_3 - P] e_1 - e_2^T S_3 e_2 + e_3^T (S_2 - S_1) e_3 - e_4^T S_2 e_4$$

Proof. Applying Schur complement to the inequalities in (12) yields (17) with equality constraints in (18). This completes the proof. □

Following the line of CCL in [20], the non-convex feasible problem described by (17) with equality constraints in (18) can be converted into the following nonlinear minimization problem:

$$\text{NMP:} \begin{cases} \text{minimize} & \text{Tr} \left(PL + Z(d_1^2 R_1 + d_{12} R_2) \right) \\ \text{subject to} & (17) \text{ and } \begin{bmatrix} P & I \\ I & L \end{bmatrix} \geq 0, \begin{bmatrix} Z & I \\ I & d_1^2 R_1 + d_{12} R_2 \end{bmatrix} \geq 0 \end{cases} \tag{19}$$

4   Numerical Examples

In this section, two numerical examples are given to demonstrate the effectiveness of the proposed method.

Example 1. *Consider the system (6), where K is supposed to be known. A and the product of B, K and C are given as*

$$A = \begin{bmatrix} 0.7 & 0 \\ 0 & 0.95 \end{bmatrix}, \quad BKC = \begin{bmatrix} -0.15 & 0 \\ -0.1 & -0.15 \end{bmatrix} \tag{20}$$

For prescribed $d_1 = 3$, we will calculate the admissible maximum upper bound (AMUB) of d_2, which ensures the asymptotic stability of system (20) subject to (4). By Theorem 1 in [21], Theorem 1 in [22] and Theorem 1 in [23], the obtained AMUBs are 9, 9 and 8, respectively. However, by using Proposition 1 in this paper, the obtained AMUB is 10, which is larger than those by [21, 22] and [23]. Moreover, for $d_1 = 0$, the AMUBs of d_2 obtained by Theorem 1 in [14] and Corollary 1 in this paper are 5 and 9, respectively. Clearly, Corollary 1 provides a larger AMUB than Theorem 1 in [14].

Example 2. *Consider an unstable batch reactor, which is taken from [14, 24]*

$$\dot{x}(t) = \begin{bmatrix} 1.38 & -0.2077 & 6.715 & -5.676 \\ -0.5814 & -4.29 & 0 & 0.675 \\ 1.067 & 4.273 & -6.654 & 5.893 \\ 0.048 & 4.273 & 1.343 & -2.104 \end{bmatrix} x(t) + \begin{bmatrix} 0 & 0 \\ 5.679 & 0 \\ 1.136 & -3.146 \\ 1.136 & 0 \end{bmatrix} u(t) \tag{21}$$

Discretizing this system with $T_s = 0.005s$ yields

$$x(k+1) = \begin{bmatrix} 1.0070 & -0.0010 & 0.0330 & -0.0278 \\ -0.0029 & 0.9788 & -0.0000 & 0.0034 \\ 0.0052 & 0.0211 & 0.9675 & 0.0288 \\ 0.0002 & 0.0211 & 0.0066 & 0.9897 \end{bmatrix} x(k) + \begin{bmatrix} 0.0000 & -0.0003 \\ 0.0281 & 0.0000 \\ 0.0060 & -0.0155 \\ 0.0060 & -0.0001 \end{bmatrix} u(k)$$

Xiong and Lam [14] studied the problem of state feedback controller design for an NCS with the plant being (21). With a logic ZOH, a stabilizing state feedback controller was derived for $1 \leq d(k) \leq 5$. However, solving the NMP in (19) with $C = I$ in this paper, for $1 \leq d(k) \leq 20$, one can get a stabilizing network-based controller as

$$u = \begin{bmatrix} 0.1669 & -0.0324 & -0.0264 & -0.0544 \\ 1.3107 & -0.0654 & 0.2062 & 0.0102 \end{bmatrix} x \tag{22}$$

This result shows clearly that under the controller (22) the asymptotic stability of the involved NCS can be ensured even if the maximum number of packet dropouts between any two consecutive updating instants of ZOH reaches 20, or even if the control input packet driving the plant is delayed by 20 sampling periods from the sampler. Connecting this controller, Figure 3 plots the initial condition response of the closed-loop system with $x_0 = [-5\ 0\ 5\ 0]^T$.

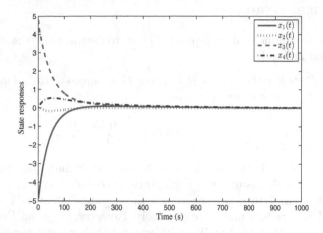

Fig. 3. State curves of the closed-loop system of Example 2

5   Conclusion

In this paper, output feedback control has been investigated for an NCS with a logic ZOH in the discrete-time domain. The closed-loop NCS has been modeled as a discrete-time linear system with an interval-like time-varying delay, whose upper bound reflects not only the maximum allowable delays induced by the network but also the maximum allowable number of packet dropouts between any two consecutive updating instants of ZOH. In order to derive a larger upper bound of the delay, a generalized finite-sum inequality has been introduced to obtain a less conservative stability criterion for the closed-loop NCS incorporating with a convex combination technique. By modifying the cone complementary algorithm, the desired network-based output feedback controllers have been designed in terms of solutions to a set of LMIs. Some numerical examples have been given to demonstrate the effectiveness of the proposed method.

References

1. Cloosterman, M., Van de Wouw, N., Heemels, W., Nijmeijer, H.: Stability of networked control systems with uncertain time-varying delays. IEEE Trans. Autom. Control 54, 1575–1580 (2009)
2. Hespanha, J.P., Naghshtabrizi, P., Xu, Y.: A survey of recent results in networked control systems. Proc. IEEE 95, 138–162 (2007)
3. Gupta, R.A., Chow, M.Y.: Networked control systems: Overview and research trends. IEEE Trans. Ind. Electron. 57, 2527–2535 (2010)
4. Yang, T.C.: Networked control system: A brief survey. IEE Proc. Control Theory Appl. 153, 403–412 (2006)
5. Zhang, X.-M., Han, Q.-L.: Event-triggered dynamic output feedback control for networked control systems. IET Control Theory Appl. 8, 226–234 (2014)

6. Zhang, X.-M., Han, Q.-L.: Network-based H_∞ filtering using a logic jumping-like trigger. Automatica 49, 1428–1435 (2013)
7. Peng, C., Han, Q.-L.: A novel event-triggered transmission scheme and control co-design for sampled-data control systems. IEEE Trans. Autom. Control 58, 2620–2626 (2013)
8. Yue, D., Tian, E., Han, Q.-L.: A delay system method for designing event-triggered controllers of networked control systems. IEEE Trans. Autom. Control 58, 475–481 (2013)
9. Zhang, L., Shi, Y., Chen, T., Huang, B.: A new method for stabilization of networked control systems with random delays. IEEE Trans. Autom. Control 50, 1177–1181 (2005)
10. Wang, Z., Ho, D.W.C., Liu, X.: Variance-constrained filtering for uncertain stochastic systems with missing measurements. IEEE Trans. Autom. Control 48, 1254–1258 (2003)
11. Zhang, W.A., Yu, L.: Output feedback stabilization of networked control systems with packet dropouts. IEEE Trans. Autom. Control 52, 1705–1710 (2007)
12. Zhang, W., Branicky, M., Phillips, S.: Stability of networked control systems. IEEE Control Syst. Magzine 21, 84–99
13. Yue, D., Han, Q.-L., Lam, J.: Network-based robust H_∞ control of systems with uncertainty. Automatica 41, 999–1007 (2005)
14. Xiong, J., Lam, J.: Stabilization of networked control systems with a logic ZOH. IEEE Trans. Autom. Control 54, 358–363 (2009)
15. Gao, H., Chen, T., Lam, J.: A new delay system approach to network-based control. Automatica 44, 39–52 (2008)
16. Zhang, X.-M., Han, Q.-L.: A delay decomposition approach to H-infinity control of networked control systems. Europ. J. Control 15, 523–533 (2009)
17. Jiang, X., Han, Q.-L.: Stability criteria for linear discrete-time systems with interval-like time-varying delay. In: Proc. Amer. Control Conf., pp. 2817–2822 (2005)
18. Shao, H., Han, Q.-L.: New stability criteria for linear discrete-time systems with interval-like time-varying delays. IEEE Trans. Autom. Control 56, 619–625 (2011)
19. Zhang, X.-M., Han, Q.-L.: A novel finite-sum inequality for stability of discrete-time linear systems with interval-like time-varying delays. In: Proc. 49th IEEE CDC, pp. 708–713 (2010)
20. Ghaoui, L., Oustry, F., Ait Rami, M.: A cone complementarity linearization algorithms for static output feedback and related problems. IEEE Trans. Autom. Control 42, 1171–1176 (1997)
21. He, Y., Wu, M., Liu, G.-P., She, J.-H.: Output feedback stabilization for a discrete-time systems with a time-varying delay. IEEE Trans. Autom. Control 53, 2372–2377 (2008)
22. Guo, Y., Li, S.: New stability criteria for discrete-time systems with interval time-varying state delay. In: Proc. 48th IEEE Conf. Decision Control, pp. 1342–1347 (2009)
23. Zhang, B., Xu, S., Zou, Y.: Improved stability criterion and its applications in delayed controller design for discrete-time systems. Automatica 44, 2963–2967 (2008)
24. Walsh, G., Ye, H.: Scheduling of networked control systems. IEEE Control Syst. Magzine 21, 57–65 (2001)
25. Gao, H., Chen, T.: New results on stability of discrete-time systems with time-varying state delay. IEEE Trans. Autom. Control 52, 328–334 (2007)
26. Peng, C., Tian, Y., Yue, D.: Output feedback control of discrete-time systems in networked environments. IEEE Trans. Syst., Man, Cybern. A, Syst., Humans 41, 185–190 (2011)

Passivity-Based Control for Fractional Order Unified Chaotic System*

Qiao Wang** and Donglian Qi

Department of Electrical Engineering, Zhejiang Uinversity, Hangzhou 310027, China
{qiao,qidl}@zju.edu.cn

Abstract. This paper concerns with the fractional order unified chaotic control based on passivity. A hybrid control strategy combined with fractional state feedback and passive control is proposed, derived from the properties of fractional calculus and the concept of passivity. The fractional chaotic system with the hybrid controller proposed can be stabilized at its equilibrium under different initial conditions. Numerical simulation results present the verification on the effectiveness of the proposed control method.

Keywords: fractional order systems, passivity, fractional calculus.

1 Introduction

Fractional order systems have received considerable attention, due to the fact that they can provide better intrinsic essence and accuracy than the classical integer order models in describing real world physical phenomena [1]. Since the restriction of the development of mathematics and physics, the fractional calculus has not been studied enough at the time it was proposed, until the last century where science and technology had a revolutionary development and many physical and biological phenomena can be given natural interpretation by the fractional order models[2, 3]. During the passed several years, there appears a lot of researches with the focus on the fractional differential equations, fractional order nonlinear system stability and control, as well as numerical solution and simulation [4–6], which consequently enable a variety of control strategies and applications for fractional order chaotic systems [7–10]. Many traditional chaotic systems have been generalized into their corresponding fractional order circumstances, where the chaos behaviors are still preserved under certain constrains on the parameters of the fractional order, such as fractional order Chua system [11], the fractional order Voltas system [12], fractional Lotka-Volterra system [13], as well as fractional order unified chaotic system. Therefore, it is important to explore effective control methods to deal with such fractional order chaotic systems. Among the literatures appeared, certain classical control

---

* This work is supported by National Natural Science Foundation of China (Grant No. 61171034) and Zhejiang Provincial Natural Science Foundation of China (Grant No. R1110443).
** Corresponding author.

M. Fei et al. (Eds.): LSMS/ICSEE 2014, Part II, CCIS 462, pp. 310–317, 2014.
© Springer-Verlag Berlin Heidelberg 2014

strategies have been introduced into the fractional order systems, including active control, linear state feedback, sliding mode control, Lyapunov method and so on [14–16]. Furthermore, inspired by the fractional order phenomena, some fractional order controllers have been proposed, one of which is the well-known fractional order PID controller with certain excellent performances [17].

Recently, passivity-based control attracts attention by the researchers [18, 19]. The concept of passivity, derived from the network theory and dissipative systems, deals with the control problem under the perspective of system energy characteristics to simplify the controller design, such that it has been applied successfully into many nonlinear systems. Since the fact that there exists certain inherent consistence between passivity and system stability, it is practicable to stabilize the nonlinear system by means of designing a controller to render the closed loop system passive, under appropriate constrains on the system [20]. Up to now, some passivity-based control methods have been introduced into chaotic systems, such as passive active control for chaotic systems, passive sliding mode control, and passive robust control and so on [19, 21]. However, most of the existed works are constrained in the scenario where only the integer order chaotic systems are considered. In this article, we proposed a hybrid control method to deal with the fractional chaotic systems, which combines the fractional order feedback and passivity-based control such that the closed loop system can be passive and then stabilized at its equilibrium.

2   Problem Formulation and Preliminaries

2.1   Fractional Calculus and Properties

Definition 1. *Let $f : [a, b] \to R$ and $f \in L^1[a, b]$, The Riemann-Liouville fractional derivative of order is defined as:*

$$D^\alpha f(t) = \frac{1}{\Gamma(n - \alpha)} \frac{d^n}{dt^n} \int_0^t \frac{f(\tau)}{(t - \tau)^{\alpha - n + 1}} d\tau \tag{1}$$

where $n - 1 < \alpha < n$ and $\Gamma(\cdot)$ is the Gamma function.

For fractional order chaotic system $0 < \alpha \leq 1$. In this paper we employ D for representing the classical integer differential $D^1 f(t) = df(t)/dt$. The fractional calculus has the following properties [2]

$$\begin{aligned} D^\alpha D^\beta f(t) &= D^{\alpha + \beta} f(t) \\ D^\alpha D^{-\alpha} f(t) &= D^0 f(t) = f(t) \\ D^\alpha [af(t) + bg(t)] &= aD^\alpha f(t) + bD^\alpha g(t) \end{aligned} \tag{2}$$

where α and β are fractional orders, a and b are real constants.

2.2   Theory of Passive Control

Passivity-based control investigates the dynamical systems in the view of system input, output and energy, through geometric nonlinear control theory for the design of nonlinear feedback systems. If the supply from external source cannot satisfy internal energy lost, then the system energy will decrease gradually and then the system will reach an asymptotical stability eventually.

Consider nonlinear system described by the equation of the form

$$\dot{x} = f(x) + g(x)u$$
$$y = h(x) \tag{3}$$

where the state space $X = R^n$, system input $U = R^m$ and system output $Y = R^m$; $f(x)$, $g(x)$ and $h(x)$ are smooth mappings. We can denote a nonnegative function $V(x) : X \to R$ the storage function of the system (3). Moreover, we call a real-valued function $s(y, u) = <y, u> = y^T u$ the supply rate, such that

$$\int_0^t s(\tau)d\tau < \infty \ \forall t \geq 0 \tag{4}$$

Definition 2. *The system (3) is said to be passive with the storage function and supply rate, if the following inequality holds*

$$V(x) - V(x_0) \leq \int_0^t y^T(\tau)u(\tau)d\tau \tag{5}$$

and it is equivalent to say that: the system is passive if there exists real constants β and $\rho > 0$ such that

$$\int_0^t y^T(\tau)u(\tau)d\tau + \beta \geq \int_0^t \rho y^T(\tau)y(\tau)d\tau \tag{6}$$

Furthermore, the system (3) can be represented by its equivalent expression called the normal form in new coordinate (z, y), where

$$\dot{z} = f_0(z) + p(z, y)y$$
$$\dot{y} = b(z, y) + a(z, y)u \tag{7}$$

Lemma 1 (Bynes [20]). *The system (3) is called a minimum phase system if $L_g h(0)$ is nonsingular and $z = 0$ is an asymptotically stable equilibrium of $f_0(z)$.*

Theorem 1 (Bynes [20]). *Suppose that the system (3) is a minimum phase system and $L_g h(0)$ is nonsingular, then there exists a feedback $u = \varphi(x)$ such that the controlled system is equivalent to be passive and then stable asymptotically at its equilibrium.*

3   Passive Control for Fractional Order Unified Chaotic System

Consider the fractional order unified chaotic system

$$D^\alpha x_1 = f_{x_1}(x) = (25\theta + 10)(x_2 - x_1)$$
$$D^\alpha x_2 = f_{x_2}(x) = (28 - 35\theta)x_1 - x_1 x_3 + (29\theta - 1)x_2, \ 0 < \alpha \leq 1 \tag{8}$$
$$D^\alpha x_3 = f_{x_3}(x) = x_1 x_2 - \frac{8+\theta}{3}x_3$$

Here we choose a fractional feedback controller $u = [u_{x_1}\ u_{x_2}\ u_{x_3}]^{\mathrm{T}}$, where

$$
\begin{aligned}
u_{x_1} &= (D^{\alpha-1} - 1)f_{x_1}(x) \\
u_{x_2} &= (D^{\alpha-1} - 1)f_{x_2}(x) + u_{\mathrm{I}} \\
u_{x_3} &= (D^{\alpha-1} - 1)f_{x_3}(x)
\end{aligned}
\tag{9}
$$

Insert (9) into (8), and we obtain the controlled system

$$
\begin{aligned}
D^\alpha x_1 &= f_{x_1}(x) + u_{x_1} = f_{x_1}(x) + (D^{\alpha-1} - 1)f_{x_1}(x) \\
D^\alpha x_2 &= f_{x_2}(x) + u_{x_2} = f_{x_2}(x) + (D^{\alpha-1} - 1)f_{x_2}(x) + u_{\mathrm{I}} \\
D^\alpha x_3 &= f_{x_3}(x) + u_{x_3} = f_{x_3}(x) + (D^{\alpha-1} - 1)f_{x_3}(x)
\end{aligned}
\tag{10}
$$

Take fractional derivative $D^{1-\alpha}$ on both sides of the equation (10), and note that $D^\alpha D^\beta = D^{\alpha+\beta}$, then we have

$$
\begin{aligned}
D^{1-\alpha}D^\alpha x_1 &= Dx_1 = D^{1-\alpha}(f_{x_1}(x) + u_{x_1}) = f_{x_1}(x) \\
D^{1-\alpha}D^\alpha x_2 &= Dx_2 = D^{1-\alpha}(f_{x_2}(x) + u_{x_2}) = f_{x_2}(x) + D^{1-\alpha}u_{\mathrm{I}} \\
D^{1-\alpha}D^\alpha x_3 &= Dx_3 = D^{1-\alpha}(f_{x_3}(x) + u_{x_3}) = f_{x_3}(x)
\end{aligned}
\tag{11}
$$

thus the system (8) is transformed into an integer differential form by the controller u.

If we design a passive controller $u_p = D^{1-\alpha}u_{\mathrm{I}}$ to render the system (11) to be passive and then stable asymptotically, then the system (8) can be stabilized under the controller u defined in equation (9), with $u_{\mathrm{I}} = D^{\alpha-1}u_p$.

Let $y = x_2, z_1 = x_1\ z_2 = x_2$ equation (11) can be rewritten as its normal form

$$
\begin{aligned}
\dot{z} &= f_0(z) + p(z,y)y \\
\dot{y} &= b(z,y) + a(z,y)u
\end{aligned}
\tag{12}
$$

where $f_0(z) = [-(25\theta + 10)z_1\quad \frac{8+\theta}{3}z_2]^{\mathrm{T}}$, $p(z,y) = [25\theta + 10\quad z_1]^{\mathrm{T}}$, $a(z,y) = 1$, and $b(z,y) = (28 - 35\theta)z_1 + (29\theta - 1)y - z_1 z_2$.

Theorem 2. *The system (12) is a minimum phase system.*

Proof. For subsystem $\dot{z} = f_0(z)$, where

$$
\begin{aligned}
\dot{z}_1 &= -(25\theta + 10)z_1 \\
\dot{z}_2 &= -\frac{8+\theta}{3}z_2
\end{aligned}
\tag{13}
$$

choose Lyapunov function $W(z) = \frac{1}{2}z_1^2 + \frac{1}{2}z_2^2$, then we obtain $\dot{W}(z) = -(25\theta + 10)z_1^2 - \frac{8+\theta}{3}z_2^2 \le 0$. thus is $z = 0$ an asymptotical equilibrium of $f_0(z)$. Since $L_g h(0) = 1$ is nonsingular, according to Lemma 1, we conclude that the system (12) is a minimum phase system.

Therefore, we have the following stability theorem

Theorem 3. *For system (12), if we choose the passive controller*

$$
u_p = a^{-1}(z,y)[-b^{\mathrm{T}}(z,y) - \frac{\partial W(z)}{\partial z}p(z,y) - ky + v]
\tag{14}
$$

where $k > 0$ and v is an external input. Then the system can be stabilized at its equilibrium.

Proof. Choose the storage function $V(z,y) = W(z) + \frac{1}{2}y^2$, then we obtain

$$
\begin{aligned}
\dot{V} &= \frac{\partial W}{\partial z}\dot{z} + y\dot{y} \\
&= \frac{\partial W}{\partial z}f_0(z) + \frac{\partial W}{\partial z}p(z,y)y + [b(z,y) + a(z,y)u_p]y
\end{aligned}
\tag{15}
$$

since the system is minimum phase [20], thus

$$
\frac{\partial W}{\partial z}f_0(z) \leq 0
\tag{16}
$$

then

$$
\dot{V} \leq \frac{\partial W}{\partial z}p(z,y)y + [b(z,y) + a(z,y)u_p]y
\tag{17}
$$

put the passive controller into inequality (17), we obtain

$$
\dot{V} \leq -ky^2 + vy
\tag{18}
$$

take integration on both side of (18), we have

$$
V(z,y) + \int_0^t ky^2(\tau)d(\tau) \leq \int_0^t v(\tau)y(\tau)d(\tau) + \beta
\tag{19}
$$

where $\beta = V(z_0, y_0)$. Since the storage function $V(z,y) \geq 0$, we obtain

$$
\int_0^t y(\tau)v(\tau)d\tau + \beta \geq \int_0^t ky^2(\tau)d\tau
\tag{20}
$$

according to definition 2, the system is passive. Furthermore, theorem 2 shows that the system (12) is minimum, and note that $L_g h(0)$ is nonsingular, thus by theorem 1, the passive controller proposed stabilizes the system asymptotically at its equilibrium.

Finally, the system (8) can be stabilized under the hybrid controller u defined in equation (10), with $u_I = D^{\alpha-1}u_p$, where is the passive controller in equation (14).

4   Simulation

There are several numerical simulation methods which have been applied broadly in nonlinear control systems, such as GL numerical method based on power series expansion, CRONE-Oustaloups approximation, Podlubny matrix approach, Adams-Bashforth-Moulton type predictor-corrector scheme and so on [2]. This paper adopts the last one for simulation.

The hybrid controller proposed in equation (19) consists of two components, the fractional order feedback and passive controller respectively. Since the fractional order feedback controller has no control parameter, the only two control

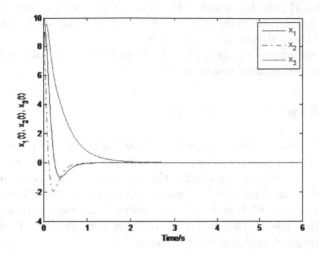

Fig. 1. State trajectories when $\alpha = 0.96$, $\theta = 0.2$, $k = 3$, $v = 0$

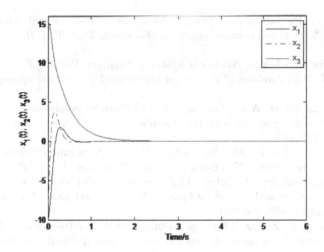

Fig. 2. State trajectories when $\alpha = 0.9$, $\theta = 0.9$, $k = 0.23$, $v = 0$

parameters, k and v, appear in the fractional order passive controller u_I . According to equation (14), we have

$$u_\mathrm{I} = D^{\alpha-1}u_p = D^{\alpha-1}[-(38 - 10\theta)x_1 - (29\theta - 1 + k)x_2 - x_1x_3 + v] \qquad (21)$$

When $\theta \in [0, 0.8)$, the fractional unified chaotic system appears as a general Lorenz system, figure 1 shows the stability of the controlled system, with $\alpha = 0.96$, $\theta = 0.2$, $k = 3$, $v = 0$, and $x_0 = [10 \quad 9 \quad 8]^\mathrm{T}$.

When $\theta \in (0.8, 1]$, the fractional unified chaotic system appears as a general Lorenz system, figure 2 shows the stability of the controlled system, with $\alpha = 0.9$, $\theta = 0.9$, $k = 0.23$, $v = 0$, and $x_0 = [-10 \ -8 \ 15]^T$.

The two simulation results above show that the hybrid controller can stabilize the fractional unified chaotic system asymptotically.

5   Conclusions

This paper deals with the control problem of the fractional order unified chaotic system, under the perspective of passivity. On the basis of the fractional calculus properties and the passive control theory, a hybrid controller is designed which can stabilize the system at its equilibrium asymptotically. The concept of passivity concerns only with system input, output and energy, other than the system description model, therefore it provides a new direction to investigate the control of the fractional nonlinear systems.

References

1. Skovranek, T., Pudlubny, I., Petras, I.: Modeling of the national economiees in state space: A fractional calculus approach. Economic Modelling 29(4), 1322–1327 (2012)
2. Petras, I.: Fractional-order Nonlinear Systems. Springer, Berlin (2011)
3. Diethelm, K.: The Analysis of Fractional Differential Equations. Springer, Berlin (2011)
4. Ahmed, E., EI-Sayed, A.M.: On some Routh-Hurwitz conditions for fractional order differential equations and their applications in Lorenz. Physics Letters A 358(2), 1–4 (2006)
5. Deng, W., Li, C., Lu, J.: Stability analysis of linear fractional differential system with multiple time delays. Nonlinear Dynamics 48(8), 409–416 (2007)
6. Sierociuk, D., Pudlubny, I., Petras, I.: Experimental Evidence of Variable-Order Behavior of Ladders and Nested Ladders. IEEE Transactions on Control Systems Technology 21(5), 459–466 (2013)
7. Qi, D.L., Yang, J., Zhang, J.L.: The stability control of fractional order unified chaotic system with sliding mode control theory. Chinese Physics B 19(7), 100506 (2010)
8. Fu, J., Yu, M., Ma, T.D.: Modified impulsive synchronization of fractional order hyperchaotic systems. Chinese Physics B 20(3), 120508 (2011)
9. Zhang, B., Pi, Y., Luo, Y.: Fractional order sliding mode control based on parameter auto-tuning velocity control of permanent magnet synchronous motors. ISA Transactions 51(9), 649–656 (2012)
10. Razminia, A., Baleanu, D.: Complete synchronization of commensurate fractional order chaotic systems using sliding mode control. Mechatronics 23(6), 873–879 (2013)
11. Lu, J.G.: Chaotic dynamics and synchronization of fractional order Chua's circuit. International Journal of Modern Physics 20(3), 3249–3259 (2005)
12. Petras, I.: Chaos in the fractional order Volta's system. Nonlinear Dynamics 57(5), 157–170 (2009)

13. Ahmed, E., EI-Sayed, A.M.: Equilibrium points, stability and numerical solutions of fractional order predator-prey and rabies models. Journal of Mathematical Analysis and Applications 325(1), 542–553 (2007)
14. Mohammad, P.A.: Finite-time chaos control and synchronization of fractional-order nonautonomous chaotic (hyperchaotic) systems using fractional nonsingular terminal sliding mode technique. Nonlinear Dynamics 69(5), 247–261 (2012)
15. Razminia, A., Baleanu, D.: Complete synchronization of commensurate fractional order chaotic systems using sliding mode control. Mechatronics 23(1), 873–879 (2013)
16. Bartolini, G., Pisano, A., Usai, E.: Digital second order slding mode control for uncertain nonlinear systems. Automatica 37(2), 1371–1377 (2010)
17. Podlubny, I.: Fractional Differential Equations. Academic Press, San Diego (1999)
18. Wu, L., Zheng, W.X.: Passivity-based sliding mode control for uncertain singular time-delay systems. Automatica 45(5), 2120–2127 (2009)
19. Dadras, S., Momeni, H.R.: Passivity-based fractional order integral sliding mode control design for uncertain fractional order nonlinear systems. Mechatronics 23(6), 880–887 (2013)
20. Byrnes, I., Aeberto, I., Willems, J.C.: Passivity, Feedback Equivalence, and the Global Stabilization of Minimum Phase Nonlinear Systems. IEEE Transactions on Automatic Control 36(2), 1228–1240 (1991)
21. Uyaroglu, Y., Emiroglu, S.: Passivity-based chaos control and synchronization of the four dimensional Lorcnz-Stenflo system via one input. Journal of Vibration and Control 56(3), 435–442 (2013)

Characteristics Analysis of Cloud Services
Based on Complex Network

Lilan Liu, Cheng Chen[*], and Tao Yu

Shanghai Key Laboratory of Intelligent Manufacturing and Robotics, Shanghai University,
Yanchangstr 149, Zhabei, Shanghai, China

Abstract. This paper researched the theory of cloud services based on complex network theory and graph theory. By virtue of the generalized network model of cooperation, a method of building cloud services network model was put forward, with regarding cloud services and their cooperative relationship as vertexs and edges in abstraction respectively. The network statistical properties are analyzed, including degree, degree distribution, shortest path length, vertex betweenness, similar matching coefficient and clustering coefficient. Using an example about fulfilling an order of LED spotlight, characteristics analysis of cloud services was suggested, to illustrate how to choose the appropriate cloud services.

Keywords: Complex network, Cloud services, Topological characteristics.

1 Introduction

The emergence of cloud manufacturing proposes a newly developed method for the transformation from manufacturing industry to service industry. Cloud manufacturing is a new networked intelligent manufacturing pattern which is service-oriented、 cost-effective and knowledge-based [1]. With the help of cloud computing, EPC network, semantic Web technology etc., Cloud manufacturing virtualizes various manufacturing resources and capabilities, operating and managing uniformly and ultimately realizes the function of accessibility readily and use on demand in the whole service of manufacturing life circle orienting all the members in the whole industry including enterprises, sale agents and end user.

In the environment of cloud manufacturing, lots of manufacturing sources and capabilities are involved which are all encapsulated as individual cloud manufacturing services (*cloud services*) through network virtualization based on knowledge. In the process of providing service to customers, choosing one or more suitable cloud service according to user requirements and provide users with reliable, efficient, low-

---

[*] Corresponding author.

Supported by the National High Technology Research and Development Program of China (Key Technology R&D Program):2012BAF12B01; the National High Technology Research and Development Program of China (863 Program): 2013AA03A112.

M. Fei et al. (Eds.): LSMS/ICSEE 2014, Part II, CCIS 462, pp. 318–330, 2014.

cost manufacturing services. A large amount of cloud services forms a complex system. In cloud manufacturing, simplify each cloud service and the relationship between different cloud services as the vertex and edge between the vertices empowered or not, thus the study of cloud services can be transformed into complex network model through that characteristics of cloud services can be analyzed by the means of topological characteristic in complex network.

2　Cloud Services and Complex Network Model

2.1　Cloud Services

Cloud Manufacturing can be treated as extend and practice of cloud computing in manufacturing industry. Cloud Manufacturing uses cloud computing, Internet of Things, semantic Web technology to virtualize the *Manufacturing Resource* and *Manufacturing capability* and encapsulate them into *Manufacturing of Cloud Services*, in the cloud manufacturing platform. And combine of multiple of *Manufacturing of Cloud Services* to form the *Manufacturing of Cloud* finally. Any enterprise or individual which distributed in different areas can visit the cloud manufacturing platform through network. The network composites cloud services according to the *Personalization of Demand* [2] (Fig.1). During this, *knowledge* plays an important role in virtualization encapsulation and display of the requirements, and provides service-oriented, cost-effective and knowledge-based networked intelligent manufacturing pattern in the process of cloud manufacturing service [1].

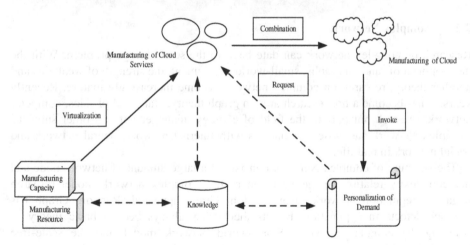

Fig. 1. The Working Principle Diagram of Cloud Services

The key of cloud manufacturing, which simultaneously support efficient management and intelligent search etc., is how to overcoming the barriers of distributional, heterogeneous manufacturing sources and capacities, and encapsulates them into the cloud manufacturing platform treating knowledge as the carrier. Cloud manufacturing can provides customers dynamically and flexibly with reliable and low-cost individual services [3].

When customers' individual services involving varies resources, the most suitable cloud services must be chosen in the cloud manufacturing platform and composited into manufacturing clouds. Cloud services can composite into various manufacturing clouds according to customers' individual requirements. How to customize the optimal service composition has been the key point in cloud manufacturing. Cloud manufacturing service composition generally can be divided into three stages: ① requirement decomposition, which decomposed customer requirements into several service modules; ②cloud services match, which aiming at the decomposed service modules to search matching cloud services in the cloud manufacturing platform; ③ cloud services composition, which composites the searched cloud services into a feasible task flow.

Cloud manufacturing service composition is the key procedure to realize effective allocation of cloud manufacturing resource. In the environment of cloud manufacturing, manufacturing enterprises provide customers with its own competitive advantages in the form of manufacturing resources and manufacturing capabilities. Usually, in order to safeguard the operation enterprise itself as well as balance the completion of manufacturing orders and high invoking ratio, requirements in cloud manufacturing are often composited together to execute a task. Multitude of cloud services thus compose a service network. Combining with the knowledge and method of complex networks to study characteristic of cloud services composition is the chief of this paper.

2.2   Complex Network

Research on complex network can date back to the seven bridge problem. With the development of random graphs, small-world experiment, the strength of weak ties and another theory, research on complex network become increasingly mature. Recently years, it has become a hot research area in graph theory, statistical physics, computer network, research hotspots in the field of ecology, management and other subjects. Complex Network has close associations with Internet network, neural network and social network in real life.

The structure of complex Network consists of a large amount of network node and the complexity relationship between the nodes. Complex network model research began in regulation network, then Random Graph appeared. Random Graph has obvious defects in application, but its theory has always been a basic theory of studying the complex network. Small-world network model and the scale-free network model has attracted a great deal of attention of complex networks researches. Through researches on complex networks, people can be visualization and quantitative for many application scenarios, and find out things' development characteristics, then predict the possible running state within a certain range.

In complex networks has a kind network named cooperative network, partnership between each basic unit is widespread in this network. Such as scientific researchers' cooperation can produce scientific papers and actors' cooperation can produce films and television programs in social network. In non-social network, many traditional Chinese medicines' cooperation can produce prescriptions to treat diseases, tourist attractions' cooperation can produce hot tourist routes which attract visitors, and bus stations' cooperation can improve the efficiency of transportation vehicle. The edges of cooperative network always stand for the cooperative relationship between elementary units. [4] This so-called affiliation network usually expressed by two-particle diagram. A kind of node is the *actor* participated in activity, event or organization, the other is the activity, event or organization called *act*. An edge connecting two points shows that two *actors* participated in the same *act*.

As shown in Fig.2, if just one type of particle is considered (just as particles in type "*actor*"), draw the single particle's projections. Two single particles of this type share the same particle in other type (just as particles in type "*act*"), with whom forms two edges. Then project the two edges as the edge of the two single particles (edges between two particles in type "*actor*"). Edges joining particles mean that participants worked with the same project. If there are multiple edges between two particles, then the more multiple edges, the more often participants work together.[5]

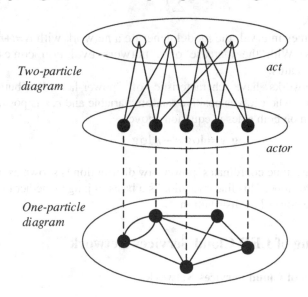

Fig. 2. Two-particle and one-particle diagram

Define a parameter h_i to represent "vertex act degree". If actor i participates in several acts, then a parameter h_i representing "vertex act degree" can be defined, which equals to the particle number that actor i connect with in two-particle diagram. In the same way, if there are several actors participate in the connection, then define a parameter T_j which means "number of nodes inside an act", which equals to the particle number that act j connect with in two-particle diagram[6]. Those actor particles constitute a complete graph due to that they have worked with. Vertex act degree and number of nodes inside an act are important properties in cooperation network which depict the important information of cooperation network.

In research of cooperation network, Barabasi and Albert[7] research on Scale-Free of cooperation network, and put forward the famous BA model: In an open system, there are new units joining in, and the total number of nodes increases continually, and probability a note form a new edge monotonously rely on the degree it has possessed. Its construction algorithm is:

(1) Increase: At the initial time, assuming there has been m_0 notes. Then every once in a while, add a new note, and connect it to m notes exist already $(m \leq m_0)$;

(2) Optimization: When connect new note to existed note i , assuming the relationship between the connection probability Π_i that note i connect to the new note and the degree of note i -- k_i, the degree of note j --k_j satisfy the following relations:

$$\Pi_i = \frac{k_i}{\sum_j k_j} \tag{1.1}$$

(3) After t times interval, the model generate a network with $N=t+m_0$ notes and m_t edges. With the increase of t, network evolution comes into scale invariant status.

Scale-free networks have characteristic of "power-law distribution", whose equation is: $y=cx^{-r}$, where x, y is positive random variable and c, r is positive constant. To take logarithm on both sides in equation above:

$$\log y = \log c - r \log x \tag{1.2}$$

In double logarithmic coordinates, power-law distribution is shown as a straight line with negative slope. This linear relation is a basis to judge whether random variable meet the power law distribution.

3   Modeling of LED Cloud Services Network

3.1   Definition of Cloud Services Network

Quantity of cloud services in cloud manufacturing platform aggregate into "cloud pool", in which stores a larger number of cloud services. How to custom low-priced, convenient services according to customer's requirements rapidly is the top priority of cloud manufacturing. The composition of cloud services can be solved by complex network theory, treating each cloud service as a note in complex network and service connections among different cloud services as edges in complex network. As shown in Fig.3, according to different mission requirements, there are different weight value of edges. This method can realize functions of cloud services composition relations, such as visual display, service composition network construction, kinetic-analysis of cloud services composition network, node control strategy and method in cloud services composition network and so on[2].

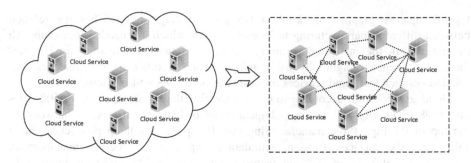

Fig. 3. Cloud services network

In the field of manufacturing, distribution, heterogeneous and massive number characteristic of manufacturing resource result in the complexity of cloud manufacturing service composition, thus a complex network among cloud services forms – a complex network model through building cloud services composition. In this paper, cooperation network is applied to cloud services composition. When building cloud services composition model, divide cloud services according to the corresponding relationship of subtask--- cloud services, thus bipartite graph between subtask T which split from ancestral task and cloud service S can be built. Then project cloud service to ingle particle network, treating each cloud service as a vertex and treating interrelation between the cloud services which participate in the same case as the edges connecting two vertex. By this way, a cloud service composition network can be built.

3.2   Cloud Services Network Building

Any network can be abstracted as a figure of node sets and edge sets: $G = (V,\ E)$, where V is a nonempty set representing node sets; E representing edge sets is a binary set consists of elements in V. When no edge in the network has direction – in other words, E is an unordered binary set which is composed of elements in V – then this network represents an undirected network; if not, this network represents a directed network. When each edge is endowed with weights, then this network represents weighting network; if not, it represents un-weighted network.

This paper conducts the study by collecting relevant information via a LEDs lighting cloud services platform. This platform has 127 LEDs lighting enterprises, including 602 cloud services involving design service, simulation service, processing service, management service, license service, marketing service and many other services which can satisfy user requirement based manufacturing tasks. Taking cloud services composition problem in this platform as a case to build a cloud services composition network. Abstract a manufacturing task into a task note, and abstract a cloud service into a service note. If one manufacturing task need to call a cloud service, then there is a relevance between this manufacturing task and cloud task, which can be abstracted into an edge connecting task notes and service notes. According to this abstract principles, a task-service network can be built, and this is a typical bipartite graph. This kind of two-particle diagram describes the relation between different manufacturing tasks and services which are needed to accomplish these tasks. In the next step, simplify the directed two-particle diagram into un-directed asymmetric figure by using this relation. In other words, transform the relation between tasks and services into the relation between the various services.

Cloud services built according to the construction theory above involves 602 notes and 8908 edges. Fig.4. depicts topological graph of LED cloud services network. In the top of the Fig., three manufacturing task (lamp design, lamp production, lamp sale) and six services (design, simulation experiment, process, management, certification, sale) comprise "manufacturing task- service" two-particle diagram. In the below of the Fig. is single-particle diagram consists of services which corporate in the same manufacturing task.

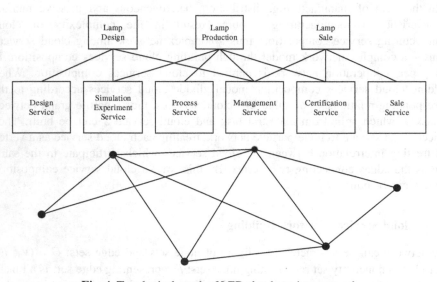

Fig. 4. Topological graph of LED cloud services network

4   Analysis of Topological Characteristics of Cloud Services Network

4.1   Degree and Degree Distribution

Network's degree and degree distribution should be focused first. In cloud services, node's $s(S_i)$ degree k_i is the amount of other nodes connecting to S_i. k_i reflects the connection degree of the node S_i and other nodes. The Network's average degree is the average of all nodes' degree.

$$k = \frac{1}{N} \sum_{i=1}^{N} k_i \ . \tag{3.1}$$

Degree distribution is the probability distribution of all nodes' degree. For the cloud service node, degree and degree distribution fully reflect the coupling of each cloud service node. Coupling is a key character to combine cloud services. Table 1 shows the top five degree cloud services.

Table 1. Degrees of the top 5 cloud services. The larger degree the cloud service has, the more other services will connect to it. Company A~E have larger degree that shows this five companies are invoked frequently.

Cloud Service	Node ID	Node Degree
Company A: Processing services of lamps and lanterns	5	560
Company B: Chip production services	23	446
Company C: Certification testing services	467	358
Company D: Power production services	215	276
Company E: Sales services of lamps and lanterns	510	154

Through the statistical analysis by *NetworkX*, in cloud services network node degree distribution $P\ (k)$ is similar to follow power-law distribution, as shown in figure 5. $P\ (k)$ is the proportion of the amount of node which one's degree is k in the total nodes. On the results found in previous research, most of the social network topology is between completely random and completely rules. The power-law distribution is exactly between exponential distribution and power distribution.

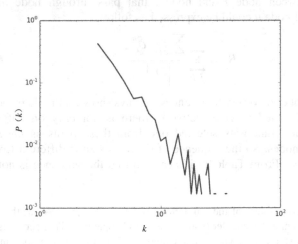

Fig. 5. Node degree distribution in double-logarithmic coordinates show $P\ (k)$ is similar to follow power-law distribution which is exactly between exponential distribution and power distribution

4.2   Shortest Path Length

In networks, shortest path length is equal to the number of edges of the path which has the least nodes among all the path from node i to node j. And the average of that is the average path length:

$$l = \frac{\sum_{i \ne j} l_{ij}}{N(N-1)}. \qquad (3.2)$$

By calculation, the average path length of LED cloud services network is 3.26. The maximal distance between any two cloud services is 4. This characteristic indicates LED cloud services network has the obvious phenomenon of small world.

In cooperative networks, there is cooperation relationship between neighbor nodes, and the relationship could be extended by nodes. Therefore in LED cloud services network, some two nodes without directly connecting by edge may participate in same one manufacturing task. And average path length reflects the length of cloud services chain on the whole.

4.3   Vertex Betweenness

Betweenness is a measure of node's centrality, and vertex betweenness is the amount of the shortest path from all vertices to all other that pass through that node. The number of vertex betweenness depicts the importance of node. Define D_{ij} as the set of shortest path between node i and node j that pass through node u. The vertex betweenness B_u of node u could be expressed as follow:

$$B_u = \sum_{i,j} \frac{\sum_{\in D_{ij}} \delta_l^u}{|D_{ij}|}. \qquad (3.3)$$

The larger number of vertex betweenness always show that node is the hub node in network. In LED cloud services network, there is not only one of shortest path between two nodes, randomly selected one from these paths as *the shortest path* between the two nodes. So the sequence of degree's size is different from the one of vertex betweenness. From Table 2., it's easy to find the sequence is not the same of that in Table 1.

Table 2. Vertex Betweenness of the top 3 cloud services. Company A has the largest degree, but its vertex betweenness is smaller than the one of Company C. That accounts for Company A connects the most other services, and Company C is the more important service than Company A. In many service combination, Company is indispensable.

Cloud Service	Node ID	Vertex Betweenness
Company C: Certification testing services	467	34678
Company A: Processing services of lamps and lanterns	5	23156
Company B: Chip production services	23	12786

4.4   Similar Matching Coefficient

Similar matching coefficient r (g) signify the connection tendency between nodes. Graph G contains N nodes and l edges, then could be defined as follow:

$$r(g) = \frac{\sum\limits_{i,j\in\varepsilon} d_i d_j / l - \left(\sum\limits_{i,j\in\varepsilon}(d_i+d_j)/l\right)^2}{\sum\limits_{i,j\in\varepsilon}\frac{1}{2}(d_i^2+d_j^2)/l - \left(\sum\limits_{i,j\in\varepsilon}\frac{1}{2}(d_i+d_j)/l\right)^2}. \tag{3.4}$$

In (3.4), $i, j \in \varepsilon$ are edges in Graph G, if r (g) > 0, then r (g) is called similar matching coefficient; if r (g) < 0, then it's non- similar matching coefficient. In LED cloud services network, r (g) $= 0.024$, belongs to similar matching coefficient. A property of this kind of network is that the node of lager degree always be connected by others easily. Such as Company A could be in collaborations frequently because of its largest degree (560).

4.5   Clustering Coefficient

Clustering coefficient c reflects nodes' ability of gathering. In LED cloud services network, clustering coefficient c_i indicates tightness of the connection between cloud service node S_i and its neighbors. Definition of c_i as follow:

$$c_i = \begin{cases} \dfrac{2n_i}{k_i \times (k_i - 1)} & k_i > 1 \\ 0 & k_i \le 1 \end{cases}. \tag{3.5}$$

And the average clustering coefficient C is defined as the average of all nodes' clustering coefficient:

$$C = \frac{1}{N}\sum_i^N c_i. \tag{3.6}$$

After analysis and calculating all the value of c_i, the lager degree node i has, the smaller c_i it has, and the smaller degree it has, the larger c_i it has. That is because of that nodes connecting to the same node may be compete with each other.

In LED cloud services network, C is equal to 0.044. For the network nodes, the smaller c_i it has, the more easily it will be replaced. This kind of node usually is not important.

5   Application of LED Cloud Services' Characteristics

Take as an example to fulfilling an order of LED spotlight. Decompose this manufacturing task into 4 subtasks: *raw materials procurement, modular design, simulating optimization, production & processing*. Combining with those aforementioned the statistical features of topological characteristics, an application is given.

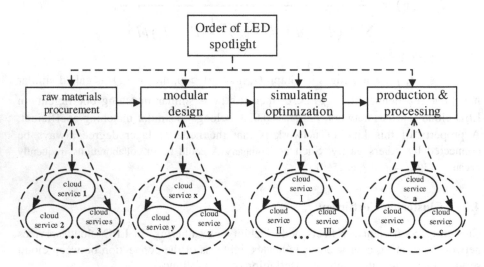

Fig. 6. Decompose the manufacturing task about fulfilling an order of LED spotlight into 4 subtasks: *raw materials procurement, modular design, simulating optimization, production & processing*. For *raw materials procurement*, cloud service **1**, **2**, **3** are candidate services. Similarly, cloud service **x**, **y**, **z** are candidate for *modular design*. Cloud service I, II, III are candidate for *simulating optimization*. Cloud service **a**, **b**, **c** are candidate for *production & processing*.

1) Calculate node degrees and degree distribution, in consideration of the coupling between nodes, try to choose the node with large degree distribution (such as the alternate contracts in table 3.).
2) In the selected nodes, to guarantee the quality of manufacturing task, choose the optimized nodes by vertex betweenness. Select the one which has larger value.
3) The shortest path length reflects the length of cloud services chain on the whole, which is positive relevant to the time for fulfill the manufacturing task. Hence, it's necessary to take the shortest path length of alternative nodes into account. Get shortest path length by *NetworkX*, and choose the smallest one prior.
4) From the point of view of the market economy, keep the clustering coefficient of alternative node as small as possible, because of that high clustering coefficient will reduce bargain chips.[13]
5) According to the optimized principle above, the best service combination as follow: cloud service **2** → cloud service **z** → cloud service I → cloud service **c**.

Table 3. The statistical attribute of Cloud service **1-3**, **x-z**, I -III, **a-b**, including degree distribution $P(k)$, vertex betweenness c_i , Clustering coefficient B_u .To find the most suitable alternate contract for each subtask, Synthesize all the statistical attributes.

Subtask item	Alternate Contracts	$P(k)$ %	c_i	B_u
raw materials procurement	cloud service **1**	18.00	1.000	32560
	cloud service **2**	12.00	0.800	39238
	cloud service **3**	9.00	0.900	28905
modular design	cloud service **x**	12.00	0.400	25032
	cloud service **y**	10.00	0.300	23456
	cloud service **z**	8.00	0.200	30976
simulating optimization	cloud service **I**	14.00	0.600	45609
	cloud service **II**	13.00	0.800	35992
	cloud service **III**	8.00	0.800	48950
production	cloud service **a**	17.00	0.300	34929
&	cloud service **b**	14.00	0.400	47672
processing	cloud service **c**	13.00	0.300	56782

6   Conclusion

This paper took the theory of cloud manufacturing as study background, put forward a LED cloud services model construction method based on complex network. By analysis of topological characteristics of cloud services network, including degree, degree distribution, shortest path length, vertex betweenness, similar matching coefficient and clustering coefficient, an application about fulfilling an order of LED spotlight explains the feature of cloud services. Due to the cloud manufacturing is still in the preliminary stage of development at the present stage, most of it remain in academic stage basically. It's difficult to collect relevant information and data, we will offer a further argument later.

References

1. Li, B.H., Zhang, L., Chai, X.D.: Introduction to Cloud Manufacturing. ZTE Communications 16(4), 05–08 (2010)
2. Tao, F., Zhang, L., Guo, H., et al.: Typical characteristics of cloud manufacturing and several key issues of cloud service composition. Computer Integrated Manufacturing Systems 17(3), 477–486 (2011)
3. Zhang, L., Luo, Y.L., Tao, F., et al.: Key technologies for the construction of manufacturing cloud. Computer Integrated Manufacturing Systems 16(11), 2510–2520 (2010)
4. He, D., Liu, Z.H., Wang, B.H.: Complex systems and complex network. Higher Education Press, Beijing (2009)
5. Wang, X.F., Li, X., Chen, G.R.: The theory and application of Complex Networks. Tsinghua University Press, Beijing (2006)

6. da F. Costa, L., Rodrigues, F.A., Travieso, G., Boas, P.R.V., et al.: Characterization of Complex Networks: A Survey of Measurements. Advances in Physics 56(1), 167–242 (2007)
7. Albert, R., Barabasi, A.L.: Statistical mechanics of complex networks. Review of Modern Physics 74, 47–97 (2002)
8. Hang, P.P., He, Y., Zhou, T., et al.: A model describing the degree distribution of collaboration networks. Submitted to Acta Phys. Sin. (1), 60–66 (2005)
9. Wu, J.S., Di, Z.R.: Complex networks in statistical Physics. Progress in Physics 24(1), 18–46 (2004)
10. Yuan, J., Zhang, N.: Analysis of corporation competition network topology. University of Shanghai for Science and Technology 29(1), 37–41 (2007)
11. Watts, D.J., Strogatz, S.H.: Collective dynamics of 'small-world' networks. Nature 393, 440–442 (1998)
12. Freeman, L.: A set of measures of centrality based upon betweenness. Socimetry 40, 35–41 (1977)
13. Yin, Y., Zhou, Z.D., Liu, Q., Long, Y.H.: Resource node selection in manufacturing grid based on complex network theory. Machine Tool & Hydraulics 39(15), 5–9 (2010)

Room Cooling Load Calculation Based on Soft Sensing[*]

Zhanpei Li, Xinyuan Luan, Tingzhang Liu, Biyao Jin, and Yingqi Zhang

Shanghai key Laboratory of Power Station Automation Technology, Shanghai University,
Shanghai 200072, China
{Liu Tingzhang,liutzh}@staff.shu.edu.cn

Abstract. The calculation of real-time dynamic room cooling load can be solved effectively by soft sensing technology based on auxiliary variable. This paper studies on the relationships between room cooling loads at different reference temperatures, and presents the system equations for cooling load calculation based on soft sensing. The undetermined coefficients in the system equations were identified via the least squares method, which reflect the magnitude relationship between the non-measurable primary variables and the auxiliary variables which can be measured accurately. Finally, commentary was presented based on the comparison between the results of simulation in Dwelling Environment Simulation Tools (DeST) and results of calculation via the equations in this paper.

Keywords: Room cooling load, central air conditioning, energy saving, soft sensing.

1 Introduction

There is enormous energy saving potential on the real-time control at the running time of central air conditioning, so that accuracy real-time building space load value is the foundation of control and supply according demands of the air conditioning system. According to the study and practice in the energy industry in recent 30 years, it is widely believed that the building energy saving is the most promising and most directly efficient way in all of the energy saving approach, as well as one of the most efficient ways to ease the energy tension and resolve the contradiction between the development of economical society and the shortfall of energy supplies [1]. Heating, ventilation and air-conditioning (HVAC) system is the major electricity consumer in an air-conditioned building; therefore, an accurate cooling load calculation method is indispensable [2].

Since the inception of air conditioning in 1902, there are many methods for the room cooling load calculation. The earliest method equations are founded on the basis of the thermal theory and architectural features [3-8]. According to these traditional cooling load calculation methods, models are complex, while modeling is difficult and real-time performance can not suffice. Difficulty on the data acquisition and

---

[*] This work is supported by The National Science Fund (61273190).

M. Fei et al. (Eds.): LSMS/ICSEE 2014, Part II, CCIS 462, pp. 331–341, 2014.

shortage of underlying data make it a challenge for the application of these methods. The usage and dependence on empirical value, estimated value and recommend value also influence the calculation accuracy. In recent years, with more attention on the energy saving at the running time of central air conditioning, the hot plot of academic research turns to the prediction methods based on data learning, the basic idea of which method is establishing the room cooling load prediction model on the base of the history statistics [9-15]. It is a large time delay process on the transformation from ambient variables to the room cooling/heating load, meanwhile the room air temperature won't stay at the design value all the time. Load prediction methods need huge history data, and have difficulty on reflecting the random dynamic factors, which both impacts the accuracy and practicability.

For many processes, the quantity that one wishes to control cannot be measured quickly or easily. In some cases a number of non-specific measurements are available and it should be possible to combine these measurements to provide good inferences for important, non-measured quantities. The soft sensing technology can realize online prediction for the process parameters which are difficult to measure online or to accurately measure, with the utilization of the process parameters that can be measured online or can be accurately measured [16]. In later 1980s, the soft sensing technology was formally proposed as academic term, and brought with research upsurge all over the world. In 1992, a report named "Contemplative Stance for Chemical Process Control" was given and played an important role in the development of soft sensing technology. With development of more than 30 years, the soft sensing technology has been widely used in the chemical industry, the metallurgical engineering, the bioengineering and so on, while the theory and practice on modeling methods have also made considerable progress [16-20].

For the limitations of traditional methods and prediction methods on real-time dynamical characteristics of room cooling load calculation, some new approaches are developed based on soft sensing technology. These methods realize the online calculation relying on modeling with auxiliary variables. References [21-22] present a new thought on air conditioning room cooling load evaluation with soft sensing technology based on auxiliary variables, which method fits well on real-time dynamic characteristics, but the accuracy needs to be increased, and the equations for cooling load calculation in references[21-22] need to be improved.

2   Soft Sensing Method for Room Cooling Load Calculation

Room cooling load is the rate of heat removal required to maintain a space at the designed temperature. It is the fundamental basis for the ascertainment of air conditioning equipment capacity, air supply volume and air supply temperature. The room cooling load is a variable that can't be measured directly. As to the traditional modeling method based on mechanism analysis and the prediction modeling method, the large demand on history data and the complex calculation make them unsuitable for the real-time cooling load calculation.

As automatic control system, the air conditioning system and the room space form a temperature closed-loop system. Various dynamic disturbances are contained in the closed-loop. The mapping relationship between the input of measurable heat extraction rate of the HAVC system and the output of room temperature covers the static characteristic and the dynamic characteristic of the load formation. Actually, the change of the room temperature is the joint action result of the heat extraction rate of the HAVC system and the room actual heat/cooling load. The actual heat extraction rate of the HAVC system will achieve balance with the room actual heating/cooling load and the energy for the change of room temperature. With the previous thinking the room cooling load can be back calculated according to the response characteristics between the heat extraction rate of the HAVC system and the room temperature. The energy balance equation among the room cooling load, heat extraction rate of the HAVC system and the heat storage was then researched as the room is treated as controlled process.

For the intermittently use of air conditioning system or the inequality between the room cooling load and the actual heat extraction rate of the HAVC system, the room air temperature won't stay at the design temperature all the time. The deviation of the actual heat extraction rate of the HAVC system from the room cooling load is then defined as heat storage. As present in reference [23], the room heat balance equation at a reference temperature is written as

$$CL(n) = HE(n) - HS(n) \tag{1}$$

where: $CL(n)$ is the current hourly room cooling load at the reference temperature, $HE(n)$ is the current hourly actual heat extraction rate of the HAVC system, $HS(n)$ is the current hourly heat storage.

Although the current hourly actual heat extraction rate of the HAVC system in equation (1) can't be measured directly with the current technology, it can be calculated by means of the air supply volume, the temperature and humidity of the supply air and return air can be measured. As written in equation (2)

$$HE(n) = f\left(Q_s, w_c, T_C, w_h, T_h\right) \tag{2}$$

where: Q_s is the flow rate of the HAVC system, w_c, T_C is the humidity/temperature of supply air, w_h, T_h is the humidity/temperature of return air.

As to the heat storage, which is also non-measureable variable, a characteristics equation needs to be established with the further derivation on the response characteristics of the room temperature. The current hourly heat storage can be managed as equation (3).

$$HS(n) = f\left(T_a\right) \tag{3}$$

where T_a is the room air temperature.

As auxiliary variables, the flow rate of the HAVC system, the humidity/temperature of the supply/return air and the room air temperature can be measured accurately. The current hourly actual heat extraction rate of the HAVC system and the current hourly heat storage can be calculated in equation (2) and equation (3), the current hourly room cooling load at the reference temperature would be calculated by the utilization of the equation (1).

3 Room Cooling Load at Different Reference Temperatures

Room cooling load is defined at a corresponding reference temperature, but the room designed temperature is always changed with different demands. It is obviously too complex to establish a group of system equations at every setting temperature, so the relationship of room cooling load at different setting temperatures is worth exploring. In all of the factors influencing the cooling load, the occupant heat gain and the corresponding cooling load changes as the people move out and in, so it is not considered in this paper.

The reference [24] presents a complete introduction of harmonic response method, which is a classical method in the room cooling load calculation. The cooling load calculation equation is described as

$$CL = CL_d + CL_f + CL_s + CL_r + CL_i \tag{4}$$

where CL is the total cooling load, CL_d is the cooling load from convection heat gain across the wall, CL_f is the cooling load from radiation across the wall, CL_s is the cooling load from transient conduction heat gain across windows, CL_r is the cooling load from solar heat gain across windows, CL_i is the cooling load from internal heat gain.

$$CL_d = \beta_d KF\left[\overline{T_Z} - T_i + \frac{\alpha_N}{K}\sum_{n=1}^{m}\frac{\Delta T_{z,n}}{v_n}COS(\omega_n\tau - \varphi_n - \varepsilon_n)\right] \tag{5}$$

$$CL_f = \beta_f KF\left(\overline{T_Z} - T_i\right) + \beta_f \alpha_N F\sum_{n=1}^{m}\frac{\Delta T_{z,n}}{v_n\mu_n}COS\left(\omega_n\tau - \varphi_n - \varepsilon_n - \varepsilon_n'\right) \tag{6}$$

$$CL_s = \beta_d KF\sum_{n=1}^{m}A_n COS(\omega_n\tau - \varphi_n) + \beta_f KF\sum_{n=1}^{m}\frac{A_n}{\mu_n}COS\left(\omega_n\tau - \varphi_n - \varepsilon_n'\right) \tag{7}$$

$$CL_r = \beta_d C_s C_n F\sum_{n=1}^{m}B_n COS(\omega_n\tau - \varphi_n) + \beta_f C_s C_n F\sum_{n=1}^{m}\frac{B_n}{\mu_n}COS\left(\omega_n\tau - \varphi_n - \varepsilon_n'\right) \tag{8}$$

$$CL_i = \beta_d Q\sum_{n=1}^{m}A_n COS(\omega_n\tau - \varphi_n) + \beta_f Q\sum_{n=1}^{m}\frac{A_n}{\mu_n}COS(\omega_n\tau - \varphi_n - \varepsilon_n') \tag{9}$$

where β_d / β_f is the ratio of convection/radiation to corresponding heat gain, K is the heat transfer coefficient of corresponding building envelope $W / (m^2 \bullet K)$, F is the area of corresponding building envelop m^2, $\overline{T_Z}$ is the outdoors comprehensive temperature, T_i is the indoor constant temperature, α_N is the heat transfer coefficient of the inner surface of building envelope $W / (m^2 \bullet K)$, $\Delta T_{z,n}$ is the n order harmonic disturbance of outdoors comprehensive temperature, v_n is the decay on n order harmonic disturbance of outdoors comprehensive, ω_n is the frequency of n order outdoors comprehensive temperature, φ_n is the initial phase angle of n order outdoors comprehensive temperature, ε_n is the phase delay of n order comprehensive temperature, μ_n is the decay on n order radiation disturbance, ε_n' is the phase delay of n order radiation disturbance, A_n is the amplitude of n order outdoor comprehensive temperature, C_s is the shading coefficient of glasses, C_n is the shading coefficient of shading equipment, B_n is the amplitude of n order solar heat gain, Q is the internal heat gain. At the condition without ventilation, the internal heat gain is constant.

At different setting temperature $T_{i,1}$ and $T_{i,2}$, the deviation of the cooling load is present as

$$CL|_{T_{i,1}} - CL|_{T_{i,2}} = (\beta_d + \beta_f)KF(T_{i,2} - T_{i,1}) \tag{10}$$

where: $CL|_{T_{i,1}} / CL|_{T_{i,2}}$ is the cooling load at the setting temperature $T_{i,1} / T_{i,2}$.

Then, the condition with ventilation is researched. Reference [25] describes the ventilation's effect on cooling load as

$$q_s = C_s Q_s \Delta T \tag{11}$$

$$q_l = C_l Q_s \Delta w \tag{12}$$

where: q_s is the sensible heat, q_l is the latent heat, C_s is the air sensible heat factor, C_l is the air latent heat factor, ΔT is the change of dry-bulb temperature, Δw is the change of absolute humidity.

Latent heat gain exists for the existence of enthalpy difference via ventilation. According to the relationship between enthalpy difference and indoor/outdoor humidity on cooling load from ventilation in reference [26], the equation (12) can be rewritten as

$$q_l = C_l Q_s \begin{cases} w_w - w_n, & \text{when} : w_w > w_n \\ 0, & \text{when} : w_w \leq w_n \end{cases} \tag{13}$$

where: w_w is the outdoor humidity, w_n is the indoor humidity.

There is no absorption and release of radiation in the transform of ventilation to cooling load, so there is no lag in the transform of this part of heat gain to cooling load. Different room air temperatures can lead to different humidity. It is assumed that $T_{i,1} > T_{i,2}$ and the corresponding humidity $w_1 > w_2$. Hence, the cooling load difference from ventilation at different designed temperatures can be written as

$$CL|_{T_{i,1}} - CL|_{T_{i,2}} = C_s Q_s (T_{i,2} - T_{i,1}) + C_l Q_s \begin{cases} w_2 - w_1, & \text{when} : w_w > w_1 \\ w_2 - w_w, & \text{when} : w_1 > w_w > w_2 \\ 0, & \text{when} : w_2 > w_w \end{cases} \tag{14}$$

The temperature and humidity can get measured, and the other variables were then treated as undetermined coefficients. According to equation (10) and equation (14), the room cooling load difference at different designed temperatures is written as

$$CL|_{T_{i,1}} - CL|_{T_{i,2}} = k_T (T_{i,2} - T_{i,1}) + k_w Q_t \begin{cases} w_2 - w_1, & \text{when} : w_w > w_1 \\ w_2 - w_w, & \text{when} : w_1 > w_w > w_2 \\ 0, & \text{when} : w_2 > w_w \end{cases} \tag{15}$$

where: k_r is the coefficient for temperature, k_w is the coefficient for humidity, Q_t is the ventilation times every hour.

By the simulation in DeST of a room in a building of Shanghai University at the setting temperature 26/25°C, the coefficient for temperature and humidity can be got. With this data the cooling load at setting temperature 24°C can be calculated, which can be reflected by the simulation as well. The max hourly relative error of the two previous cooling load is 1.22E-05. The equation (15) is obviously effective.

4   Room Heat Storage Calculation

Reference [24-25,27] introduces the weighting-factor method for cooling load calculation in which method the relationship between heat storage and temperature is written as

$$\Delta T(n) = \frac{1}{g_0}\left[HS(n) + p_1 HS(n-1) + p_2 HS(n-2)\right] + ... + g_1 \Delta T(n-1) + g_2 \Delta T(n-2) + ...] \quad (16)$$

where: $HS(n-i)$ is the heat storage at previous i hour, $\Delta T(n-i)$ is the deviation of air temperature to reference value at previous i hour, g_i / p_i is the weighting factors of corresponding temperature differences and heat storages.

For the sake of convenient calculation, the references [25,27] rewrite the equation (16) as a 3-order system, as equation

$$HS(n) = -g_0 \Delta T(n) + g_1 \Delta T(n-1) + g_2 \Delta T(n-2) + g_3 \Delta T(n-3) - p_1 HS(n-1) - p_2 HS(n-2)$$
$$(18)$$

Since the heat storage value also can't be measured, the previous hourly heat storage should be replaced by some variables that can be measured. As the method gives the following equation

$$HS(n-1) = -g_0 \Delta T(n-1) + ... + g_3 \Delta T(n-4) - p_1 HS(n-2) - p_2 HS(n-3) \quad (18)$$

At last, the relationship between the current hourly heat storage and deviation of air temperature to reference value can be written as

$$HS(n) = b_0 \Delta T(n) + b_1 \Delta T(n-1) + b_2 \Delta T(n-2) + ... \quad (19)$$

where: b_i is the coefficient for corresponding deviation of air temperature from reference value.

A 4-order system was also assumed in reference [21]. The influence that the different temperature exert on cooling load before 3 hours was so little that it can be ignored as B, as in the following equation

$$HS(n) = b_0 \Delta T(n) + b_1 \Delta T(n-1) + b_2 \Delta T(n-2) + b_3 \Delta T(n-3) + B \quad (20)$$

It has been proved in the simulation that the approximate method like equation (20) is too simple. In the work of this paper, we reconsider and rewrite it.

DeST is dwelling environment simulation tools which can realize the simulation on room cooling load [28]. We build a model of a building in Shanghai University and get the hourly room cooling load as well as the room-base temperature of cooling season from the simulation of a room in the building. According to these data in July and August, we can get two groups of undetermined coefficients values b_i in (19) from 0-order to one 99-order. Then we can calculate the room cooling load in cooling season. The average relative error can be calculated as

$$w = \sum_{i=1}^{m}(CL_f - CL_j) / \sum_{i=1}^{m}CL_f \tag{21}$$

where: w is the average relative error, CL_f is the cooling load got in simulation, CL_j is the cooling load got in calculation. m = 744 is the length of data set, as there are 31 days counted in one month. The last day of May and the first day of October were added to June and September, correspondingly, for easy counting.

Then the curve of average relative error is drawn in Fig. 1

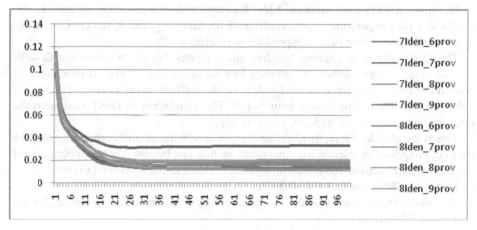

Fig. 1. Average relative error in different system equations

In fig.1, there are 8 plots with different color, corresponding to the cooling load got with different data groups. The symbol "7Iden_6prov" means that the undetermined coefficients values were got with the set of data in July and the data in June were used in proving the effectiveness of the method, so as the other plots and symbols. It is present in the figure 1 that the value of average relative error falls rapidly at 1-order system, and practically arrives at the steady state at 19-order system. After that, the valve of the average relative error stays at the same level. So it is credible that 19-order is reasonable for the system equation (19), not 3-order.

The equation for an acceptable value of current heat storage calculation at reference temperature can be wrote as

$$HS(n) = b_0\Delta T(n) + b_1\Delta T(n-1) + b_2\Delta T(n-2) + ... + b_{19}\Delta T(n-19) \tag{22}$$

5   Identification of Undetermined Coefficients in System Equations

On the condition of central air conditioning closed, the actual heat extraction rate is zero, so the equation (1) can be rewritten as

$$CL(n) = -HS(n) \qquad (23)$$

The equation (15) presents the relationship between room cooling load at a temperature and the cooling load at the reference temperature. Equation (22) presents the method to calculate the current heat storage. The current cooling load can be calculate by equation (23) and then used in (15) for different cooling loads at different temperatures.

In equation (15) and equation (22), the undetermined coefficients are: k_T the coefficient for temperature, k_w the coefficient for humidity, and b_i the coefficient for corresponding deviation of air temperature to reference value.

The Dest will run at a setting building environment. Set the conditions with 40W as max value of lighting heat disturbance in total index and 1200W as max value of equipment heat disturbance in total index, while the min value is zero. The setting value of ventilation times every hour is 0.5. The simulation in DeST runs with the reference temperature 26 •and 25•, successively.

The ambient/indoor humidity, the air temperature without air conditioning and the room cooling load at reference temperatures in July can be got. Via the application of the least square method, the undetermined coefficients in the equation (15) is

K_T=0.04600958, and K_w=0.07480946.

The undetermined coefficients in equation (22) at reference temperature 25• are present in Table 1.

Table 1. Undetermined coefficient

b0	1.1314	b5	0.02556	b10	-0.0076	b15	-0.0552
b1	-0.2182	b6	0.02269	b11	-0.0507	b16	0.05017
b2	-0.3171	b7	0.00148	b12	-0.0188	b17	0.03902
b3	-0.2345	b8	0.00028	b13	-0.0403	b18	0.34872
b4	-0.1288	b9	0.00587	b14	-0.0465	b19	-0.4555

6   Evaluation and Conclusion

Set a new condition with 60W as max value of lighting heat disturbance in total index and 1000W as max value of equipment heat disturbance in total index, while the min value is zero. The setting value of ventilation times every hour is 0.5, which is also the recommend value. The simulation in DeST can be completed with the reference temperature 24°C.

Via the simulation in DeST, the room cooling load CL_f at the reference temperature 24°C can be got. According to equations (15), (22) and (23), the room cooling load CL_j can be calculated at the temperature 24°C, as shown in figure 2.

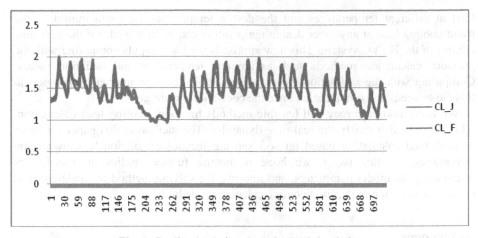

Fig. 2. Cooling load via calculation and simulation

Comparing these two groups of data, the biggest value in hourly relative error is 11.1%, as present in fig.3, the average relative error is 2.37%.

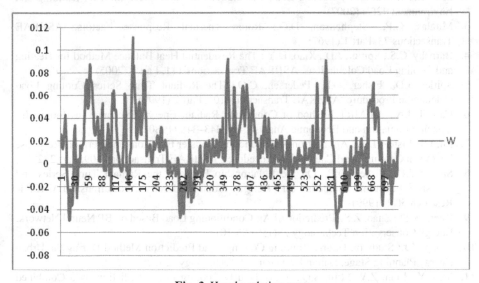

Fig. 3. Hourly relative error

The room cooling load can be expressed as the difference between the heat extraction rate of the HAVC system and the heat storage. The heat extraction rate of the HAVC system and the heat storage can be real-time calculated based on the auxiliary variables. With previous data groups, the configuration parameters which reflect the relationship between the non-measurable cooling load and the measurable auxiliary variables can be got via identification methods. Then the real-time room cooling load at the current reference temperature can be calculated with the real-time measuring auxiliary variables. For the linear relationship between the room cooling load at different temperatures and the design temperature/the room humidity, the room cooling load at any other design temperature can be managed, for the real-time control of the HAVC system. This new approach need less inputs comparing with the previous calculation methods such as ambient temperature and radiation values. Comparing with the simulation tools such as DeST, the new approach proposed in this paper needs less computation once the coefficients were got.

We need more accuracy and feasible methods for room cooling load calculation, which should also satisfy the real-time dynamics. The method in this paper for room cooling load calculation based on soft sensing technology obviously shows some advantages. By this paper, we hope to inspire further studies on the factors influencing the model parameters, and improve the solving method for undetermined coefficients, for a more accurate online calculation method.

References

1. Jiang, Y.: Current Building Energy Consumption in China and Effective Energy Efficiency Measures. HV&AC 35(5) (2005)
2. Mui, K.W., Wong, T.: Cooling Load Calculations in Subtropical Climate. Building and Environment 42(7) (2007)
3. Mitalas, G.P., Stephenson, D.G.: Room Thermal Response Factors. ASHRAE Transactions 73, Part 1 (1967)
4. Barnaby, C.S., Spitler, J.D., Xiao, D.Y.: The Residential Heat Balance Method for Heating and Cooling Load Calculations. ASHRAE Transactions 111, Part 1 (2005)
5. Spitler, J.D., Fisher, D.E., Pedersen, C.O.: The Radiant Time Series Cooling Load Calculation Procedure. ASHRAE Transactions 103, Part 2 (1997)
6. Carroll, J.A.: An MRT Method of Computing Radiant Energy Exchange in Rooms. In: System Simulation and Economic Analysis, pp. 343–348 (1980)
7. Lu, I.S., Fisher, D.E.: Application of Conduction Transfer Functions and Periodic Response Factors in Cooling Load Calculation Procedures. ASHRAE Transactions 109, Part 2 (2004)
8. Spitler, J.D., Fisher, D.E.: On the Relationship Between the Radiant Time Series and Transfer Function Methods for Design Cooling Load Calculations. HVAC&R Research 5(2) (1999)
9. Chen, W.D., Zhao, Z.S.: Prediction of Air Conditioning Load Based on BP Neural Network. Energy Conservation Technology 28(159) (2010)
10. Wang, Z.J.: Study on Hourly Building Cooling Load Prediction Method During the Urban Energy Planning Stage. Dalian University of Technology (2010)
11. Yao, Y., Lian, Z.W., Liu, S.Q., et al.: Hourly Cooling Load Prediction by a Combined Forecasting Model Based on Analytic Hierarchy Process. International Journal of Thermal Sciences 43(11) (2004)
12. Li, Q., Meng, Q.L., Cai, J.J., et al.: Applying Support Vector Machine to Predict Hourly Cooling Load in the Building. Applied Energy 86(10) (2009)
13. Zhang, Y., Lu, N.: Demand-side Management of Air Conditioning Cooling Loads for Intra-hour Load Balancing. In: 2013 IEEE PES Innovative Smart Grid Technologies Conference (2013)
14. Lu, N.: An Evaluation of the HVAC Load Potential for Providing Load Balancing Service. IEEE Transactions on Smart GRID 3(3) (2012)

15. Guo, Y., Ehans, N., Ko, J., et al.: Hourly Cooling Load Forecasting Using Time-indexed ARX Models with Two-stage Weighted Least Squares Regression. Energy Conversion and Managementl 80, 46–53 (2014)
16. Mcavoy, T.J.: Contemplative Stance for Chemical Process Control. Automatica 28(2) (1992)
17. Sun, Y.: Research on Adaptive Correction of Soft sensors and High-temperature Fields Soft Sensing. Central South University, Hunan (2012)
18. Arafa, A.M., Seddik, K.G., Sultan, A.K., et al.: A Feedback Soft Sensing Based Access Scheme for Cognitive Radio Networks. IEEE Transactions on Wireless Communications 12(7) (2013)
19. Yan, D., Tang, J., Zhao, L.J.: Soft Sensor Approach Based on Feature Extraction and Extreme Learning Machines. Control Engineering of China 20(1) (2013)
20. Qiao, Z.L., Zhang, L., Zhou, J.X., et al.: Soft Sensor Modeling Method Based on Improved CPSO-LSSVM and its Applications. Chinese Journal of Scientific Instrument 35(1) (2014)
21. Yao, L.X., Liu, T.Z.: Modeling of Real-time Air Conditioning Load Based on Room Temperature Response Characteristic. Journal of System Simulation 23(10) (2011)
22. Liu, T.Z., Gong, A.H., Yao, L.X.: A Real-time Experimental Method for Room Cooling Load Calculations. Advanced Materials Research, 433–440, 4490–4495 (2012)
23. Cao, S.W.: The room heat process and air conditioning load. Shanghai Scientific and Technological Literature Publishing House, Shanghai (1991)
24. Zhao, R.Y., Fan, C.Y., Xue, D.H., et al.: Air Conditioning. China Architecture & Building Press, Beijing (2008)
25. ASHRAE: ASHRAE Handbook 2009. American Society of Heating, Refrigerating and Air-Conditioning Engineers, Inc., Atlanta (2009)
26. Yu, X.P., Fu, X.Z.: Calculation Procedure of Fresh Air Cooling Loads in Hot Summer and Cold Winter Region. Journal of Chengdu Textile College 19(2) (2002)
27. Kerrisk, J.F., Hunn, B.D., Schnurr, N.M., et al.: The Custom weighting-factor method for Thermal Load Calculations in the DOE-2 Computer program. ASHRAE Transactions 87(2), 569–584 (1981)
28. Jiang, Y.: Building Environmental System Simulation and Analysis DeST. China architecture & building press, Beijing (2006)

Trajectory Tracking of Nonholonomic Mobile Robots via Discrete-Time Sliding Mode Controller Based on Uncalibrated Visual Servoing

Gang Wang, Chaoli Wang, Xiaoming Song, and Qinghui Du

University of Shanghai for Science and Technology, 516 Jun Gong Road,
200093 Shanghai, China
2010wanggang@gmail.com, clclwang@126.com, miko_song@163.com,
39652297@qq.com

Abstract. This paper considers the problem of trajectory tracking of nonholonomic mobile robots based on uncalibrated visual servoing. A prerecorded image sequence or a video taken by the pin-hole camera is used to define a desired trajectory for the mobile robot. First, a novel discrete-time model is present based on visual servoing. And then the discrete-sliding mode controller is designed for the model associated with uncertain parameter. The asymptotic convergence of the tracking errors is proved rigorously. Finally, simulation results confirm the effectiveness of the proposed methods.

Keywords: Nonholonomic, trajectory tracking, discrete-time, visual servoing.

1 Introduction

The problem of control nonholonomic mobile robots has caused great interests due to its wide range of applications in medical, agriculture, industry and so on in the last three decades. Controlling nonholonomic mobile robots is a nontrivial problem for many reasons even the simple structure which will be investigated here. The stabilization problem cannot be solved by many methods of classical linear system for the fact nonholonomic system fails to meet the three necessary conditions of the theorem of Brockett (1983) [1]. The purpose of this note is to study the problem of trajectory tracking of nonholonomic mobile robots. To solve this problem, many scholars has done a lot of relevant research in this area such as continuous sliding mode method [2], discrete-time sliding mode [3], backstepping technique [4] and dynamic feedback linearization [5] etc. In addition, many constraints also have been considered such as input saturation, finite-time tracking, input and communication delays and so on. For the purpose of controlling the robots, it is always supposed that all the states of the robot are obtained accurately. However this hypothesis may be not held all the time due to some inevitable issues includes uncertain disturbances and accuracy of measurements. A useful approach to overcome those difficulties is exploiting the pin-hole camera to get the position information which is needed by the controller.

M. Fei et al. (Eds.): LSMS/ICSEE 2014, Part II, CCIS 462, pp. 342–350, 2014.

Visual servoing as a simple and important sensor has been an increasing interest in controlling mobile robots recently. From review of the previous work, the clear drawback of monocular camera system is that the depth information cannot be obtained directly. In the reference [6], Chen et al. developed an adaptive tracking controller via Lyapunov-based method, and the adaptive update law is designed to compensate for the unknown and time-varying depth parameter. In [7], a novel adaptive torque tracking controller has been designed based on backstepping method and Lyapunov-based stability with uncalibrated visual parameters and unknown disturbances in the dynamic system. Since a local state and input transformations has been applied to change the camera-object visual servoing kinematic model to the uncertain chained form system, the controller is semiglobal which means the control law presents singularity for some situations. And Wang et al. [8] adopted Lyapunov technique and Barbalat lemma to craft a dynamic feedback controller that the adaptive update law was not needed to estimate the unknown camera parameter. However, the above proposed approaches are only concerned with the continuous-time control input which may cause unavoidable errors due to the discretization before applying it to the actual robots.

In this paper, the discrete-time tracking problem of nonholonomic mobile robots is considered. The main contribution is twofold. Firstly, a novel discrete-time model is present based on uncalibrated visual servoing. Secondly, a discrete-sliding mode controller is designed for this model which is different from the available methods.

The organization of this paper is as follows. In Section 2, the pin-hole camera model is introduced to describe the camera-robot system. And the discrete-time model is present. In Section 3, the discrete-time sliding mode control law is designed. And the asymptotic convergence of the tracking errors is proved rigorously. Section 4 illustrates the proposed method via simulation. Some concluding remarks are offered in the last section.

2   System Configuration and Problem Statement

The nonholonomic mobile robot considered here is mobile robot of type (2,0) (see [9] for more details) shown in Fig.1. Assume a pin-hole camera is placed on the ceiling and the robot plane and the camera plane is parallel. There are three coordinate frames, namely the inertial frame X-Y-Z, the image frame u-o-v and the coordinate system attached to the camera X_1-Y_1-Z_1. Suppose that X_1-Y_1 plane of the camera frame is parallel with the u-v of the image coordinate plane. (p_x,p_y) denotes a projection of the optical center of the camera on the X-Y plane, the coordinate of the original point of the camera frame with respect to the image frame is defined by (O_{c1},O_{c2}), and (x, y) is the coordinate of the mass center P of the robot with respective to X-Y plane. Consider that (x_m,y_m) is the coordinate of (x, y) relative to the image frame. θ represents the angle between the heading direction of the robot and the X axis. Pinhole camera model yields [8].

$$
\begin{bmatrix} x_m \\ y_m \end{bmatrix} = \begin{bmatrix} \alpha_1 & 0 \\ 0 & \alpha_2 \end{bmatrix} R \left(\begin{bmatrix} x \\ y \end{bmatrix} - \begin{bmatrix} p_x \\ p_y \end{bmatrix} \right) + \begin{bmatrix} O_{c1} \\ O_{c2} \end{bmatrix}, \tag{1}
$$

where α_1, α_2 are the unknown constants that depend on depth formation, focus length, and scale factors along x axis and y axis, respectively.

$$
R = \begin{bmatrix} \cos\theta_0 & \sin\theta_0 \\ -\sin\theta_0 & \cos\theta_0 \end{bmatrix}, \tag{2}
$$

θ_0 denotes the positive, anticlockwise rotation angle of the image frame system with respect to the inertial frame.

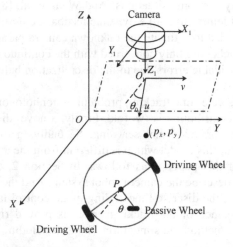

Fig. 1. Camera-robot system configuration

The system to be controlled is robot of type (2,0), whose body is of symmetric shape and the center of mass is at the geometric center P of the body. The robot consists of two driving wheels mounted on the same axis and a passive front wheel which could prevent the robot from turning over. We assume the driving wheels purely roll and do not slip, and the motion of the passive wheel can be ignored in the kinematics of the mobile robot. Thus kinematic of the mobile robot considered by many researchers can be expressed in accordance with the frame defined in Fig.1 as follows:

$$
\begin{cases} \dot{x} = v\cos\theta, \\ \dot{y} = v\sin\theta, \\ \dot{\theta} = \omega. \end{cases} \tag{3}
$$

where v and ω denote the linear velocity and angular velocity of the mobile robot, respectively.

In the image frame, the kinematic model can be deduced by (1) and (3)

$$\begin{cases} \dot{x}_m = \alpha_1 v \cos(\theta - \theta_0), \\ \dot{y}_m = \alpha_2 v \sin(\theta - \theta_0), \\ \dot{\theta} = \omega. \end{cases} \tag{4}$$

In the following, we make the same assumption as in [3].

Assumption 1. The sample time T is small enough to make the variations of $\cos(\theta - \theta_0)$ and $\sin(\theta - \theta_0)$ ignored. Simultaneously, considering a zero-order hold for control input v and ω and integrating equations (4) from kT to $(k+1)T$, then the following approximated discrete-time model can be obtained.

$$\begin{cases} x_{m,k+1} = x_{m,k} + T\alpha_1 v_k \cos(\theta_k - \theta_0), \\ y_{m,k+1} = y_{m,k} + T\alpha_2 v_k \sin(\theta_k - \theta_0), \\ \theta_{k+1} = \theta_k + T\omega_k. \end{cases} \tag{5}$$

In order to simplify the following analysis, the next assumption is made.

Assumption 2. Assume that θ_0 is known, and $\alpha_1 = \alpha_2 = \alpha$ are unknown. But there exists two known positive constants $\bar{\alpha}, \hat{\alpha}_{max}$ such that the following conditions hold:

$$\alpha = \bar{\alpha} + \hat{\alpha}, |\hat{\alpha}| \leq \hat{\alpha}_{max}.$$

With some abuse of notation, we replace $\theta - \theta_0$ with θ, x_m with x and y_m with y, then (5) can be rewritten as

$$\begin{cases} x_{k+1} = x_k + T\alpha v_k \cos\theta_k, \\ y_{k+1} = y_k + T\alpha v_k \sin\theta_k, \\ \theta_{k+1} = \theta_k + T\omega_k. \end{cases} \tag{6}$$

The primary control objective is to design a tracking controller to let the mobile robot track a desired trajectory defined by a prerecorded sequence of images or a video, generated by a reference robot whose equation of motion is similar as (6) as

$$\begin{cases} x_{k+1}^d = x_k^d + T\alpha v_k^d \cos\theta_k^d, \\ y_{k+1}^d = y_k^d + T\alpha v_k^d \sin\theta_k^d, \\ \theta_{k+1}^d = \theta_k^d + T\omega_k^d. \end{cases} \tag{7}$$

where v_d and ω_d denote the linear velocity and angular velocity of the reference mobile robot, respectively. And in practical application $x_k^d, y_k^d, \theta_k^d, v_k^d, \omega_k^d$ are available all time.

To facilitate the subsequent closed-loop error system development and stability analysis, the auxiliary errors signal is defined as

$$\begin{cases} e_k^x = x_k - x_k^d, \\ e_k^y = y_k - y_k^d, \\ e_k^\theta = \theta_k - \theta_k^d. \end{cases} \tag{8}$$

Then the tracking error dynamics can be obtained based on (6) and (7)

$$\begin{cases} e_{k+1}^x = e_k^x + T\alpha v_k \cos\theta_k - T\alpha v_k^d \cos\theta_k^d, \\ e_{k+1}^y = e_k^y + T\alpha v_k \sin\theta_k - T\alpha v_k^d \sin\theta_k^d, \\ e_{k+1}^\theta = e_k^\theta + T\omega_k - T\omega_k^d. \end{cases} \tag{9}$$

In addition, the following auxiliary variables are defined

$$a_k := v_k \cos\theta_k, b_k := v_k \sin\theta_k, \tag{10}$$

then equations (9) can be reduced to

$$\begin{cases} e_{k+1}^x = e_k^x + T\alpha(a_k - v_k^d \cos\theta_k^d), \\ e_{k+1}^y = e_k^y + T\alpha(b_k - v_k^d \sin\theta_k^d), \\ e_{k+1}^\theta = e_k^\theta + TT(\omega_k - \omega_k^d). \end{cases} \tag{11}$$

3   Discrete-Time Sliding Mode Controller Design

To finish the control objective of this paper, the following sliding surface is introduced

$$\begin{aligned} S_k^x = e_{k+1}^x - e_k^x + \gamma_x e_{k-1}^x = 0, \\ S_k^y = e_{k+1}^y - e_k^y + \gamma_y e_{k-1}^y = 0, \end{aligned} \tag{12}$$

where $\gamma_x, \gamma_y \in (0, 0.25]$. The principle of this choice is to make sure the roots of the subsequent polynomials

$$z^2 - z + \gamma_x = 0, z^2 - z + \gamma_y = 0, \tag{13}$$

inside the unit circle. Hence whenever a sliding mode is achieved on it, the position tracking errors are asymptotically convergent.

The achievement of (12) can be guaranteed by the variables a_k and b_k respectively. Since a_k and b_k are not independent, the subsequent condition must be guaranteed:

$$v = \sqrt{a_k^2 + b_k^2}, \theta_k = a\tan 2\left(\frac{b_k}{a_k}\right). \tag{14}$$

And the following sliding surface is introduced

$$S_k^\theta = a\tan 2\left(\frac{b_k}{a_k}\right) - \theta_k = 0. \tag{15}$$

If the achievement of (15) can be guaranteed, then the second condition of (14) is satisfied. In addition, it can be easily concluded that $a_k \to v_k^d \cos\theta_k^d, b_k \to v_k^d \sin\theta_k^d$ for $e_k^x \to 0, e_k^y \to 0$ from (11). So if the condition (15) is imposed, the orientation error e_k^θ tends to its desired value provided the position tracking errors e_k^x, e_k^y vanish.

Theorem: A sliding motion on the surface (12) and (15) can be enforced by the following control law:

$$\omega_k = \frac{1}{T}\left[a\tan 2\left(\frac{b_{k+1}}{a_{k+1}}\right) - \theta_k\right],\tag{16a}$$

$$v_k = \sqrt{(a_k)^2 + (b_k)^2}\tag{16b}$$

with

$$a_k = v_k^d \cos\theta_k^d + a_{1,k} + a_{2,k}, b_k = v_k^d \sin\theta_k^d + b_{1,k} + b_{2,k},\tag{17}$$

where

$$a_{1,k} = -\frac{\gamma_x}{T\tilde{\alpha}}e_{k-1}^x,$$

$$a_{2,k} = \begin{cases} \lambda_x \dfrac{\dfrac{|S_{k-2}^x|}{T} - \hat{\alpha}_{max}|a_{1,k}|}{\tilde{\alpha} + \hat{\alpha}_{max}}, & \text{if } \dfrac{|S_{k-2}^x|}{T} > \hat{\alpha}_{max}|a_{1,k}| \\[4mm] -\dfrac{\tilde{\alpha}a_{1,k}}{\tilde{\alpha} + \tilde{\alpha}}, & \text{if } \dfrac{|S_{k-2}^x|}{T} \le \hat{\alpha}_{max}|a_{1,k}| \end{cases}\tag{18a}$$

and

$$b_{1,k} = -\frac{\gamma_y}{T\tilde{\alpha}}e_{k-1}^y,$$

$$b_{2,k} = \begin{cases} \lambda_y \dfrac{\dfrac{|S_{k-2}^y|}{T} - \hat{\alpha}_{max}|b_{1,k}|}{\tilde{\alpha} + \hat{\alpha}_{max}}, & \text{if } \dfrac{|S_{k-2}^y|}{T} > \hat{\alpha}_{max}|b_{1,k}| \\[4mm] -\dfrac{\tilde{\alpha}b_{1,k}}{\tilde{\alpha} + \tilde{\alpha}}, & \text{if } \dfrac{|S_{k-2}^y|}{T} \le \hat{\alpha}_{max}|b_{1,k}| \end{cases}\tag{18b}$$

and $\lambda_x, \lambda_y \in (-1,1)$, $\tilde{\alpha}$ is the approximation of $\hat{\alpha}$ performed by a neural network (see [10] for more details, here omitted), which means the tracking errors will converge to zero.

Proof: Substitute (16a) to (6), we have

$$\theta_{k+1} = a\tan 2\left(\frac{b_{k+1}}{a_{k+1}}\right).\tag{19}$$

According to (19), the sliding surface (15) is achieved. Based on our previous analysis and (16b), we can conclude that a_k, b_k will be consistent with their definition (17). Considering the sliding surface S_k^x and replacing (17) in equations (11), gives

$$
\begin{aligned}
S_k^x &= e_{k+1}^x - e_k^x + \gamma_x e_{k-1}^x \\
&= T\alpha(a_k - v_k^d \cos\theta_k^d) + \gamma_x e_{k-1}^x \\
&= T[(\bar{\alpha} + \hat{\alpha})a_{2,k} + \hat{\alpha}a_{1,k}]
\end{aligned}
\tag{20}
$$

The following condition $\left|S_k^x\right| < \left|S_{k-2}^x\right|$ is imposed to ensure the asymptotic convergence of the position tracking errors. Noting that

$$
\left|S_k^x\right| = T\left|(\bar{\alpha} + \hat{\alpha})a_{2,k} + \hat{\alpha}a_{1,k}\right| \le T(\bar{\alpha} + \hat{\alpha}_{\max})\left|a_{2,k}\right| + T\hat{\alpha}_{\max}\left|a_{1,k}\right|
\tag{21}
$$

when $\dfrac{\left|S_{k-2}^x\right|}{T} > \hat{\alpha}_{\max}\left|a_{1,k}\right|$, from (18a) we have

$$
\begin{aligned}
&T(\bar{\alpha} + \hat{\alpha}_{\max})\left|a_{2,k}\right| + T\hat{\alpha}_{\max}\left|a_{1,k}\right| - \left|S_{k-2}^x\right| \\
&= (\lambda_x - 1)\left|S_{k-2}^x\right| < 0
\end{aligned}
\tag{22}
$$

Comparing (21) and (22), it is straightforward to get $\left|S_k^x\right| < \left|S_{k-2}^x\right|$.

When $\dfrac{\left|S_{k-2}^x\right|}{T} \le \hat{\alpha}_{\max}\left|a_{1,k}\right|$, the sliding mode condition cannot be imposed exactly and the approximation of $\hat{\alpha}$ is used. Replace $\hat{\alpha}$ by $\tilde{\alpha}$, the approximation is obtained from (18a)

$$
S_k^x = T[(\bar{\alpha} + \hat{\alpha})a_{2,k} + \hat{\alpha}a_{1,k}] \approx T[(\bar{\alpha} + \tilde{\alpha})a_{2,k} + \tilde{\alpha}a_{1,k}] = 0
\tag{23}
$$

Note that this approximation is used only inside the sector. So the sliding surface S_k^x can be imposed based on the above proof.

Due to the proof that the sliding surface S_k^y can be imposed by the controller (17) is similar with S_k^x, here omitted since limited space. Here we complete the proof of the theorem.

4   Simulation

Suppose the reference robot initial states $(0,0,0)$, the actual robot initial states $(0.5,0.5,0)$, $v_d = 3, \omega_d = \pi/2, \alpha = 0.5, \bar{\alpha} = 0.45, \hat{\alpha}_{\max} = 0.1$, Sample time $T = 0.1\text{sec}$, $\gamma_x = 0.2, \gamma_y = 0.2, \lambda_x = 0.5, \lambda_y = 0.5$. the controller is chosen as (16a) and (16b). The simulation results are shown in Fig 2~5.

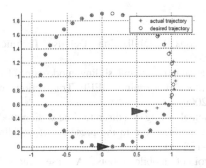

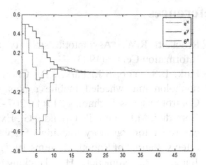

Fig. 2. The trajectory of the robot

Fig. 3. History of the tracking errors

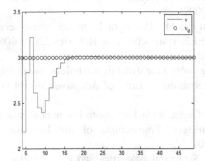

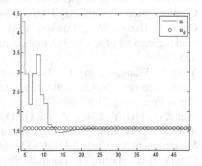

Fig. 4. The control input v

Fig. 5. The control input ω

5   Conclusions

A discrete-time sliding mode controller is proposed for the tracking problem of nonholonomic mobile robot based on uncalibrated visual servoing. The asymptotic convergence of the tracking errors is proved rigorously. And our theoretical results have been confirmed by the simulation results. Our future work is extending the results to other uncertain nonholonomic systems and uncertain dynamic of the mobile robot.

Acknowledgments. This paper was partially supported by The National Natural Science Foundation (61374040), Key Discipline of Shanghai (S30501), Scientific Innovation program (13ZZ115), Graduate Innovation program of Shanghai (54-13-302-102).

References

1. Rrockett, R.W.: Asymptotic stability and feedback stabilization. Defense Technical Information Center (1983)
2. Park, B.S., Yoo, S.J., Park, J.B., Choi, Y.H.: Adaptive neural sliding mode control of nonholonomic wheeled mobile robots with model uncertainty. IEEE Transactions on Control Systems Technology 17(1), 207–214 (2009)
3. Corradini, M.L., Leo, T., Orlando, G.: Experimental testing of a discrete-time sliding mode controller for trajectory tracking of a wheeled mobile robot in the presence of skidding effects. Journal of Robotic Systems 19(4), 177–188 (2002)
4. Jiang, Z.P., Nijmeijer, H.: Tracking control of mobile robots: A case study in backstepping. Automatica 33(7), 1393–1399 (1997)
5. Oriolo, G., De Luca, A., Vendittelli, M.: WMR control via dynamic feedback linearization: design, implementation, and experimental validation. IEEE Transactions on Control Systems Technology 10(6), 835–852 (2002)
6. Chen, J., Dixon, W.E., Dawson, D.M., McIntyre, M.: Homography-based visual servo tracking control of a wheeled mobile robot. IEEE Transactions on Robotics 22(2), 406–415 (2006)
7. Yang, F., Wang, C., Jing, G.: Adaptive tracking control for dynamic nonholonomic mobile robots with uncalibrated visual parameters. International Journal of Adaptive Control and Signal Processing 27(8), 688–700 (2013)
8. Wang, C., Mei, Y., Liang, Z., Jia, Q.: Dynamic feedback tracking control of nonholonomic mobile robots with unknown camera parameters. Transactions of the Institute of Measurement and Control 32(2), 155–169 (2010)
9. Campion, G., Bastin, G., Dandrea-Novel, B.: Structural properties and classification of kinematic and dynamic models of wheeled mobile robots. IEEE Transactions on Robotics and Automation 12(1), 47–62 (1996)
10. Corradini, M.L., Fossi, V., Giantomassi, A., Ippoliti, G., Longhi, S., Orlando, G.: Minimal resource allocating networks for discrete time sliding mode control of robotic manipulators. IEEE Transactions on Industrial Informatics 8(4), 733–745 (2012)

A Systematic Fire Detection Approach Based on Sparse Least-Squares SVMs

Jingjing Zhang[1], Kang Li[2], Wanqing Zhao[1], Minrui Fei[3], and Yigang Wang[4]

[1] Cardiff School of Engineering, Cardiff University, Cardiff, CF24 3AA, U.K.
[2] School of Electronics, Electrical Engineering and Computer Science,
The Queen's University of Belfast, Belfast, BT9 5AH, U.K.
[3] Shanghai Key Laboratory of Power Station Automation Technology,
School of Mechatronics and Automation, Shanghai University,
Shanghai, 200072, China
[4] Department of Information, Ningbo Institute of Technology, Zhejiang University,
Ningbo, 315010, China

Abstract. In this paper, a systematic approach adopting sparse least-squares SVMs (LS-SVMs) is proposed to automatically detect fire using vision-based systems with fast speed and good performance. Within this framework, the features are first extracted from input images using wavelet analysis. The LS-SVM is then trained on the obtained dataset with global support vectors (GSVs) selected by a fast subset selection method, in the end of which the classifier parameters can be directly calculated rather than updated during the training process, leading to a significant saving of computing time. This sparse classifier only depends on the GSVs rather than all the patterns, which helps to reduce the complexity of the classifier and improve the generalization performance. Detection results on real fire images show the effectiveness and efficiency of the proposed approach.

1 Introduction

Elaborate empirical studies have shown that least-squares support vector machines (LS-SVMs) can achieve good generalization performance on various classification problems [1–3] and it has been widely applied in many areas such as medicine, economy, etc [4–6]. Among these areas, application involving videos and images is an important and hot topic, since the image input for machine is an important way for data collection and analysis the same way as the eyes do to humans. Fire detection is a typical application area. Several vision-based fire detection papers have appeared in the literature [7–9]. The features of the fire region in terms of color, shape, temporal, motion and information have been studied to improve the detection accuracy. However, the complexity of the system will increase and computing speed will reduce with more information being added to verify the candidate fire pixels. Moreover, such scene-dependant information may deteriorate the generalization performance of the system.

To tackle this problem, a data-driven method, namely support vector machine (SVM) has been adopted. Recently, the application of SVMs in vision-based

M. Fei et al. (Eds.): LSMS/ICSEE 2014, Part II, CCIS 462, pp. 351–362, 2014.

fire detection, which can improve the reliability, accuracy and robustness with good generalization performance has drawn a lot of attention. Moreover, a SVM does not require heuristic features to be determined as in [7, 8]. In [10, 11], the fire-colored pixels and moving pixels were detected. Non-fire pixels were then removed using temporal luminance variation. After these two steps, the remaining pixels were used for fire classification by SVMs using features extracted with a Daubechies wavelet. The wavelet transform was also adopted to extract the features for a SVM classifier. SVMs were also adopted in [12] to gain a low generalization error rate with some more universal features suggested. Experimental results on different styles of smoke in different scenes have shown the proposed algorithm was reliable and effective. Although the application of SVMs does improve the performance of fire detection, the training of a SVM classifier is time-consuming. This is due to solving a convex quadratic programming of SVM itself. Thus, the least-squares support vector machines [13, 14] was proposed to cope with this problem. In LS-SVMs, a two-norm cost function is adopted together with equality, instead of inequality, constraints to obtain a linear set of equations rather than a Quadratic Programming (QP) problem to be solved in the dual space in SVMs. These modifications to the problem formulation implicitly correspond to a ridge regression formulation with binary targets ± 1. As a result, LS-SVMs can overcome the high computational complexity issue for conventional SVMs. However, the main drawback of LS-SVM is the non-sparse issue where all training patterns tend to be the support vectors (SVs), limiting its application in fire detection.

In this paper, a systematic approach is developed which also enables the recently developed sparse LS-SVM training method [15] to be used for constructing the classifier for fire detection. Firstly, the information for fire detection is selected by a three level decomposition of Daubechies wavelet, based on which, the features for identifying the fire can be extracted and the training dataset for a sparse LS-SVM can thus be established. Then, the sparse LS-SVM classifier will be trained with a subset selection method in optimizing the same cost function of the conventional LS-SVMs. Compared to the conventional LS-SVMs, this sparse classifier is constructed based on selected global support vectors (GSVs), like the SVs in SVMs, with subset selection method. The selected GSVs from the continuous input space ensure the low complexity of the classifier and further improve the generalization performance. The speed is improved through the employment of a fast subset selection method, with less GSVs being selected and all classifier parameters being calculated only at the end of the training. The performance of the proposed approach will be tested on real fire images and further verified through visualized detection results.

2   Fire Detection with LS-SVMs

The information contained in the pixels of fire image can be extracted as features to be used in the classification methods, like least-squares SVMs. Thus based on the previously proposed methods for LS-SVMs and fire detection methods from the literature, here a new structure for fire detection with LS-SVM is formed. In the new structure, a LS-SVM is adopted to implement classification on the fire dataset, the features of which are extracted using a wavelet transformation. Then the fire detection can be automatically conducted with the trained classifier. The detailed structure is shown in Fig. 1.

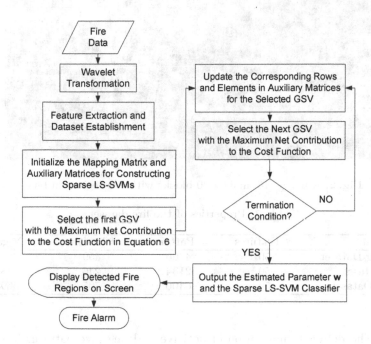

Fig. 1. The structure of proposed fire detection with a sparse classifier of LS-SVMs

2.1 Feature Selection and Training Dataset Establishment

In the proposed structure, fire videos are first transformed into frames. Suppose the image size is $m \times n$ pixels and there are k frames in the video sequences. Then the number of pixels from the video will be $m \times n \times k$, which might result in a huge dataset to be processed. So here, the block processing technique is adopted. Every frame is divided into 32×40 blocks, each of which includes 8×10 pixels, as shown in Fig. 2. Each block will be taken as one data pattern.

In this paper, five videos taken from [16] are used to form the fire dataset. The images from those are all 256×400 in size. Five images are separately selected from the five videos to form the whole dataset. The blocks with fire are labeled manually as positive and non-fire blocks are labeled as negative for the five images and the total number of blocks in the whole dataset is therefore 6400. That is to say, there are 6400 data patterns in the whole dataset. Then a third of the data patterns (2134 patterns) are randomly selected from the whole dataset as the test dataset and the rest (4266 patterns) are taken as the training

Fig. 2. A image [16] in 32 × 40 blocks with fire regions in blue

Table 1. Properties of the fire dataset

	# Features	# Patterns	# Positive	# Negative
Training Dataset	50	4266	459	3807
Test Dataset	50	2134	246	1888
Whole Dataset	50	6400	705	5695

dataset. The detailed information of positive and negative patterns in the fire dataset are listed in Table 1. The ratio of positive to negative patterns in the whole dataset is 1:8.

To visualize the fire detection results, a separate frame is randomly selected from the five videos as shown in Fig. 2, which is also divided into 32 × 40 blocks and labeled of the fire blocks manually.

The feature extraction is conducted with a wavelet transform. The wavelet transform [17] provides the frequency of the signals and the time associated with those frequencies. In the continuous wavelet transform (CWT), the input signal $f(t)$ is decomposed into a series of frequency components with a set of wavelet coefficients as

$$[W_\psi, f](a, b) = \frac{1}{\sqrt{a}} \int_{-\infty}^{\infty} f(t) \overline{\psi(\frac{t-b}{a})} dt \tag{1}$$

where $\psi(a, b)$ is the mother wavelet function, based on which all the child wavelets are derived by shifting with the shifting coefficient b and scaling with the scaling coefficient a. If parameters a and b are discrete values, then this kind of transformation is the discrete wavelet transform (DWT) [18, 19]. The input signal is decomposed through filters with different cutoff frequencies at different scales. The DWT is realized by successive lowpass and highpass filtering of the

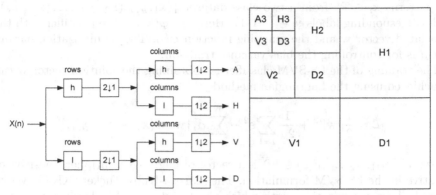

(a) A one level decomposition of two dimensional DWT

(b) A three level decomposition of
two dimensional DWT

Fig. 3. Decomposition of two dimensional DWT

input signal. In the two dimensional DWT, four coefficients are produced instead
of two in one dimensional DWT: approximation coefficients A and detailed coefficients H (horizontal), V (vertical) and D (diagonal) as shown in Fig. 3(a). In
multi-level decomposition, the approximation part will be further decomposed
into four components. For example, a three level decomposition is shown in Fig.
3(b).

Daubechies wavelet [20] is the most commonly used DWT. It is a family of
orthogonal wavelets defining a discrete wavelet transform and characterized by
a maximal number of vanishing moments for some given support. In [21], the
Daubechies second order moment has been evaluated as efficient in fire detection.
So in the proposed approach, it is also adopted for extracting features from the
images. Similar to [21], the three level of wavelets decomposition is used to obtain
the coefficients and five features are computed for each sub-band in Fig. 3(b):
arithmetic mean, standard deviation, skewness, kurtosis and entropy, resulting in
50 features in total as listed in Table 1. Based on the patterns preprocessed with
blocking method and the features extracted with the Daubechies wavelet, the fire
dataset is finally formed, the detailed information of which is demonstrated in
Table 1. Then the sparse LS-SVM classifier can be trained on the training dataset
and evaluated to show the generalization performance on the test dataset.

2.2 Sparse LS-SVMs

The problem of least-squares support vector machines is formulated as

$$\min_{\mathbf{w}, \varepsilon_i} \frac{1}{2} ||\mathbf{w}||^2 + \frac{1}{2\mu} \sum_{i=1}^{N} \varepsilon_i^2,$$

$$\text{subject to} \quad \varepsilon_i = y_i - \mathbf{w} \cdot \boldsymbol{\phi}(\mathbf{x}_i),$$

(2)

for patterns $\mathbf{x}_i \in \Re^n$ from a two class dataset $\{(\mathbf{x}_1, y_1), (\mathbf{x}_2, y_2), \ldots, (\mathbf{x}_N, y_N)\}$ with corresponding labels $y_i \in \{-1, 1\}$. Here, $\mathbf{w} \cdot \phi(\mathbf{x}_i)$ is the classifier with the associated vector $\mathbf{w}$ and the mapping function $\phi(\cdot)$. The regularization parameter μ is for controlling the bias-variance trade-off.

The training of the LS-SVM classifier for obtaining the solution vector $\mathbf{w}$ can be achieved using the Lagrangian method

$$\mathcal{L} = \frac{1}{2}||\mathbf{w}||^2 + \frac{1}{2\mu}\sum_{i=1}^{N}\varepsilon_i^2 - \sum_{i=1}^{N}\alpha_i\{\mathbf{w} \cdot \phi(\mathbf{x}_i) + \varepsilon_i - y_i\}, \tag{3}$$

where $\boldsymbol{\alpha} = (\alpha_1, \alpha_2, \ldots, \alpha_N) \in \Re^N$ is a vector of Lagrange multipliers, positive or negative in the LS-SVM formulation. The Karush-Kuhn-Tucker (KKT) system is applied for optimization, with which the training of the classifier equals to solving the following equation

$$\mathbf{M}\boldsymbol{\alpha} = \mathbf{y}, \tag{4}$$

where $\mathbf{M} = \mathbf{K} + \mu\mathbf{I}$ is a definite symmetric matrix and $\mathbf{K}(\mathbf{x}_i, \mathbf{x}_j) = \phi(\mathbf{x}_i) \cdot \phi(\mathbf{x}_j)$ is known as kernel function. Solving this simpler set of linear equations in (4) instead of the complex QP problem in SVMs makes the resultant LS-SVMs superior to SVMs in terms of computational efficiency. However, this improvement causes a vital drawback of LS-SVMs, a non-sparse classifier. That is because the support values α_i are proportional to the errors for the input patterns, most of which are non-zero. This is to say, the final output of the LS-SVMs classifier depends on almost all the training patterns, resulting in a non-sparse classifier.

Since the conventional LS-SVM is non-sparse, a new sparse classifier was proposed in [15]. In this paper, the sparse classifier is now examined to build up the classifier for vision-based fire detection based on the previously selected features. To initialize the sparse classifier, an alternative solution to the conventional one is suggested. Here, assume that the mapping function is known a *priori* and given as

$$\phi(\mathbf{x}_i) = [\varphi_1(\mathbf{x}_i), \varphi_2(\mathbf{x}_i), \ldots, \varphi_m(\mathbf{x}_i)]^{\mathrm{T}},$$
$$\varphi_k(\mathbf{x}_i) = \exp\{-\sigma(\mathbf{x}_i - \mathbf{s}_k)^{\mathrm{T}}(\mathbf{x}_i - \mathbf{s}_k)\}; \tag{5}$$
$$k = 1, \ldots, m,$$

where m is the dimension of the mapped high-dimensional space $\mathcal{H}^m$, σ is the width and $\mathbf{s}_k \in \Re^n, (k = 1, 2, \ldots, m)$ are some data vectors from continuous input space, which are not necessarily confined to the training patterns.

Accordingly, the original optimization problem in (2) can be reformulated as the following regularized problem

$$\min_{\mathbf{w}} J(\mathbf{w}) = \frac{1}{2}||\mathbf{w}||^2 + \frac{1}{2\mu}\sum_{i=1}^{N}(y_i - \mathbf{w} \cdot \phi(\mathbf{x}_i))^2. \tag{6}$$

Here the idea of ridge regression [22] is adopted to solve the objective instead of using Lagrangian method in the conventional approach. The parameter estimate of $\mathbf{w}$ can be given as in (7), considering that the gradient of (6) with respect to the parameter $\mathbf{w}$ has to be equal to zero

$$\hat{\mathbf{w}} = (\mathbf{\Phi}^{T}\mathbf{\Phi} + \mu\mathbf{I})^{-1}\mathbf{\Phi}^{T}\mathbf{y}, \tag{7}$$

where

$$\mathbf{\Phi} = [\varphi_1, \varphi_2, \ldots, \varphi_m],$$
$$\varphi_k = [\varphi_k(\mathbf{x}_1), \varphi_k(\mathbf{x}_2), \ldots, \varphi_k(\mathbf{x}_N)]^{T} \in \Re^{N}. \tag{8}$$

Each column φ_k (called as regressor) in the whole mapping matrix $\mathbf{\Phi}$ corresponds to one mapping dimension of the high-dimensional space for all the input patterns. It is worth noting that solution (7) optimizes the same objective function (2) as in conventional LS-SVMs.

With the computed solution vector $\hat{\mathbf{w}}$ in (7), for a new test data vector $\mathbf{x}$ in the input space, its decision value can be determined by

$$f(\mathbf{x}) = \hat{\mathbf{w}} \cdot \phi(\mathbf{x}) = \sum_{k=1}^{m} w_k \varphi_k(\mathbf{x}). \tag{9}$$

Based on this new solution, the sparseness can be achieved by selecting the dimension m of the high-dimensional space, i.e. the columns of mapping matrix $\mathbf{\Phi}$, with a fast subset selection method similar as in [23, 24]. Now with the definition of a recursive matrix $\mathbf{R} \in \Re^{N \times N}$ of the form

$$\mathbf{R} \triangleq \mathbf{I} - \mathbf{\Phi}(\mathbf{\Phi}^{T}\mathbf{\Phi} + \mu\mathbf{I})^{-1}\mathbf{\Phi}^{T}, \tag{10}$$

and auxiliary matrix $\mathbf{A}$ and vector $\mathbf{B}$ including elements defined as

$$a_{k+1,i} = \phi_{k+1}^{T}\mathbf{R}_k\phi_i,$$
$$b_{k+1} = \mathbf{y}^{T}\mathbf{R}_k\phi_{k+1}, \tag{11}$$

where $k = 0, 1, \ldots, m-1$ and $i = 1, 2, \ldots, N$, the net contribution of the $(k+1)th$ added regressor to the cost function will be

$$\Delta J_{k+1} = \frac{1}{2\mu} \frac{(b_{k+1})^2}{a_{k+1,k+1} + \mu}. \tag{12}$$

For each selection step, the regressor which results in the maximum net contribution of the cost function will be selected. This selection only changes the columns of mapping matrix without deleting any input patterns, thus keeping the information contained in the input patterns, superior to other sparse methods by deleting some input patterns during training. The selection process will be ended when some termination condition is met, for example, predefined number m ($m \ll N$) of GSVs. Finally, each element $\hat{\mathbf{w}}_{m,i}$, $i = 1, 2, \ldots, m$ in the estimated vector $\hat{\mathbf{w}}_m$ is calculated directly with the values from the auxiliary matrix $\mathbf{A}$ and vector $\mathbf{B}$ as

$$\hat{\mathbf{w}}_{m,i} = \frac{b_i}{a_{i,i} + \mu} - \frac{1}{\mu} \sum_{j=i+1}^{m} \frac{a_{j,i}b_j}{a_{j,j} + \mu}. \tag{13}$$

Table 2. Cost values on fire dataset with various number of GSVs

# GSVs	Minimum J	Running Time (s)	Training Acc. (%)	Test Acc. (%)
1	1.5906×10^3	0.3615	89.24	88.47
5	1.1933×10^3	1.0509	90.83	89.13
10	966.3615	2.0551	93.95	93.25
15	915.8171	3.1085	94.30	93.67
20	889.9646	3.9200	95.54	93.63
30	857.4315	5.6070	94.70	93.96
40	835.3370	6.8930	94.73	94.38
50	817.3197	8.9410	94.80	94.28
60	803.5796	10.7773	94.77	94.28
80	780.6155	14.4710	94.73	94.33
100	765.9533	18.1485	94.87	94.24
125	752.0591	22.7409	94.89	94.14
150	741.3625	27.4526	95.08	94.24
175	732.9921	32.2084	95.19	94.19
200	725.3695	36.5419	95.26	94.19
250	714.7755	46.2379	95.29	94.24
300	707.6249	55.2485	95.29	94.38
350	702.3870	65.2093	95.34	94.38
400	698.2766	73.4476	95.36	94.38
450	694.8493	83.4306	95.38	94.38
500	692.1020	93.1268	95.38	94.38
1000	679.7927	204.4456	95.43	94.42
2000	674.7748	499.6815	95.43	94.42
4266	673.5321	1.6357×10^3	95.43	94.42

It's worth to mention that the calculation of the estimated vector $\hat{\mathbf{w}}_m$ only at the end of the subset selection further helps saving the training time.

3   Fire Detection Results

In this section, the performance of the proposed systematic approach will be tested in terms of the accuracy and speed by applying the sparse LS-SVM classifier described in Section 2 to detect fire from images, which are established after feature extraction as in Table 1. Also, the superiority of the sparse LS-SVM classifier will also be verified comparing with conventional LS-SVMs for the proposed vision-based fire detection system.

Firstly, suppose GSVs for constructing the final classifier in the LS-SVM are selected from the input patterns and the hyperparameters including parameter of the regularization μ and kernel parameter σ are set the same as those in conventional solution. Given the values of hyperparameter pairs (μ, σ) as $(0.5, 0.01)$, the values of the cost function, the classification accuracy on training dataset and test dataset with different number of selected GSVs on fire dataset are listed in Table 2.

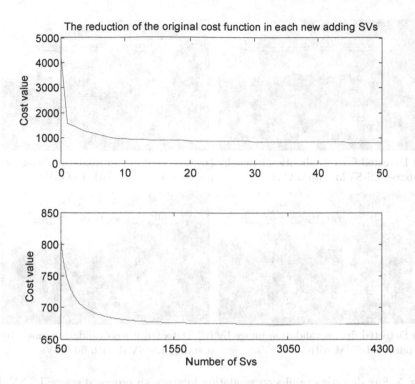

Fig. 4. The convergence curve of the cost value with the number of GSVs in fire detection

From Table 2, it can be first found that the cost value decreases as more GSVs are included in the classifier. Meanwhile, it can be noticed that the test accuracy does not change too much after 15 GSVs being selected, even though the optimal cost function values keep decreasing. Thus, in practice, 15 GSVs are selected to construct the final sparse classifier considering acceptable performance and the complexity of the classifier. As a matter of fact, reducing the complexity of the classifier can always degrade the training performance, but this is not true for the generalization performance when certain range of GSV numbers are included in the classifier as a low complexity classifier with less GSVs can usually help avoiding overfitting.

Alternatively, the change in the cost value with adding more GSVs into the classifier is shown in Fig. 4. The upper figure shows the variation of cost value when the first 50 GSVs being selected and the lower figure shows that for more than 50 GSVs being selected. The figure also shows that the decreasing rate of cost value is very small after 50 GSVs being selected, which means GSVs selected after that have less contribution to the reduction of the cost value. That is to say, the trained classifier only depends on certain important patterns from the training dataset.

For comparison, the conventional LS-SVM was also operated on the same dataset as above. The training accuracy of the obtained non-sparse classifier

(a) Detected fire on validation image by a sparse LS-SVM with 5 GSVs

(b) Detected fire on validation image by a sparse LS-SVM with 15 GSVs

(c) Detected fire on validation image by a sparse LS-SVM with 30 GSVs

(d) Detected fire on validation image by a sparse LS-SVM with 50 GSVs

Fig. 5. Fire detection results on validation image with proposed sparse LS-SVM

consisted of 4266 SVs is 95.92% with the training time of 151.0223 seconds and the validation accuracy on test dataset is 94.56%. However, from the results in Table 2, the proposed sparse LS-SVM classifier can obtain the training accuracy of 94.30% in 3.1085 seconds with 15 GSVs, and the test accuracy of 93.67%. This shows that for this fire detection application, the sparse LS-SVM classifier can achieve competitive generalization performance while also using less support vectors and being trained in shorter time, compared with the conventional LS-SVM.

Similar results can be seen from the visualization of the validation image, shown in Fig. 5. In these images, the logo of the video from which it was taken are interestingly viewed as the background noise to test the noise tolerance of the proposed approach. There are four sub-figures separately demonstrating the detection results on the same validation image with sparse LS-SVMs constructed from 5, 15, 30 and 50 GSVs. While the sparse classifier only depended on 5 GSVs, significant background blocks were incorrectly identified as fire regions compared to that manual labeled in Fig. 2. With the number of the GSVs increased to 15, 30 and 50, the false alarm reduced with more fire regions in the center of the fire detected, which can also be reflected by the increased detection accuracies 91.72%, 95.00%, 95.16% and 95.31%. It is also noticed that, the detection accuracy of the trained classifiers with 15, 30 and 50 GSVs did not change significantly, also shown as the detected fire regions are very similar to each other in Fig. 5(b), Fig. 5(c) and Fig. 5(d). This verifies that the number of 15 GSVs is acceptable for constructing the sparse LS-SVM with comparable performance in this practical fire detection application.

4 Conclusions and Future Work

This paper has proposed a systematic fire detection approach using data driven classification method to automatically detect fire from images. Instead of using the time-consuming SVMs and non-sparse least-squares SVMs, a sparse LS-SVM classifier was adopted in order to simultaneously speed up the training and reduce the model complexity. Within the framework of this new fire detection approach, the wavelet analysis was first used for feature extraction and a fast subset selection method was then applied for constructing sparse LS-SVM classifiers through selecting global support vectors. The application of this sparse classifier on real fire images have demonstrated that, the proposed approach was able to detect fire with high accuracy while also using less support vectors.

However, it was also noticed that in the application of fire detection, the ratio of positive to negative patterns in the whole dataset is 1:8, which is an unbalanced dataset. Although the fire detection results were generally satisfactory, the unbalanced dataset problem may still produce a strong bias towards the majority class with a consequence of increased false-negative rate and reduced classification accuracy. Common approaches for dealing with unbalanced datasets can be categorized into two groups. One aims to re-sample the dataset, such as randomly under-sampling and over-sampling. The other usually modifies the algorithm, such as adding weights on different classes. Both strategies could potentially be investigated under the proposed approach in the future.

Acknowledgement. This work was partially supported by the UK RCUK grant EP/G042594/1, Chinese Scholarship Council (CSC) and the International Corporation Project of Ningbo under Grant 2013D10009.

References

1. Burges, C.J.C.: A tutorial on support vector machines for pattern recognition. Data Mining and Knowledge Discovery 2(2), 121–167 (1998)
2. Gestel, T.V., Suykens, J.A.K., Baesens, B., Viaene, S., Vanthienen, J., Dedene, G., Moor, B.D., Vandewalle, J.: Benchmarking least squares support vector machine classifiers. Machine Learning 54(1), 5–32 (2004)
3. Baesens, B., Viaene, S., Gestel, T.V., Suykens, J.A.K., Dedene, G., Moor, B.D., Vanthienen, J.: An empirical assessment of kernel type performance for least squares support vector machine classifiers. In: Proceedings of Fourth IEEE International Conference on Knowledge-Based Intelligent Engineering Systems and Allied Technologies, vol. 1, pp. 313–316 (2000)

4. Tarabalka, Y., Fauvel, M., Chanussot, J., Benediktsson, J.A.: SVM- and MRF-based method for accurate classification of hyperspectral images. IEEE Geoscience and Remote Sensing Letters 7(4), 736–740 (2010)
5. Polat, K., Gnes, S.: Breast cancer diagnosis using least square support vector machine. Digital Signal Processing 17(4), 694–701 (2007)
6. Gestel, T.V., Suykens, J.A.K., Baestacns, D.E., Lambrechts, A., Lanckriet, G., Vandaele, B., Moor, B.D., Vandewalle, J.: Financial time series prediction using least squares support vector machines within the evidence framework. IEEE Transactions on Neural Networks 12(4), 809–821 (2001)
7. Healey, G., Slater, D., Lin, T., Drda, B., Goedeke, A.D.: A system for real-time fire detection. In: Proceeding of IEEE Computer Society Conference on Computer Vision and Pattern Recognition, pp. 605–606 (1993)
8. Phillips III, W., Shah, M., Lobo, N.D.V.: Flame recognition in video. Pattern Recognition Letters 23(1-3), 319–327 (2002)
9. Celik, T., Demirel, H.: Fire detection in video sequences using a generic color model. Fire Safety Journal 44(2), 147–158 (2009)
10. Ko, B.C., Cheong, K.H., Nam, J.Y.: Fire detection based on vision sensor and support vector machines. Fire Safety Journal 44(3), 322–329 (2009)
11. Cheong, K.H., Ko, B.C., Nam, J.Y.: Vision sensor-based fire monitoring system for smart home. In: Proceeding of International Conference on Ubiquitous Information Technology and Applications, B, pp. 1453–1463 (2007)
12. Yang, J., Chen, F., Zhang, W.: Visual-based smoke detection using support vector machine. In: Proceeding of Fourth IEEE International Conference on Natural Computation, vol. 4, pp. 301–305 (2008)
13. Suykens, J.A.K., Vandewalle, J.: Least squares support vector machine classifiers. Neural Processing Letters 9(3), 293–300 (1999)
14. Jia, J., Wang, H.Q., Hu, Y., Ma, Z.F.: Fire smoke recognition algorithm based on least squares support vector machine. Computer Engineering 2, 091 (2012)
15. Zhang, J., Li, K., Irwin, G.W., Zhao, W.: A regression approach to LS-SVM and sparse realization based on fast subset selection. In: Proceeding of 10th World Congress on Intelligent Control and Automation, Beijing, China, pp. 612–617 (2012)
16. Töreyin, B.U.: Fire video dataset, http://signal.ee.bilkent.edu.tr/VisiFire/ (accessed July 13, 2014)
17. Chui, C.K.: An introduction to wavelets, vol. 1. Academic Press (1992)
18. Mallat, S.G.: A theory for multiresolution signal decomposition: The wavelet representation. IEEE Transactions on Pattern Analysis and Machine Intelligence 11(7), 674–693 (1989)
19. Shensa, M.J.: The discrete wavelet transform: wedding the a trous and mallat algorithms. IEEE Transactions on Signal Processing 40(10), 2464–2482 (1992)
20. Daubechies, I.: Orthonormal bases of compactly supported wavelets. Communications on Pure and Applied Mathematics 41(7), 909–996 (1988)
21. Gubbi, J., Marusic, S., Palaniswami, M.: Smoke detection in video using wavelets and support vector machines. Fire Safety Journal 44(8), 1110–1115 (2009)
22. Hoerl, A.E., Kennard, R.W.: Ridge regression: Biased estimation for nonorthogonal problems. Technometrics 12(1), 55–67 (1970)
23. Li, K., Peng, J.X., Irwin, G.W.: A fast nonlinear model identification method. IEEE Transactions on Automatic Control 50(8), 1211–1216 (2005)
24. Li, K., Peng, J.X., Bai, E.W.: A two-stage algorithm for identification of nonlinear dynamic systems. Automatica 42(7), 1189–1197 (2006)

Sensorless Vector Control of PMSM in Wide Speed Range

Tao Yan, Jun Liu, and Haiyan Zhang

Shanghai Dianji University, NO.690 Jiangchuan Road, Minhang District, Shanghai, China
yt.sh69@163.com, liujun@sdju.edu.cn, haiyok@126.com

Abstract. In order to achieve the sensorless vector control of PMSM in wide speed range, a hybrid control mode strategy is presented, which included sliding mode observer (SMO) and high frequency injection (HFI). At medium or high speed, sliding mode observer method that was based on fundamental wave model is applied to estimate of speed and position of PMSM. While at low speed, for avoiding the shortcomings of SMO, it had to switch to the HFI method. Firstly, the application of SMO method was achieved and the speed limit of SMO method is calculated, and it is as basis for switching region of the speed of the hybrid mode. The simulation results show that the hybrid mode can reduce the buffet in the procedure of switching of algorithm effectively. And it achieves the control of PMSM in wide speed range.

Keywords: PMSM, Sensorless Vector Control, SMO, HIF, Wide Speed Range.

1 Introduction

With the development of microelectronics technology, PMSM received more and more attention and was used widely. In high dynamic performance servo control system of PMSM, position or speed of the rotor was indispensable feedback information. The traditional method is that the mechanical position sensors were installed in the PMSM. But the cost of the system will increase and reliability and stability will lower [1].

Sensorless control strategy is based on the fundamental wave model of PMSM. And the voltage and current of the motor stator windings are sampled; estimate and extract the position and speed of rotor, through certain mathematical method. There are a lot of related mathematical methods; but direct back emf [2] or flux estimation method, or various types of state observer [3-6] and model reference adaptive method [7-8], are based on estimate of back emf. However, at low or zero speed range, signal to noise ratio of the back emf is too low or even zero, which is the inevitable. Through high frequency injection method[9-10], the rotor position information can be extracted from the feedback frequency signal at low speed.

In the paper, a hybrid control model strategy is established, using linear weighted average approach to achieve a smooth transition of the two methods. And simulation experiments validate the feasibility of the hybrid mode.

M. Fei et al. (Eds.): LSMS/ICSEE 2014, Part II, CCIS 462, pp. 363–369, 2014.
© Springer-Verlag Berlin Heidelberg 2014

2   Sliding Mode Observer

2.1   Mathematical Principle of SMO

The mathematical model of PMSM in the $\alpha\beta$-reference frame is as follows:

$$\begin{cases} \dfrac{di_\alpha}{dt} = -\dfrac{R}{L}i_\alpha - \dfrac{1}{L}K_e e_\alpha + \dfrac{1}{L}u_\alpha \\ \dfrac{di_\beta}{dt} = -\dfrac{R}{L}i_\beta - \dfrac{1}{L}K_e e_\beta + \dfrac{1}{L}u_\beta \end{cases} \tag{1}$$

Where $u_\alpha, u_\beta, i_\alpha, i_\beta, e_\alpha, e_\beta$ are voltage, current, component of back emf in the $\alpha\beta$ axis. And R, L, K_e are motor phase resistance, phase inductance, and back emf coefficient. Correlations of the back emf in (1) can be represented as:

$$\begin{cases} e_\alpha = -K_e \omega_r \sin\theta_r \\ e_\beta = -K_e \omega_r \sin\theta_r \end{cases} \tag{2}$$

Where ω_r is the speed of the rotor; θ_r is the position angle of the rotor.

The back emf contains information of speed and position of the rotor from Eq. (2). So based on mathematical model, SMO equation can be structured.

$$\begin{cases} \dfrac{d\hat{i}_\alpha}{dt} = -\dfrac{R}{L}\hat{i}_\alpha + \dfrac{u_\alpha}{L} - \dfrac{K}{L}sign\left(\overline{i_\alpha}\right) \\ \dfrac{d\hat{i}_\beta}{dt} = -\dfrac{R}{L}\hat{i}_\beta + \dfrac{u_\beta}{L} - \dfrac{K}{L}sign\left(\overline{i_\beta}\right) \end{cases} \tag{3}$$

Where $\hat{i}_\alpha, \hat{i}_\beta$ are observed current; $\overline{i_\alpha}, \overline{i_\beta}$ are observed current error; K is coefficient of sliding mode observer; and $sign(x)$ is sign function, which can be represented as follow:

$$sign(x) = \begin{cases} 1, x > 0 \\ 0, x = 0 \\ -1, x < 0 \end{cases} \tag{4}$$

The section of sliding mode is defined as:

$$s_\alpha = \overline{i_\alpha}, s_\beta = \overline{i_\beta} \tag{5}$$

Use sliding mode variable structure (SMVSC) of function switching:

$$u = u_{eq} + Ksign\big(s(x)\big) = e + Ksign\big(s(x)\big) \tag{6}$$

When the system is in sliding mode, we have $s(x) = 0, \dfrac{d}{dt}s(x) = 0$. And after a finite time interval, there is $\overline{i}_\alpha = 0, \overline{i}_\beta = 0$. We make $\dfrac{d}{dt}\overline{i}_\alpha = 0, \dfrac{d}{dt}\overline{i}_\beta = 0$; we can have:

$$\begin{cases} u_{eq\alpha} = \left(Ksign\left(\overline{i}_\alpha\right)\right)_{eq} = e_\alpha \\ u_{eq\beta} = \left(Ksign\left(\overline{i}_\beta\right)\right)_{eq} = e_\beta \end{cases} \tag{7}$$

The current switch error signal contains the information of back emf. We can obtain estimated value of back emf from the signal, using a low pass filter that filters out the high frequency signal. Therefore, estimated value is:

$$\begin{cases} \hat{e}_\alpha = \dfrac{\omega_s}{s+\omega_s}e_\alpha \\ \hat{e}_\beta = \dfrac{\omega_s}{s+\omega_s}e_\beta \end{cases} \tag{8}$$

Therefore, we obtain the position angle and estimated value of speed of the rotor:

$$\begin{cases} \hat{\theta}_r = \arctan\left(-\dfrac{\hat{e}_\alpha}{\hat{e}_\beta}\right) \\ \hat{\omega}_r = \dfrac{d\,\hat{\theta}_r}{dt} \end{cases} \tag{9}$$

For the phase lag that is caused by the low pass filter, it has to estimate the position angle of the rotor for phase compensation. And depending on the operating speed ω_r, we obtain the relative displacement angle $\Delta\theta$, as noted in Fig. 1. So the final estimated value of position angle is:

$$\hat{\theta}_r = \arctan\left(-\dfrac{\hat{e}_\alpha}{\hat{e}_\beta}\right) + \Delta\theta \tag{10}$$

$\Delta\theta$ is given by $\Delta\theta = \arctan\left(\dfrac{\omega_r}{\omega_s}\right)$.

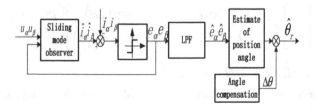

Fig. 1. Schematic diagram of SMO

2.2   Simulation Analysis

Fig. 2 and 3 is respectively the speed waveforms when rated speed is 1000rpm, 100rpm. As can be seen from the figures, when the rotational speed starts to rise from 0, there is a great buffet on the waveform. And when rotational speed reaches the command speed, it exhibits good convergence property. With the speed reducing, the error increases. When rotational speed is 100rpm, the output waveform has been unable to converge, which means that estimation of speed is failed.

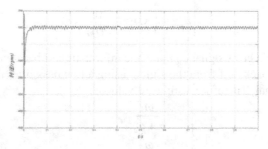

Fig. 2. Waveform of speed when Rated speed is 1000rpm

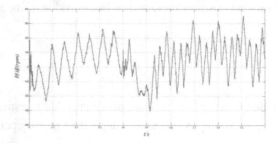

Fig. 3. Waveform of speed when Rated speed is 100rpm

3   High Frequency Injection

When high frequency voltage signal is injected, the response current is also high frequency signal. In this case, the value of the stator resistance and rotating EMF voltage can be negligible. The voltage equation of salient PMSM is:

$$\begin{bmatrix} u_\alpha \\ u_\beta \end{bmatrix} = \begin{bmatrix} L + \Delta L \cos(2\theta) & \Delta L \sin(2\theta) \\ \Delta L \sin(2\theta) & L - \Delta L \cos(2\theta) \end{bmatrix} \begin{bmatrix} pi_\alpha \\ pi_\beta \end{bmatrix} \tag{11}$$

And there's $L = (L_d + L_q)/2, \Delta L = (L_d - L_q)/2$.

After the injection of high frequency sinusoidal voltage, the voltage space vector is generated in the motor.

$$\begin{bmatrix} u_{\alpha i} \\ u_{\beta i} \end{bmatrix} = U_i \begin{bmatrix} \cos(\omega_i t) \\ \sin(\omega_i t) \end{bmatrix} = U_i e^{j\omega_i t} \tag{12}$$

Where U_i is amplitude of the high frequency voltage signal; ω_i is angular frequency of high frequency voltage signal, and $\omega_i \gg \omega_r$. The high frequency current response is:

$$\begin{bmatrix} \dfrac{di_{\alpha i}}{dt} \\ \dfrac{di_{\beta i}}{dt} \end{bmatrix} = \begin{bmatrix} L + \Delta L \cos(2\theta) & \Delta L \sin(2\theta) \\ \Delta L \sin(2\theta) & L - \Delta L \cos(2\theta) \end{bmatrix}^{-1} \begin{bmatrix} u_{\alpha i} \\ u_{\beta i} \end{bmatrix} \tag{13}$$

So

$$\begin{bmatrix} i_{\alpha i} \\ i_{\beta i} \end{bmatrix} = \begin{bmatrix} I_{ip} \cos\left(\omega_i t - \dfrac{\pi}{2}\right) + I_{in} \cos\left(2\theta - \omega_i t + \dfrac{\pi}{2}\right) \\ I_{ip} \sin\left(\omega_i t - \dfrac{\pi}{2}\right) + I_{in} \sin\left(2\theta - \omega_i t + \dfrac{\pi}{2}\right) \end{bmatrix} \tag{14}$$

Where $I_{ip} = \dfrac{U_i}{\omega_i} \left[\dfrac{L}{L^2 - \Delta L^2} \right]$, $I_{in} = \dfrac{U_i}{\omega_i} \left[\dfrac{-\Delta L}{L^2 - \Delta L^2} \right]$.

There are positive and negative sequence component in the response current. Only the negative component contained the position information of rotor. We can obtain the negative component through band pass filter that can filters fundamental harmonic current and low frequency current, synchronous frame filter that can filters positive component.

$$i_{in} = I_{in} e^{j\left(2\theta - 2\omega_i t + \frac{\pi}{2}\right)} e^{j\omega_i t} = I_{in} e^{j\left(2\theta - \omega_i t + \frac{\pi}{2}\right)} \tag{15}$$

We can have the rotor position error signal by using heterodyne:

$$\varepsilon = i_{\alpha in} \cos\left(2\hat{\theta} - \omega_i t\right) + i_{\beta in} \sin\left(2\hat{\theta} - \omega_i t\right) = I_{in} \sin\left(2\hat{\theta} - 2\theta\right) \tag{16}$$

When $\Delta\theta$ is small enough, we have $\varepsilon \approx 2I_{in}\Delta\theta$.

A tracking error signal can be obtained by using heterodyne method. Only when error signal approaches zero, it can guarantee that estimation of angle could approach true value.

4   Hybrid Control Mode Strategy

In order to achieve a smooth switch of two methods, at a certain speed range, the estimations are calculated mean value, which could make estimation closer to the true value. In the simulation results before the first section, you can see that the hybrid mode can be set: when less than 20% of rated speed, use high frequency injection; and when more than 40% of rated speed, use back emf estimation method; at 20% ~ 40% of the rated speed, the hybrid algorithm could be used for estimating the value by mean linear scale processing. The estimated mean value of position angle is shown as:

$$\hat{\theta}_{hyb} = \begin{cases} \hat{\theta}_{inj}, (k_\omega < 0.2) \\ \dfrac{k_\omega - 0.2}{0.4 - 0.2} \bullet \hat{\theta}_{emf} + \dfrac{0.4 - k_\omega}{0.4 - 0.2} \bullet \hat{\theta}_{inj}, (k_\omega \in [0.2, 0.4]) \\ \hat{\theta}_{emf}, (k_\omega > 0.4) \end{cases} \tag{17}$$

5   Simulations and Verification

Figure 6 shows the output waveform when speed accelerates from 0 to 800rpm; and figure 7 shows the speed error waveform. As can be seen from the simulation results, in the acceleration process, the error increases firstly; and then decreases. After the speed reaches 600rpm (the system is in the hybrid mode), estimation error is further reduced. It indicates that the hybrid mode has high accuracy during acceleration. When at 1.8s, the load increases from 3Nm to 6Nm by the sudden; accordingly, speed waveform appears buffeting and restores homeostasis quickly.

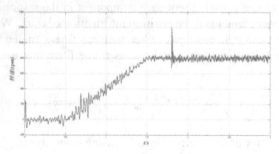

Fig. 4. Waveform of speed when accelerated from 0 to 800rpm

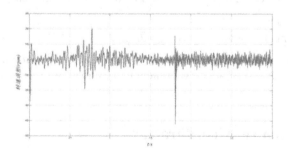

Fig. 5. Waveform of speed error

6   Conclusions

At first, based on the principle of sliding mode observer algorithm, it achieved a research of sensorless PMSM vector control. And for the disadvantages of sliding mode observer at low speed state that it cannot estimate rotor position angle and speed accurately. Be compensated by a high-frequency voltage injection method, which is based on the rotation principle of extrinsic motivation. Finally, in order to achieve a smooth switching of two estimation methods, a hybrid observer is designed, which is based on the principles of linear weighted average rate in a certain speed range. Simulation results show that the hybrid model can achieve sensorless control of PMSM effectively in wide speed range, which will provide a strong theoretical basis for further PMSM's applications and experiments.

References

1. Widyan, M.S., Hanitsch, R.E.: High-power density rapid-flux permanent sinusoidal three-phase three slot four-pole electrical generator. Int. J. Electr. Power Energy Syst., 1221–1227 (2012)
2. Fabio, G., Rosario, M., Cosimo, R., Giuseppe, R.G.: Back EMF Sensorless-Control Algorithm for High-Dynamic Performance PMSM. IEEE Transaction on Industrial Electronics 57, 2092–2100 (2010)
3. Accetta, A., Cirrincione, M., Pucci, M.: TLS EXIN based neural sensorless control of a high dynamic PMSM. Control Engineering Practice, 725–732 (2012)
4. Hassan, A.A., El-Sawy, A.M., Mohamed, Y.S., Shehata, E.G.: Sensorless sliding mode torque control of an IPMSM drive based on active flux concept. Alexandria Engineering Journal, 1–9 (2012)
5. Shehata, E.G.: Speed sensorless torque control of an IPMSM drive with online stator resistance estimation using reduced order EKF. Electrical Power and Energy Systems, 378–386 (2013)
6. Sanath, A., Tyrone, T., Hieu, T., Victor, S.: Unknown input sliding mode functional observers with application to sensorless control of permanent magnet synchronous machines. Journal of the Franklin Institute, 107–128 (2013)
7. Fan, S.C., Luo, W.Q., Zou, J.X., Zheng, G.: A Hybrid Speed Sensorless Control Strategy for PMSM Based on MRAS and Fuzzy Control. In: 2012 IEEE 7th International Power Electronics and Motion Control Conference, Harbin, pp. 2976–2980 (2012)
8. Frederik, M.D.B., Sergeant, P., Jan, A.M.: A sensorless PMSM drive using modified high-frequency test pulse sequences for the purpose of a discrete-time current controller with fixed sampling frequency. Mathematics and Computers in Simulation, 367–381 (2010)
9. Osama, A.M., Khan, A.A.: A Wavelet Filtering Scheme for Noise and Vibration Reduction in High-frequency Signal Injection-Based Sensorless Control of PMSM at Low Speed. IEEE Transactions on Energy Conversion, 250–260 (2012)
10. Aydogmus, O., Sünter, S.: Implementation of EKF based sensorless drive system using vector controlled PMSM fed by a matrix converter. Electrical Power and Energy Systems, 736–743 (2012)

Pitch Angle Control for Improving the Low Voltage Ride-Through Based on DFIG

Ruming Li, Tianyu Liu*, Qinghua Zhu, and Li Zhang

School of Electrical Engineering, Shanghai Dian Ji University,
Shanghai 200240, China
liuty@sdju.edu.cn

Abstract. Pitch angle control for wind generation is one of the most important segments in wind turbine. Wind turbine can capture the max wind energy which is called Maximum Power Point Tracking through regulating the pitch angle and then the output power of wind turbine is the best. On the base of it, one new method of pitch angle is put forward in this paper. When the grid was failure, we can reduce the output power through regulating the pitch angle in order to improve the performance of LVRT of the wind turbine. When the grid voltage is under the sudden dip, the rotor speed can be limited to reduce the engine and the sudden disconnection of wind turbine and grid can be confined. The pitch angle control is described in this paper in a short time when the emergency power supply is discussed so that we can keep the coordination between the wind turbine and the grid.

Keywords: Wind power, pitch angle control, LVRT.

1   Introduction

Recently, wind power is one of the fastest segments of clean-energy industry all over the world. It is not only related to the development of energy strategy for many countries, but also the wind power itself contains many advantages that the coals and nuclear don't have.

With the interconnection capacity of wind power higher, the dip of the grid voltage may case the wind farms disconnect with the grid suddenly. It may break the balance of voltage between the grid and the wind farm. So, when the grid has a dip on voltage, we don't want the turbine disconnect the grid and also transmit power to it. So it puts forward to high quality to the wind power and the power system. All in all, the technique of low voltage ride-through (LVRT) is paid much attention to the wind power integration [1]. In order to improve the ability of LVRT in wind farms, a new method of pitch angle control for improving the Low-Voltage Ride-Through based on DFIG is described and its novelty and feasibility are put forward in theory.

---

* Corresponding author.

M. Fei et al. (Eds.): LSMS/ICSEE 2014, Part II, CCIS 462, pp. 370–376, 2014.

2   Mathematic Model of DFIG

2.1   The Principles of DFIG

The DFIG is similar to wound rotor induction motor in structure and it has two windings of rotors and stators. When the wind speed that is variable and then cause the generator speed n, the frequency of rotor current can make the stator current constant. And it should be satisfied with the equation:

$$f_1 = pf_m + f_2$$

In the equation f1 is stator current frequency, and it is the same with the grid because of connecting with it; fm is rotor machinery frequency, and $f_m = \dfrac{n_m}{60}$ (n_m is mechanical speed of generator); p is the pole pair of generator; f2 is rotor current frequency.

2.2   Dynamic Model of DFIG

The mathematic model of DFIG is an equation contained by time-variable coefficients which are complicated. In order to simplify the process of calculating, we take the rotor and stator into the conversation of coordinates.

In the following discussion, the mathematic model of DFIG in the condition of Three-phase static coordinate system and Two phase synchronous speed rotating coordinate system, stator uses the routine of generator and the stator current is positive which is flowing; rotor uses the routine of electromotor and the rotor current is positive which is inflowing.

We hold a hypothesis to emphasize the key questions that the DFIG is used in the wind power system.

2.2.1   Mathematic Model in abc Three-Phase Static Coordinate System

a_s, b_s, c_s of Three-phase winding of the stator is fixed in the space and has a difference of 120 degrees each other. The axis of rotor winding is rotated with the rotor and a hypothesis is hold that rotor a_r and stator a_s has an electric angle which is θr. Θr is a variable angular displacement in the space.

1. the voltage equations

According to Kirchoff's law and Lenz's law, the voltage equations of stator loop and rotor loop can be expressed as:

$$u = Ri + d\varphi / dt \tag{1}$$

voltage equations of Three-phase stator winding

$$\begin{cases} U_{a_s} = -R_s i_{a_s} + \dfrac{d\varphi_{a_s}}{dt} \\[2mm] U_{b_s} = -R_s i_{b_s} + \dfrac{d\varphi_{b_s}}{dt} \\[2mm] U_{c_s} = -R_s i_{c_s} + \dfrac{d\varphi_{c_s}}{dt} \end{cases} \tag{2}$$

voltage equations of Three-phase rotor winding

$$\begin{cases} U_{a_r} = -R_r i_{a_r} + \dfrac{d\varphi_{a_r}}{dt} \\[2mm] U_{b_r} = -R_r i_{b_r} + \dfrac{d\varphi_{b_r}}{dt} \\[2mm] U_{c_r} = -R_r i_{c_r} + \dfrac{d\varphi_{c_r}}{dt} \end{cases} \tag{3}$$

In these equations, $U_{a_s}, U_{b_s}, U_{c_s}, U_{a_r}, U_{b_r}, U_{c_r}$ are the voltage instantaneous value of stator and rotor and the index s and represent stator and rotor; $i_{a_s}, i_{b_s}, i_{c_s}, i_{a_r}, i_{b_r}, i_{c_r}$ are the phrases current instantaneous value of stator and rotor; $\varphi_{a_s}, \varphi_{b_s}, \varphi_{c_s}, \varphi_{a_r}, \varphi_{b_r}, \varphi_{c_r}$ are the phase winding magnetic chain and R_S, R_r are the equivalent of resistance.

2. the flux-linkage equations

In the equation (1), the full magnetic chain of every resistance is the sum of self-induction magnetic chain and mutual inductance magnetic chain by other resistances and it can be expressed as

$$\begin{bmatrix} \varphi_s \\ \varphi_r \end{bmatrix} = \begin{bmatrix} L_{ss} & L_{sr} \\ L_{rs} & L_{rr} \end{bmatrix} \begin{bmatrix} i_s \\ i_r \end{bmatrix} = Li \tag{4}$$

In the equation, L_{ss} is stator leakage inductance; L_{rr} is rotor leakage inductance; L_{sr} is mutual inductance magnetic chain which rotor works on stator; L_{rs} mutual inductance magnetic chain which stator works on rotor; ψ_s is self-induction magnetic chain of stator; ψ_r is self-induction magnetic chain of stator; i_s is the stator current and i_r is the rotor current.

3.the equations of torque

Put the equation (4) into equation (3), it can be got

$$u = Ri + L\frac{di}{dt} + L\frac{di}{dt}i \tag{5}$$

In abc three-phase static coordinate system, electromagnetic torque equation of DFIG can be expressed as

$$T_c = 0.5\, p\, i^T \begin{bmatrix} 0 & \dfrac{dL_{s_r}}{d\theta_r} \\ \dfrac{dL_{r_s}}{d\theta_r} & 0 \end{bmatrix} i \tag{6}$$

In the equation, T_c is the magnetic torque of generator and p is number of pole-pairs.

4. the equation of motion

Provided that T_m is the input mechanical torque of prime motor, magnetic torque T_e and rotational inertia J can keep balance to get the torque equation as

$$T_m = T_e + \frac{J}{n_p}\frac{dw_r}{dt} \tag{7}$$

3   Existing Technique of LVRT

Low Voltage Ride-Through (LVRT) refers to when the voltage net in wind farm has a decline on voltage the wind farm could connect the grid and supply some reactive power until the grid is recovered and then ride through the time of low voltage [2].

In wind power system, under the situation of the balance and unbalance of grid voltage fault in principle the operational control of DFIG can be improved by GSC and RSC. But the control effect is controlled by the converters. And so the hardware protection measures should be added to through the fault time under the background of serious grid fault. Nowadays many protection schemes are put forward with [3], such as the hardware protection in stator side; the Crowbar technology on direct current side; the Crowbar technology on active current side; the pitch angle technology. These technologies of LVRT and Crowbar make up another research hot in wind power system.

4   Pitch Angle Control

4.1   The Principle of Pitch Angle Control in Turbines

Compared with turbines of the fixed pitch angle control, the turbines of variable pitch angle control are featured by better power smoothing control performance and flexible in grid connection and then they are widely used in the wind power system [4]. According to the role of variable pitch angle control system in variable pitch angle, it can be described as four models as control of starting, control of lacking of power, control of rated power and control of stopping. Variable speed and variable pitch angle are two main power regulation ways in wind turbines and also be the research hot at home and abroad [5]. When the wind speed is under the rated speed, the variable speed pattern (regulate the speed of rotor of generator) is chosen to capture the most wind power; when the wind speed is over the rated speed, the variable pitch angle control model (regulate the pitch angle) is used to balance the power between the wind turbines and generators.

1. Control of Starting

When the variable pitch angle turbines in the situation of stopping and feathering state, in order to ensure the security of turbines the angle of turbine is regulated to 90 degree and then the blade and the airflow has a degree just 0. So airflow has no effect on blade and in that time the blade is the same to a spoiling flap. When the wind speed increases gradually and reaches to the cut-in wind speed (the starting speed), we could control institution of variable pith to regulate the pitch angle in order to decrease it (in the direction of 0 degree). In order to ensure the wind turbines to connect the grid smoothly and decrease the impact to it, we can control the rotor speed of the generator nearby the synchronous speed with a period of time to seek to connect the grid.

2. Control of Lacking of Power

When the wind turbines connect the grid and the indeed speed is under the rated speed, the power of generator is less than rated power and the turbines are in the state of low power situation, the situation of wind turbines is called control of lacking of power. In order to improve the output power and capture the largest wind power, the maximum power point tracking is used (MPPT). In the situation of lacking of power, the pattern of MPPT controls the pitch angle nearby 0 degree. And then we can regulate the stator current by motor-side converter to control the speed of generator to track the most power of wind turbines.

3. Control of Rated Power

When the wind turbines connect to the grid, wind speed is over than rated speed and less than cut-out wind speed, the output power of turbines is over than rated power and then the wind turbines go into the pattern of rated power. The capacity of converter, mechanical stress and capacity of generator is limited and the pitch angle control is introduced. At that time the pitch angle is needed to enlarge to decrease the power of turbines to keep balance between the wind power and the rated power.

4. Control of Stopping

When the wind turbines connect to the grid and the wind speed increases gradually, according to the principle of pitch angle, the pitch angle is increased in order to decrease the power of turbines. When the pitch angle is nearby 90 degree and at the same time the wind speed also increases and the wind speed is over than the cut-out wind speed, the wind turbines will stop by braking. Then the blade is in the process of feathering state and out of the grid. Because of the wind speed variable constantly, the capturing power of turbines will be variable because of not using pitch angle control and pattern of stopping state. When the wind speed is more and more big, the turbines will capture the higher power. When the turbines capture power beyond its ability of itself, the turbines may be blow down. And it may cause a bad effect on the grid and even may lead to the grid to paradise.

4.2   The Principle of Pitch Angle Control to Achieve LVRT

Pitch angle can be changed between 0 degree and 90 degrees because of the pitch-controlled system and then the output power can be controlled. The capacity of generator can be controlled using the method of pitch angle in order to have little effect on the grid to have a high quality of electricity. The rotor of DFIG is determined by the difference of output power of turbines and output power of DFIG. When the grid has a great dip on voltage, the input power of turbines keep fixed, the power of DFIG into the grid is decreased and then the unbalanced power will cause the jump of rotor of generator. At that time, the pitch angle should be increased to decrease the input power of turbine in order to hinder the jump of rotor of generator. The typical method of pitch angle is expressed as figure 1 [6]. When the grid is normal, the DFIG capture the most power in variable speed constant frequency according to the pitch angle θ to capture the most power in variable speed constant frequency. When the grid has a dip, the pitch angle control should be taken immediately. According to the limit of power to calculate the index C_{p_lim} , we can get the reference of pitch angle through check the tables. Through decreasing the input power of turbine, the grid fault can be adopted [7].

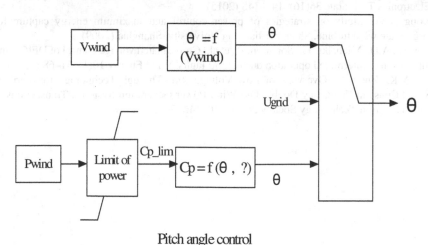

Pitch angle control

Fig. 1. Pitch angle control

5   Conclusion

Pitch angle control for improving the Low-Voltage Ride-Through based on DFIG is put forward with in this paper. Firstly, the mathematic model of DFIG is described to derive the model and the existing method of LVRT is analyzed. In addition, the principle of pitch angle control and through it to achieve LVRT is represented. Last, the emergency power supply when the grid has a dip is taken.

Overview it, a new method for achieving LVRT is put forward with. It has an advantage that it can guard against the jump of rotor of generator when the grid is fault. But it may add a new pitch angle system to cause complexity and the high odds of fault. When the grid is normal, the pitch angle system is supplied by the grid directly. When the grid faults, it has to supply by the emergency power supply. It is a complex program.

At last, considering the complexity of the model, it can't make an experiment through software and test bed. It is only testified in theory and principle. And it is the task in the following research and study.

References

1. Thet, A.K., Saitoh, H.: Pitch control for improving the low-voltage ride-through of wind farm. In: Transmission and Distribution Conference and Exposition: Asia and Pacific, T and D Asia 2009 (December 16, 2009)
2. Liu, Z.Y.: Technique of smart grid, pp. 132–138. China Electric Power Press, Beijing (2010)
3. Zhou, P.: Investigation of Low Voltage Ride-Through technique of Duobly-Fed Induction Generator based wind power generation systems. Zhe Jiang University, Hangzhou (2011)
4. Tian, Q.: Research on wind turbine pitch control system based on fuzzy PID. Modern Electronics Technique 36(16), 146–148 (2013)
5. Kong, Y.G.: Study on strategies of power control and maximum energy capture for large-scale wind turbine. Shanghai Jiaotong University, Shanghai (2009)
6. Hansen, A.D., Michalke, G., Sorensen, P., et al.: Co-ordinated voltage control of DFIG wind turbines in uninterrupted operation during grid faults. Wind Energy 10(1), 51–68 (2007)
7. He, Y.K., Zhou, P.: Overview of Low Voltage Ride-Through Technology for Variable Speed Constant Frequency Doubly Fed Wind Power Generation Systems. Transactions of China Electron Technology Society 24(9), 140–146 (2009)

A New Method for the Control of the Conditioning Temperature of Hoop Standard Granulator[*]

Yuan Xue[1], Jianguo Wu[1], Lei Qin[1], Jin Liu[1], and Kun Zhang[2,3]

[1] School of Electrical Engineering of Nantong University, Nantong, Jiangsu 226019
[2] School of Electronics, Electrical Engineering and Computer Science,
Queen's University Belfast, United Kingdom
[3] School of Mechatronic Engineering & Automation, Shanghai University, Shanghai 200072
wu.jg@ntu.edu.cn

Abstract. The conditioning temperature is one of the important parameters of the hoop standard granulator production system. It can be influenced by temperature system with nonlinear, time-varying and hysteresis characteristics and feed quantity. This paper creates a control algorithm which combines disturbance observer and fuzzy PID to control the temperature modulation. In this way, feed quantity of the hoop standard granulator is also seen as a part of the interference. By constructing disturbance observer, this paper predicts the disturbance on temperature system and variations of parameters, so as to suppress the effect of interference on the system. Meanwhile this paper introduces fuzzy PID for adaptive PID tuning parameters to achieve optimal parameters. Finally, the numerical simulation on temperature system has strong adaptability and robustness.

Keywords: hoop standard granulator, temperature control, disturbance observer-fuzzy_PID, Matlab simulation.

1   Introduction

Hoop standard granulator is production equipment which can produce higher economic added value in feed production modern enterprise. The conditioning temperature is a very important parameter in the process of granulating, which is not the same for different types and formulations of the feed. For example ,feed containing sensitive formulas generally requires conditioning temperature less than 60℃, livestock and poultry feed is roughly in the conditioning temperature of 75°C~ 90°C and aquatic feed is in 85°C~100°C[1]. For each feed formulation, it has the best conditioning temperature. If the conditioning temperature can be controlled in the temperature of best value or of a narrow range with small amplitude fluctuations, product quality requirements can be met to the greatest degree. But in the real production, due to the relatively complex production scene, such as unstable gas pressure provided by boiler, disturbances generated by the

---

[*] Corresponding author.

M. Fei et al. (Eds.): LSMS/ICSEE 2014, Part II, CCIS 462, pp. 377–388, 2014.

steam flow and temperature and different volume of feed, the actual temperature system has three characteristics: nonlinear, time-varying, hysteresis. It is difficult to ensure that stable temperature is in the vicinity of the optimum value. Thus that affects the quenching and tempering effect and even causes machines blocked. The staffs need to constantly adjust temperature, wasting a lot of raw materials. For such a lagging system, it is difficult for the traditional conventional PID control strategy to achieve satisfactory results of the temperature control[2]. Although the fuzzy PID has better effect, there are certain fluctuations. Because of disturbance observer compensating for interference prediction, this paper proposes a control algorithm by means of combining the fuzzy PID with the disturbance observer, giving full play to the advantages of both. Kinds of interference in temperature compensation system can be predicted by constructing the disturbance observer, while the best parameters of PID control can be obtained by introducing fuzzy PID and modifying parameters.

2   Conditioning Temperature Control System Design

In the course of working, hoop standard granulator exudes heat out. High-temperature steam continuously adds replenishes moisture and heat for the hoop standard granulator to ensure that conditioning temperature is controlled within a certain range. So conditioning temperature control of the hoop standard granulator is a temperature dynamic balance system.

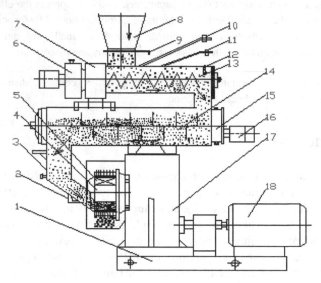

1-rack,2-magnet,3-blanking chute,4-loopdie,5-press roll,6-feed motor,7-feed reducer, 8-hopper, 9-blanking gate,10-electric steam flow control value,11-electric level flow control value,12-temperature sensor,13-feeder,
14-conditioner,15-reducer of condition,16-condition motor,17-main transmission box,18-main motor of granulation

Fig. 1. The schematic diagram of hoop standard granulator

As is seen hoop standard granulator from the structure diagram (Figure 1) , powdery materials mixed with steam is divided into two levels. The first level is a mixture of warm-up session. In the feeder, the powder materials from the hopper down are mixed with high temperature steam from steam pipe, which is seen as part of mixing preheating process. The temperature of the mixed material rises at a faster rate according to the amount of steam controlled by the steam valve. The second stage is fully mixing process. Through the cage and scraper clockwise pushing materials, the paste materials are fully stirred in the temperature of the desired range. In this mixing process, the conditioning temperature gradually stabilizes around the set value with the changing degree of temperature approaching zero. The final temperature is the conditioning temperature in the hoop standard granulator.

Currently, the shortcomings of widely used PID temperature control method is to make conditioning temperature of the object with great delay and inertia[3].This paper designs the diagram of this conditioning temperature control system ,which is shown in Figure 2.

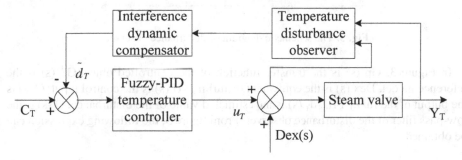

Fig. 2. The schematic diagram of refining temperature control system

In figure 2, C_T is conditioning temperature input, u_T is the fuzzy PID output, d_T is the estimates of the interference, Dex is feed quantity, external disturbance, etc., Y_T is the system output to the steam regulating valve.

3 Disturbance Observer Design

3.1 The Principle of Disturbance Observer

The basic concept of the disturbance observer is that errors, produced by variations of parameters from external interference, the actuator and the ideal mathematical model, can be effectively predicted, and interference can be suppressed by introducing the equivalent compensation[4]. For hoop standard granulator, feed quantity is also seen as a disturbance. The disturbance observer structure diagram is shown in Figure 3.

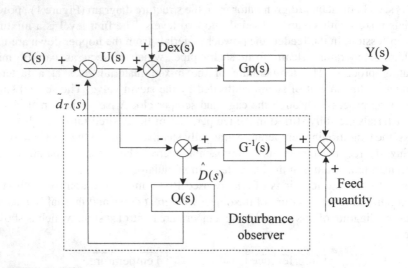

Fig. 3. The schematic diagram of disturbance observer

In Figure 3: Gp (s) is the transfer function of the controlled object.G^{-1} (s) is the reference model. Dex (s) is the equivalent disturbing. U (s) is the control input. C (s) is the output of the fuzzy PID. d_T (s) is the estimated value of the disturbance. Q (s) is the low-pass filter of the disturbance observer. From figure 3, the following expression can be obtained:

$$Y(s) = G_{cy}(s)C(s) + G_{dy}(s)D_{ex}(s) + G_{ny}(s)N(s) \qquad (1)$$

where,

$$Gcy(s) = \frac{Y(s)}{C(s)} = \frac{Gp(s)G(s)}{G(s) + (Gp(s) - G(s))Q(s)} \qquad (2)$$

$$Gdy(s) = \frac{Y(s)}{D(s)} = \frac{Gp(s)G(s)(1 - Q(s))}{G(s) + (Gp(s) - G(s))Q(s)} \qquad (3)$$

$$Gny(s) = \frac{Y(s)}{N(s)} = \frac{Gp(s)Q(s)}{G(s) + (Gp(s) - G(s))Q(s)} \qquad (4)$$

3.2   Low-Pass Filter Q (s) Design

Assume that fq is the cutoff frequency of low-pass filter Q (s).Where f <= fq, it is in the low frequency. Then if |Q (jw) | = 1, then Gcy = G (s), Gdy = 0, Gny = 1.Where f> fq, it is in the high frequency. Then if |Q (jw) | = 0, then Gcy = Gp (s), Gdy = Gp (s), Gny = 0, where the impact of the high-frequency measurement error N (s) on the system can be reduced. Therefore, Q (s) design is a particularly important component in designing disturbance observer. Firstly, in order to make Q (s) G-1 (s) is positive, Q (s) relative order should be greater than or equal to 1/G (s) relative order [5]; Secondly, Q (s) bandwidth design should consider the balance of robust stability and inhibitory ability of disturbance observer [6]. Considering dynamic change of parameters from the system; the controlled object Gp (s) can be expressed as follows

$$Gp(s) = G(s)(1+\Delta(s)M(s)) \tag{5}$$

Δ (s) represents the transfer function of the variable and satisfies $|| \Delta$ (s) $|| \infty << 1$, M (s) represents the transfer function of a fixed balance. According to the robust stability theorem [7] [8], the stable condition of the disturbance observer is

$$||\Delta(s)Q(s)||_\infty <<1 \tag{6}$$

Q (s) can be designed to achieve required stability. Q (s) should be selected reasonably according to the specific needs of control system. A common design is that Q (jw) is equal to the slope on the right side 1-Q (jw) slope in the high frequency region. In the frequency domain, | Q (jw) | and the slope of the low-frequency region | 1-Q (jw) | is

$$\frac{L_m \, |Q(\, jw)\, |}{L_m(w)} \approx -r \tag{7}$$

$$\frac{L_m \, |1-Q(jw)|}{L_m(w)} \approx N-r-1 \tag{8}$$

The optimal is obtained through equal slope:

$$r = \frac{N+1}{2} \tag{9}$$

In addition to the robustness of disturbance observer, interference suppression must also be considered. So the cut-off frequency of Q(s) is an important factor. In this system fq is chosen between system error signal frequency and the frequency of noise measurement. Based on the analysis of the robustness and interference suppression, the features of hoop standard granulator conditioning temperature system are taken into account. The low-pass filter uses the following form:

$$Q(s) = \frac{3\tau + 1}{\tau^3 s^3 + 3\tau^2 \tau s^2 + 3\tau s + 1}$$ (10)

4   Fuzzy PID Controller Design

This study designs a fuzzy PID controller for the system. The fuzzy controller can establish an accurate mathematical model for complex objects like hoop standard granulator. The fuzzy controller is designed based on experience about control objects without mathematical model of the controlled system. And according to artificial control rules and the organization control decisions, the size of the control quantity is determined. It belongs to the nonlinear control of nonlinear systems which require relatively fast response and small overshoot. It is not sensitive to process parameters, and namely it has very strong robustness and overcomes the influence of nonlinear factors[9]. However, it does not have the overall effect of the traditional PID control, so steady-state error can not be controlled in the ideal level. Combined PID control and fuzzy control, fuzzy PID controller is designed to achieve temperature control. The control circuit is shown in Figure 4.

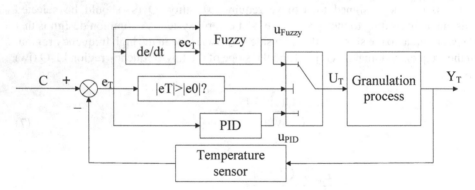

Fig. 4. Fuzzy PID temperature control loop

4.1   PID Control Strategy

When absolute value of the error about conditioning temperature is less than 5℃, the PID control algorithm, in which the output of controller is u_{PID}, is used to achieve precise control and improve the stability and accuracy of the system. PID control formula is

$$u(t) = K_p [e(t) + \frac{1}{T_i} \int_0^t e(t)dt + T_d \frac{de(t)}{dt}] \tag{11}$$

The formula (11) converts into a digital PID control algorithm output:

$$u(k) = K_p \{e(k) + \frac{T}{T_i} \sum_{i=0}^k e(k) + \frac{T_d}{T}[e(k) - e(k-1)]\} \tag{12}$$

At time k-1, the formula (11) becomes:

$$u(k-1) = K_p \{e(k-1) + \frac{T}{T_i} \sum_{i=0}^{k-1} e(k) + \frac{T_d}{T}[e(k-1) - e(k-2)]\} \tag{13}$$

At time k, the output of the PID controller is

$$u_{PID} = u_{PID}(k-1) + K_p[e(k) - e(k-1)] + K_i e(k) + K_d[e(k) - 2e(k-1) + e(k-2)] \tag{14}$$

Formula (14) is the operation amount of the PID controller, its output is U_{PID}, T is the sampling period, $Ki = KpT / Ti$, $Kd = KpTd / T$.

4.2   Fuzzy Control Strategy

When the error is greater than 5°C, the fuzzy control algorithm is adopted. Fuzzy controller uses double input variables and single output variable, with the controller input being the error E, the change rate of temperature error being Ec and the controller output being U_{Fuzzy}. The signals collected are the exact. It requires blurring the exact and then fuzzy inference.

4.2.1   Input and Output of the Fuzzy Controller

4.2.1.1   Input Linguistic Variables
Temperature deviation: E_T, the corresponding domain: [-5, 5]
    Fuzzy subset: {NB, NM, NS, ZE, PS, PM, PB}
    Temperature deviation change rate: EC_T, the corresponding domain: [-5, 5]
    Fuzzy subset: {NB, NM, NS, ZE, PS, PM, PB}

4.2.1.2   Output Linguistic Variables
Steam valve control signal: U_{Fuzzy}, the corresponding domain: [-6, 6]
    Fuzzy subset: {NB, NM, NS, ZE, PS, PM, PB}

4.2.2   Membership Function Choice

This paper uses the most simple triangle membership function to indicate the temperature control loop temperature deviation E_T, temperature deviation change rate EC_T and the steam valve control signal U_{Fuzzy}. These membership function are shown in Figure 5, 6, 7 respectively.

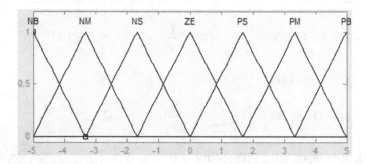

Fig. 5. Membership functions of temperature deviation E_T

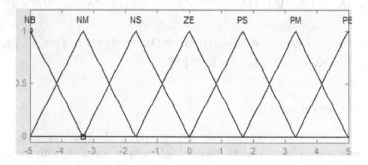

Fig. 6. Membership functions of temperature deviation rate EC_T

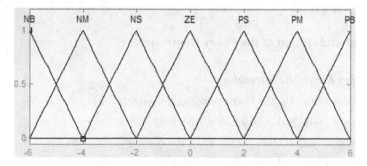

Fig. 7. Membership functions of steam valve control signal U_{Fuzzy}

4.2.3   Fuzzy Knowledge Base Establishment

The knowledge base of fuzzy controller is composed of two parts: fuzzy database and fuzzy rule base summed up by experts.

4.2.3.1   Fuzzy Database

Fuzzy Logic Control database include: division of blurred space, conversion way from the exact to the blurred, conversion factors and membership fuzzy sets. That is to say, the exact value, which have been blurred, is deposited into fuzzy database.

4.2.3.2   Fuzzy Rule Base

Fuzzy control system is described by a series of experts' knowledge. According to operating experience of experts in the field of production and processing, this paper summarizes 49 fuzzy control rules. The IF-THEN statement expressing forms are used to get the control rule table of manipulated variable U_{Fuzzy} fuzzy, as is shown in table 1.

Table 1. Fuzzy control rules

U_{Fuzzy}		F_T						
		NB	NM	NS	ZE	PS	PM	PB
	NB	PB	PB	PM	PM	PS	ZE	ZE
	NM	PB	PB	PM	PS	PS	ZE	NS
	NS	PM	PM	PM	PS	ZE	NS	NS
EC	ZE	PM	PM	PS	ZE	NS	NM	NM
	PS	PS	PS	ZE	NS	NS	NM	NM
	PM	PS	ZE	NS	NM	NM	NM	NB
	PB	ZE	ZE	NM	NM	NM	NB	NB

4.2.4   Fuzzy Inference Engine

According to fuzzy relationship summarized by the fuzzy rule, this paper uses Ma Dani (Mamdani) fuzzy reasoning method to obtain the membership function values of domain elements u corresponding U_{Fuzzy}, and conduct defuzzification using weighted average method to obtain u_{Fuzzy} ,which can control the opening of the steam valve , suppress the disturbance of the temperature control system and enhance the speed and stability. The fuzzy controller is forward reasoning strategy based on data driven, one by one judging each control rule, until the condition is satisfied, and otherwise continuing to search.

5   Simulation Results

Actual production requirement is that temperature stability should be modulated as soon as possible in the optimal setting temperature, so as to ensure certain steam quantity to produce high quality products. So that is a target tracking problem. This paper sets the target signal simulation 2 as the step signal, which is equivalent to the optimal temperature set in advance. In the control process the interference signal (feed quantity of the granulator system) and the noise signal whose amplitude is 0.01 are added. The time constant of the low-pass filter is 0.002 (determined by analysis of the experimental data). Compared with traditional PID control and fuzzy PID control, time of the setting value is short and fluctuation is small .The simulation structure is shown in Figure 8.

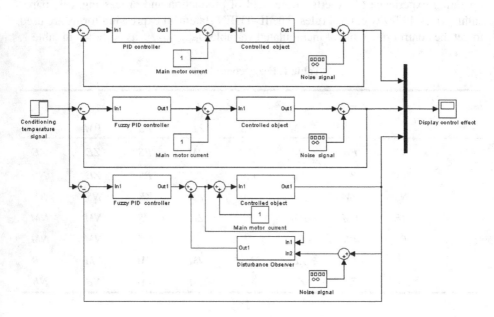

Fig. 8. The simulation diagram of the hoop standard granulator conditioning temperature

From the simulation results shown in Figure 9, it can be seen: the traditional PID control has a relatively large fluctuations, larger overshoot, and instability; fuzzy PID control is better than traditional PID. With the shorter period of time achieving the set value, there is still a certain degree of fluctuation, which is not in conformity with the requirements of the temperature control. Temperature fluctuations cause steam quantity changes, resulting in unstable moisture content of products, production of low quality products, and even leading to blocking machine. Compared with two methods, due to high accuracy from the disturbance observer impacting on hoop standard granulator temperature system, the combination control of the disturbance observer and fuzzy PID makes the time of conditioning longer than PID control. And from the diagram, the combination control can better track desired signals with higher stability in order to stabilize the temperature in the set value. Based on the above condition, the production can be improved without blocking machines.

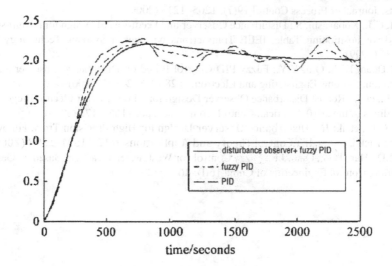

Fig. 9. The schematic diagram of simulation results

6   Conclusions

The control algorithm combining fuzzy PID and disturbance observer is applied to the conditioning temperature control system of hoop standard granulator with nonlinear, time-varying and hysteresis characteristics. The disturbance observer estimates external interference and changes in its parameters and feed quantity-another important parameter. By predicting the amount of compensation, the control algorithm suppresses the influence under the control of the process. The conditioning temperature of the hoop standard granulator can be controlled effectively combined with PID controller of fuzzy self-tuning parameters.

Finally, by building a simulation platform simulation experiments are carried out. Compared with the traditional PID and fuzzy PID, the new control method proposed in this paper is proved better. It is significant for practical engineering to apply this method to debugging the actual hoop standard granulator system.

References

1. Zhang, K., Fei, M.R., Zhang, P.J., Wu, J.G., Hu, Z.J.: Application Study on Intelligent Control of a Class of Time Delay Systems with Parameter Uncertainty. Chinese Journal of Scientific Instrument 35(6), 1394–1401 (2014)
2. Li, C.S., Wang, Y.N., E J Q.: Optimization Design of Intelligent Nonlinear PI Controller Based on One Degree with Time Delay. Control and Decision 3, 103–106 (2007)

3. Utkin, V.I., et al.: Sliding Mode Control Design Principles and Applications to Electric Drives. IEEE Transactions on Industrial Electronics 40(1), 23–36 (1993)
4. Chen, X.S., et al.: Disturbance Observer Based Multi-variable Control of Ball Mill Grinding Circuits. Journal of Process Control 19(7), 1205–1213 (2009)
5. Kempf, C.J., Kobayashi, S.: Disturbance Observer and Feedforward Design for a High-speed direct-drive Positioning Table. IEEE Transactions on Control Systems Technology 7(5), 513–526 (1999)
6. Zi, B., Duan, B.Y., Qiu, Y.Y.: Fuzzy PID Control Based on Disturbance Observer and its Application. Systems Engineering and Electronics 28(6), 892–895 (2006)
7. Xuan, F., et al.: Robust Disturbance Observer Design for a Power-assist Electric Bicycle. In: Proceedings of the 2010 American Control Conference, pp. 1166–1171 (2010)
8. Thum, C.K., et al.: H_∞ Disturbance Observer Design for High Precision Track Following in Hard Disk Drivers. IET Control Theory and Applications 3(12), 1591–1598 (2009)
9. Bai, M.D., Han, H.G., Qiao, J.F.: Fuzzy Control for Wastewater Treatment Based on Genetic Algorithm. Control Engineering of China 16(1), 46–49 (2009)

Integrated IMC-ILC Control System Design
for Batch Processes

Qinsheng Li, Li Jia, and Tian Yang

Shanghai Key Laboratory of Power Station Automation Technology,
Department of Automation, College of Mechatronics Engineering and Automation,
Shanghai University, 200072 Shanghai, China

Abstract. Considering conventional iterative learning control (ILC) is actually
an open-loop control approach within each batch, which cannot guarantee the
control performance of batch process when uncertainties and disturbances exist,
an integrated iterative learning control scheme is presented in this paper. The
proposed approach systematically integrates continuous-time information along
with time-axis and discrete-time information along with batch-axis into one
uniform frame, namely an internal model control (IMC)based PID control along
time-axis, while the optimal ILC along batch-axis. As a result, the operation
policy of batch process leads to superior tracking performance and better
robustness compared with conventional ILC strategy. An illustrative example is
exploited to verify the effectiveness of the investigated approach.

Keywords: batch process, integrated learning control, internal model control,
two-dimension control.

1   Introduction

Batch processes have the characteristic of repetition, which makes both technologies
and applications of iterative learning control (ILC) are possible to be widely used in
the optimization control of batch processes in the past decade [1]. Lee et al.presented
the quadratic criterion based iterative learning control (QILC) approach for tracking
control of batch processes based on a linear time-varying tracking error transition
model [2]. Xiong and Zhang presented a recurrent neural network based ILC scheme
for batch processes where the filtered recurrent neural network prediction errors from
previous batches are added to the model predictions for the current batch and
optimization is performed based on the updated predictions [3]. However, the
conventional ILC algorithms only use historical batch data, resulting in the ILC is
open loop control scheme in each batch, which cannot guarantee the tracking control
performance of batch process when uncertainties and disturbances exist. Hence, two-
dimension control strategy in which the information of batch-axis and time-axis are
both utilized synchronously for controller synthesis of batch process is required.
Rogers firstly employed two-dimension (2D) theory to solve above problem [4].

M. Fei et al. (Eds.): LSMS/ICSEE 2014, Part II, CCIS 462, pp. 389–396, 2014.

Recently, in order to enhance the robustness and stability, 2D control of batch processes is becoming new research focus presented in a number of literatures. From a 2D system point of view, based on a 2D cost function defined over a single-cycle or multi-cycle prediction horizon, Shi J et al. proposed two ILC schemes, referred respectively as single-cycle and multi-cycle generalized 2D predictive ILC (2D-GPILC) schemes, which are formulated in the GPC framework for batch processes [5]. Liu et al. combined the internal model control (IMC) with ILC to deal with uncertain time delay for linear batch processes [6]. Based on the integrated model, which integrates the learning ability of ILC into the prediction model of model predictive control (MPC), Mo S et al. developed a 2D dynamic matrix control (2D-DMC) algorithm for batch processes to against the 2D interval uncertainty[7]. Based on neuro-fuzzy model (NFM) and QILC, Jia L et al. proposed an integrated iterative learning control strategy with model identification and dynamic R-parameter to improve tracking performance and robustness for batch processes [8].

Based on 2D control viewpoint, and motivated by the previous works [8], an integrated iterative learning control strategy for batch process is proposed in this paper. This strategy integrates an internal model control (IMC) based PID control along time-axis and QILC along batch-axis into one uniform framework. Simulation results show that the control profile demonstrates better tracking performance and robustness compared with conventional QILC.

2   Integrated ILC Scheme for Batch Processes

The framework of the proposed integrated iterative learning control strategy is comprised of two parts: the iterative learning control working as a feedforward controller and IMC-based PID controller playing as a real-time feedback controller, as shown in Fig.1, where $u_k^{ILC}(t)$ is ILC control action variable, and $u_k^{PID}(t)$ is feedback control action variable. Batch length is defined as t_f. Here the batch length t_f is divided into n equal intervals. Here k is the batch index and t is the discrete batch time. $y_d(t_f)$ is desired product quality, $y_k(t)$ is the corresponding product quality of integrated control action, $u_k(t) = u_k^{ILC}(t) + u_k^{PID}(t)$, $y_k^m(t) = \hat{y}_k$ is the predicted output of neuro-fuzzy model with model modification algorithms from batch to batch [8]. $e_k(t)$ denotes the error between the measured output and the desired product quality. $G(s)$ represents the batch process, and $C(s)$ denotes the PID controller .

Define that U_k is the input sequence of kth batch, $U_k = U_k^{ILC} + U_k^{PID}$, here U_k^{ILC} and U_k^{PID} are the corresponding ILC sequence and feedback control sequence, respectively. Y_k and $\hat{Y}_k$ are the corresponding product quality sequence and predicted product quality sequence. In this study, U_k, Y_k and $\hat{Y}_k$ are stored to eliminate the model-plant mismatch. Based on the historical batch information, the optimization controller can find an updating mechanism for the new batch input sequence U_k^{ILC} ,using improved quadratic criterion based iterative learning control method. Next batch, this procedure is repeated to make the product qualities asymptotically converge towards the desired product qualities $y_d(t_f)$ at the batch end.

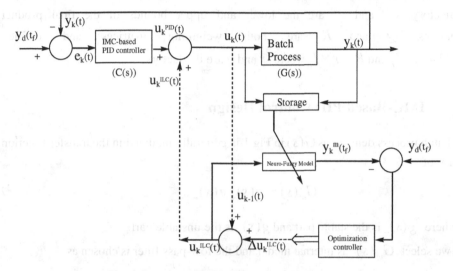

Fig. 1. Integrated iterative learning control system

3   ILC Design

As discussed above, the proposed integrated iterative learning control action can be described as

$$U_k = U_k^{ILC} + U_k^{PID}. \tag{1}$$

Considering that the objective is to design a learning algorithm to implement the control trajectory U_k so that the product qualities at the batch end convergence asymptotically towards the desired product qualities. Thus, instead of using the whole reference sequences for product qualities, which is less important for batch processes product quality control, only the desired product quality $y_d(t_f)$ at the end of each batch is used. Here the following quadratic objective function is employed:

$$\min J(U_{k+1}^{ILC}, k+1) = \left\| y_d(t_f) - \hat{y}(U_{k+1}^{ILC}, t_f) \right\|_Q^2 + \left\| U_{k+1}^{ILC} - U_k \right\|_R^2. \tag{2}$$

s.t.

$$u^{low} \leq u_{k+1}(t) \leq u^{up}$$
$$y^{low} \leq y_{k+1}(t_f) \leq y^{up} \quad . \tag{3}$$

where u^{low} and u^{up} are the lower and upper bounds of the input trajectory, y^{low} and y^{up} are the lower and upper bounds of the final product qualities, Q and R are both weighting matrices and defined as $Q = q \times I_n$ and $R = r \times I_n$ where r and q are the factors.

4   IMC-Based PID Control Design

A batch process denoted as $G(s)$ in Fig.1 is generally modeled in the transfer function form of

$$G_m(s) = g(s)_+ g(s)_- . \tag{4}$$

where $g(s)_-$ is the stable part and $g(s)_+$ is the unstable part.

If we select $G_m(s)$ as internal model and the low-pass filter is chosen as

$$F(s) = \frac{1}{(\lambda s + 1)^n}, \lambda \succ 0 \quad . \tag{5}$$

It follows from the IMC theory that the IMC should be designed as

$$C_{IMC}(s) = g(s)_-^{-1} F(s) \quad . \tag{6}$$

Furthermore, according to IMC theory, we can design PID controller via

$$C(s) = \frac{C_{IMC}(s)}{1 - G_M(s)C_{IMC}(s)} = \frac{g(s)_-^{-1}}{(\lambda s + 1)^n - g(s)_+} . \tag{7}$$

Where λ is the adjustable parameter for controller tuning.

5   Convergence Analysis

For convenience of convergence analysis, define two errors according to following conditions respectively

1) Condition one: conventional iterative learning control system without the part of feedback controller

$$E_1^k(s) = Y_d(s) - Y_k^{ILC}(s) = Y_d(s) - U_k^{ILC}(s)G(s). \tag{8}$$

2)   Condition two: the proposed integrated iterative learning control system

$$E_2^k(s) = Y_d(s) - Y_k(s) = \frac{Y_d(s) - U_k^{ILC}(s)G(s)}{1 + G(s)C(s)} = \frac{E_1^k(s)}{1 + G(s)C(s)} \tag{9}$$

In order to obtain smaller tracking error using the proposed integrated iterative learning control system, compared with conventional iterative learning control strategy, we can derive that the following inequality should holds.

$$\left\| E_2^k(s) \right\|_\infty \le \left\| E_1^k(s) \right\|_\infty. \tag{10}$$

Furthermore, from eq. (9), the designed feedback controller $C(s)$ satisfies

$$\left\| (1 + G(s)C(s))^{-1} \right\|_\infty \le 1. \tag{11}$$

Substituting eq (7) into ineq (11) yields

$$\left\| \frac{(\lambda s + 1)^n g(s)_- - G_m(s)}{(\lambda s + 1)^n g(s)_- - G_m(s) + G(s)} \right\|_\infty \le 1. \tag{12}$$

Form ineq (12), if $G(s) = G_m(s) = g(s)_-$ holds,   we can obtain

$$\left\| 1 - \frac{1}{(\lambda s + 1)^n} \right\|_\infty \le 1. \tag{13}$$

Furthermore, form ineq. (13), if $n = 1$, we can get

$$\left\| \frac{\lambda s}{\lambda s + 1} \right\|_\infty \le 1. \tag{14}$$

Remark 1. Clearly, the ineq. (14) always holds, as $\lambda \succ 0$. This is to say, under the above discussed assumptions, the ineq. (11) always holds, which can guarantee the ineq (10) always holds. From the convergence analysis of the conventional ILC system for batch process presented and proved in our previous works[9], we can derive $\left\| E_1^k(s) \right\|_\infty \to 0$,as $k \to \infty$,   combined   with   the   ineq.(10),namely $\left\| E_2^k(s) \right\|_\infty \le \left\| E_1^k(s) \right\|_\infty$,and it is clear that we can derive $\left\| E_2^k(s) \right\|_\infty \to 0$, as $k \to \infty$.

6   Examples

This case study is a typical nonlinear batch reactor, in which a first-order irreversible exothermic reaction $A\xrightarrow{k_1} B \xrightarrow{k_2} C$ takes place [8], [9]. This process can be described as follow:

$$\dot{x}_1 = -4000\exp(-2500/T)x_1^2$$
$$\dot{x}_2 = 4000\exp(-2500/T)x_1^2 - 6.2\times10^5\exp(-5000/T)x_2$$

where T denotes reactor temperature, x_1 and x_2 are, respectively, the reactant concentration.

In this simulation, the reactor temperature is first normalized using $T_d = (T - T_{min})/(T_{max} - T_{min})$, in which T_{min} and T_{max} are respectively 298 (K) and 398 (K). T_d is the control variable which is bounded by $0 \le T_d \le 1$ and $x_2(t)$ is the output variable. The control objective is to manipulate the reactor temperature T_d to control concentration of B at the end of the batch $x_2(t_f)$.

The independent random signal with uniform distribution between [0, 1] are used to simulate 30 input-output data for training purpose and another 20 input-output data for testing purpose. And the robustness of the proposed method is evaluated by introducing Gaussian white noise with 0.002^2 standard deviation to the output variable of the process. In the simulation, the model of six fuzzy rules is chosen. The following parameters in batch process are assumed: $r = 1$, $q = 10$, $U_0 = \vec{0}$, and $y_d(t_f) = 0.6100$. 3% noisy is put into the quality output of the 5th batch.

The batch process is linearized as $G_m(s) = \dfrac{0.002447}{s+1.846}$, from eq.(7), if $n=1$, we can obtain

$$C(s) = \frac{1.846+s}{0.002447\lambda s}$$

Obversly, this is a PI controller. The parameters are $k_p = \dfrac{1}{0.002447\lambda}$ and $k_i = \dfrac{1.846}{0.002447\lambda}$. λ is the unique adjustable parameter for controller tuning. In this case, λ is chosen as $\lambda = 29000$.

The control trajectory and product quality trajectory are shown in Fig.2 and Fig.3, respectively. From Fig.4, we can observe that the investigated integrated iterative learning control strategy has faster convergence rate along batch-axis and better tracking performance against disturbances compared with conventional Q-ILC method.

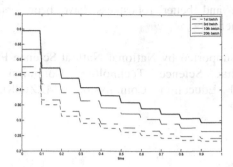

Fig. 2. Control trajectory of integrated iterative learning controller

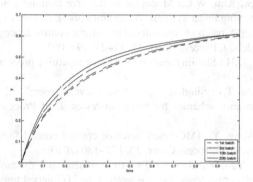

Fig. 3. Product quality trajectory of integrated iterative learning control system

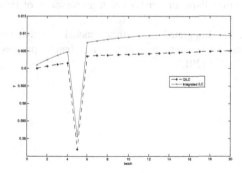

Fig. 4. The quality at the end of each batch

7   Conclusion

In this paper, an integrated iterative learning control strategy is proposed, which can utilize the information of batch-axis and time-axis synchronously for controller design of batch processes. The results on a simulated batch reactor show that the superior tracking performance and better robustness have been obtained compared to conventional Q-ILC method.

Acknowledgments. Supported by National Natural Science Foundation of China (61374044), Shanghai Science Technology Commission (12510709400), Shanghai Municipal Education Commission (14ZZ088), Shanghai talent development plan.

References

1. Lee, J.H., Lee, K.: Iterative learning control applied to batch processes: An overview. Control Eng. Pract. 15(10), 1306–1318 (2007)
2. Lee, J.H., Lee, K.S., Kim, W.C.: Model-based iterative learning control with a quadratic criterion for time-varying linear systems. Automatica 36(5), 641–657 (2000)
3. Xiong, Z., Zhang, J.: A batch-to-batch iterative optimal control strategy based on recurrent neural network models. J. Process Contr. 15, 11–21, 19 (2005)
4. Rogers, E., Owens, D.H.: Stability analysis for linear repetitive processes. LNCIS, vol. 175. Springer, Heidelberg (1992)
5. Shi, J., Gao, F., Wu, T.J.: Single-cycle and multi-cycle generalized 2D model predictive iterative learning control schemes for batch processes. J. of Process Contr. 17, 715–727 (2007)
6. Liu, T., Gao, F., Wang, Y.: IMC-based iterative optimal control for batch processes with uncertain time delay. J. of Process Contr. 20, 173–180 (2010)
7. Mo, S., Wang, L., Yao, Y., Gao, F.: Two-time dimensional dynamic matrix control for batch processes with convergence analysis against the 2D interval uncertainty. J. of Process Contr. 22, 899–914 (2012)
8. Jia, L., Yang, T., Qiu, M.: An integrated iterative learning control strategy with model identification and dynamic R-parameter for batch processes. J. of Process Contr. 23, 1332–1341 (2013)
9. Jia, L., Shi, J., Qiu, M.: Integrated neuro-fuzzy model and dynamic R-parameter based quadratic criterion-iterative learning control for batch process control technique. Neurocomptuing 98, 24–33 (2012)

Decentralized Control for Power Systems Components Based-on Nonlinear Differential-Algebraic Equations Subsystem Model

Qiang Zang[1,2], Ying Zhou[3], Ping Mei[1], Baichao Zheng[1], and Kaifeng Zhang[1]

[1] School of Information and Control Engineering, Nanjing University
of Information Science & Technology, Nanjing 210044, China
[2] Key Laboratory of Measurement and Control of Complex Systems
of Engineering, Ministry of Education, Southeast University,
Nanjing 210096, China
[3] College of Automation, Nanjing University of Posts and Telecommunications,
Nanjing 210003, China

Abstract. Components of power systems essentially belong to a special class of nonlinear differential-algebraic equations subsystem, whose index is one and interconnection is locally measurable. In this paper, the decentralized control problem is discussed using inverse systems method for such class of power systems components. Firstly, the definition of α-order integral right inverse system is presented. Secondly, a recursive algorithm is proposed to judge whether the controlled component is invertible. Then a physically feasible α-order integral right inverse system is constructed with which the controlled component is made linearization and decoupled, so that linear control theory and methods can be applied. Finally, a decentralized excitation controller is designed for one synchronous generator within multi-machine power systems based on the proposed method in this paper. The simulation results demonstrate the effectiveness of the proposed control scheme.

Keywords: Power systems components, Differential-algebraic equations subsystems, Decentralized control, Inverse systems.

1 Introduction

For the safe operation under various disturbance and change of structure and load, the nonlinear control of power systems components, including synchronous generator and FACTS, etc., has been well-studied. In most existing research, the model of controlled component is usually described by nonlinear Ordinary Differential Equations (ODE), and great progress has been made based-on such systems description[1-3]. But nonlinear ODE model can hardly describe the various complex characteristics of components, especially the nonlinear algebraic constraint relation. Some research has realized such problem and begun to study the control of components based-on nonlinear Differential Algebraic Equations (DAE) model [4-9].

M. Fei et al. (Eds.): LSMS/ICSEE 2014, Part II, CCIS 462, pp. 397–406, 2014.

In most existing results based-on power system DAE model, the mutual influence between components and the rest of the power systems is not taken into account and components are treated as isolated systems. However, components are essentially nonlinear DAE subsystems within power systems. Some preliminary research has been done for components nonlinear DAE systems, for instance see [10-12] and the references therein. A so-called component structural model(i.e., nonlinear DAE systems model) is established, which can describe the complex characteristics and meet the need of component decentralized control design[10-11]. The particularities of components nonlinear DAE systems model are analyzed, including its characteristics of index 1 and interconnection measurability[12].

Among various nonlinear control methods, the inverse systems method holds an important position in linearization and decoupling control. The research about inverse systems control of DAE systems can be backed to [13,14], where invertibility of the continuous-time and discrete-time linear DAE systems are studied respectively in [13] and [14]. The invertibility of affine nonlinear DAE systems is studied in [15]. Furthermore, the invertibility of general nonlinear DAE systems is studied in [16].

In this paper, the decentralized control problem is considered using inverse systems method for the power systems components whose model is described by nonlinear DAE subsystems. The method can be divided into two steps: At first step to judge the invertibility of component nonlinear DAE subsystem. If it does, then to construct the inverse systems that can be realized physically, with which the decoupling and linearization of the composed systems is achieved. At second step, various linear control theories and methods can be used to the composed systems such that the desired performance can be obtained.

2   Systems Description and Problem Formulation

For power systems composed by N components, the $i^{th}(i=1,\cdots,N)$ component is described by following nonlinear differential algebraic equations[7-8]:

$$\begin{cases} \dot{x}_i = f_i(x_i,z_i,u_i) \\ g_i(x_i,z_i,\overline{v}_i) = 0 \\ y_i = h_i(x_i,z_i,\overline{v}_i) \end{cases} \quad (1)$$

where $x_i \in N_0^i \subset R^{n_i}, z_i \in M_0^i \subset R^{l_i}, u_i \in L_0^i \subset R^{m_i}, y_i \in K_0^i \subset R^{m_i}, \overline{v}_i \in S_0^i \subset R^{s_i}$ are differential variable, algebraic variable, manipulated input, controlled output and interconnection input respectively. $f_i \in R^{n_i}, g_i \in R^{l_i}, h_i \in R^{m_i}$ are smooth map. Obviously controlled component i is a nonlinear DAE subsystem within power systems, meanwhile the other components $j(j \neq i)$ and AC grid compose the rest of the power systems.

For the sake of simplicity, we will omit the subscript i of (1) in remainder. Without loss of generality, suppose component (1) bears compatible initial conditions $X_0 = (x(t_0), z(t_0), u(t_0), y(t_0), \bar{v}(t_0)) \triangleq (x_0, z_0, u_0, y_0, \bar{v}_0) \in U_0$, i.e., $g(x_0, z_0, \bar{v}_0) = 0$ [3], where $U_0 = [N_0, M_0, L_0, K_0, S_0]$.

Component nonlinear DAE subsystem (1) bears following particularities [12]:

P1. The Jacobian matrix of $g(x, z, \bar{v})$ with respect to z has full rank on U_0:

$$rank(\frac{\partial g}{\partial z}) = l \tag{2}$$

i.e., (1) is of index one.

P2. The interconnection input $\bar{v}$ and its sufficient order derivatives are locally measurable and bounded.

From the functional view, nonlinear DAE subsystem (1) can be regarded as an operator (marked by Θ) which maps the manipulated input $u(t)$ and interconnection input $\bar{v}(t)$ to control output $y(t)$, i.e.,

$$y = \Theta(u, \bar{v}) \tag{3}$$

Definition 1. Suppose there exist system Σ_α which bears input-output mapping relationship: $\quad \hat{y} = \bar{\theta}_\alpha(\varphi, \bar{v}) \quad$ where $\quad$ the $\quad$ input $\varphi(t) = (\varphi_1, \cdots, \varphi_m)^T = r^{(\alpha)} = (r_1^{(\alpha_1)}, \cdots, r_m^{(\alpha_m)})^T$ is a continuous vector. For nonlinear DAE subsystem1, if $u(t) = \hat{y}(t)$ we have $y^{(\alpha)} = \varphi$ where $y^{(\alpha)} = (y_1^{(\alpha_1)}, \cdots, y_m^{(\alpha_m)})^T$ (i.e., $y_i^{(\alpha_i)} = \varphi_i, i = 1, \cdots, m$), then the system Σ_α is the α-order integral right inverse system and (1) is called invertible.

3   Invertibility of Component Nonlinear DAE Subsystem

Now we will present a recursive algorithm with which to judge the invertibility of component nonlinear DAE subsystem (1).

From (2) we know that $\left(\dfrac{\partial g}{\partial z}\right)^{-1}$ on U_0. At first we use following operator

$$E_\xi(F) \triangleq \frac{\partial F}{\partial \xi}\bigg|_{g(x,z,\bar{v})=0} = \frac{\partial F}{\partial \xi} - \frac{\partial F}{\partial z}\left(\frac{\partial g}{\partial z}\right)^{-1}\frac{\partial g}{\partial \xi} \tag{4}$$

to denote the Jacobian matrix of vector function $F(x,z,u,y,\cdots,y^{(k)},\overline{v},\cdots,\overline{v}^{(k)})$ respect to some variable $\xi \in (x,u,\overline{v})$ under the algebraic constraint $0 = g(x,z,u,\overline{v})$. The detailed procedure of algorithm are as follows.

Step 1. Definite

$$h_0 = y - h(x,z,\overline{v}) = 0 \tag{5}$$

To derivative on h_0 and we get

$$\hat{h}_1(x,z,u,y,\dot{y},\overline{v},\dot{\overline{v}}) = \frac{d}{dt}h_0 = E_x(h_0)\dot{x} + \frac{\partial h_0}{\partial y}\dot{y} + E_{\overline{v}}(h_0)\dot{\overline{v}} \tag{6}$$

Accordingly we get point $X_0^1 = (x_0,z_0,u_0,y_0,\dot{y}_0,\overline{v}_0,\dot{\overline{v}}_0)$. Suppose matrix $\dfrac{\partial \hat{h}_1}{\partial u}$ has constant rank $\gamma_1 \geq 0$ at some neighborhood of X_0^1. If $\gamma_1 = m$, then algorithm stops; If $\gamma_1 < m$, the algorithm enters next step.

Step $k(k = 2,\cdots)$: Suppose until to step k, we get a list of integers $\gamma_1,\cdots,\gamma_{k-2},\gamma_{k-1}$, vector $h_{k-1} = \begin{bmatrix} H_{k-2} \\ \hat{h}_{k-1} \end{bmatrix}$ and point X_0^{k-1}. Similarly, we have

$rank \dfrac{\partial h_{k-1}}{\partial u} = \gamma_{k-1}$ at a neighborhood of X_0^{k-1}. Obviously $\gamma_{k-1} \geq \gamma_{k-2}$, so we can pick $\gamma_{k-1} - \gamma_{k-2}$ rows from $\hat{h}_{k-1}$, which is denoted as $\hat{h}_{k-1,1}$ and satisfies

$rank \begin{bmatrix} \dfrac{\partial H_{k-2}}{\partial u} \\ \dfrac{\partial \hat{h}_{k-1,1}}{\partial u} \end{bmatrix} = \gamma_{k-1}$. Let $H_{k-1} = \begin{bmatrix} H_{k-2} \\ \hat{h}_{k-1,1} \end{bmatrix}$ and the other rows of $\hat{h}_{k-1}$ are

denoted as $\hat{h}_{k-1,2}$, then there exist a neighborhood of X_0^{k-1} and a smooth mapping $\lambda_{k-1}(x,z,u,y,\cdots,y^{(k-1)},\overline{v},\cdots,\overline{v}^{(k-1)})$ such that

$$\frac{\partial \hat{h}_{k-1,2}}{\partial u} = \lambda_{k-1}(\cdot)\frac{\partial H_{k-1}}{\partial u} \tag{7}$$

If $E_u(\hat{h}_{k-1,2}) = 0$, we need only to set $\lambda_{k-1}(\cdot) = 0$. Definite

$$\hat{h}_k = \hat{h}_k(x, z, u, y, \cdots, y^{(k)}, \overline{v}, \cdots, \overline{v}^{(k)}) =$$

$$\left[E_x(\hat{h}_{k-1,2}) - \lambda_{k-1}(\cdot)E_x(H_{k-1}) \right]\dot{x} + \sum_{i=0}^{k-1}\left[\frac{\partial \hat{h}_{k-1,2}}{\partial y^{(i)}} - \lambda_{k-1}(\cdot)\frac{\partial H_{k-1}}{\partial y^{(i)}} \right]y^{(i+1)}$$

$$+\left[E_{\overline{v}}(\hat{h}_{k-1,2}) - \lambda_{k-1}(\cdot)E_{\overline{v}}(H_{k-1}) \right]\dot{\overline{v}} + \sum_{i=1}^{k-1}\left[\frac{\partial \hat{h}_{k-1,2}}{\partial \overline{v}^{(i)}} - \lambda_{k-1}(\cdot)\frac{\partial H_{k-1}}{\partial \overline{v}^{(i)}} \right]\overline{v}^{(i+1)} \qquad (8)$$

and definite recursively

$$h_k(x, z, u, y, \cdots, y^{(k)}, \overline{v}, \cdots, \overline{v}^{(k)}) = \begin{bmatrix} H_{k-1} \\ \hat{h}_k \end{bmatrix} = 0 \qquad (9)$$

then we get the point $X_0^k = (x_0, z_0, u_0, y_0, \cdots, y_0^{(k)}, \overline{v}_0, \cdots, \overline{v}_0^{(k)})$. If matrix $\dfrac{\partial h_k}{\partial u}$ has constant rank γ_k at a neighborhood of X_0^k. If $\gamma_k = m$, then algorithm stops ; If $\gamma_k < m$, then algorithm enters next step.

Definition 2. The relative order θ for the component nonlinear DAE subsystem (1) is the least positive integer k such that $\gamma_k = m$ or $\theta = \infty$ if $\gamma_k < m$ for all $k = 1, 2, \cdots$.

Theorem 1. Consider the nonlinear DAE subsystem (1) with relative order θ. If $\theta < \infty$, then component nonlinear DAE subsystem (1) is invertible.

Proof. From the recursive algorithm, we can get a neighborhood of X_0 and following equation at the θth step

$$h_\theta(x, z, u, y, \cdots, y^{(\theta)}, \overline{v}, \cdots, \overline{v}^{(\theta)}) = 0 \qquad (10)$$

where h_θ satisfies $rank(\dfrac{\partial h_\theta}{\partial u}) = \gamma_\theta = m$. By the Implicit Function Theory, equation (10) can determine one unique solution for u

$$u = h_\theta^{-1}(x, z, y, \cdots, y^{(\theta)}, \overline{v}, \cdots, \overline{v}^{(\theta)}) \qquad (11)$$

Replace y with r in (11), we can get

$$u = h_\theta^{-1}(x, z, r, \cdots, r^{(\theta)}, \overline{v}, \cdots, \overline{v}^{(\theta)}) \qquad (12)$$

where $r = (r_1, \cdots, r_m)^T$ and suppose the highest order derivative and lowest order derivative of $r_i (i = 1, \cdots, m)$ are $r_i^{(\beta_i)}$ and $r_i^{(\alpha_i)}$ respectively. Definite

$$\overline{v}_k = (\overline{v}^T, \cdots, (\overline{v}^{(k)})^T)^T, k = 0, 1, \cdots, \theta$$
$$\varphi = (\varphi_1, \cdots, \varphi_m)^T = (r_1^{(\alpha_1)}, \cdots, r_m^{(\alpha_m)})^T,$$
$$\xi_i = (r_i^{(\beta_i)}, r_i^{(\beta_i+1)}, \cdots, r_i^{(\alpha_i-1)})^T, i = 1, \cdots, m \tag{13}$$

and construct following systems $\overline{\Sigma}_\alpha$:

$$\dot{\xi}_i = A_i \xi_i + B_i \varphi_i, i = 1, \cdots, m$$
$$\hat{y} = h_\theta^{-1}(x, z, \xi, \varphi, \overline{v}_\theta) \tag{14}$$

where

$$A_i = \begin{pmatrix} 0 & 1 & & \\ & \ddots & \ddots & \\ & & \ddots & 1 \\ & & & 0 \end{pmatrix}_{(\alpha_i-\beta_i)\times(\alpha_i-\beta_i)}, B_i = \begin{pmatrix} 0 \\ \vdots \\ 0 \\ 1 \end{pmatrix}_{(\alpha_i-\beta_i)\times 1}, \xi = (\xi_1^T, \cdots, \xi_m^T)^T \tag{15}$$

Similar to [13], it can be shown that (14) is a α-order integral right inverse system of component nonlinear DAE subsystem (1) realized by state feedback and dynamic compensation, where $\alpha = (\alpha_1, \cdots, \alpha_m)$. This completes the proof.

4   Decentralized Excitation Control of Synchronous Generator

Excitation control is studied for one synchronous generator within multi-machine power systems. The mathematical model of synchronous generator excitation control is described by following nonlinear DAE subsystem [10]:

$$\dot{\delta} = \omega - \omega_0$$
$$\dot{\omega} = \frac{\omega_0}{H}\{P_m - \frac{D}{\omega_0}(\omega - \omega_0) - \left[E_q' + (x_q - x_d')I_d\right]I_q\}$$
$$\dot{E}_q' = \frac{1}{T_{d0}'}\left(E_f - E_q' - (x_d - x_d')I_d\right) \tag{16}$$
$$0 = g(x, z, \overline{v})$$

where

$$g(x,z,\overline{v}) = \begin{pmatrix} P_t - [E_q' + (x_q - x_d')I_d]I_q + r_a(I_d^2 + I_q^2) \\ \theta_U - \delta + \operatorname{arc} ctg \dfrac{x_q I_q - r_a I_d}{E_q' - x_d' I_d - r_a I_q} \\ I_t - \sqrt{I_d^2 + I_q^2} \\ Q_t - E_q' I_d + x_q I_q^2 + x_d' I_d^2 \end{pmatrix} \tag{17}$$

The differential variables $x = (\delta, \omega, E_q', P_H)^T$ are relative power angle between G1 and G4, rotate speed deviation of G1, q-axis transient potential and the high pressure mechanical power respectively, the algebraic variables $z = (P_t, \theta_U, I_d, I_q)^T$ are active power, the angle of voltage, the d-axis current and the q-axis current respectively, and the interconnection input $\overline{v} = (I_t, Q_t)^T$ are the generator stator current and the reactive power respectively. The manipulated input $u = E_f$ are induction electromotive force. The others are the systems parameters. The controlled output y is chosen as voltage:

$$y = h(\cdot) = V_t = \frac{\sqrt{P_t^2 + Q_t^2}}{I_t} \tag{18}$$

From (12) we can get the explicit solution of manipulated control E_f as follows:

$$E_f = \frac{T_{d_0}'}{-I_q - c_1 I_d} \{ \frac{1}{P_t} [\frac{1}{I_t} \sqrt{P_t^2 + Q_t^2} (I_t^2 \dot{y}_1 + \dot{I}_t \sqrt{P_t^2 + Q_t^2}) - Q_t \dot{Q}_t] + b_1 \dot{I}_t + c_1 \dot{Q}_t \} + E_q' + (x_d - x_d')I_d \tag{19}$$

According to (14), now we can construct the 1-order integral right inverse systems of synchronous generator (16), with which the linearization and decouple is achieved.

The simulation is conducted based on a two-area four-machine power systems and the parameters of each generator and transformer can be seen in [18]. The simulation is based-on MATLAB and the results are shown in Fig.1.

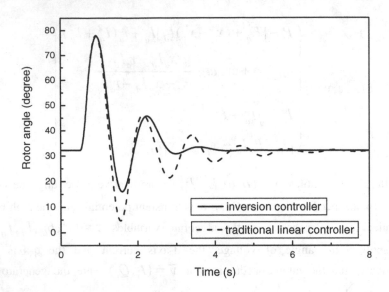

(a) Relative power angle between G1 and G4

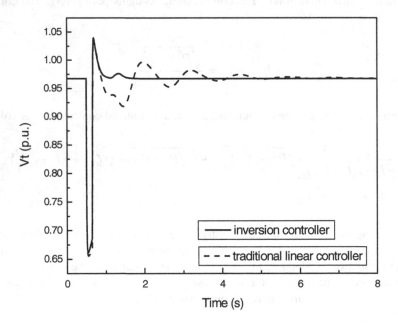

(b) Load bus voltage of G1

Fig. 1. Simulation Results of G1

As shown in Fig.1, both inter-area oscillation and inner-area oscillation are improved dramatically when generator G1 equipped excitation coordinative inverse controller.

5   Conclusion

Power systems components in essence belong to a special class of nonlinear DAE subsystems, which are of index one and the interconnection input is locally bounded and measurable. For power systems component described by such class of nonlinear DAE subsystem, the decentralized control problem is discussed using inverse systems method. Compared with the traditional linear controller, better closed-loop performance can be achieved using the control method proposed in this paper.

Acknowledgments. This work is supported by National Natural Science Foundation of China (61004001; 61104103; 61304089; 51177019); Natural Science Foundation of Jiangsu Province of China (BK2011826; BK20131000); Open Project of Key Laboratory of Measurement and Control of Complex Systems of Engineering, Ministry of Education, Southeast University (MCCSE2012A07); State Grid Corporation of China, Major Projects on Planning and Operation Control of Large Scale Grid (SGCC-MPLG022-2012).

References

1. Lu, Q., Sun, Y., Xu, Z.: Decentralized nonlinear optimal excitation control. IEEE Trans. Power Systems 11(4), 1957–1962 (1996)
2. Wang, M.H., Sun, Y.Z., Song, Y.H.: Backstepping design of modulation controller for multi-infeed HVDC systems. Proceedings of the CSEE 25(23), 7–11 (2005)
3. Yan, W., Wu, W.S., Hua, Z.M.: The direct feedback linearization of SSSC nonlinear control. Proceedings of the CSEE 23(3), 65–68 (2003)
4. Hill, D., Mareels, I.: Stability theory for differential/algebraic systems with application to power systems. IEEE Trans. Circuits and Systems 37(11), 1416–1423 (1990)
5. Wang, J., Chen, C.: Nonlinear control of differential algebraic model in power systems. Proceedings of the CSEE 21(8), 15–18 (2001)
6. Zang, Q., Dai, X.Z.: Output feedback stabilization control for nonlinear differential-algebraic equation systems. Acta Automatica Sinica 35(9), 1244–1248 (2009)
7. Xu, G.H., Wang, J., Chen, C.: Design of nonlinear controller for ac/dc power system based on differential algebraic models. Proceedings of the CSEE 25(7), 52–57 (2005)
8. Li, C.W., Liu, Y.H., Chen, T.J.: Feedback control for nonlinear singular systems with application to power systems: An inverse system method. Control Theory & Application 24(5), 799–802 (2007)
9. Ríos, A., López, M.: Order-reduction strategy of non-linear differential-algebraic equation models with application on power systems. Electric Power Components & Systems 40(15), 1690–1707 (2012)
10. Dai, X.Z., Zhang, K.F.: Interface concept and structural model of complex power systems. Proceedings of the CSEE 27(7), 7–12 (2007)

11. Zhang, K.F., Dai, X.Z., Qi, H., et al.: Analysis and application of component structural model of complex power systems. Proceedings of the CSEE 27(13), 24–28 (2007)
12. Dai, X.Z., Zhang, K.F., Zang, Q.: Nonlinear decentralized control method of component of power systems based on structural model. Proceedings of the CSEE 28(22), 15–22 (2008)
13. Tan, S.H., Vandewalle, J.: Inversion of singular systems. IEEE Trans. on Circuits and Systems 35(5), 583–587 (1998)
14. EI-Tohami, M., Lovassnagy, V., Powers, D.L.: On minimal order inverse of discrete-time descriptor systems. International Journal of Control 41(4), 991–1004 (1985), Wang, J., Liu, X.P.:The invertibility of affine nonlinear singular systems. Acta Automatica Sinica 24(2), 254–257(1998)
15. Dai, X.Z., Zang, Q., Zhang, K.F.: Construction of inverse system for nonlinear differential-algebraic equations subsystem. Acta Automatica Sinica 35(8), 1094–1100 (2009)
16. Zhang, K.F., Dai, X.Z.: ANN inversion excitation control method for multi-machine power systems. Automation of Electric Power Systems 27(21), 23–29 (2003)
17. Kundur, P.: Power system stability and control. McGraw Hill (1994)

Event-Triggered State Estimation for Complex Systems with Randomly Nonlinearities and Time-Varying Delay

Yushun Tan, Jinliang Liu, and Yuanyuan Zhang

Department of Applied Mathematics, Nanjing University of Finance and Economics,
Nanjing, Jiangsu 210046, P.R. China
tyshun994@163.com

Abstract. The event-triggered state estimation is investigated for a class of complex networks system with randomly nonlinearities. A novel event-triggered scheme is proposed, which can reduce the information communication burden in the network. Considering the effect of transmission delay, a time-varying delayed system model is constructed. Attention of this paper is focused on the analysis and design of a reliable estimator for the complex networks through the available output measurements under event-triggered scheme. In order to design the state estimator, a Lyapunov functional approach and the linear matrix inequality technique are employed. A sufficient condition is obtained in which the estimator error dynamics is exponential asymptotically stable, and a state estimator of considered complex networks can be achieved by solving some linear matrix inequalities. Finally, a numerical example is provided to illustrate the effectiveness of the proposed method.

Keywords: event-triggered scheme, complex networks, time-varying delay, randomly occurring.

1 Introduction

There exist many complicated systems which can be modeled as complex networks in our daily life and the natural world, such as power grids in engineering application, neural networks in biology and human relationship networks in social life. In the past decades, a lot of researches about complex networks have been attracted from different working area, see e.g. [1,2,3,4,5] and reference therein.

Dynamic analysis for complex networks has become a hot issues and experienced an increasing attentions due to their theoretical importance and as well as the extensive applications of these systems in many fields. Different kinds of issue have been extensively investigated for complex networks, such as, stabilization, pinning control, synchronization and state estimation. As one of the mostly investigated dynamical behavior, some effective conclusions have been obtained[3,5,6,8,9,7]. For example, in [6], by employing the Lyapunov functional

M. Fei et al. (Eds.): LSMS/ICSEE 2014, Part II, CCIS 462, pp. 407–418, 2014.

method combined with the stochastic analysis approach, some desired state estimators were designed for a class of uncertain stochastic network.

However, how to reduce communication requirements becomes an important problem. In [10], this problem was addressed for a networked control systems, where the maximum time allowed to elapse was obtained to guarantee the stability of the system. But their approach leads to an inherently periodic transmission. Recently, event-triggered techniques for control design, advocating the use of actuation only when some function of the system state exceeds a threshold, have been paid an increased attention in the literature [11,12,14,13,15]. Event-triggered provides a useful way of determining when the sampling action is carried out, which guarantees that only really necessary state signal will be sent out to the controller. Thus, the amount of the sent state signals is relatively little. Compared with periodic sampling method, the event-triggered method has the following advantages: (1) closer in nature to the way a human behaves as a controller, which only samples when necessary; (2) reduction in the release times of the sensor and then the burden of the network communication; (3) reduction in the computation cost of the controller and the occupation of the sensor and the actuator.

To the best of authors knowledge, there are no papers to deal with the event-triggered state estimation for complex networks with randomly occurring nonlinearities and time-varying delay up to now. Complex networks system with time-varying delays still remains as a challenging problem. Motivated by the above discussion, in this paper, we consider the state estimation problem for the time-varying delayed complex networks with stochastic nonlinearities. The main purpose is to estimate the complex networks states through output sampled measurement based on event-triggered control, which can provide a significant method in reducing burden of information communication. By constructing a Lyapunov function and using stochastic analysis technique, a desirable estimator of complex network system is designed by solving some linear matrix inequalities.

The rest of this paper is organized as follows: in Section 2, considering event-triggered scheme, a state estimation model for complex network model with randomly occurring nonlinearities is provided. In Section 3, by employing the Lyapunov stability theory, a sufficient condition in term of linear matrix inequalities (LMIs) is established and the design of the estimator is derived. In Section 4, a numerical example is presented to demonstrate the effectiveness of the results achieved. Finally, a conclusion is drawn in Section 5.

2   Model and Preliminaries

Considering the following the stochastic complex network consisting of N coupled nodes with time-varying delayed:

$$\dot{x}_i(t) = \delta(t)Af_1(x_i(t)) + (1-\delta(t))Bf_2(x_i(t)) + \sum_{j=1}^{N} g_{ij}\Gamma_1 x_j(t) + \sum_{j=1}^{N} \Gamma_2 x_j(t-\tau(t)) \quad (1)$$

where $x_i(t) = (x_{i1}(t), x_{i2}(t), \ldots, x_{in}(t))^T$ is the state vector of the ith node, $f_1(x_i(t))$ and $f_2(x_i(t))$ are nonlinear vector valued functions satisfying certain conditions given later. A and B are constant matrix with appropriate dimensions, Γ_1 and Γ_2 are the inner coupling matrices of the network from the vertical and the horizontal directions, respectively. $G = (g_{ij}) \in R^{N \times N}$ is the coupling the configuration matrix denoting the topological structure of the complex network, if there is a connection between node i and $j(i \neq j)$, then $g_{ij} = g_{ji} = 1$, otherwise, $g_{ij} = g_{ji} = 0$, the diagonal elements of the matrix G are defined as $g_{ij} = -\sum_{j=1,j \neq i}^{N} g_{ij}(i = 1, 2, \cdots, N)$. The positive integer $\tau(t)$ denotes time-varying delay that satisfies $\tau_1 \leq \tau(t) \leq \tau_2$ with $0 \leq \tau_1 \leq \tau_2$ being known positive integers. $\delta(t)$ is Bernoulli distributed with sequence governed by $Prob\{\delta(t) = 1\} = \delta_0, Prob\{\delta(t) = 0\} = 1 - \delta_0$, where δ_0 is known constant.

Assumption 1. The nonlinear vector-valued functions $f_1(\cdot)$ and $f_2(\cdot)$ are continuous and satisfy [7]

$$[f_1(u) - f_1(v) - \Xi_1(u - v)]^T [f_1(u) - f_1(v) - \Xi_2(u - v)] \leq 0$$
$$[f_2(u) - f_2(v) - \Xi_3(u - v)]^T [f_2(u) - f_2(v) - \Xi_4(u - v)] \leq 0 \tag{2}$$

Remark 1. On the basis assumption 1, we can get

$$\begin{bmatrix} x_i(t) \\ f_1(x_i(t)) \end{bmatrix}^T \begin{bmatrix} \Omega_{11} & \Omega_{12} \\ \Omega_{21} & \Omega_{22} \end{bmatrix} \begin{bmatrix} x_i(t) \\ f_1(x_i(t)) \end{bmatrix} \geq 0,$$

$$\begin{bmatrix} x_i(t) \\ f_2(x_i(t)) \end{bmatrix}^T \begin{bmatrix} \bar{\Omega}_{11} & \bar{\Omega}_{12} \\ \bar{\Omega}_{21} & \bar{\Omega}_{22} \end{bmatrix} \begin{bmatrix} x_i(t) \\ f_2(x_i(t)) \end{bmatrix} \geq 0$$

where

$$\Omega_{11} = \frac{\Xi_1^T \Xi_2 + \Xi_2^T \Xi_1}{2}, \quad \Omega_{21}^T = \Omega_{12} = -\frac{\Xi_1^T + \Xi_2^T}{2},$$
$$\bar{\Omega}_{11} = \frac{\Xi_3^T \Xi_4 + \Xi_4^T \Sigma_3}{2}, \quad \bar{\Omega}_{21} = \bar{\Omega}_{12} = -\frac{\Xi_3^T + \Xi_4^T}{2}$$

With the matrix Kronecker product , we can rewrite the networks (2) in the following compact form:

$$\dot{x}(t) = \delta(t)(I_n \otimes A)F_1(x(t)) + (1 - \delta(t))(I_N \otimes B)F_2(x(t))$$
$$+ (G \otimes \Gamma_1)x(t) + (G \otimes \Gamma_2)x(t - \tau(t)) \tag{3}$$

where

$$x(t) = \begin{bmatrix} x_1(t) \\ x_2(t) \\ \vdots \\ x_N(t) \end{bmatrix}; \quad F_1(x(t)) = \begin{bmatrix} f_1(x_1(t)) \\ f_1(x_2(t)) \\ \vdots \\ f_1(x_N(t)) \end{bmatrix}; \quad F_2(x(t)) = \begin{bmatrix} f_2(x_1(t)) \\ f_2(x_2(t)) \\ \vdots \\ f_2(x_N(t)) \end{bmatrix}$$

In order to make use of the networks in practice, it is necessary to estimate the nodes states through available network output. Suppose that the output $y(t) \in \mathbb{R}^m$ of the complex network (3) is of the form:

$$y(t) = Cx(t) \tag{4}$$

where C is a known constant matrix.

In this paper, the measurement output is sampled before it enters the estimator. Based on the zero-order hold and the sampling technique, the actual output in system (2) can be described as

$$\tilde{y}(t) = y(t_k h) = Cx(t_k h), \quad t \in [t_k h, t_{k+1} h) \tag{5}$$

Remark 2. In most systems, we know that periodic sampling mechanism may lead to sending many unnecessary signals thought the network, which will decrease the efficiency of communication. Therefore, it is significant to introduce a mechanism to determine which sampled signal should be sent out or not. In this sense, we propose an event-triggered mechanism which has some advantages over exiting ones, and the sensor data is transmitted only when some function of the system exceeds a threshold. Thus, the event-triggered sampling scheme is effective way because it can reduce the burden of the network communication and the computation cost of the estimator.

Here, similar with [14,15], the event generator is constructed between the sensor and the estimator, which is used to determine whether the newly sampled data $x(t)$ to be send out to the estimator by using the following judgement algorithm:

$$[x((k+j)h) - x(kh)]^T W [x((k+j)h) - x(kh)] \le \rho x((k+j)h) W x((k+j)h) \tag{6}$$

Where W is a symmetric positive definite matrix, $j = 1, 2, \cdots$, and $\rho \in [0,1)$. $x(kh)$ is the previous transmitted sensor data. Only when the current sampled sensor measurements $x((k+j)h)$ and the latest transmitted sensor measurements $x(kh)$ variate the specified threshold (6), the current sampled sensor measurements $x((k+j)h)$ can be transmitted by the event generator and sent into the estimator. When $\rho = 0$, it is obvious that for all the sampled state $x((k+j)h)$ do not satisfy the inequality, thus the event-triggered scheme reduces to a periodic release scheme.

Remark 3. The sensor measurement are sampled at time kh by sampler with a given period h. The next sensor measurement is at time $(k+1)h$. Suppose that the release times are $t_0 h, t_1 h, t_2 h, \cdots$, it is easily seen that $t_i h = t_{i+1}h - t_i h$ denotes the release period of event generator in (6).

When the sampled data has been transmitted by the event generator, we suppose that the time-varying delay in network communication is $\tau(k)(k = 1, 2, \cdots,)$. The output $\tilde{y}(t)$ can be rewritten as

$$\tilde{y}(t) = y(t_k h) = Cx(t_k h), \quad t \in [t_k h + \tau_k, t_{k+1} h + \tau_{k+1}) \tag{7}$$

Based on the available sampled measurement $\tilde{y}(t)$, the following state estimator is adopted:

$$\begin{cases} \dot{\hat{x}}(t) = G_1 \hat{x}(t) + G_2 \hat{x}(\tau - \tau(t)) + K(\tilde{y}(t) - \hat{y}(t)), \\ \hat{y}(t) = C\hat{x}(t). \end{cases} \tag{8}$$

where $\hat{x}(t)$ is estimator state vector, $\hat{y}(t)$ is estimator measurement vector, and K is feedback gain matrix to be designed.

Similar to [14,16,17], for technical convenience, consider the following two cases:

Case 1: If $t_k h + h + \bar{\tau} \geq \tau_{k+1} h + \tau_{k+1}$, where $\bar{\tau} = \max\{\tau_k\}$, define a function $d(t)$ as

$$d(t) = t - \tau_k h, \quad for \quad t \in [t_k h + \tau_k, t_{k+1} h + \tau_{k+1}) \tag{9}$$

obviously,

$$\tau_k \leq d(t) \leq (t_{k+1} - \tau_k)h + \tau_{k+1} \leq h + \bar{\tau} \tag{10}$$

Case 2: If $t_k h + h + \bar{\tau} < t_{k+1} h + \tau_{k+1}$, $[t_k h + \tau_k, t_k h + h + \bar{\tau})$, considering the following intervals:

$$[t_k h + ih + \bar{\tau}, \tau_k h + ih + h + \bar{\tau})$$

since $\tau_k < \bar{\tau}$, it can be shown that d_M exists such that $t_k h + d_M h + \bar{\tau} < t_{k+1} h + \tau_{k+1} \leq t_k h + d_M h + h + \bar{\tau}$ and $x(t_k h)$ and $x(t_k h + ih)$ with $i = 1, 2, \cdots, d_M$ satisfy (6), it also can be seen that

$$[t_k h + \tau_k, t_{k+1} h + \tau_{k+1}) = I_1 \cup I_2 \cup I_3. \tag{11}$$

where

$$\begin{cases} I_1 = [\tau_k h + \tau_k, \tau_k h + h + \bar{\tau}) \\ I_2 = \bigcup_{i=1}^{d_M - 1}\{I_2^i\} = \bigcup_{i=1}^{d_M - 1}[t_k h + ih + \bar{\tau}, \tau_k h + ih + h + \bar{\tau}) \\ I_3 = [\tau_k h + d_M h + \bar{\tau}, t_{k+1} h + \tau_{k+1}) \end{cases} \tag{12}$$

Define

$$d(t) = \begin{cases} t - t_k h & \text{for } t \in I_1 \\ t - t_k h - ih & \text{for } t \in I_2^i \\ t - t_k h - d_M h & \text{for } t \in I_3 \end{cases} \tag{13}$$

We can easily show that

$$\begin{cases} \tau_k \leq d(t) \leq h + \bar{\tau} & \text{for } t \in I_1 \\ \tau_k \leq \bar{\tau} \leq d(t) \leq h + \bar{\tau} & \text{for } t \in I_2^i \\ \tau_k \leq \bar{\tau} \leq d(t) \leq h + \bar{\tau} & \text{for } t \in I_3 \end{cases} \tag{14}$$

where the third row in (14) holds because $t_{k+1} h + \tau_{k+1} \leq \tau_k h + (d_M + 1)h + \bar{\tau}$. Therefore, for $t \in [\tau_k h + \tau_k, \tau_{k+1} h + \tau_{k+1})$, $0 \leq \tau_k \leq h + \bar{\tau} = \tau_M$, that is $\tau(t) \in [0, \tau_M]$.

In case 1, for $t \in [t_k h + \tau_k, t_{k+1} h + \tau_{k+1})$, define $e_k(t) = 0$. In case 2, define

$$e_k(t) = \begin{cases} 0 & \text{for } t \in I_1 \\ x(t_k h) - x(t_k h + ih) & \text{for } t \in I_2^i \\ x(t_k h) - x(t_k h + d_M h) & \text{for } t \in I_3 \end{cases} \tag{15}$$

From the definition of $e_k(t)$ and $d(t)$, the triggering algorithm can be rewritten as

$$e_k^T(t)\Omega e_k(t) \leq \sigma x^T(t - d(t))\Omega x(t - d(t)) \tag{16}$$

Then by setting the estimation error $x(t) - \hat{x}(t) = e(t)$, the error dynamics of the estimation can be obtained from (3) and (8) as follows:

$$
\begin{aligned}
\dot{e}(t) = {} & \delta(t)(I_N \otimes A)F_1(x(t)) + (1 - \delta(t))(I_N \otimes B)F_2(x(t)) + (G \otimes \Gamma_1 - KC)e(t) \\
& + (G \otimes \Gamma_2)e(t - \tau(t)) + KCx(t) - KCx(t - d(t)) - KCe_k(t)
\end{aligned} \tag{17}
$$

Denoting $\bar{x}(t) = \left[x^T(t)\ e^T(t) \right]^T$, $I_A = I_N \otimes A$ and $I_B = I_N \otimes B$, we can get the following augmented system from (3) and (17)

$$
\begin{aligned}
\dot{\bar{x}}(t) = {} & \delta(t)\bar{I}_A F_1(H\bar{x}(t)) + (1 - \delta(t))\bar{I}_B F_2(H\bar{x}(t)) + \bar{A}\bar{x}(t) \\
& + \bar{B}\bar{x}(t - \tau(t)) + \bar{C}\bar{x}(t - d(t)) + \bar{D}e_k(t)
\end{aligned} \tag{18}
$$

where

$$
\bar{A} = \begin{bmatrix} G \otimes \Gamma_1 & 0 \\ KC & G \otimes \Gamma_1 - KC \end{bmatrix}, \quad \bar{B} = \begin{bmatrix} G \otimes \Gamma_1 & 0 \\ 0 & G \otimes \Gamma_1 \end{bmatrix}, \quad \bar{C} = \begin{bmatrix} 0 & 0 \\ -KC & 0 \end{bmatrix}
$$

$$
\bar{D} = \begin{bmatrix} 0 \\ KC \end{bmatrix}, \bar{I}_A = \begin{bmatrix} I_A \\ I_A \end{bmatrix}, \quad \bar{I}_B = \begin{bmatrix} I_B \\ I_B \end{bmatrix}, \quad \bar{H}^T = \begin{bmatrix} I \\ 0 \end{bmatrix}
$$

Introduce a new vector

$$
\begin{aligned}
\zeta^T(t) = [& F_1^T(H\bar{x}(t)) \quad F_2^T(H\bar{x}(t)) \quad \bar{x}^T(t) \quad \bar{x}^T(t - \tau_m) \quad \bar{x}^T(t - \tau(t)) \\
& \bar{x}^T(t - \tau_M) \quad \bar{x}^T(t - d_M) \quad \bar{x}^T(t - d(t)) \quad e_k^T(t)]
\end{aligned}
$$

And let

$$
\mathscr{A}_1 = \begin{bmatrix} \bar{I}_A & 0 & \bar{A} & 0 & \bar{B} & 0 & 0 & \bar{C} & \bar{D} \end{bmatrix}, \quad \mathscr{A}_2 = \begin{bmatrix} 0 & \bar{I}_B & \bar{A} & 0 & \bar{B} & 0 & 0 & \bar{C} & \bar{D} \end{bmatrix}
$$

We can rewrite (18) as following:

$$
\dot{\bar{x}}(t) = \delta(t)\mathscr{A}_1\zeta(t) + (1 - \delta(t))\mathscr{A}_2\zeta(t) \tag{19}
$$

Before giving the main results, the following lemmas are essential in establishing our main results.

Lemma 1. [19] For any $x, y \in R^n$ and positive definite matrix $Q \in R^{n \times n}$, the following inequality holds:

$$
2x^T y \leq x^T Q x + y^T Q^{-1} y \tag{20}
$$

Lemma 2. [20,21] For any constant positive matrix $R \in \mathbb{R}^{n \times n}$, $\tau_1 \leq \tau(t) \leq \tau_2, x(t) \in R^n, R > 0$ and vector function $\dot{x}(t) : [-\tau_2, \tau_1] \to \mathbb{R}^n$, such that the following integration is well defined, then the following inequality is holds:

$$
(\tau_1 - \tau_2) \int_{t-\tau_2}^{t-\tau_1} \dot{x}^T(s)R\dot{x}(s) \leq \begin{bmatrix} x(t-\tau_1) \\ x(t-\tau(t)) \\ x(t-\tau_2) \end{bmatrix}^T \begin{bmatrix} -R & R & 0 \\ * & -2R & R \\ * & * & -R \end{bmatrix} \begin{bmatrix} x(t-\tau_1) \\ x(t-\tau(t)) \\ x(t-\tau_2) \end{bmatrix} \tag{21}
$$

Lemma 3. [22,23] Ω_1, Ω_2 and Ω are matrices with appropriate dimensions, $\tau(t)$ is a function of t and $\tau_1 \leq \tau(t) \leq \tau_2$, then

$$(\tau(t) - \tau_1)\Omega_1 + (\tau_2 - \tau(t))\Omega_2 + \Omega < 0 \tag{22}$$

if and only if the following two inequalities hold:

$$(\tau_2 - \tau_1)\Omega_1 + \Omega < 0, \quad (\tau_2 - \tau_1)\Omega_2 + \Omega < 0 \tag{23}$$

3 Main Results

In this section, we will design a desirable estimator with form (18) for complex networks systems with random nodes based on event-triggered control. We first present some sufficient conditions for the augmented system (19).

Theorem 1. Suppose that assumption 1 holds, for given scalars $0 \leq \tau_1 \leq \tau_2$ and the estimator gain matrix K in (7), the augmented system (19) is exponential asymptotically stable if there exist matrices $P > 0, Q_1 > 0, Q_2 > 0, R_1 > 0, R_2 > 0$ and $M, N, \Lambda_1, \Lambda_2$ with appropriate dimensions, such that the following matrix inequalities hold for $s = 1, 2$

$$\begin{bmatrix} \Phi_{11} + \Gamma + \Gamma^T & * & * \\ \Phi_{21} & \Phi_{22} & * \\ \Phi_{31}(s) & * & -R_2 \end{bmatrix} < 0 \tag{24}$$

where

$$\Phi_{11} = \begin{bmatrix} \Lambda_1 \otimes I_n & * & * & * & * & * & * & * & * \\ 0 & \Lambda_2 \otimes I_n & * & * & * & * & * & * & * \\ \Pi_{31} & \Pi_{32} & \Pi_3 & * & * & * & * & * & * \\ 0 & 0 & R_1 & -R_1 - Q_1 & * & * & * & * & * \\ 0 & 0 & \bar{B}^T P & 0 & 0 & * & * & * & * \\ 0 & 0 & 0 & 0 & 0 & -Q_2 & * & * & * \\ 0 & 0 & 0 & 0 & 0 & 0 & -R_3 - Q_3 & * & * \\ 0 & 0 & R_3 + \bar{C}^T P & 0 & 0 & 0 & R_3 & -R_3 + \Omega & * \\ 0 & 0 & \bar{D}^T P & 0 & 0 & 0 & 0 & 0 & -\Omega \end{bmatrix}$$

$$\Phi_{21} = \begin{bmatrix} \theta_1 R_1 \bar{I}_A & 0 & \theta_1 R_1 \bar{A} & 0 & \theta_1 R_1 \bar{B} & 0 & 0 & -\theta_1 R_1 \bar{C} & -\theta_1 R_1 \bar{D} \\ \theta_2 R_2 \bar{I}_A & 0 & \theta_2 R_2 \bar{A} & 0 & \theta_2 R_2 \bar{B} & 0 & 0 & -\theta_2 R_2 \bar{C} & -\theta_2 R_2 \bar{D} \\ \theta_3 R_3 \bar{I}_A & 0 & \theta_3 R_3 \bar{A} & 0 & \theta_3 R_3 \bar{B} & 0 & 0 & -\theta_3 R_3 \bar{C} & -\theta_3 R_3 \bar{D} \\ 0 & \theta_{10} R_1 \bar{I}_B & \theta_{10} R_1 \bar{A} & 0 & \theta_{10} R_1 \bar{B} & 0 & 0 & -\theta_{10} R_1 \bar{C} & -\theta_{10} R_1 \bar{D} \\ 0 & \theta_{20} R_2 \bar{I}_B & \theta_{20} R_2 \bar{A} & 0 & \theta_{20} R_2 \bar{B} & 0 & 0 & -\theta_{20} R_2 \bar{C} & -\theta_{20} R_2 \bar{D} \\ 0 & \theta_{30} R_3 \bar{I}_B & \theta_{30} R_3 \bar{A} & 0 & \theta_{30} R_3 \bar{B} & 0 & 0 & -\theta_{30} R_3 \bar{C} & -\theta_{30} R_3 \bar{D} \end{bmatrix}$$

$$\Pi_{22} = diag\{-R_1, -R_2, -R_3, -R_1, -R_2, -R_3\}$$

$$\Pi_3 = -R_3 + Q_1 + Q_2 + Q_3 + \bar{A}^T P + P \bar{A} + H^T (\Lambda_1 \otimes \Omega_{11}) H + H^T (\Lambda_2 \otimes \bar{\Omega}_{11}) H$$

$$\Pi_{31} = H^T (\Lambda_1 \otimes \Omega_{21})^T + \delta_0 P \bar{I}_A, \Pi_{31} = H^T (\Lambda_2 \otimes \bar{\Omega}_{21})^T + \delta_{10} P \bar{I}_B$$

$$\Psi_{31}(1) = \sqrt{\delta}N^T, \Psi_{31}(2) = \sqrt{\delta}M^T, \tau_{21} = \tau_M - \tau_m, \delta_{10} = 1 - \delta_0$$

$$\Omega = \begin{bmatrix} W & 0 \\ 0 & 0 \end{bmatrix}, \Gamma = [0\ 0\ 0\ N\ -N + M\ -M\ 0\ 0\ 0]$$

$$M^T = [M_1^T\ M_2^T\ M_3^T\ M_4^T\ M_5^T\ M_6^T\ M_7^T\ M_8^T\ M_9^T]$$

$$M^T = [N_1^T\ N_2^T\ N_3^T\ N_4^T\ N_5^T\ N_6^T\ N_7^T\ N_8^T\ N_9^T]$$

$$\theta_1 = \tau_1\sqrt{\delta_0}, \theta_2 = \sqrt{\tau_{21}\delta_0}, \theta_3 = d_M\sqrt{\delta_0}$$

$$\theta_{10} = \tau_1\sqrt{\delta_{10}}, \theta_{20} = \sqrt{\tau_{21}\delta_{10}}, \theta_{30} = d_M\sqrt{\delta_{10}}$$

Proof: Construct the following Lyapunov functional candidate:

$$V(t, \bar{x}(t)) = \sum_{i=1}^{3} V_i(t, \bar{x}(t)) \tag{25}$$

where

$$V_1(t, \bar{x}(t)) = \bar{x}^T(t)P\bar{x}(t)$$

$$V_2(t, \bar{x}(t)) = \int_{t-\tau_m}^{t} \bar{x}^T(s)Q_1\bar{x}(s)ds + \int_{t-\tau_M}^{t} \bar{x}^T(s)Q_2\bar{x}(s)ds + \int_{t-d_M}^{t} \bar{x}^T(s)Q_3\bar{x}(s)ds$$

$$V_3(t, \bar{x}(t)) = \tau_m \int_{t-\tau_m}^{t} \int_{s}^{t} \dot{\bar{x}}^T(v)R_1\dot{x}(v)dvds + \int_{t-\tau_M}^{t-\tau_m} \int_{s}^{t} \dot{\bar{x}}^T(v)R_2\bar{x}(v)dvds$$

$$+ d_M \int_{\tau-d_M}^{t} \int_{s}^{t} \dot{\bar{x}}^T(v)R_3\bar{x}(v)dvds$$

and $P > 0, Q_1 > 0, Q_2 > 0, R_1 > 0, R_2 > 0$ are matrices to be determined.

Taking the derivative of $V_i(t, \bar{x}(t)), (i = 1, 2, 3)$ along the trajectory of system (18) and taking expectation on $\mathcal{L}V(t, \bar{x}(t))$, then similar to the proof in [7] and recalling Lemma 1-3, the proof can be completed.

Theorem 1 established analysis results for the complex networks(3), but it can not be used to design estimator K directly for the condition (24) are not linear matrix inequalities. The following Theorem 2 gives the estimator gain matrix K.

Theorem 2. For some given constants $0 \leq \tau_1 \leq \tau_2$ and $0 \leq \tau_m \leq \tau_M$, the augmented systems (19) is exponential asymptotically stable if there exist $P = diag\{\mathbf{P}_1, \mathbf{P}_2\} > 0$, $Q_1 = diag\{\mathbf{Q}_1, \mathbf{Q}_1\} > 0$, $Q_2 = diag\{\mathbf{Q}_2, \mathbf{Q}_2\} > 0$, $Q_3 = diag\{\mathbf{Q}_3, \mathbf{Q}_3\} > 0$, $R_1 = diag\{\mathbf{R}_1, \mathbf{R}_1\} > 0$, $R_2 = diag\{\mathbf{R}_2, \mathbf{R}_2\} > 0$, $R_3 = diag\{\mathbf{R}_3, \mathbf{R}_3\} > 0$, $M_k = diag\{\mathbf{M}_k, \mathbf{M}_k\}$, $N_k = diag\{\mathbf{N}_k, \mathbf{N}_k\}(k = 1, 2\cdots, 9)$, Λ_1 and Λ_2 with appropriate dimensions so that the following linear matrix inequalities hold for given $\varepsilon_j > 0(j = 1, 2, 3)$

$$\begin{bmatrix} \Phi_{11} + \Gamma + \Gamma^T & * & * \\ \tilde{\Phi}_{21} & \tilde{\Phi}_{22} & * \\ \Phi_{31}(s) & * & -R_2 \end{bmatrix} < 0 \tag{26}$$

where

$$\tilde{\Phi}_{21} = \begin{bmatrix} \theta_1 P\bar{I}_A & 0 & \theta_1 P\bar{A} & 0 & \theta_1 P\bar{B} & 0 & 0 & \theta_1 P\bar{C} & \theta_1 P\bar{D} \\ \theta_2 P\bar{I}_A & 0 & \theta_2 P\bar{A} & 0 & \theta_2 P\bar{B} & 0 & 0 & \theta_2 P\bar{C} & \theta_2 P\bar{D} \\ \theta_3 P\bar{I}_A & 0 & \theta_3 P\bar{A} & 0 & \theta_3 P\bar{B} & 0 & 0 & \theta_3 P\bar{C} & \theta_3 P\bar{D} \\ 0 & \theta_{10} P\bar{I}_B & \theta_{10} P\bar{A} & 0 & \theta_{10} P\bar{B} & 0 & 0 & \theta_{10} P\bar{C} & \theta_{10} P\bar{D} \\ 0 & \theta_{20} P\bar{I}_B & \theta_{20} P\bar{A} & 0 & \theta_{20} P\bar{B} & 0 & 0 & \theta_{20} P\bar{C} & \theta_{20} P\bar{D} \\ 0 & \theta_{30} P\bar{I}_B & \theta_{30} P\bar{A} & 0 & \theta_{30} P\bar{B} & 0 & 0 & \theta_{30} P\bar{C} & \theta_{30} P\bar{D} \end{bmatrix}$$

$$\tilde{\Phi}_{22} = diag\{\tilde{R}_1, \tilde{R}_2, \tilde{R}_3, \tilde{R}_1, \tilde{R}_2, \tilde{R}_3\} \quad \tilde{R}_j = \varepsilon_j^2 R_j - 2\varepsilon_j P, \quad (j = 1, 2, 3),$$

$$P\bar{I}_A = \begin{bmatrix} \mathbf{P}_1 I_A \\ \mathbf{P}_2 I_A \end{bmatrix}, \quad P\bar{I}_B = \begin{bmatrix} \mathbf{P}_1 I_B \\ \mathbf{P}_2 I_B \end{bmatrix}, \quad P\bar{A} = \begin{bmatrix} \mathbf{P}_1(G \otimes \Gamma_1) & 0 \\ YC & \mathbf{P}_2(G \otimes \Gamma_1) - YC \end{bmatrix},$$

$$P\bar{B} = \begin{bmatrix} \mathbf{P}_1(G \otimes \Gamma_2) & 0 \\ 0 & \mathbf{P}_2(G \otimes \Gamma_2) \end{bmatrix}, \quad P\bar{C} = \begin{bmatrix} 0 & 0 \\ -YC & 0 \end{bmatrix}, \quad P\bar{D} = \begin{bmatrix} 0 \\ -YC \end{bmatrix}.$$

and the other symbols are defined in Theorem 1. Moreover, the desired state estimator K in (8) can be obtained by $K = \mathbf{P}_2^{-1} Y$ if (26) is true.

Proof: Performing a congruence transformation of diagonal matrix

$$diag\{I, PR_1^{-1}, PR_2^{-1}, PR_3^{-1}, PR_1^{-1}, PR_2^{-1}, PR_3^{-1}, I\}$$

to (24), then setting $Y = \mathbf{P}_2 K$ and considering the following inequality:

$$-PR_j^{-1}P \leq \varepsilon_j^2 R_j - 2\varepsilon_j P \quad (j = 1, 2, 3).$$

Similar with [7], we can easily obtained (26) state estimator $K = \mathbf{P}_2^{-1} Y$.

4   Numerical Results

In this section, a numerical example is given to verify the effectiveness of the proposed control techniques for estimation of complex networks with time-varying delays.

Considering the following continuous complex network consisting of five coupled nodes:

$$\dot{x}_i(t) = \delta(t) A f_1(x_i(t)) + (1 - \delta(t)) B f_2(x_i(t))$$

$$+ \sum_{j=1}^{N} g_{ij} \Gamma_1 x_j(t) + \sum_{j=1}^{N} \Gamma_2 x_j(t - \tau(t)), (i = 1, 2, 3, 4, 5)$$

where

$$x_i(t) = \begin{bmatrix} x_{i1}(t) \\ x_{i2}(t) \end{bmatrix}, \quad A = \begin{bmatrix} 0.2 & -0.2 \\ -0.4 & 0.3 \end{bmatrix}, \quad B = \begin{bmatrix} 0.2 & -0.25 \\ -0.35 & 0.3 \end{bmatrix}$$

The coupling configuration matrix and the inner-coupling matrix are assumed to be

$$G = \begin{bmatrix} -15 & 12.01 & 0 & 0 & 0.01 \\ 12.01 & -15 & 0 & 0 & 0 \\ 0.01 & 0.02 & -16 & 0 & 0.02 \\ 0.02 & 0.01 & 0 & -16 & 0.01 \\ 0 & 0 & 0.01 & 0.01 & -14 \end{bmatrix}, \quad \Gamma_1 = \begin{bmatrix} 1 & 0 \\ 0 & 1 \end{bmatrix}, \quad \Gamma_2 = 0.1\Gamma_1.$$

The network nodes activation function is described as follows:

$$f_1(x_i(t)) = \begin{bmatrix} 0.4x_{i1}(t) - tanh(0.3x_{i2}(t)) \\ 0.9x_{i2}(t) - tanh(0.7x_{i1}(t)) \end{bmatrix}, \quad f_2(x_i(t)) = \begin{bmatrix} 0.3x_{i1}(t) - tanh(0.2x_{i2}(t)) \\ 0.8x_{i2}(t) - tanh(0.6x_{i1}(t)) \end{bmatrix}$$

Suppose $C = \begin{bmatrix} 0.2 & -0.5 & 0.2 & 0 & 0.2 & -0.6 & 0.2 & 0 & -0.7 & 0.2 \end{bmatrix}$, $\delta_0 = 0.7$, and $\Xi_i, (i = 1, 2, 3, 4)$ in Assumption 1 are chosen as follows: $\Xi_1 = diag\{0.1, 0.2\}$, $\Xi_2 = diag\{-0.15, -0.25\}$, $\Xi_3 = diag\{-0.1, -0.2\}$, $\Xi_4 = diag\{0.25, 0.35\}$.

Table 1. $\tau_1 = 0.1$, $h = 0.05$, $\varepsilon_1 = \varepsilon_2 = \varepsilon_3 = 0.1$

ρ	0	0.1	0.2	0.5
The upper bound of τ_2	1.37	1.08	0.63	0.41
Trigger times	200	138	74	52
Data transmission	100%	69%	37%	26%

In here, the communication delays satisfy $\tau_1 \leq \tau(t) \leq \tau_2$, and assume $\tau_1 = 0.1$, $h = 0.05$, $\rho = 0.2$, $\varepsilon_1 = \varepsilon_2 = \varepsilon_3 = 0.1$. By applying Theorem 2, it can be obtained the upper bound allowable delay $\tau_2 = 0.63$. In fact, for different values of ρ given, different results can be obtained, which are described in Table 1. It can be seen that the larger the ρ, the smaller trigger times and the maximum allowable delay, which are reasonable.

Then combining (26) and $K = P^{-1}Y$, by employing LMI Toolbox in LMIs (26), the desired estimator parameters and triggered matrix can be obtained as follows:

$$K^T = 10^{-2} \begin{bmatrix} -6.67 & 16.78 & -6.58 & 0.01 & -6.6 & 19.9 & -6.66 & -0.03 & 22.85 & -6.49 \end{bmatrix}$$

$$W = 10^{-3} diag\{1.3, \quad 1.3, \quad 1.4, \quad 1.4, \quad 1.4, \quad 1.4, \quad 1.4, \quad 1.4, \quad 1.5, \quad 1.5\}$$

Simulation results are shown in Fig.1 which shows that the output error $e(t)$ with $\rho = 0.2$. From Fig.1 we can see that the designed estimator performs well. Moreover, the times of data transmission under event-triggered scheme much small number than time-triggered periodic communication scheme from Table 1.

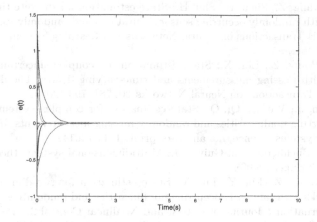

Fig. 1. Estimation error trajectory $e(t)$ of (17) with $\rho = 0.2$

5 Conclusion

In order to reduce the computation load, this paper has investigated a event-triggered state estimation problem for the complex networks system. Under the event-triggering scheme, the sampled sensor measurements information will be transmitted to the controller only when it meets a specified triggering condition. By using Lyapnnov functional approach and stochastic techniques, sufficient conditions is obtained in which the estimator error dynamics is exponential asymptotically stable, and a novel state estimator for the stochastic complex networks systems can be achieved by solving some linear matrix inequalities. Finally, a numerical example is provided to show the effectiveness of the even-triggered scheme.

Acknowledgment. The authors would like to acknowledge the Natural Science Foundation of China (No.11226240), the Natural Science Foundation of Jiangsu Province of China (No.BK2102469).

References

1. Ding, D., Wang, Z., Dong, H., Shu, H.: Distributed H_∞ state estimation with stochastic parameters and nonlinearities through sensor networks: The finite-horizon case. Automatica 48, 1575–1585 (2012)
2. Gao, H., Lam, J., Chen, G.: New criteria for synchronization stability of general complex dynamical networks with coupling delays. Physics Letters A 360, 263–273 (2006)
3. Liu, Y., Wang, Z., Liang, J., Liu, X.: Synchronization and state estimation for discrete-time complex networks with distributed delays. IEEE Transactions on Systems, Man, and Cybernetics, Part B: Cybernetics 38, 1314–1325 (2008)
4. Xiang, L., Liu, Z., Chen, Z., Chen, F., Yuan, Z.: Pinning control of complex dynamical networks with general topology. Physica A: Statistical Mechanics and its Applications 379, 298–306 (2007)

5. Ding, D., Wang, Z., Shen, B., Shu, H.: State estimation for discrete-time complex networks with randomly occurring sensor saturations and randomly varying sensor delays. IEEE Transactions on Neural Networks and Learning Systems 23, 725–736 (2012)
6. Liang, J., Wang, Z., Liu, X.: State estimation for coupled uncertain stochastic networks with missing measurements and time-varying delays: The discrete-time case. IEEE Transactions on Neural Networks 20, 781–793 (2009)
7. Liu, J., Cao, J., Wu, Z., Qi, Q.: State estimation for complex systems with randomly occurring nonlinearities and randomly missing measurements. International Journal of Systems Science, no. ahead-of-print, 1–11 (2014)
8. Bose, N.K., Buchberger, B., Guiver, J.: Multidimensional systems theory and applications. Springer (2003)
9. Liang, J., Wang, Z., Liu, Y., Liu, X.: State estimation for two-dimensional complex networks with randomly occurring nonlinearities and randomly varying sensor delays. International Journal of Robust and Nonlinear Control 24, 18–38 (2014)
10. Seto, D., Lehoczky, J.P., Sha, L., Shin, K.G.: On task schedulability in real-time control systems. In: 17th IEEE Real-Time Systems Symposium, pp. 13–21. IEEE (1996)
11. Sijs, J., Lazar, M.: On event based state estimation. In: Majumdar, R., Tabuada, P. (eds.) HSCC 2009. LNCS, vol. 5469, pp. 336–350. Springer, Heidelberg (2009)
12. Tabuada, P.: Event-triggered real-time scheduling of stabilizing control tasks. IEEE Transactions on Automatic Control 52, 1680–1685 (2007)
13. Wang, X., Lemmon, M.D.: Event design in event-triggered feedback control systems. In: 47th IEEE Conference on Decision and Control, CDC 2008, pp. 2105–2110. IEEE (2008)
14. Hu, S., Yue, D.: Event-triggered control design of linear networked systems with quantizations. ISA Transactions 51, 153–162 (2012)
15. Hu, S., Yue, D.: Event-based H_∞ filtering for networked system with communication delay. Signal Processing 92, 2029–2039 (2012)
16. Yue, D., Tian, E., Han, Q.L.: A delay system method for designing event-triggered controllers of networked control systems. IEEE Transactions on Automatic Control 58, 475–481 (2013)
17. Liu, J., Yue, D.: Event-triggering in networked systems with probabilistic sensor and actuator faults. Information Sciences 240, 145–160 (2013)
18. Mao, X.: Exponential stability of stochastic delay interval systems with markovian switching. IEEE Transactions on Automatic Control 47, 1604–1612 (2002)
19. Wang, Y., Xie, L., de Souza, C.: Robust control of a class of uncertain nonlinear systems. Systems & Control Letters 19, 139–149 (1992)
20. Han, Q., Yue, D.: Absolute stability of lur'e systems with time-varying delay. IET Control Theory & Applications 1, 854–859 (2007)
21. Peng, C., Tian, Y.: Delay-dependent robust stability criteria for uncertain systems with interval time-varying delay. Journal of Computational and Applied Mathematics 214, 480–494 (2008)
22. Yue, D., Tian, E., Zhang, Y.: A piecewise analysis method to stability analysis of linear continuous/discrete systems with time-varying delay. International Journal of Robust and Nonlinear Control 19, 1493–1518 (2009)
23. Yue, D., Tian, E., Zhang, Y., Peng, C.: Delay-distribution-dependent stability and stabilization of t–s fuzzy systems with probabilistic interval delay. IEEE Transactions on Systems, Man, and Cybernetics, Part B: Cybernetics 39, 503–516 (2009)

Integrated Iterative Learning Control Strategy for Batch Processes

Li Jia[1], Tian Yang[1], and Min-Sen Chiu[2]

[1] Shanghai Key Laboratory of Power Station Automation Technology,
Shanghai University, Shanghai 200072, China
[2] Department of Chemical and Biomolecular Engineering,
National University of Singapore, Singapore, 117576, Singapore
jiali@staff.shu.edu.cn

Abstract. An integrated iterative learning control strategy based on time-varying perturbation models for batch processes is proposed in this paper. A linear perturbation model is firstly obtained in order to control the perturbation variables rather than the actual process variables themselves. Next, an integrated control strategy which combines ILC with real-time feedback control is used to control the perturbation model. It leads to superior tracking performance and better robustness against disturbance and uncertainty. Lastly, the effectiveness of the proposed method is verified by examples.

Keywords: batch process, integrated learning control, iterative learning control, Time-varying perturbation model.

1 Introduction

Batch processes have been used increasingly in the production of low volume and high value added products, such as special polymers, special chemicals, pharmaceuticals, and heat treatment processes for metallic or ceramic products [1]. With strong nonlinearity and dynamic characteristics, optimal control of batch processes is more challenging than that of continuous processes and thus it needs new non-traditional techniques.

Batch processes have the characteristic of repetition, and thus iterative learning control (ILC) can be used in the optimization control of batch processes [2-4]. After its initial development for industrial robot [5], ILC et al has been increasingly practiced for batch processes with repetitive natures to realize perfect tracking and control optimization [6, 7]. However, ILC is actually an open-loop control from the view of a separate batch because the feedback-like control just plays role between different batches. In most reported results, only the batch-to-batch performance is taken for consideration but not the performance of real-time feedback. As a result, it is difficult to guarantee the performance of the batch process when uncertainties and disturbances exist. Therefore, an integrated optimization control system is required to derive the maximum benefit from batch processes, in which the performance of

M. Fei et al. (Eds.): LSMS/ICSEE 2014, Part II, CCIS 462, pp. 419–427, 2014.

time-axis and batch-axis are both analyzed synchronously. Rogers firstly employed two-dimension (2D) theory to solve above-mentioned problem [8]. Li et al presented an ILC strategy for 2D time-invariant linear repetitive systems with fixed time delays [9]. Chin et al proposed a two-stage iterative learning control technique by using the real-time feedback information to modify the ILC parameters for independent disturbance rejection [10]. Liu et al combined the internal model control (IMC) with the ILC to deal with uncertain time delay for linear batch processes [11]. Recently, Wang et al proposed an advanced ILC-based PI control for MIMO batch processes to hold robust stability based on a 2D system formulation [12]. Gao's research group did a series of 2D optimization control-based research for batch processes [13, 14]. However, most reported results assume that the batch process is linear in order to simplify the design procedure [15]. Furthermore, the analysis on the stability and robustness of batch process is more troublesome.

Motivated by previous works, an integrated iterative learning control strategy based on time-varying perturbation models for batch processes is proposed in this paper. A linear perturbation model is firstly obtained in order to control the perturbation variables rather than the actual process variables themselves. Then, an integrated control strategy which combines ILC with real-time feedback control was used to control the perturbation model.

2   Process Descriptions

With strong nonlinearity and dynamic characteristics, the optimal control of batch process is more complex than that of continuous processes. In batch process operation practice, it is quite useful to look at the changes of variables away from their nominal trajectories, rather than the actual process variables themselves [16]. In this section, a linear perturbation model will be firstly obtained by linearizing a nonlinear process with respect to the nominal trajectories.

Consider a 2D nonlinear batch process described by the following nonlinear function

$$y_k(t) = f_t(u_k(t)) + v_k(t) \tag{1}$$

$$\begin{aligned} u^{low} \le u_k(t) \le u^{up} \\ y^{low} \le y_k(t) \le y^{up} \end{aligned} \tag{2}$$

where t and k denote time and cycle indices, respectively. $u_k(t)$, $y_k(t)$ and $v_k(t)$ are, respectively, the control input, the output and the measurement noise at time t in k-th cycle, and $f_t(\cdot)$ represents the nonlinear function between $u_k(t)$ and $y_k(t)$. u^{low} and u^{up} are the lower and upper bounds of the input sequence, y^{low} and y^{up} are the lower and upper bounds of the final product qualities.

Eq.(1) can be rewritten in matrix form as

$$\mathbf{Y}_k = \mathbf{F}(\mathbf{U}_k) + \mathbf{V}_k \tag{3}$$

where $\mathbf{U}_k$ and $\mathbf{V}_k$ denote the input sequence and the noise sequence of $k-th$ batch, respectively. $\mathbf{Y}_k$ is the corresponding product quality sequence. $\mathbf{F}(\cdot)$ represents the nonlinear static function between $\mathbf{U}_k$ and $\mathbf{Y}_k$.

By linearizing the nonlinear batch process model described by Eq. (3) with respect to $\mathbf{U}_s$ around the nominal trajectories $(\mathbf{U}_s, \mathbf{Y}_s)$, the following expression can be obtained

$$\mathbf{Y}_k = \mathbf{Y}_s + \left.\frac{\partial \mathbf{F}(\mathbf{U}_k)}{\partial \mathbf{U}_k}\right|_{\mathbf{U}_s} (\mathbf{U}_k - \mathbf{U}_s) + \mathbf{W}_k + \mathbf{V}_k \tag{4}$$

where, $\mathbf{W}_k$ is the model errors due to the linearization.

Define $\mathbf{G}_s = \left.\dfrac{\partial \mathbf{F}(\mathbf{U}_k)}{\partial \mathbf{U}_k}\right|_{\mathbf{U}_s}$, then Eq.(4) can be rewritten as

$$\mathbf{Y}_k - \mathbf{Y}_s = \mathbf{G}_s (\mathbf{U}_k - \mathbf{U}_s) + \mathbf{d}_k \tag{5}$$

where $\mathbf{d}_k = \mathbf{W}_k + \mathbf{V}_k$, and is supposed to be bounded by a certain small positive constant, namely $\|\mathbf{d}_k\| < B_d$.

Thus, the linearized perturbation model can be obtained from Eq.(5) as

$$\bar{\mathbf{Y}}_k = \mathbf{G}_s \bar{\mathbf{U}}_k + \mathbf{d}_k \tag{6}$$

where, $\bar{\mathbf{U}}_k = \mathbf{U}_k - \mathbf{U}_s$ and $\bar{\mathbf{Y}}_k = \mathbf{Y}_k - \mathbf{Y}_s$ are the perturbation variables of the control and product quality variables, respectively.

3   Itergrated Iterative Learning Control Strategies Based on Perturbation Models

The formulation of the proposed integrated iterative learning control system based on perturbation model is depicted in Fig. 1. Batch length is defined as t_f. Here the batch length t_f is divided into L equal intervals. k and t are same as those defined in Section 2, $\bar{y}_d(t)$ is the deviated desired product quality defined as $\bar{y}_d(t) = y_d(t) - y_s(t)$, where $y_d(t)$ is desired product quality. $\bar{u}_k(t)$, $\bar{y}_k(t)$ and $d_k(t)$ are input variable, output variable and disturbance of the perturbation model at time t in k-th cycle, respectively. $\bar{u}_k^{ILC}(t)$ and $\bar{u}_k^{time}(t)$ are ILC control action variable and PIDC control action variable, and $\bar{u}_k(t) = \bar{u}_k^{ILC}(t) + \bar{u}_k^{time}(t)$. $\bar{e}_k(t)$ represents the error between $\bar{y}_k(t)$ and $\bar{y}_d(t)$. G_s denotes the batch process. Based on the information from previous batch, the optimization controller can find an updating mechanism for the input sequence $\bar{\mathbf{U}}_k^{ILC}$ of the new batch using improved iterative optimal control law derived by the rigorous mathematic proof method discussed shortly. At next batch, this procedure is repeated to let the product qualities asymptotically converge towards $\bar{y}_d(t_f)$ at the batch end.

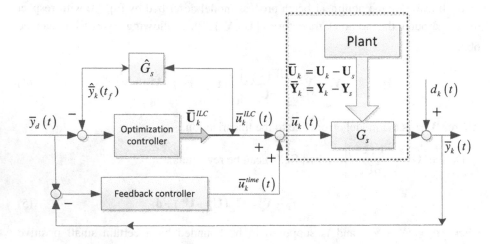

Fig. 1. Integrated Iterative Learning Control System Based on Perturbation Model

As discussed above, the proposed integrated learning optimization control action can be described as

$$\bar{\mathbf{U}}_k = \bar{\mathbf{U}}_k^{ILC} + \bar{\mathbf{U}}_k^{time} \qquad (7)$$

where $\bar{\mathbf{U}}_k^{time}$ is the control sequence from time-axis control action.

The error between the desired product qualities and the measured output is defined by

$$e_k(t) = y_d(t) - y_k(t) \qquad (8)$$

Thus

$$\begin{aligned}
\bar{e}_k(t) &= \bar{y}_d(t) - \bar{y}_k(t) \\
&= (y_d(t) - y_s(t)) - (y_k(t) - y_s(t)) \\
&= y_d(t) - y_k(t) \\
&= e_k(t)
\end{aligned} \qquad (9)$$

Owing to the model-plant mismatch, the process output may not be same as the one predicted by the model. The offset between the measured output and the model prediction is termed as model prediction error defined by

$$\hat{e}_k(t) = y_k(t) - \hat{y}_k(t) \qquad (10)$$

where $\hat{y}_k(t) = y_s(t) + \hat{\bar{y}}_k(t)$.

Then, according to Eq.(10) and $\hat{y}_k(t) = y_s(t) + \hat{\bar{y}}_k(t)$, we can get

$$
\begin{aligned}
\hat{e}_k(t) &= y_k(t) - \hat{y}_k(t) \\
&= \left(y_k(t) - y_s(t) \right) - \left(\hat{y}_k(t) - y_s(t) \right) \\
&= \bar{y}_k(t) - \hat{\bar{y}}_k(t)
\end{aligned}
\tag{11}
$$

The objective is to design a learning algorithm to implement the control trajectory $\bar{\mathbf{U}}_k$ such that the product quality asymptotically converges to the desired product qualities along the batch axes. Thus, instead of using the whole reference sequences for product qualities, which is less important for batch processes product quality control, only the desired product quality $y_d\left(t_f\right)$ at the end of a batch is used.

The batch-to-batch dynamic parameters based quadratic criterion-iterative optimization problem can be formulated as:

$$
\min J(\bar{\mathbf{U}}_{k+1}^{ILC}, k+1) = \left\| \bar{y}_d\left(t_f\right) - \hat{\bar{y}}\left(\bar{\mathbf{U}}_{k+1}^{ILC}, t_f\right) \right\|_{\mathbf{Q}}^2 + \left\| \Delta\bar{\mathbf{U}}_{k+1}^{ILC} \right\|_{\mathbf{R}_{k+1}}^2
\tag{12}
$$

where $\Delta\bar{\mathbf{U}}_{k+1}^{ILC} = \bar{\mathbf{U}}_{k+1}^{ILC} - \bar{\mathbf{U}}_k$, $\mathbf{Q}$ and $\mathbf{R}_{k+1}$ are both weighting matrices defined as $\mathbf{Q} = q \times \mathbf{I}_L$ and $\mathbf{R}_{k+1} = r_{k+1} \times \mathbf{I}_L$, where r_{k+1} and q are both positive real numbers.

The solution to the constrained optimization problem in Eq. (12) can be easily solved using classical mathematic method or intelligent algorithm such as sequential quadratic programming (SQP) algorithm [17], particle swarm optimization (PSO) [18], and genetic algorithm (GA) [19]. The condition that the optimization algorithm can accurately find out the best solution is assumed in this work in order to develop the first of its kind that guarantees the convergence of control policy with the proposed batch-wise varying quadratic-criterion derived from a rigorous proof. In this paper we use SQP to solve this optimization problem. That is to say, we assume that the best control solution can be guaranteed by using SQP optimization algorithm.

In this paper, the nonlinear batch process model in Eq. (6) is identified by NFM [20]. Without loss of generality, we assume that the prediction errors of the model converge to a very small region along batch axes.

As a sum-up, the procedure to design the integrated iterative learning control system can be described by five steps as follows:

Step 1 Modeling the prediction model based on historical batch operation date points.

Step 2 Obtaining a linear perturbation model by linearizing a nonlinear object with respect to the nominal trajectories according to Eq.(4). Let k= 1 and initialize $\bar{\mathbf{U}}_1$, $\mathbf{Q}$ and $\mathbf{R}_k$.

Step 3 Update control input sequence $\bar{\mathbf{U}}_k^{ILC}$ in batch-axis according to Eq.(12). Let t= 1.

Step 4 Compute $\bar{u}_k^{time}(t)$ and $\bar{u}_k(t)$ at the t-th time of the k-th batch, and then $u_k(t)$ and the corresponding output $y_k(t)$ can be obtained. Update the parameters of prediction model according to Eq.(13).

Step 5 If $t < T$, set $t = t + 1$ and go back to step 4, else set $k = k + 1$ and go back to Step 3.

4   Example

The proposed method is implemented to a classical batch process proposed by Kwon and Ecans [21] in this section. The mechanism model of the batch process describing polymerization process is as follow:

$$\frac{dx_1}{dt} = f_1 = \frac{\rho_0^2 \rho}{M_m}(1-x_1)^2 \exp(2x_1 + 2\chi x_1^2) A_m \exp\left(-\frac{E_m}{uT_{ref}}\right)$$

$$\frac{dx_2}{dt} = f_2 = \frac{f_1 x_2}{1+x_1}\left[1 - \frac{1400x_2}{A_w \exp(B/uT_{ref})}\right] \tag{34}$$

$$\frac{dx_3}{dt} = f_3 = \frac{f_1}{1+x_1}\left[\frac{A_w \exp(B/uT_{ref})}{1500} - x_3\right]$$

where $\rho = \frac{1-x_1}{r_1+r_2 T_c} + \frac{x_1}{r_3+r_4 T_c}$, $\rho_0 = r_1 + r_2 T_c$ and $T_c = uT_{ref} - 273.15$, x_1 is the conversion, $x_2 = x_n/x_{nf}$ and $x_3 = x_w/x_{wf}$ are dimensionless number average and weight average chain lengths (NACL and WACL), respectively, $u = T/T_{ref}$ is the control variable which is bounded in the interval $[0.93486, 1.18539]$; T is the absolute temperature of the reactor; T_c is the temperature in degrees Celsius; A_w and B are coefficients in the relation between the WACL and temperature obtained from experiments; A_m and E_m are the frequency factor and activation energy, respectively, of the overall monomer reaction; the constants $r_1 - r_4$ are density-temperature corrections, and M_w and χ are the monomer molecular weight and polymer-monomer interaction parameter, respectively. Table 1 lists the reference value used to obtain the dimensionless variables, as well as the values of the reactor parameters. The final time, t_f was fixed to be 313 min, the initial value of the states used were $x_1(0) = 0$, $x_2(0) = 1$, $x_3(0) = 1$, and the desire value is $y_d = (0.8 \quad 1 \quad 1)$.

Table 1. Parameter values for the batch polymerization process

Parameters	Values
A_m	4.266×10^5 m^3 / (kmol s)
A_w	0.033 454
B	4364 K
E_m	10 103.5 K
M_m	104 kg/kmol
r_1	0.9328×10^3 kg/m^3
r_3	1.0902×10^3 kg/m^3
r_2	$-0.879\,02$ kg/(m^3 °C)
r_4	-0.59 kg/(m^3 °C)
T_{ref}	399.15 K
t_f	313 min
x_{nf}	700
x_{wf}	1500
χ	0.33

In the simulation, we choose NFM model as the prediction model, and three different NFM models are firstly constructed to represent the mapping relationship of $u \rightarrow x_1, u \rightarrow x_2, u \rightarrow x_3$, respectively. 50 independent random signals are used to simulate 30 batches data for training purpose and another 20 batches data for testing purpose. In addition, the robustness of the proposed method is evaluated by introducing Gaussian white noise with 0.5% standard deviation to the output variable of the process. After models training, test the average errors of prediction models are $5.4657 \times 10^{-3}, 1.1393 \times 10^{-2}$ and 8.8480×10^{-3} [22].

The control object is to drive system output $\mathbf{y} = \begin{pmatrix} x_1 & x_2 & x_3 \end{pmatrix}$ approximating to $\mathbf{y}_d^f = \begin{pmatrix} 0.8 & 1 & 1 \end{pmatrix}$, and the object function of this example is as follow

$$J(\mathbf{U}_{k+1}, k+1) = \left\| \tilde{\mathbf{e}}_{k+1}^f \right\|_Q^2 + \left\| \Delta \mathbf{U}_{k+1} \right\|_R^2 = \left\| \tilde{e}_{k+1,1}^f \right\|_{q_1}^2 + \left\| \tilde{e}_{k+1,2}^f \right\|_{q_2}^2 + \left\| \tilde{e}_{k+1,3}^f \right\|_{q_3}^2 + \left\| \Delta \mathbf{U}_{k+1} \right\|_R^2 \quad (35)$$

where, $\mathbf{Q} = \begin{pmatrix} q_1 & 0 & 0 \\ 0 & q_2 & 0 \\ 0 & 0 & q_3 \end{pmatrix}$, $\mathbf{R} = r \cdot \mathbf{I}_T$ is dynamic parameter :

$$r = \tau_1 \left(1 - \exp\left(-\tau_2 \sum_{k=k_0}^{K} \sum_{i=1}^{3} \left| \bar{\hat{e}}_{k,i}^{f} \right| \Big/ (K - k_0) \right) \right) \tag{36}$$

Other parameters in this simulation are : $\tau_1 = 1 \times 10^5$, $\tau_2 = 0.1$, $q_1 = 0.1$, $q_2 = 1 \times 10^4$, $q_3 = 1 \times 10^4$, $\alpha = 0.2$, $\beta = 0.01$. Time-axis choose a simple P controller, and $k_p = [0.01, 0.1, 0.1]$. The batch reactor is linearized as $G_1(s) = 6.48/(s + 8.274)$, $G_2(s) = 1.827/(s + 3.023)$ and $G_3(s) = 1.394/(s + 1.949)$.

The proposed integrated ILC is compared with traditional ILC [22] and Zhang's method [16] as given in table 2. Clearly, the proposed integrated learning algorithm has better performance.

Table 2. 15-th batch final output error value based on three control systems under disturbance

methods	output x_1	output x_2	output x_3
Integrated control	3.8×10^{-3}	1.8×10^{-3}	1.67×10^{-3}
ILC[22]	8.87×10^{-2}	1.6×10^{-4}	4.7×10^{-3}
Zhang[16]	4.5×10^{-3}	1.12×10^{-2}	1.15×10^{-2}

5   Conclusion

An optimal control strategy for batch processes nonlinear 2D system was presented in this paper. First of all, obtaining a linear perturbation model by linearizing a nonlinear object with respect to the nominal trajectories, and thus the perturbation variables were used as controlled variables rather than the actual process variables themselves. Then, an integrated control strategy which combines ILC with real-time feedback control was used to control the perturbation model. Thus, the complex nonlinear process was turned into a linearization process for easily researching, controlling, and then obtaining better control effects.

Acknowledgments. Supported by National Natural Science Foundation of China (61374044), Shanghai Science Technology Commission (12510709400), Shanghai Municipal Education Commission (14ZZ088), Shanghai talent development plan.

References

1. Bonvin, D.: Optimal operation of batch reactors: A personal view. Journal of Process Control 8, 355–368 (1998)
2. Xiong, Z.H., Zhang, J., Wang, X., Xu, Y.M.: Integrated tracking control strategy for batch processes using a batch-wise linear time-varying perturbation model. Control Theory & Applications 1, 178–188 (2007)

3. Lu, N., Gao, F.: Stage-based process analysis and quality prediction for batch processes. Industrial and Engineering Chemistry Research 44, 3547–3555 (2005)
4. Lee, J.H., Lee, K.S.: Iterative learning control applied to batch processes: An overview. Control Engineering Practice 15, 1306–1318 (2007)
5. Arimoto, S., Kawamura, S., Miyazaki, F.: Bettering operation of robots by learning. J. Robotic Systems 1, 123–140 (1984)
6. Lee, K.S., Chin, I.S., Lee, H.J., Lee, J.H.: Model predictive control technique combined with iterative learning control for batch processes. AIChE J. 45, 2175–2187 (1999)
7. Chien, C.J., Liu, J.S.: P2type iterative learning controller for robust output tracking of nonlinear time varying systems. Int. J. Control 64, 319–334 (1996)
8. Rogers, E., Owens, D.H.: Stability analysis for linear repetitive processes. LNCIS, vol. 175. Springer, Heidelberg (1992)
9. Li, X.D., Chow, T.W.S., Ho, J.K.L.: 2-D system theory based iterative learning control for linear continuous systems with time delays. IEEE Transactions on Circuits and Systems–I: Regular Papers 52, 1421–1430 (2005)
10. Chin, I., Qin, S.J., Lee, K.S., Cho, M.: A two-stage iterative learning control technique combined with real-time feedback for independent disturbance rejection. Automatica 40, 1913–1922 (2004)
11. Liu, T., Gao, F., Wang, Y.: IMC-based iterative learning control for batch processes with uncertain time delay. Journal of Process Control 20, 173–180 (2010)
12. Wang, Y., Liu, T., Zhao, Z.: Advanced PI control with simple learning set-point design: Application on batch processes and robust stability analysis. Chemical Engineering Science 71, 153–165 (2012)
13. Shi, J., Gao, F., Wu, T.: From Two-dimensional Linear Quadratic Optimal Control to Iterative Learning Control Paper 1. Two-dimensional Linear Quadratic Optimal Controls and System Analysis. Industrial & Engineering Chemistry Research 45, 4603–4616 (2006)
14. Shi, J., Gao, F., Wu, T.: From Two-dimensional Linear Quadratic Optimal Control to Iterative Learning Control Paper 2. Iterative Learning Controls for Batch Processes. Industrial & Engineering Chemistry Research 45, 4617–4628 (2006)
15. Jia, L., Yang, T., Chiu, M.: An Integrated Iterative Learning Control Strategy with Model Modification and Dynamic R-Parameter for Batch Processes. Journal of Process Control 23, 1332–1341 (2013)
16. Xiong, Z.H., Zhang, J.: Product Quality Trajectory Tracking in Batch Processes Using Iterative Learning Control Based on Time-Varying Perturbation Models. American Chemical Society 42, 6802–6814 (2003)
17. Teng, Z.B., Zhang, S.Y.: A successive quadratic programming algorithm with global and superlinear convergence properties. Journal of UEST of China 30, 103–106 (2001)
18. Jia, L., Cheng, D.S.: Particle swarm optimization algorithm based iterative learning algorithm for batch processes. Control Engineering of China 18, 341–344 (2011)
19. Mou, J.H., Su, S.M.: A hybrid genetic algorithm for constrained optimization. Computer Simulation 26, 184–187 (2009)
20. Jia, L., Shi, J.P., Chiu, M.: Neuro-Fuzzy based dynamic Quadratic Criterion-Iterative Learning Control for Batch Process Control Technique. Transactions of the Institute of Measurement and Control 35, 92–101 (2013)
21. Kwon, Y.D., Evans, L.B.: A Coordinate Transformation Method for the Numerical Solution of Nonlinear Minimum-Time Control Problems. AIChE J. 21, 1158–1164 (1975)
22. Jia, L., Shi, J.P., Chiu, M.: Integrated Neuro-Fuzzy Model and Dynamic R-parameter based Quadratic Criterion-Iterative Learning Control for Batch Process Control Technique. Neurocomputing 98, 24–33 (2012)

Almost Sure Consensus of Multi-agent Systems over Deterministic Switching Agent Dynamics and Stochastic Topology Jumps

Yang Song[1,2], Zhenhong Yang[1], and Taicheng Yang[3]

[1] Dept. of Automation, Shanghai University, 200072 Shanghai, China
[2] Shanghai Key Laboratory of Power Station Automation Technology,
200072 Shanghai, China
{y_song,yangzhh}@shu.edu.cn
[3] Dept. of Engineering and design, University of Sussex, Brighton,
BN1 9QT, UK
T.C.Yang@sussex.ac.uk

Abstract. In this paper, consensus of multi-agent systems under both determinstic switching agent dynamics and stochastic topology jumps is investigated. The analysis relies on the fact that these two switching signals are independent. First, a necessary and sufficient condition for achieving a consensus under time-invariant agent dynamics and fixed communication topology is presented. Based on this result, the almost sure consensus condition under deterministic switching agent dynamics and stochastic topology jumps is investigated. It is shown that almost sure consensus can be reached if determinstic switching signal satisfies minimum average dwell time constraint.

Keywords: Multi-agent systems, Consensus, Markov switching, Dwell time.

1 Introduction

Recent years have witnessed an increasing attention on cooperative control of MASs partly due to its broad applications in many areas including formation control [1], distributed sensor network [2], flocking [3], attitude alignment [4] and etc. Among them, consensus is a fundamental problem. Consensus refers to all the agent reach an agreement using the local information, which is determined by the underlying network topology.

Consensus problem has a long-standing tradition in automatic theory. Reference [5] firstly applies the thoughts of consensus from statistics to sensor network information fusion, uncovering the research of consensus problem in control theory and application. In the pioneering work [6], a general framework is introduced to analyze the consensus of multi-agent system by using graph theory. Many existing literatures study the consensus problem of MASs whose agent dynamics are single or double integrators. See, for example, reference [7] and [8]. In practice, many systems cannot be described by simple single- or double-integrator dynamics, so recently more and

M. Fei et al. (Eds.): LSMS/ICSEE 2014, Part II, CCIS 462, pp. 428–438, 2014.

more efforts are devoted to consensus of MASs with general linear dynamics. For instance, reference [9] proposed a distributed observer-type consensus controller based on relative output measurements to address the consensus problem of multi-agent systems with general linear node dynamics and the synchronization of complex networks in a unified framework. In [10], several fully distributed adaptive protocols requiring neither global information about the communication graph nor the upper bound of the leader's control is designed to solve consensus problem of multi-agent systems with general linear dynamics. However, most existing literatures only assume that the agent dynamics are time-invariant. In fact, the dynamic of each agent can be changed due to, i.e., mode changes. To the best of our knowledge, only literature [18] investigates the output feedback control of MASs whose agent dynamic is subject to ergodic Markov chains.

Another interesting topic of MASs is consensus under time-varying topologies. The time-varying nature of a communication network is originated from the fact that two agents may lose or generate a link due to, for example, obstacles, transmission failures, and limited communication range. Motivated by this fact, the consensus problem of MASs under time-varying network topology has been actively addressed in the literature. For example, the consensus problem of first- or second-order MASs has been considered in [7] and [11] respectively. For high-order MASs, reference [12]-[14] have addressed the problem under fast-switching, frequently connected, and jointly connected networks, respectively. Reference [15] and [16] consider the consensus problems for multi-agent systems with switching topologies governed by a Markov process, showing that the effect of switching topologies on consensus is determined by the union of topologies associated with the positive recurrent states of the Markov process. In [17], both deterministic and Markov switching topologies are considered for consensus problem of high-order multi-agent systems.

This paper considers consensus of general linear MASs under both time-varying agent dynamics and switching network topologies. We assume that there exists a deterministic sync signal to change the dynamics of all the agents simultaneously while the topology jumping is subject to a Markov process. A distributed observer-type protocol is constructed to solve the problem. Firstly, a necessary and sufficient condition for achieving a consensus under time-invariant agent dynamics and fixed network topology is presented. Based on this result, the almost sure consensus condition under deterministic switching agent dynamics and stochastic topology jumps is investigated. It is shown that almost sure consensus can be reached if agent switching signal satisfies minimum average dwell time constraint.

The rest part of this paper is organized as follows. Section 2 gives some basic concepts and results of graph theory. Section 3 gives problem formulation and a necessary and sufficient for achieving consensus under time-invariant agent dynamics and fixed communication topology. Section 4 presents a sufficient almost sure consensus condition to reach a consensus over deterministic switching agent dynamics and Markovian jumping topologies. Finally, Section 5 concludes the whole paper.

Notations: Let $\mathbb{R}^{n \times n}$ be the set of $n \times n$ real matrices. The superscript T means the transpose for real matrices. $I_n \in \mathbb{R}^{n \times n}$ represents the identity matrix. Denoted by **1** a column vector with all entries equal to one and its dimension is appropriate. For a vector x, let $\|x\|$ denotes its 2-norm. A matrix is Hurwitz if all of its eigenvalues have negative real parts. The matrix inequality $A > (\geq)B$ means that $A - B$ is positive (semi-)definite. The Kronecker product, denoted by $\otimes$, facilitates the manipulation of matrices by the following properties: (1) $(A \otimes B)(C \otimes D) = AC \otimes BD$; (2) $(A \otimes B)^T = A^T \otimes B^T$.

2   Graph Theory

In this section, some basic concepts and results of graph theory are introduced. For more details, readers can please refer to ([12–14]).

We use a digraph $\mathcal{G} = (\mathcal{V}, \mathcal{E})$ to describe the information exchange between agents, where $\mathcal{V} = \{1, ..., N\}$ is a finite nonempty set of nodes (i.e. agents) and $\varepsilon \subset \mathcal{V} \times \mathcal{V}$ is the set of edges. An edge (i, j) in ε means that agent j can obtain information form agent i. A sequence of edges (i_1, i_2), (i_2, i_3), ..., (i_{k-1}, i_k) with $(i_{j-1}, i_j) \in \mathcal{E}$, $\forall j \in \{2, ..., k\}$ is called a directed path form agent i_1 to agent i_k. A digraph is said to be connected (or contain a directed spanning tree) if there is a node $i \in \mathcal{V}$ such that there exists a direct path form this node to the rest. Digraph $\mathcal{G}$ can also be described by a weighted adjacency matrix $\mathcal{A} = [a_{ij}] \in \mathbb{R}^{N \times N}$ such that $a_{ij} > 0$ if $(j, i) \in \varepsilon$ and $a_{ij} = 0$, otherwise. $\mathcal{G}$ is undirected if $\mathcal{A}^T = \mathcal{A}$.

The Laplacian matrix $\mathcal{L} = [\mathcal{L}_{ij}] \in \mathbb{R}^{N \times N}$ of the graph $\mathcal{G}$ is defined as $\mathcal{L}_{ii} = \sum_{j \neq i} a_{ij}$, $\mathcal{L}_{ii} = -a_{ii}$. Clearly, $\mathcal{L}$ is symmetric if the graph is undirected.

Lemma 1. *([9])* All the eigenvalues of $\mathcal{L}$ have nonnegative real parts. Zero is an eigenvalue of $\mathcal{L}$, with **1** as the right eigenvector.

Lemma 2. *([17])* The algebraic multiplicity of the zero eigenvalues is 1 if and only if the given graph $\mathcal{G}$ is connected.

3   Problem Formulation

Consider the following MAS with time-invariant agent dynamics and fixed topology

$$\dot{x}_i = A x_i + B u_i$$

$$y_i = Cx_i, \ 1 \le i \le N, \tag{1}$$

where $x_i, u_i \in \mathbb{R}^n$ and $y_i \in \mathbb{R}^p$ are the state, input and output of the agent i, respectively. All $A^{[j]}$ are not Hurwitz (if $A^{[j]}$ is Hurwitz, the consensus can be reached by just setting $u_i(k) = 0$). Moreover, it is assumed that (A, B, C) is stabilizable and detectable.

The communication topology among agents is modeled as a directed graph $\mathcal{G} = (\mathcal{V}, \mathcal{E}, \mathcal{A})$. For the MAS (1), only relative output measurements

$$\zeta_i = \sum_{j=1}^{N} a_{ij}(y_i - y_j), \ 1 \le i \le N, \tag{2}$$

are assumed to be available to the agent i, where a_{ij} are the (i, j) elements of adjacency matrix $\mathcal{A}$. An observer-type consensus protocol is proposed as

$$\dot{v}_i = (A + BK) + F\left(\sum_{j=1}^{N} a_{ij}C(v_i - v_j) - \zeta_i\right), \tag{3}$$

$$u_i = Kv_i$$

where $v_i \in \mathbb{R}^n$ is the protocol state, $i = 1, \ldots, N$, $F \in \mathbb{R}^{q \times n}$ and $K \in \mathbb{R}^{q \times n}$ are constant matrices to be determined.

Define $\zeta_i := [x_i^T; v_i^T]^T$, $\zeta := [\zeta_1^T; \ldots; \zeta_N^T]^T$. Then, from (1)-(3), the collective dynamic ζ is written as

$$\dot{\zeta} = [I_N \otimes \mathcal{J} + \mathcal{L} \otimes \mathcal{H}]\zeta, \tag{4}$$

where $\mathcal{J} = \begin{bmatrix} A & BK \\ 0 & A + BK \end{bmatrix}$, $\mathcal{H} = \begin{bmatrix} 0 & 0 \\ -FC & FC \end{bmatrix}$, $\mathcal{L}$ is the Laplacian matrix of $\mathcal{G}$.

Let us partition $\mathcal{L}$ into the following from:

$$\mathcal{L} = \begin{bmatrix} l^{11} & L^{12} \\ L^{21} & L^{22} \end{bmatrix}, \tag{5}$$

where $l^{11} \in \mathbb{R}$ and $L^{22} \in \mathbb{R}^{(N-1) \times (N-1)}$. Introduce the following error variable:

$$\delta := \zeta_R - 1_{N-1} \otimes \zeta_1, \tag{6}$$

where $\zeta_R = [\zeta_2^T; \ldots; \zeta_N^T]^T \in \mathbb{R}^{(N-1) \times (N-1)}$. Then, the so-called disagreement dynamics can be written as:

$$\dot{\delta} = [I_N \otimes \mathcal{J} + (L^{22} - 1_{N-1}L^{12}) \otimes \mathcal{H}]\delta := \Pi\delta. \tag{7}$$

Readers can refer to [17] for detail derivation of (7). It is obvious that the MAS (1) can reach a consensus if Π is Hurwitz. The following lemma gives the condition.

Lemma 3. Given an arbitrary digraph $\mathcal{G}$, the matrix Π in (7) is Hurwitz if and only if $\mathcal{G}$ is connected and all matrices $A + BK$, $A + \lambda_i FC$, $i = 2,...,N$, are Hurwitz, where λ_i are the nonzero eigenvalues of the Laplacian matrix $\mathcal{L}$ associated with $\mathcal{G}$.

Proof: By *Theorem 3.2* of [17], Π is Hurwitz if and only if $\mathcal{J} + \lambda_i \mathcal{H}$ is Hurwitz for all $i = 2,...,N$. It is easy to check that $\mathcal{J} + \lambda_i \mathcal{H}$ is similar to

$$\begin{bmatrix} A + \lambda_i FC & 0 \\ -\lambda_i FC & A + BK \end{bmatrix}, \ i = 2,...,N.$$

Then, Π is Hurwitz if matrices $A + BK$, $A + \lambda_i FC$, $i = 2,...,N$ are stable. $\square$

As (A, B) is stabilizable, it can always find a K satisfing $A + BK$ is Hurwitz. Matrix F can be obtained by the following instruction.

If there exists a $P > 0$ such that

$$P(A + \lambda_i FC) + (A + \lambda_i FC)^T P < 0, \tag{8}$$

then $A + \lambda_i FC$ is Hurwitz. Let $F = -P^{-1}C^T$, then (8) becomes

$$PA + A^T P - 2\operatorname{Re}(\lambda_i)C^T C < 0.$$

P can be acquired by solving the above linear matrix inequality (LMI).

Next, we assume there exists a sync signal which can transform the dynamics of all the agents simultaneously. Then the dynamics of agent i takes the following form:

$$\dot{x}_i = A^{[\gamma(t)]}x_i + B^{[\gamma(t)]}u_i,$$

$$y_i = C^{[\gamma(t)]}x_i, \ 1 \leq i \leq N, \tag{9}$$

where $\gamma(t) \in \mathcal{M} = \{1, 2,...,m\}$ is an exogenous piecewise constant function. m can be regarded as different working modes of the agents. Let $D(0, t)$ denote the number of deterministic switches that happened in the interval $[0, t)$. τ_γ is the average dwell time of deterministic switching during $[0, t)$, which satisfies that $D(0, k) \leq D_0 + t/\tau_\gamma$, where $D_0 \geq 0$ is a chatter bound.

The communication topology is represented by a directed switching graph $\mathcal{G}_{\sigma(t)} = (\mathcal{V}, \mathcal{E}_{\sigma(t)}, \mathcal{A}_{\sigma(t)})$, where $\sigma(t)$ takes values in a finite set $\mathcal{P} = \{1, 2, \ldots, p\}$ and is subjects to a irreducible Markov process with stationary transition probabilities

$$\Pr\{\sigma(t + h) = j \mid \sigma(t) = i\} = \lambda_{ij} h + o(h), \ i \neq j,$$

where $h > 0$ and $\lambda_{ij} > 0$ is the transition rate form mode i at time t to mode j at time $t + h$. Let $\lambda_{ii} = -\sum_{j=1, j \neq i}^{p} \lambda_{ij}$ and define $\Lambda = [\lambda_{ij}]_{p \times p}$. The matrix Λ is called the transition rate matrix of the Markov process $\sigma(t)$. From the irreducibility of $\sigma(t)$, it exponentially converges to a unique invariant distribution $\pi = [\pi_1 \cdots \pi_p]^T$ which can be obtained by solving the equation $0 = \Lambda^T \pi$.

Under these settings, the information available to the agent i is modified as:

$$\zeta_i = \sum_{j=1}^{N} a_{\sigma(t)}^{ij} (y_i - y_j), \tag{10}$$

and the consensus protocol is proposed as

$$\dot{v}_i = (A^{[\gamma(t)]} + B^{[\gamma(t)]} K^{[\gamma(t)]}) + F^{[\gamma(t)]} \left(\sum_{j=1}^{N} a_{ij} C^{[\gamma(t)]} (v_i - v_j) - \zeta_i \right).$$

$$u_i = K^{[\gamma(t)]} v_i \tag{11}$$

Then, the disagreement dynamics is obtained as

$$\delta = [I_N \otimes \mathcal{J}^{[\gamma(t)]} + (L_{\sigma(t)}^{22} - 1_{N-1} L_{\sigma(t)}^{12}) \otimes \mathcal{H}^{[\gamma(t)]}] \delta := \Pi_{\sigma(t)}^{[\gamma(t)]} \delta \tag{12}$$

where $\mathcal{J}^{[\gamma(t)]}$ and $\mathcal{H}^{[\gamma(t)]}$ can be obtained analogously from (4).

Definition 1. The MAS (9) is said to reach an almost sure consensus under a deterministic agent switching $\gamma(t)$ and a Markov topology jumping $\sigma(t)$ if $\Pr\left\{ \lim_{t \to \infty} \delta = 0 \right\} = 1$ holds for any initial conditions.

Assumption 1. The digraph set $\{\mathcal{G}_1, ..., \mathcal{G}_p\}$ contains r connected graphs, where r is an integer satisfying $1 \le r \le p$. Without loss of generality, let $\{\mathcal{G}_1, ..., \mathcal{G}_r\}$ and rest be the sets of connected and disconnected digraphs, respectively.

We now proposed an algorithm for the MAS (9) with protocol (11).

Algorithm 1. For an arbitrary $\bar{\gamma} \in \mathcal{M}$

1. Choose matrix $K^{[\bar{\gamma}]}$ such that $A^{[\bar{\gamma}]} + B^{[\bar{\gamma}]}K^{[\bar{\gamma}]}$ is Hurwitz;
2. Solve the following LMI to get a $P^{[\bar{\gamma}]} > 0$:

$$P^{[\bar{\gamma}]}A^{[\bar{\gamma}]} + (P^{[\bar{\gamma}]}A^{[\bar{\gamma}]})^T - 2\lambda^+(C^{[\bar{\gamma}]})^T C^{[\bar{\gamma}]} < 0,$$

where $\lambda^+ := \min_{k \in \mathcal{P}, \lambda_i(\mathcal{L}_k) \ne 0} \operatorname{Re} \lambda_i(\mathcal{L}_k) > 0$.

Let us assume that $\sigma(t) = i$ and $\gamma(t) = j$ for some $t > 0$. Then by *Lemma 3*, if i is less than or equal to r, $\Pi_i^{[j]}$ is Hurwitz, otherwise $\Pi_i^{[j]}$ is not. Based on this fact, one can find two constants $a_i^{[j]} > 0$ and $\delta_i^{[j]} > 0$ satisfying that

$$\left\| e^{\Pi_i^{[j]}t} \right\| \le \begin{cases} e^{a_i^{[j]} - \delta_i^{[j]}t}, & 1 \le i \le r \\ e^{a_i^{[j]} + \delta_i^{[j]}t}, & r < i \le p \end{cases} \tag{13}$$

where $1 \le j \le m$. We can also define $\bar{a}_i := \max_{1 \le j \le m} a_i^{[j]}$, $\bar{a} := \max_{1 \le i \le p} \bar{a}_i$.

4   Main Result

Theorem 1. Suppose Assumption 1 hold. Then, the MAS (9) with protocol (11) reach an almost sure consensus under deterministic agent switching $\gamma(t)$ and Markovian network switching $\sigma(t)$ if there exists average dwell time τ_γ such that

$$\eta^{[j]} > \eta, \forall j \in \mathcal{M}, \tau_\gamma \ge \frac{\bar{a}}{\min_j \eta^{[j]} - \eta}$$

where $\eta^{[j]} = \sum_{i=1}^{p} \bar{a}_i \pi_i \lambda_{ii} - \sum_{i=1}^{r} \delta_i^{[j]} \pi_i + \sum_{i=r+1}^{p} \delta_i^{[j]} \pi_i$, η is a positive constant.

Proof: If $\gamma(t) = j \in M$ is invariant in time interval $[\tau, t)$, we can define $\Phi^{[j]}(t, \tau) := \int_\tau^t e^{F_{\sigma(t)}^{[j]} t} dt$. Let $t_1, t_2, \ldots$ be the time instant at which $\gamma(t)$ switches. Denote $\Psi(t, 0)$ as the transition matrix of (12). Then, for any t satisfying $0 = t_0 < t_1 < \cdots < t_k \leq t < t_{k+1}$, $\Psi(t, 0)$ satisfies

$$\Psi(t, 0) = \Phi^{[\gamma(t_k)]}(t, t_k) \prod_{s=0}^{k-1} \Phi^{[\gamma(t_s)]}(t_{s+1}, t_s),$$

which can be further obtained as:

$$\ln \left\| \Psi(t, 0) \right\| \leq \ln \left\| \Phi^{[\gamma(t_k)]}(t, t_k) \right\| + \sum_{s=0}^{k-1} \ln \left\| \Phi^{[\gamma(t_s)]}(t_{s+1}, t_s) \right\| \tag{14}$$

Rearrange the terms in (14) by using (13),

$$\ln \left\| \Psi \right\| \leq \sum_{i=1}^{p} a_i^{[\gamma(t_k)]} N_i(t, t_k) + \sum_{i=1}^{p} \sum_{s=0}^{k-1} a_i^{[\gamma(t_s)]} N_i(t_{s+1}, t_s)$$

$$- \sum_{i=1}^{r} \delta_i^{[\gamma(t_k)]} T_i(t, t_k) - \sum_{i=1}^{r} \sum_{s=0}^{k-1} \delta_i^{[\gamma(t_s)]} T_i(t_{s+1}, t_s) \tag{15}$$

$$+ \sum_{i=r+1}^{p} \delta_i^{[\gamma(t_k)]} T_i(t, t_k) + \sum_{i=r+1}^{p} \sum_{s=0}^{k-1} \delta_i^{[\gamma(t_s)]} T_i(t_{s+1}, t_s)$$

where $N_i(t, \tau)$ and $T_i(t, \tau)$ represent the discontinuous and cumulative dwell time of graph $\mathcal{G}_i$ in the interval $[\tau, t)$, respectively. By rearranging the terms in the sum of (15), we obtain

$$\ln \left\| \Psi \right\| \leq \sum_{i=1}^{p} a_i^{[\gamma(t_k)]} N_i(t, t_k) + \sum_{i=1}^{p} \sum_{s=0}^{k-1} a_i^{[\gamma(t_s)]} N_i(t_{s+1}, t_s) - \sum_{i=1}^{r} \sum_{j=1}^{m} \delta_i^{[j]} T_i^{[j]}(t)$$

$$+ \sum_{i=r+1}^{p} \sum_{j=1}^{m} \delta_i^{[j]} T_i^{[j]}(t)$$

Note that the visits for which $\sigma(t_s^-) = \sigma(t_s^+)$ are counted twice in the summation of $N_i(t, t_k) + \sum_{s=0}^{k-1} N_i(t_{s+1}, t_s)$. Precisely,

$$N_i(t, t_k) + \sum_{s=0}^{k-1} N_i(t_{s+1}, t_s) = N_i(t, 0) + D_i(t, 0) \tag{16}$$

where $D_i(t,0)$ is the number of agent dynamics switching occurrence during $[0,t)$ when the graph is in $\mathcal{G}_i$. Note that $\sum_{i=1}^{p} D_i(t,0) = D(t,0)$. Then

$$\ln\|\Psi\| \le \bar{a}D(t,0) + \sum_{i=1}^{p} \bar{a}_i N_i(t,0) - \sum_{i=1}^{r}\sum_{j=1}^{m} \delta_i^{[j]} T_i^{[j]}(t) + \sum_{i=r+1}^{p}\sum_{j=1}^{m} \delta_i^{[j]} T_i^{[j]}(t)$$

Since $\sigma(t)$ does not depend on $\gamma(t)$ and the Markov chain is unique, it holds that

$$E[N_i(t,\tau)] = \pi_i - \pi_i\lambda_{ii}(t-\tau) \tag{17}$$

$$E[T_i(t,\tau)] = \pi_i(t-\tau) \tag{18}$$

$$E[T_i^{[j]}] = r^{[j]}(t)\pi_i t \tag{19}$$

By the definition of τ_γ and exploit the ergodic law of large numbers, one obtains with probability 1,

$$\limsup_{t\to\infty} \frac{1}{t}\ln\|\Psi(t,0)\| \le \frac{\bar{a}}{\tau_\gamma} - \sum_{i=1}^{p} \bar{a}_i\pi_i\lambda_{ii} - \sum_{i=1}^{r}\pi_i\sum_{j=1}^{m} \delta_i^{[j]}\bar{r}^{[j]} + \sum_{i=r+1}^{p} \pi_i\sum_{j=1}^{m} \delta_i^{[j]}\bar{r}^{[j]}$$

where $\bar{r}^{[j]} = \limsup_{t\to\infty} r^{[j]}(t) \le 1$. It is also clear that

$$\sum_{j=1}^{p} \bar{r}^{[j]} \ge 1 \tag{20}$$

Use the condition of the Theorem,

$$\eta + \frac{\bar{a}}{\tau_\gamma} \le \min_j \eta^{[j]} \le \sum_{j=1}^{m} \bar{r}^{[j]} \min_j \eta^{[j]} \le \sum_{j=1}^{m} \bar{r}^{[j]}\eta^{[j]}$$

$$= \sum_{j=1}^{m} \bar{r}^{[j]}\sum_{i=1}^{p} \bar{a}_i\pi_i\lambda_{ii} - \sum_{i=1}^{r}\pi_i\sum_{j=1}^{m} \delta_i^{[j]}\bar{r}^{[j]} + \sum_{i=r+1}^{p} \pi_i\sum_{j=1}^{m} \delta_i^{[j]}\bar{r}^{[j]}$$

Since $\bar{a}_i\pi_i\lambda_{ii} < 0$, $\forall i$, and recalling (20),

$$\eta + \frac{\bar{a}}{\tau_\gamma} \le \sum_{i=1}^{p} \bar{a}_i\pi_i\lambda_{ii} - \sum_{i=1}^{r}\pi_i\sum_{j=1}^{m} \delta_i^{[j]}\bar{r}^{[j]} + \sum_{i=r+1}^{p} \pi_i\sum_{j=1}^{m} \delta_i^{[j]}\bar{r}^{[j]} \tag{21}$$

It results form (21) that

$$\limsup_{t\to\infty} \frac{1}{t}\ln\|\Psi(t,0)\| \le -\eta, \text{ a.s., } \forall\gamma(t) \in \mathcal{M}.$$

This completes the proof.   □

5   Conclusion

Almost sure consensus of multi-agent system under both switching agent dynamics and stochastic topology jumping has been investigated in this paper. By introducing the notions of average dwell time, a sufficient AS consensus condition is proposed The analysis is carried out by the fact that the switching of agent dynamics and network topology are independent. The future work includes stochastic switching of agent dynamics.

Acknowledgments. This paper is supported by Shanghai Natural Science Fund (13ZR1416300).

References

1. Fax, J.A., Murray, R.M.: Information flow and cooperative control of vehicle formations. IEEE Trans. on Automatic Control 49, 1465–1476 (2004)
2. Olfati-Saber, R., Shamma, J.S.: Consensus filters for sensor networks and distributed sensor fusion. In: Proc. IEEE Conf. Decision Control, European Control Conf., pp. 6698–6703. IEEE, Seville (2005)
3. Olfati-Saber, R.: Flocking for multi-agent dynamic systems: Algorithms and theory. IEEE Trans. on Automatic Control 51, 401–420 (2006)
4. Lawton, J.R., Beard, R.W.: Synchronized multiple spacecraft rotations. Automatica 38, 1359–1364 (2002)
5. Benediktsson, J.A., Swain, P.H.: Consensus theoretic classification methods. IEEE Transactions on Systems, Man and Cybernetics 22, 688–704 (1992)
6. Jadbabaie, A., Lin, J., Morse, A.S.: Coordination of groups of mobile autonomous agents using nearest neighbor rules. IEEE Trans. on Automatic Control 48, 988–1001 (2003)
7. Olfati-Saber, R., Fax, J.A., Murray, R.M.: Consensus and cooperation in networked multi-agent systems. Proceedings of the IEEE 95, 215–233 (2007)
8. Ren, W., Beard, R.W.: Distributed Consensus in Multi-vehicle Cooperative Control: Theory and Applications. Springer, Heidelberg (2008)
9. Li, Z., Duan, Z., Chen, G., Huang, L.: Consensus of multiagent systems and synchronization of complex networks: A unified viewpoint. IEEE Transactions on Circuits and Systems I: Regular Papers 57, 213–224 (2010)
10. Li, Z., Ren, W., Liu, X., Xie, L.: Distributed consensus of linear multi-agent systems with adaptive dynamic protocols. Automatica 49, 1986–1995 (2013)
11. Gao, Y., Zuo, M., Jiang, T., Du, J., Ma, J.: Asynchronous consensus of multiple second-order agents with partial state information. International Journal of Systems Science 44(5), 966–977 (2013)
12. Kim, H., Shim, H., Back, J., Seo, J.H.: Consensus of output-coupled linear multi-agent systems under fast switching network: Averaging approach. Automatica 49, 267–272 (2013)
13. Wang, J., Cheng, D., Hu, X.: Consensus of multi-agent linear dynamic systems. Asian Journal of Control 10, 144–155 (2008)
14. Ni, W., Cheng, D.: Leader-following consensus of multi-agent systems under fixed and switching topologies. Systems & Control Letters 59, 209–217 (2010)

15. Miao, G., Xu, S., Zou, Y.: Necessary and sufficient conditions for mean square consensus under Markov switching topologies. International Journal of Systems Science 44, 178–186 (2013)
16. You, K.Y., Li, Z.K., Xie, L.H.: Consensus condition for linear multi-agent systems over randomly switching topologies. Automatica 49, 3125–3132 (2013)
17. Vengertsev, D., Kim, H., Seo, J.H., Shim, H.: Consensus of output-coupled high-order linear multi-agent systems under deterministic and Markovian switching networks. International Journal of Systems Science, http://dx.doi.org/10.1080/00207721.2013.835884
18. Wang, B.C., Zhang, J.F.: Distributed output feedback control of Markov jump multi-agent systems. Automatica 49, 1397–1402 (2013)

An Effective Integral Quadratic Constraint Construction Approach for Stability Analysis of a Class of Systems with Time Delays

Min Zheng[1,2], Zheng Mao[2], and Kang Li[3]

[1] School of Mechatronic Engineering and Automation, Shanghai University
[2] Shanghai Key Laboratory of Power Station Automation Technology,
Shanghai 200072, China
[3] School of Electronics, Electrical Engineering and Computer Science, Queens
University Belfast, Belfast, UK

Abstract. This paper proposes an effective method for the construction of Integral Quadratic Constraints(IQC) based on Quadratic Separation(QS) framework for the stability analysis of a wide range of systems with time delays. An unified framework for both the QS framework and the Lyapunov theory is proposed, an effective IQCs construction method for Lyapunov candidates is thus obtained. This method is then applied to the stability analysis of linear systems with time delays. Stability criteria are established based on the proposed approach. Numerical examples are finally provided to show its effectiveness.

1 Introduction

As an alternative tool to Lyapunov theory for stability analysis, IQCs provide a general framework for the stability analysis of feedback interconnections of linear time invariant(LTI) plants with uncertainty blocks. Many well-known results, such as robust stability analysis for systems with structured time-varying uncertainties can be viewed as a special case of the IQC approach [1][2]. To improve the flexibility of this approach, multipliers are used to select proper variables for the partitioning. However, the applicability of many results are limited by the computational difficulties at that time. In the 1970s, IQCs were explicitly used by Yakubovich to treat the stability problem of systems with nonlinearities. A unifying framework was introduced in [1], and since then research on IQCs has mainly focused on finding suitable multipliers[3-6]. Recently, Peaucelle and Ariba etc. [5][6] proposed a novel QS framework for feedback connected systems with implicit linear transformation. The merit of this framework lies on not only its effectiveness in dealing with time varying operators but also obtaining LMIs conditions directly without using KYP lemma. The associated fundamental concept is the topological separation proposed by Safonov [3], in which it states that internal signals of a multi variable feedback connection of two systems F and G are unique and bounded under external disturbances if and only if the graph of F is topologically separated from the inverse graph of G. However, as mentioned in

M. Fei et al. (Eds.): LSMS/ICSEE 2014, Part II, CCIS 462, pp. 439–448, 2014.
© Springer-Verlag Berlin Heidelberg 2014

[5], finding such topological separator is tricky in general. The main goal of this paper is to develop a new quadratic separator's, i.e. IQC's construction method, and extends the fundamental stability theorem 1 in [6] to more general forms.

Notation: $diag\{\cdots\}$ denotes the block diagonal matrix. Matrices, if not explicitly stated, are assumed to have compatible dimensions for algebraic operations. L_2^m denotes the space of R^m valued, square summable (integrable) functions defined on time interval $(-\infty, \infty)$, and L_{2e}^m denotes the extension of the space L_2^m, which consists of functions whose time truncation lie in L_2^m. Introduce a truncation operator P_T, which leaves a function unchanged on the interval $[0, T]$ and gives the value zero on $(T, \infty]$. For all measurable functions $u(t) \in L_2^m$, define the norm $\|u(t)\|_2 = \{\int_{-\infty}^{\infty} u^T(t)u(t)dt\}^{1/2}$, then this norm corresponds to the inner product $<>$ for $(u, v \in L_2^m)$ defined as $<u, v> = \int_{-\infty}^{\infty} u^T(t)v(t)dt$.

2 Stability Analysis Based on Quadratic Separation Framework and Lyapunov Theory

2.1 Recap of QS Stability Analysis

Consider the following feedback configuration.

$$E(z - \bar{z}) = Gw \tag{1}$$

$$w - \bar{w} = \nabla z \tag{2}$$

where E is real valued, possibly non-square full column rank matrix; G is real valued, possibly non-square matrix; z, w are internal variables, $\bar{z}, \bar{w}$ are external inputs and ∇ is a linear operator from L_{2e} to L_{2e}.

Two signals $w \in L_2^m$ and $v \in L_2^l$ are said to satisfy the *IQC* defined by Π, if

$$\begin{aligned}
\delta_\Pi(v, w) :&= \left\langle \begin{bmatrix} v \\ w \end{bmatrix}, \Pi \begin{bmatrix} v \\ w \end{bmatrix} \right\rangle \\
&= \int_{-\infty}^{\infty} \begin{bmatrix} v(t) \\ w(t) \end{bmatrix}^T \Pi \begin{bmatrix} v(t) \\ w(t) \end{bmatrix} dt \\
&= \int_{-\infty}^{\infty} \begin{bmatrix} \hat{v}(jw) \\ \hat{w}(jw) \end{bmatrix}^* \Pi(jw) \begin{bmatrix} \hat{v}(jw) \\ \hat{w}(jw) \end{bmatrix} dw \\
&\geq 0
\end{aligned}$$

where the operator Π is referred to as the *multiplier* of the quadratic form δ_Π. To facilitate the development, the multiplier Π is often a block decomposed into the form $\begin{bmatrix} \Pi_{11} & \Pi_{12} \\ \Pi_{12}^* & \Pi_{22} \end{bmatrix}$ and the dimensions of Π_{ij} confirms to those of v and w (see [11]).

The feedback interconnection of G and ∇ is *well-posed* if the map $(w, z) \Rightarrow (\bar{w}, \bar{z})$ defined by equations (1)(2) has a causal inverse on L_{2e}.

The feedback interconnection of G and ∇ is *stable* if the interconnection is well-posed and if the map $(w, z) \Rightarrow (\bar{w}, \bar{z})$ has a bounded inverse, i.e. there exists a constant $\gamma > 0$ such that

$$\left\| \begin{matrix} w_T \\ z_T \end{matrix} \right\|_2 \leq \gamma \left\| \begin{matrix} \bar{w}_T \\ \bar{z}_T \end{matrix} \right\|_2$$

for $\forall T \geq 0$, $\forall (\bar{w}, \bar{z}) \in L_{2e}$.

Theorem 1 [6]: The feedback interconnection (1)(2) is stable if there exists a symmetric matrix Π satisfying both conditions

$$\left[E - G \right]^{\perp^T} \Pi \left[E - G \right]^{\perp} > 0 \tag{3}$$

$$\forall \xi \in L_{2e}, \forall T > 0, \left\langle \begin{bmatrix} 1 \\ P_T \nabla \end{bmatrix} \xi, \Pi \begin{bmatrix} 1 \\ P_T \nabla \end{bmatrix} \xi \right\rangle \leq 0 \tag{4}$$

This theorem suggests that the proof of stability include two conditions: a matrix inequality (3) related to the lower block and an inner product (4) that states IQCs on the upper one. Basically, inequality (4) which forms IQCs, is derived from definitions and characteristics on different operators relating to the matrix ∇. Then, inequality (3) provides the stability condition of the interconnection.

2.2   Lyapunov Candidates Induced IQCs Approach

The two stability conditions (3) an (4) are similar to the two stability conditions of Lyapunov theory and one is positive definite and the other is negative definite. Based on this observation, consider the internally stability of system (3) and (4), let $\bar{z} = \bar{w} = 0$, $\Omega = -\Pi$, and notice the facts that $\left[E - G \right] * \begin{bmatrix} z(t) \\ w(t) \end{bmatrix} = 0$, then, (3) and (4) can be rewritten as

$$\dot{V}_{IQC} = \begin{bmatrix} z(t) \\ w(t) \end{bmatrix}^T \Omega \begin{bmatrix} z(t) \\ w(t) \end{bmatrix} < 0 \tag{5}$$

$$\forall t > 0, V_{IQC} = \int_{-\infty}^{t} \begin{bmatrix} z(v) \\ w(v) \end{bmatrix}^T \Omega \begin{bmatrix} z(v) \\ w(v) \end{bmatrix} dv \geq 0 \tag{6}$$

It should be noted that the internally stability can be viewed as zero input response, and the equal sign of (5) (6) is established if and only if states $z(t) = w(t) = 0$. Therefore, V_{IQC} in (6) can be regarded as a Lyapunov candidate, and the $\dot{V}_{IQC}$ in (5) is the corresponding derivative function.

Based on the above discussion, an unified version of stability theorem for feedback system (3) with (4) encompass both IQC theory and Lyapunov theory can be established as follows.

Theorem 2: The feedback interconnection system

$$Ez(t) = Gw(t), \quad w(t) = \nabla z(t) \tag{7}$$

is stable if there exists a symmetric matrix Ω and a continuous function $V_{comp}(t)$ satisfying both conditions($\forall t; z, w \neq 0$)

$$\dot{V}(z, w) = \dot{V}_{comp}(t) + \dot{V}_{IQC}(t) < 0 \tag{8}$$

$$V(z, w) = V_{comp}(t) + V_{IQC}(t) > 0 \tag{9}$$

where $V_{IQC}(t)$ and $\dot{V}_{IQC}(t)$ are defined as (5) and (6) respectively.

Remark 1: Theorem 2 meets the two basic conditions of Lyapunov theory, therefore the result is confirmed. Obviously, when the additional term $V_{comp}(t)$ is zero, Theorem 2 becomes the IQC based framework, on the other hand, if the IQC based part $V_{IQC}(t)$ is set to zero, Theorem 2 becomes the traditional Lyapunov theory.

From the relationship between IQC and Lyapunov candidate, we propose the following IQC construction method.

Consider system (7), if there exists a positive definite Lyapunov function $V(t) = \sum_{i=1}^{l} V_i(t)$, and for term $V_i(t) \geq 0$, $\dot{V}_i(t)$ is the derivative along the solution of system (7) with respect to t, and has the following quadratic form

$$\dot{V}_i(t) = \begin{bmatrix} z_i(t) \\ w_i(t) \end{bmatrix}^T \begin{bmatrix} \Omega_{i11} & \Omega_{i12} \\ \Omega_{i21} & \Omega_{i22} \end{bmatrix} \begin{bmatrix} z_i(t) \\ w_i(t) \end{bmatrix} \tag{10}$$

then, the alternative IQC condition can be constructed as follows:

Define operator $\nabla_i : L_{2e} \to L_{2e}, z_i(t) \to w_i(t)$. An integral quadratic constraint for the operator ∇_i is given by the following inequality,$\forall T > 0, \forall z_i \in L_{2e}^n$ and for a symmetric matrix Ω_i (i.e. multiplier),

$$\left\langle \begin{bmatrix} 1_n \\ \nabla_i 1_n \end{bmatrix} z_i, \Omega_i \begin{bmatrix} 1_n \\ \nabla_i 1_n \end{bmatrix} z_i \right\rangle_{P_T} \geq 0 \tag{11}$$

Remark 2: The above IQCs are directly from Lyapunov candidates, and mostly IQCs are related to Lyapunov candidates, however, different from the Lyapunov approach, IQC based approach has a few advantages in the frequency domain. For example in [12][13], the delay element is defined as

$$\phi(s, \tau) = \left\{ \begin{array}{ll} \frac{e^{-\tau s}-1}{\bar{\tau} s} & s \in C, s \neq 0 \\ -\frac{\tau}{\bar{\tau}} & s = 0 \end{array} \right.$$

and the set $\Phi := \{\lambda \phi(jw, \tau) | w \in R_e^+, \tau \in [0, \bar{\tau}], \lambda \in [0, 1]\}$. According to the analysis from frequency domain, the disk $D(q; r)$ with (q=0.251,r=0.749) or (q=0.327, r=0.726) can cover this set with minimum radius rather than unit disk with (q=0, r=1), which may lead to more conservative results in delay systems. Based on this property, [12] proposed the following IQC Lemma.

Lemma 1: An IQC for the operator $\nabla_b : x(t) \mapsto \int_{t-h_m}^{t} x(u)du$ is given by the following inequality $\forall T > 0, \forall x \in L_{2e}, \forall Q > 0$,

$$\left\langle \begin{bmatrix} 1 \\ \nabla_b \end{bmatrix} x, \Pi \begin{bmatrix} 1 \\ \nabla_b \end{bmatrix} x \right\rangle < 0 \tag{12}$$

where $\Pi = \begin{bmatrix} (q^2 - r^2)h_m^2 Q & -qh_m Q \\ -qh_m Q & Q \end{bmatrix}$, (q, r) are two appropriate reals, for example $q = 0.25, r = 0.75$.

Remark 3: It is worth pointing out that when $q = 0, r = 1$, this IQC corresponds to the traditional Lyapunov functional term

$$V(x_t) = h_m \int_{t-h_m}^{t} \int_{v}^{t} x(s)^T Q x(s) ds dv$$

with its derivative

$$\dot{V}(x_t) = \begin{bmatrix} x(t) \\ \int_{t-h_m}^{t} x(s)ds \end{bmatrix}^T \begin{bmatrix} h_m^2 Q & 0 \\ 0 & -Q \end{bmatrix} \begin{bmatrix} x(t) \\ \int_{t-h_m}^{t} x(s)ds \end{bmatrix}$$

Obviously, this Lyapunov approach based condition is more conservative than Lemma 1, and gives tighter constraint with smaller radius ($0.75 < 1$).

3   Stability of Time-Delay Systems

In this section, we will apply this approach to time delay systems to show the effectiveness of the proposed framework.

Consider the following time delay system:

$$\begin{aligned} \dot{x}(t) &= Ax(t) + A_d x(t - h(t)), \forall t > 0 \\ x(t) &= \Phi(t), \quad \forall t \in [-h_M, 0] \end{aligned} \tag{13}$$

where $x(t) \in R^n$ is the state; $\Phi(t)$ is the initial function; $A \in R^{n \times n}$ and $A_d \in R^{n \times n}$ are constant matrices.

3.1   Constant Delay Case

To reformulate the original time delay system as an interconnected system that can be fit into the framework of quadratic separation, the following operators are needed.

Define the following operators:

$$\begin{aligned} \nabla_1 &: x(t) \to \int_{-\infty}^{t} x(\theta)d\theta \\ \nabla_2 &: x(t) \to x(t - h) \\ \nabla_3 &: \dot{x}(t) \to x(t) - x(t - h) \end{aligned}$$

For the operators $\nabla_i (i = 1, 2, 3)$, $\forall T > 0, \forall x \in L_{2e}, \forall Q > 0$, define

$$\left\langle \begin{bmatrix} 1 \\ \nabla_i \end{bmatrix} x, \Pi_i \begin{bmatrix} 1 \\ \nabla_i \end{bmatrix} x \right\rangle > 0 \tag{14}$$

where

$$\Pi_1 = \begin{bmatrix} 0 & Q \\ Q & 0 \end{bmatrix}$$

$$\Pi_2 = \begin{bmatrix} Q & 0 \\ 0 & -Q \end{bmatrix}$$

$$\Pi_3 = \begin{bmatrix} (r^2 - q^2)h^2 Q & qhQ \\ qhQ & -Q \end{bmatrix}$$

To rewrite the delay system $\dot{x}(t) = Ax(t) + A_d x(t - h)$ as the form of (7), let

$$z(t) = \begin{bmatrix} \dot{x}(t) \\ x(t) \\ \dot{x}(t) \end{bmatrix}, w(t) = \begin{bmatrix} x(t) \\ x(t - h) \\ x(t) - x(t - h) \end{bmatrix}$$

then,

$$E = I, \nabla = diag\{\nabla_1, \nabla_2, \nabla_3\}, G = \begin{bmatrix} A & A_d & 0 \\ I & 0 & 0 \\ A & A_d & 0 \end{bmatrix}$$

Theorem 3: The system (13) with constant delay $h(t) = h$ is asymptotically stable if there exist matrices $P > 0, Q > 0, R > 0$, such that the following LMI holds

$$\Gamma = \begin{bmatrix} \Gamma_{11} & \Gamma_{12} \\ * & \Gamma_{22} \end{bmatrix} < 0 \tag{15}$$

with

$$\begin{aligned} \Gamma_{11} &= A^T P + PA + Q + (r^2 - q^2)h^2 A^T RA \\ &\quad + hq(A^T R + RA) - R \\ \Gamma_{12} &= PA_d + (r^2 - q^2)h^2 A^T RA_d + hqRA_d \\ &\quad - hqA^T R + R \\ \Gamma_{22} &= -Q + (r^2 - q^2)h^2 A_d^2 RA_d - hq(RA_d + A_d^2 R) - R \end{aligned}$$

Proof. For the operators $\nabla_i, (i = 1, 2, 3)$, and the IQCs (15), the overall multiplier Π can be easily constructed as block diagonal matrix $\Pi = diag\{\Pi_1, \Pi_2, \Pi_3\}$. In order to satisfy the condition (5) to guarantee the stability of system, define the vector $\xi(t)$ to fit for the overall multiplier Π

$$\xi(t) = \begin{pmatrix} z_1(t) \\ w_1(t) \\ z_2(t) \\ w_2(t) \\ z_3(t) \\ w_3(t) \end{pmatrix} = \begin{bmatrix} A & A_d \\ I & 0 \\ I & 0 \\ 0 & I \\ A & A_d \\ I & -I \end{bmatrix} \begin{pmatrix} x(t) \\ x(t - h) \end{pmatrix} = \aleph \mu(t)$$

for $\xi^T \Pi \xi = \mu^T \aleph^T \Pi \aleph \mu$, then as long as $\Gamma = \aleph^T \Pi \aleph < 0$ holds, which guarantees (5) and implies system (13) with delay h is stable. Further, notice the form of $\mu = \begin{pmatrix} x(t) \\ * \end{pmatrix}$, $\dot{V}_{IQC} \leq -\epsilon\|x(t)\|^2, (\exists\epsilon > 0)$ as long as the LMIs (15) holds, which implies that system (13) is also asymptotically stable. This completes the proof.

Remark 4: To avoid finding the orthogonal of $[E - G]$, condition (5) instead of (3) is applied. Deduction entirely follow the framework of IQC based approach. The equivalent LKFs can be given as

$$V_{IQC} = V_1(t) + V_2(t) + V_3(t)$$

$$= \int_{-\infty}^{t} \begin{pmatrix} \dot{x}(s) \\ x(s) \end{pmatrix}^T \Pi_1 \begin{pmatrix} \dot{x}(s) \\ x(s) \end{pmatrix} ds$$

$$+ \int_{-\infty}^{t} \begin{pmatrix} \dot{x}(s) \\ x(s-h) \end{pmatrix}^T \Pi_2 \begin{pmatrix} \dot{x}(s) \\ x(s-h) \end{pmatrix} ds$$

$$+ \int_{-\infty}^{t} \begin{pmatrix} \dot{x}(s) \\ x(s) - x(s-h) \end{pmatrix}^T \Pi_3 \begin{pmatrix} \dot{x}(s) \\ x(s) - x(s-h) \end{pmatrix} ds$$

and terms V_1, V_2 are equivalent to the traditional LKFs $x^T P x$ and $\int_{t-h}^{t} x(s)^T Qx(s)ds$ respectively. The merit to introduce term V_3 based on IQC is demonstrated in the following benchmark example, which also shows the benefit of bridging the relations between IQC based approach and Lyapunov approach.

3.2   Time Varying Delay Case

The time varying delay case is also considered. The related theorem can be stated as follows:

Theorem 4: The system (13) with time varying delay $0 \leq h_m \leq h(t) \leq h_M, \mu_m \leq \dot{h}(t) \leq \mu_M$ is asymptotically stable if there exist positive matrices $Q_1, Q_2 \in R^{2n \times 2n}$ and matrices $P, R_i(i = 1, 2, 3) \in R^{3n \times 3n}$, such that the following LMIs hold

$$E^T P E > 0, \ E^T R_i E > 0, (i = 1, 2, 3)$$

$$\Gamma_1 = \begin{bmatrix} \mathcal{C}^T \Gamma_{01} \mathcal{C} & \begin{bmatrix} 0_{5n \times 3n} \\ \bar{B}_M^T \Phi \end{bmatrix} \\ * & -\Phi \end{bmatrix} < 0,$$

$$\Gamma_2 = \begin{bmatrix} \mathcal{C}^T \Gamma_{02} \mathcal{C} & \begin{bmatrix} 0_{5n \times 3n} \\ \bar{B}_s^T \Phi \end{bmatrix} \\ * & -\Phi \end{bmatrix} < 0 \qquad (16)$$

with

$$E = \begin{bmatrix} I & 0 \\ 0 & I \\ I & 0 \end{bmatrix}, \mathcal{A} = \begin{bmatrix} A & 0 \\ 0 & A \\ 0 & I \end{bmatrix},$$

$$\Gamma_{01} = \begin{bmatrix} \Xi_{11} & \Xi_{12} & \Xi_{13} & \Xi_{14} \\ * & \Xi_{22} & 0 & 0 \\ * & * & \Xi_{33} & \Xi_{34} \\ * & * & * & \Xi_{44} \end{bmatrix}, \Gamma_{02} = \begin{bmatrix} \Xi_{11} & \Xi_{12} & \Xi_{13} & \Psi_{14} \\ * & \Xi_{22} & 0 & 0 \\ * & * & \Xi_{33} & \Psi_{34} \\ * & * & * & \Psi_{44} \end{bmatrix}$$

$$\mathcal{C} = \begin{bmatrix} I_n & 0_{2n \times 6n} \\ \mathcal{A} & 0_{2n \times 4n} \; A_d \; 0_n \\ 0_{6n \times 2n} & I_{6n \times 6n} \end{bmatrix}$$

$$\bar{B}_M = \begin{bmatrix} A_d & 0 \\ 0 & (1-\mu_M)A_d \\ 0 & 0 \end{bmatrix}, \bar{B}_s = \begin{bmatrix} A_d & 0 \\ 0 & (1-\mu_m)A_d \\ 0 & 0 \end{bmatrix}$$

$$\Phi = (h_a/2)^2 R_1 + \delta^2 R_2 + \alpha R_3, h_a = \frac{h_M + h_m}{2}$$

$$\delta = \frac{h_M - h_m}{2}, \alpha = (r^2 - q^2)h_a^2, \beta = h_a q$$

$$\Xi_{11} = \mathcal{A}^T P E + E^T P \mathcal{A} + Q_1 + Q_2 - E^T R_1 E + \mathcal{A}^T \Phi \mathcal{A}$$
$$- E^T R_3 E + \beta(\mathcal{A}^T R_3 E + E^T R_3 \mathcal{A})$$

$$\Xi_{12} = E^T R_1 E$$

$$\Xi_{13} = E^T R_3 E - \beta \mathcal{A}^T R_3 E$$

$$\Xi_{14} = E^T P B_M + \mathcal{A}^T \Phi B_M + \alpha \mathcal{A}^T R_3 B_M + \beta E^T R_3 B_M$$

$$\Xi_{22} = -Q_1 - E^T R_1 E$$

$$\Xi_{33} = -E^T R_2 E - E^T R_3 E$$

$$\Xi_{34} = E^T R_2 E - \beta E^T R_3 B_M$$

$$\Xi_{44} = -(1-\mu_M)Q_2 - E^T R_2 E$$

$$\Psi_{14} = E^T P B_s + \mathcal{A}^T \Phi B_s + \alpha \mathcal{A}^T R_3 B_s + \beta E^T R_3 B_s$$

$$\Psi_{34} = E^T R_2 E - \beta E^T R_3 B_s$$

$$\Psi_{44} = -(1-\mu_m)Q_2 - E^T R_2 E$$

Remark 5: Theorem 2 gives the flexibility in choosing V function, making the IQC based framework similar as Lyapunov approach. The IQCs and Lyapunov functions can easily be constructed mutually. For example, for Lyapunov function $V = x^T P x$, its derivative $\dot{V} = x^T P \dot{x} + \dot{x}^T P x$, can be reformulated as $\dot{V} = \begin{bmatrix} \dot{x} \\ x \end{bmatrix}^T \begin{bmatrix} 0 & P \\ P & 0 \end{bmatrix} \begin{bmatrix} \dot{x} \\ x \end{bmatrix}$, and this follows the form of (8) in Theorem 2. Then operator ∇_1 and multiplier Π_1 can be constructed consequently.

4   Numerical Examples

To illustrate the effectiveness of our results, this section will give the following numerical examples.

Example 1   Consider the system (13) with

$$A = \begin{bmatrix} -2 & 0 \\ 0 & -0.9 \end{bmatrix}, A_d = \begin{bmatrix} -1 & 0 \\ -1 & -1 \end{bmatrix}$$

Table 1. Upper bounds h for different (q,r)

Π_3 (q,r)	$(0,1)$	$(0.249, 0.751)$	$(0.327, 0.726)$
Theorem3	4.472	5.835	6.164

For this well-known example, if h is constant delay, the maximum allowable upper bounds are obtained by using Theorem 3 with different parameters (q,r). The results are shown in Table 1. when $(q = 0, r = 1)$, the maximum $h = 4.472$, which is almost the best result obtained from LKF without using the decomposition techniques or augmentation approach. When more restrictive constraint parameters $(q = 0.327, 0.726)$ are used in Theorem 3, a fairly good result is presented as $h = 6.164$, almost recovers the true stability margin (6.172), the accuracy reaches 99.87%. Remarkably, this result obtained only by using three 2×2 decision matrices (P,Q,R), has much less than the existing results for the same degree of conservatism.

Example 2 Consider the time varying delay case of example1. For this example, many results were obtained in this literature. To demonstrate the effectiveness of our result Theorem 4, results are compared with those in [6][7][11].For various μ, the maximal allowable delays computed are shown in Table 2. From Table 2, we observe that the Theorem 4 with $q = 0.327, r = 0.726$ give better stability margins than [6][7][11]. Conservatism is reduced thanks to the combination of augmented derivative state technique,discrete average of delay interval, proper selected parameters of q and r.

Table 2. Upper bounds h for different μ

μ	0	0.1	0.5	0.8
Theorem 2 [6]	5.12	4.08	2.52	2.15
IQC analysis [11]	6.11	4.71	2.28	1.68
Theorem 3[7]	6.11	4.79	2.68	1.95
Th.4(0.249,0.751)$(h_m = 0)$	5.83	4.99	4.98	4.97
Th.4(0.327,0.726)$(h_m = 0)$	6.16	5.05	5.04	5.01
Th.4(0.327,0.726)$(h_m = 5)$	-	5.92	5.89	5.81

The simulation results illustrate the effectiveness of our theorems.

5 Conclusions

In this paper, a novel stability analysis approach were proposed by combining IQCs with traditional Lyapunov function. An effective QS approach is thus presented from Lyapunov function accordingly. Theorem 2 in this paper provides an flexible framework for stability analysis of various systems. Two well-known systems, linear time delay system and delayed neural networks, have been particularly investigated in this new framework. Finally, three widely used examples are given to show the effectiveness of the results.

Acknowledgements. This work is supported by Shanghai Science Technology Commission No. 14ZR1414800.

References

1. Megretski, A., Rantzer, A.: System analysis via integral quadratic constraints. IEEE Trans. Automatic Control 42(6), 819–830 (1997)
2. Scherer, C.W., Emre Kose, I.: Robustness with dynamic IQCs: An exact state-space characterization of nominal stability with applications to robust estimation. Automatica 44, 1666–1675 (2008)
3. Safonov, M.G.: Stability and robustness of multivariable feedback systems. MIT Press, Cambridge (1980)
4. Rantzer, A.: On the Kalman-Yakubovich-Popov lemma. Sys. and Contr. Lett. 28, 7–10 (1996)
5. Peaucelle, D., Arzelier, D., Henrion, D., Gouaisbaut, F.: Quadratic separation for feedback connection of an uncertain matrix and an implicit linear transformation. Automatica 43, 795–804 (2007)
6. Ariba, Y., Gouaisbaut, F., Johansson, K.: Stability interval for time-varying delay systems. In: 49th IEEE Conf. on Decision and Contr., Atlanta, GA, USA, pp. 1017–1022 (2010)
7. Ariba, Y., Gouaisbaut, F.: Input-output framework for robust stability of time-varying delay systems. In: 48th IEEE Conf. on Decision and Contr., Shanghai, China, pp. 274–279 (2009)
8. Meinsma, G., Shrivastava, Y., Fu, M.: A dual formulation of mixed μ and on the loselessness of (D,G)-scaling. IEEE Trans. Automatic Control 42(7), 1032–1036 (1997)
9. Iwasaki, T., Hara, S.: Well-posedness of feedback systems: Insights into exact robustness analysis and approximate computations. IEEE Trans. Automatic Control 43(5), 619–630 (1998)
10. Iwasaki, T., Shibta, G.: LPV system analysis via quadratic separator for uncertain implicit systems. IEEE Trans. Automatic Control 46(8), 1195–1207 (2001)
11. Kao, C., Rantzer, A.: Stability analysis of systems with uncertain time-varying delays. Automatica 43, 959–970 (2007)
12. Ariba, Y., Gouaisbaut, F., Peaucelle, D.: Stability analysis of time-varying delay systems in quadratic separation framework. In: Int. Conf. on Mathematical Problems in Engineering, Aerospace and Sciences. Italie, Genoa (2008)
13. Zhang, J., Knospe, C., Tsiotras, P.: Toward less conservative stability analysis of time-delay systems. In: Proc. of the 38th Conf. on Decision & Contr. Phoenix Arizona, USA (1999)
14. Fu, M., Li, H., Niculescu, S.: Robust stability and stabilization of time-delay systems via integral quadratic constraint approach. In: Dugard, L., Verriest, E.I. (eds.) Stability and Control of Time-delay Systems. LNCIS, vol. 228, pp. 101–116. Springer, Heidelberg (1998)
15. Zhang, H., Gong, D., Cheng, B., Liu, Z.: Synchronization for coupled neural networks with interval delay: A novel augmented Lyapunov-Krasovskii functional method. IEEE Trans. Neural Netw. Learning Sys. 24(1), 58–70 (2013)
16. Sun, J., Liu, G., Chen, J., Rees, D.: Improved delay-range-dependent stability criteria for linear systems with time-varying delays. Automatica 46(2), 466–470 (2010)
17. Park, P., Ko, J., Jeong, C.: Reciprocally convex approach to stability of systems with time-varying delays. Automatica 47(1), 235–238 (2011)

An Online Recursive Identification Method over Networks with Random Packet Losses

Dajun Du[1], Lili Shang[1], and Wanqing Zhao[2]

[1] Shanghai Key Laboratory of Power Station Automation Technology,
School of Mechatronical Engineering and Automation,
Shanghai University, Shanghai, 200072, China
ddj@shu.edu.cn, sll713@126.com
[2] Cardiff School of Engineering, Cardiff University, Cardiff, CF24 3AA, U.K.
wqzhao007@gmail.com

Abstract. Under the network environment, data packet losses bring an undesirable effect on parameter identification. To solve this problem, an online recursive identification method over networks with random packet losses is proposed. In this algorithm, the Bernoulli processes is firstly employed to describe the character of data packet losses. Considering random packet losses, an improved recursive least squares (RLS) algorithm is then presented by using Givens transformation, where the intermediate matrix can be updated recursively. Simulation results show that the proposed algorithm is able to significantly enhance the accuracy of parameter estimation, and the estimated variance becomes smaller as the number of iterations increases.

Keywords: network environment, data packet losses, RLS, Givens transformation.

1 Introduction

Network control systems (NCSs) are real-time feedback control systems, where the information among the sensor, controller and the actuator are exchanged via a communication network [1]. Although NCSs bring the great advantages such as easy installation and maintenance, great flexible, and so on, it also inevitably introduces data packet dropout due to the unreliability of communication networks. A lot of works for NCSs have been reported [2], but they usually assumed that the model of controlled plant is known.

Some studies have been reported on the system identification under the network environment. These results can be roughly classified into two categories according to whether the model is updated in real time by using the new data, i.e., offline identification and online identification under the network environment.

With respect to the offline identification under the network environment, the system model is identified by using the complete input/output observation data. For example, using the cubic spline interpolation to compensate lost data, a

M. Fei et al. (Eds.): LSMS/ICSEE 2014, Part II, CCIS 462, pp. 449–458, 2014.

fast recursive algorithm is employed to identify off-line the system model [3]. Considering the network-induced delay, a covariance function based method that relies on the second order statistical properties of the output signal is proposed for estimating the parameters [4].

Recently, some online identification methods are developed for NCSs. In practice, online identification is useful, which can dynamically tune the system model. For example, considering randomly missing measurements, a recursive algorithm for parameter estimation by the modified Kalman filter-based algorithm is proposed [5]. Considering randomly missing outputs, unknown model parameters are identified by a Kalman filter-based method, and the adaptive controller is then designed based on the identification model [6]. Considering data packet dropout, a modified adaptive Kalman filter is proposed to estimate the unknown parameters [7].

In this paper, an online recursive identification method over networks with random packet losses is proposed. It not only can overcome the effectiveness of data packet losses but also can be updated recursively by using Givens transformation. The paper is organized as follows. Section 2 gives problem formulation for system identification over networks. On-line estimation method of model parameters under the network environment is presented in Section 3. Simulation results are presented in Section 4, followed by the conclusion in Section 5.

2   Problem Formulation

Consider the linear discrete system by the following difference equation

$$y_k = \sum_{i=1}^{n_y} a_i y_{k-i} + \sum_{j=1}^{n_u} b_j u_{k-j} + v_k, \tag{1}$$

where u_k and y_k are the system input and output at sampling instant k, respectively, v_k is a white noise with zero mean and variance σ^2, a_i and b_i are the unknown coefficients of the system to be estimated, n_y and n_u are the corresponding maximal lags, and $m = n_y + n_u$.

Defining the n-component vectors $Y_n = [y_1, y_2, \cdots, y_n]^T \in \Re^{n \times 1}$, $\Theta = [a_1, \ldots, a_{n_y}, b_1, \ldots, b_{n_u}]^T \in \Re^{m \times 1}$ and $V_n = [v_1, v_2, \cdots, v_n]^T \in \Re^{n \times 1}$, (1) can be re-written in matrix form

$$Y_n = \Phi_n \Theta + V_n, \tag{2}$$

where $\Phi_n = [\varphi_1, \ldots, \varphi_n]^T \in \Re^{n \times m}$ with row vectors $\varphi_k^T = [y_{k-1}, y_{k-2}, \cdots, y_{k-n_y}, u_{k-1}, u_{k-2}, \cdots, u_{k-n_u}], k = 1, \ldots, n.$

The identified problem under investigation is shown in Fig. 1, where u_k and y_k are the system input and measured output, and $\bar{y}_k$ and $\bar{u}_k$ are the input of the identifier and actuator, respectively. The identifier and controller are connected to the plant via communication networks, where packet losses are inevitably introduced. The packet loss is modeled as the Bernoulli process α_k with the following probability distribution law:

$$\Pr\{\alpha_k = 1\} = E\{\alpha_k\} = \rho, \ \Pr\{\alpha_k = 0\} = 1 - E\{\alpha_k\} = 1 - \rho,$$
$$Var\{\alpha_k\} = E\{(\alpha_k - \rho)^2\} = \rho(1 - \rho) = \beta^2, \tag{3}$$

where $\alpha_k = 1$ if the packet transmits successfully, and 0 otherwise. The known positive constant $0 < \rho < 1$ indicates the probability that the packet will be transmitted successfully while β denotes the variance of α_k. Thus, at the kth sampling instant, the measured output signals received by the identifier can be described as

$$\bar{y}_k = \alpha_k y_k + (1 - \alpha_k) y_{k-1}. \tag{4}$$

Similarly, the input signals that the actuator receives can be expressed as

$$\bar{u}_k = \alpha_k u_k + (1 - \alpha_k) u_{k-1}. \tag{5}$$

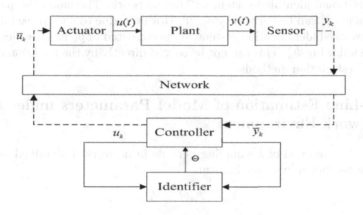

Fig. 1. Networked identification and control system

Remark 1. Note that α_k in (4) is either 1 or 0, and it represents the data packet status in the communication channel. Therefore, the identifier is subject to undesirable packet losses (i.e., incomplete information), which may significantly degrade the system identification performance or even cause the system unconvergence.

For the identification model in (2), the following cost function is considered

$$J_n = \|W_n (Y_n - \Phi_n \Theta)\|_2^2, \tag{6}$$

where $W_n = diag[w^{n-1}, w^{n-2}, \cdots, 1]$, w is a forgetting or exponential weighting factor and $0 << w < 1$.

Since the matrix $W_n \Phi_n$ is $n \times m$ and the vector $W_n Y_n$ is $n \times 1$, there exists an $n \times n$ orthogonal matrix Q_n [8] such that

$$Q_n^T W_n \Phi_n = \begin{pmatrix} R_n \\ \mathbf{0} \end{pmatrix} \tag{7}$$

$$Q_n^T W_n Y_n = \begin{pmatrix} \breve{y}_n \\ \tilde{y}_n \end{pmatrix} \tag{8}$$

where R_n is an $m \times m$ upper triangular matrix, $\mathbf{0}$ is an $(n-m) \times m$ matrix of zeros, $\breve{y}_n$ is an $m \times 1$ vector, and $\tilde{y}_n$ is an $(n-m) \times 1$ vector. The least-squares estimate to minimize (6) is given by

$$\Theta = R_n^{-1} \breve{y}_n, \tag{9}$$

$$J_n = \tilde{y}_n^T \tilde{y}_n. \tag{10}$$

Remark 2. If there are no data packet losses, i.e., $\bar{y}_k \equiv y_k$, it is the same as the traditional identified system without network. Therefore, the parameter estimation in (2) can be derived by[8], [9]. However, due to data packet dropouts in the networked identification system, Φ_n is consisted of $\bar{y}_k$ in (4) which in turn yields the following $\bar{\Phi}_n$. This can not be solved directly by the traditional on-line parameter estimation methods.

3   On-Line Estimation of Model Parameters under the Network Environment

After the time interval of k sampling periods in networked identified system, a model can be shown in a matrix form

$$\bar{Y}_n = \bar{\Phi}_n \Theta + V_n, \tag{11}$$

where $\bar{Y}_n = \begin{bmatrix} \bar{y}_1 & \bar{y}_2 & \cdots & \bar{y}_n \end{bmatrix}^T$,
$\bar{\Phi}_n = \begin{bmatrix} \bar{\varphi}_1 & \bar{\varphi}_2 & \cdots & \bar{\varphi}_n \end{bmatrix}^T$, $\bar{\varphi}_k^T = \begin{bmatrix} \bar{y}_{k-1} & \bar{y}_{k-2} & \cdots & \bar{y}_{k-n_y} & \bar{u}_{k-1} & \bar{u}_{k-2} & \cdots & \bar{u}_{k-n_u} \end{bmatrix}$, $k = 1, \ldots, n$. For the matrix $W_n \bar{\Phi}_n$ and the vector $W_n \bar{Y}_n$, there exists an orthogonal matrix $\bar{Q}_n$ such that $\bar{Q}_n W_n \bar{\Phi}_n = \begin{bmatrix} \bar{R}_n \\ 0 \end{bmatrix}$, $\bar{Q}_n W_n \bar{Y}_n = \begin{bmatrix} \breve{y}_n \\ \tilde{y}_n \end{bmatrix}$. At the $n+1^{th}$ sampling instant, a new system output $\bar{y}_{n+1}$ arrives at the identifier, the corresponding $\bar{\Phi}_{n+1}$ and $\bar{Y}_{n+1}$ becoming $\begin{bmatrix} w\bar{R}_n \\ \bar{\varphi}_{n+1}^T \end{bmatrix}$, $\begin{bmatrix} w\breve{y}_n \\ \bar{y}_{n+1} \end{bmatrix}$. There exists an orthogonal matrix $\bar{Q}_{n+1}$ such that

$$\bar{Q}_{n+1} \begin{bmatrix} w\bar{R}_n \\ \bar{\varphi}_{n+1}^T \end{bmatrix} = \begin{bmatrix} \bar{R}_{n+1} \\ 0 \end{bmatrix}, \bar{Q}_{n+1} \begin{bmatrix} w\breve{y}_n \\ \bar{y}_{n+1} \end{bmatrix} = \begin{bmatrix} \breve{y}_{n+1} \\ \tilde{y}_{n+1} \end{bmatrix} \tag{12}$$

where $\bar{\varphi}_{n+1}^T = \begin{bmatrix} \bar{y}_n & \bar{y}_{n-1} & \cdots & \bar{y}_{n-n_y+1} & \bar{u}_{n+1} & \bar{u}_n & \cdots & \bar{u}_{n-n_u+1} \end{bmatrix}$.

By partitioning the orthogonal matrix $\bar{Q}_{n+1}$, the recursive formula of parameter identification can be given by the following Theorem 1.

Theorem 1. *If the parameter vector $\hat{\Theta}_n$ is known a priori the recursive relation between $\hat{\Theta}_{n+1}$ and $\hat{\Theta}_n$ can be expressed by*

$$\hat{\Theta}_{n+1} = \hat{\Theta}_n + \frac{h_{n+1}}{f_{n+1}} e_{n+1}, \tag{13}$$

where $a_{n+1} = \bar{R}_n^{-T} \bar{\varphi}_{n+1}/w$, $f_{n+1} = \sqrt{1 + a_{n+1}^T a_{n+1}}$, $h_{n+1} = \bar{R}_n^{-1} a_{n+1}/w f_{n+1}$, $e_{n+1} = \bar{y}_{n+1} - \bar{\varphi}_{n+1}^T \hat{\Theta}_n$.

Proof: The orthogonal matrix $\bar{Q}_{n+1}$ can be expressed by the following partitioned matrix

$$\bar{Q}_{n+1} = \begin{bmatrix} S & g \\ z^T & \delta \end{bmatrix} \tag{14}$$

where S is $n \times n$, g and z are $n \times 1$, and δ is a scalar. Substituting (14) into (12), yields

$$\hat{\Theta}_{n+1} = \left[wS\bar{R}_n + g\bar{\varphi}_{n+1}^T \right]^{-1} \left[wS\bar{\bar{y}}_n + g\bar{y}_{n+1} \right]. \tag{15}$$

According to $\bar{Q}_{n+1}\bar{Q}_{n+1}^T = \mathbf{I}$, it is obvious that $g = -\frac{1}{\delta}Sz$. Substituting g into (15) thus gives

$$\hat{\Theta}_{n+1} = \left[\left(w\bar{R}_n - \frac{1}{\delta}z\bar{\varphi}_{n+1}^T \right) \right]^{-1} \left[w\bar{\bar{y}}_n - \frac{1}{\delta}z\bar{y}_{n+1} \right]. \tag{16}$$

Using (14) and $\bar{\Phi}_{n+1}$, the bottom partition of the result gives $z = -\frac{\delta \bar{R}_n^{-T} \bar{\varphi}_{n+1}}{w}$. Defining $a_{n+1} = \frac{\bar{R}_n^{-T} \bar{\varphi}_{n+1}}{w}$, then z becomes $z = -\delta a_{n+1}$. Substituting z into (16) produces

$$\hat{\Theta}_{n+1} = \left[\left(w\bar{R}_n + a_{n+1}\bar{\varphi}_{n+1}^T \right) \right]^{-1} \left[w\bar{\bar{y}}_n + a_{n+1}\bar{y}_{n+1} \right]. \tag{17}$$

Using the matrix inverse lemma [17], (17) can be re-written as

$$\hat{\Theta}_{n+1} = \left[w^{-1}\bar{R}_n^{-1} - \frac{w^{-1}\bar{R}_n^{-1} a_{n+1} a_{n+1}^T}{1 + a_{n+1}^T a_{n+1}} \right] \left[w\bar{\bar{y}}_n + a_{n+1}\bar{y}_{n+1} \right]. \tag{18}$$

Defining $f_{n+1}^2 = 1 + a_{n+1}^T a_{n+1}$, $h_{n+1} = \frac{w^{-1}\bar{R}_n^{-1} a_{n+1}}{f_{n+1}}$, (18) becomes

$$\hat{\Theta}_{n+1} = \bar{R}_n^{-1}\bar{\bar{y}}_n + \frac{h_{n+1}}{f_{n+1}} \left(\bar{y}_{n+1} - \bar{\varphi}_{n+1}^T \bar{R}_n^{-1}\bar{\bar{y}}_n \right). \tag{19}$$

Defining $e_{n+1} = \bar{y}_{n+1} - \bar{\varphi}_{n+1}^T \hat{\Theta}_n$, (19) can be computed further as

$$\hat{\Theta}_{n+1} = \hat{\Theta}_n + \frac{h_{n+1}}{f_{n+1}} e_{n+1}.$$

This completes the proof.

In Theorem 1, the upper triangular matrix $\bar{R}_{n+1}$ can be updated by [8], [10].

Online recursive identification method over networks with random packet losses can now be summarized as follows.

Step 1: Initialization: Set Θ_0 to some small value, forgetting factor w ($0 << w < 1$), the packet loss rate of input and output ρ ($0 < \rho < 1$), $R = diag(\varepsilon, \ldots, \varepsilon)$, ε is a small positive value (e.g. 10^{-6}), the variance of system noise σ^2, and the iteration index $k = 1$.

Step 2: Generate input and output data set: at the kth step, $\bar{y}(k)$ and $\bar{u}(k)$ are calculated by (4) and (5), and $\bar{\varphi}_{k+1}^T$ is produced.

Step 3: Update parameter estimation: Θ_{k+1} is updated according to (13), and $\bar{R}_{k+1}$ is updated by (12).

Step 4: The procedure is terminated when some criterion is satisfied, Otherwise, set $k = k + 1$, and go to step 2.

The whole procedure of the proposed method is shown in Fig.2.

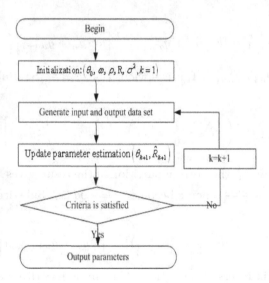

Fig. 2. The process of RLS

4   Simulation Examples

To show the effectiveness of the proposed algorithm, two examples were carried out using MATLAB.

Consider the following network system:

$$\bar{y}_k = \frac{B(z)}{A(z)}\bar{u}_k + v_k$$

where $\bar{u}_k = \alpha_k u_k + (1 - \alpha_k)u_{k-1}$, $u(k)$ is an M sequence input signal, $\bar{y}_k = \alpha_k y_k + (1 - \alpha_k)y_{k-1}$, $A(z) = 1 + a_1 z^{-1} + a_2 z^{-2} = 1 + 1.35z^{-1} - 0.75z^{-2}$, $B(z) =$

$b_1 z^{-1} + b_2 z^{-2} = 0.214 z^{-1} + 0.428 z^{-2}$, $\Theta = [a_1, a_2, b_1, b_2]^T = [\theta_1, \theta_2, \theta_3, \theta_4]^T = [1.35, -0.75, 0.214, 0.428]^T$, $\{v_k\}$ is a white noise sequence with variance $\sigma^2 = 0.25$.

Some parameters are set as $w = 1$, $\hat{\Theta}_0 = [0, 0, 0, 0]^T$, and the parameter estimation error is defined as

$$\delta_k = \frac{\left\| \hat{\theta}_k - \theta \right\|}{\|\theta\|}$$

To compare the identification performance under the different network environment, two experiments with different dropout rate have been carried out, i.e., $\rho = 0.1$ and $\rho = 0.4$.

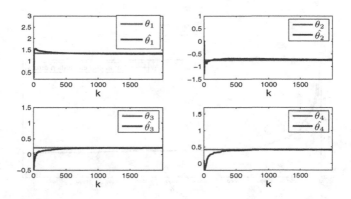

Fig. 3. The real parameter Θ and parameter estimation $\hat{\Theta}$ with $\rho = 0.1$

1) Under $\rho = 0.1$, it is seen clearly from Fig. 3 that the parameter identification results become better with the k increasing. The estimated parameters reach the real parameters after $k = 500$. Fig.4 shows that the estimated $\hat{y}$ is almost the same as the required y. It is obvious from Fig. 5 that the estimated variance σ_k is very small. It indicates that although there exists data dropout, the identification performance is reasonably good.

2) Under $\rho = 0.4$, it is also found from Fig. 6 that the estimated parameters reach gradually the real parameter with the k increasing. Fig.7 shows that the estimated $\hat{y}$ nearly coincides with y. Fig. 8 shows obviously that the estimated variance σ_k is very small. It again confirms that although there exists data dropout, the identification performance is acceptable.

Compared with the identification performance under two data dropouts, it is found that although the identification performance of the second case ($\rho = 0.4$) is slightly worse than that of the first case, the estimated parameters of the seconde case are still reasonably good. The effectiveness of the proposed method has thus been demonstrated.

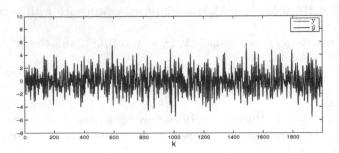

Fig. 4. The system outputs y and $\hat{y}$ with $\rho = 0.1$

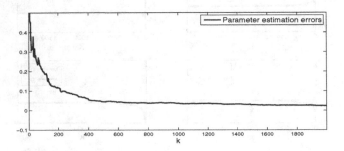

Fig. 5. The estimated variance σ_k $(\rho = 0.1)$

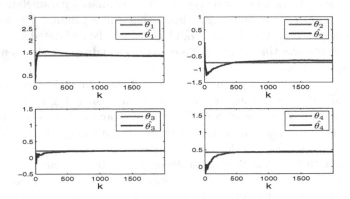

Fig. 6. The real parameter Θ and parameter estimation $\hat{\Theta}$ with $\rho = 0.4$

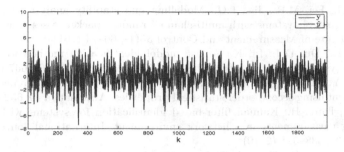

Fig. 7. The system outputs y and $\hat{y}$ with $\rho = 0.4$

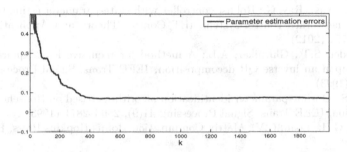

Fig. 8. The estimated variance σ_k ($\rho = 0.4$)

5   Conclusion

The paper has proposed an online recursive identification method over networks with random packet losses. In this algorithm, the Bernoulli processes was firstly employed to describe the character of data packet losses. An improved recursive least squares (RLS) algorithm was then presented where the intermediate matrix can be updated recursively. Simulation results from two simulated examples with different dropout rate demonstrated the effectiveness of the proposed method. Future work may include the convergence analysis for the online recursive identification method by considering the impact of network environment.

Acknowledgments. This work was supported in part by the National Science Foundation of China under Grant No. 61104089, project of Science and Technology Commission of Shanghai Municipality under Grant No. 11jc1404000, 12JC1404201 and Shanghai Rising-Star Program under Grant No. 13QA1401600.

References

1. Zhang, L., Gao, H., Kaynak, O.: Network-induced constraints in networked control systems-A survey. IEEE Transactions on Industrial Informatics 9(1), 403–416 (2013)

2. Du, D.J., Fei, M.R., Jia, T.G.: Modelling and stability analysis of MIMO networked control systems with multi-channel random packet losses. Transactions of the Institute of Measurement and Control 35(1), 66–74 (2011)

3. Fei, M.R., Du, D.J., Li, K.: A fast model identification method for networked control system. Applied Mathematics and Computation 205, 658–667 (2008)

4. Irshad, Y., Mossberg, M., Soderstrom, T.: System identification in a networked environment using second order statistical properties. Automatica 49, 652–659 (2013)

5. Shi, Y., Fang, H.: Kalman filter-based identification for systems with randomly missing measurements in a network environment. International Journal of Control 83(3), 538–551 (2010)

6. Shi, Y., Fang, H., Yan, M.: Kalman filter-based adaptive control for networked systems with unknown parameters and randomly missing outputs. International Journal of Roubst and Nonlinear Control 19(18), 1976–1992 (2009)

7. Sharma, A.K., Ray, G.: Robust controller with state-parameter estimation for uncertain networked control system. IET Control Theory and Applications 6(18), 2775–2784 (2012)

8. Alexander, S.T., Ghirnikar, A.L.: A method for recursive least squares filtering based upon an inverse QR decomposition. IEEE Trans. Signal Processing 41(1), 20–30 (1993)

9. Liu, Z.S.: On-line parameter identification algorithms based on Householder transformation. IEEE Trans. Signal Processing 41(9), 2863–2871 (1993)

10. Golub, G.H., Loan, C.V.: Matrix Computation. Johns Hopkins Press, Baltimore

Output Feedback Reliable H_∞ Control for Networked Control System

Zhou Gu[1,2], Shumin Fei[1], Baochun Zhuang[2], and Guangtao Shi[2]

[1] School of Automation, Southeast University,
Nanjing 210096, P.R. China
[2] College of Mechanical & Electronic Engineering,
Nanjing Forestry University, Nanjing, 201137, P.R. China
gzh1808@163.com

Abstract. This paper investigates the static output feedback control for networked control systems with actuator failure. The failure is composed of two parts, the linear deficiency of the control gain and the nonlinearity varying with the control input. Based on the more general actuator fault model, linear matrix inequality (LMI) optimization approach is used to design the reliable H_∞ output feedback control. Finally, an example is provided to demonstrate the design method.

Keywords: Actuator failure, Networked control system, Reliable control.

1 Introduction

Networked control systems (NCSs) are feedback control systems wherein the control loops are closed via real-time networks [1,2,3,4]. The past decade has witnessed continued research interest on various aspects of networked control systems, since this new type of information transmission reduces system wiring, eases maintenance and diagnosis, and increases system agility, which makes NCS a promising structure for control systems. Nevertheless, the introduction of networks also brings some new problems and challenges, such as network-induced delay, packet dropout, network scheduling and quantization problems, etc..

Most of the references concerning with NCSs assumed that the actuator/sensor operates without any flaws, i.e., the data transmit to the receiving device without any deviation, which is, unfortunately, not true in the practical systems. Due to the aging, external disturbance, etc., the actuator/sensor failure becomes a common problem [5,6]. It is worth pointing out that it plays a key role to develop a reasonable fault model in analyzing and synthesizing reliable control systems. Taking the actuator fault for example, researchers modeled the actuator fault as

$$u^F(t) = \Phi u(t) \tag{1}$$

where $u(t)$ is the true signal of the control input to be sent to the actuator device, $u^F(t)$ is the real signal of the receiver with a certain fault, $\Phi = diag\{\phi_1, \cdots, \phi_m\}$ is

M. Fei et al. (Eds.): LSMS/ICSEE 2014, Part II, CCIS 462, pp. 459–467, 2014.
© Springer-Verlag Berlin Heidelberg 2014

a fault factor which reflects the gain missing of the actuator. In [7,8], is defined to satisfy $\phi_i \in \{0, 1\}$, if $\phi_i = 0$, it means a complete failure, otherwise, $\Phi_i = 1$ denotes the device works normally. In \cite [9,10], the authors let $\phi \in [0\ 1]$, which depicts the gain of the faulty device varies from 0 to 1. In [11,12], ϕ obeys a Gaussian distribution to characterize a random fault. Most of the existed references regard the faulty device as a linear gain missing. However, the transmission characteristic of the faulty device does not always present multiplicative fault, sometimes it adhered with a certain nonlinearity. To the best of our knowledge, few authors dealt with this problem, which motivates us to further investigate the problem.

In this paper, the closed-loop NCS with the actuator failure presented in Section 2, in which the failure is composed of linear and nonlinear parts. The reliable control design method is provided in Section 3. Section 4 presents the design results and simulations. Finally, the study's findings are summarized in Section 5.

2   Problem Formulation

In this paper, we will study the reliable control for networked control system with the actuator failures , in which the plant is given by a continuous-time linear model of the form

$$\mathcal{P} : \begin{cases} \dot{x}(t) = Ax(t) + Bu(t) + D\omega(t) \\ y(t) = C_1 x(t) \\ z(t) = C_2 x(t) \\ x(t) = \varphi(t) \quad t \in [-\eta_2, 0) \end{cases} \tag{2}$$

where $x(t) \in \mathbb{R}^n$ is the state vector, $y(t) \in \mathbb{R}^{q_1}$ is the controlled output vector, $z(t) \in \mathbb{R}^{q_2}$ is output vector, $u(t) \in \mathbb{R}^m$ is the control input, and the process noise $\omega(t) \in \mathbb{R}^d$ including model uncertainties and external plant disturbance belongs to $l_2[0\ \infty]$. A, B, C_1, C_2, and D are constant matrices with appropriate dimensions.

Here we do the following assumptions,   which is widely adopted in NCS, that

- sensors are clock-driven; controllers and actuators are event-driven;
- the sampling data are hold by ZOH before the new event updates.
- the data are transmitted over the network by a single packet in every control period.

Under the above assumptions together with our previous work [13,14], the system (2) can be further written as

$$\dot{x}(t) = Ax(t) + Bu(i_k h) + D\omega(t) \tag{3}$$

for $t \in [i_k h + \tau_{i_k}, i_{k+1} h + \tau_{i_k+1})$, where h is the sensor sampling period, $i_k(k = 1, 2, 3, \cdots)$ is some non-negative integer and $i_k h$ is the sensor sampling instant. It can be obviously see that $\cup_{k=1}^{\infty} [i_k h + \tau_{i_k}, i_{k+1} h + \tau_{i_k+1}) = [0\ \infty)$.

We consider the following actuator fault model

$$u^F(t) = \Xi_1 u(t) + g(u(t)) \tag{4}$$

Where $0 < \Xi_1 = diag\{e_1, e_2, \cdots, e_m\} \leq I$ and   the   vector   function
$g(u(t)) = [g_1(u_1(t)), g_2(u_2(t)), \cdots, g_m(u_m(t))]^T$. For each channels of the control
input, it satisfies

$$|g_i(u_i(t))| \leq \alpha_i u_i(t) \tag{5}$$

where $0 \leq \alpha_i$ and $e_i + \alpha_i \leq 1$ for$i \in \mathcal{I} = \{i | i = 1, 2, \cdots, m\}$.
From (5), it follows

$$g^T(u(t))g(u(t)) \leq u^T(t)\Xi_2 u(t) \tag{6}$$

where$\Xi_2 = diag\{\alpha_1, \alpha_2, \cdots, \alpha_m\}$.

Remark 1. Let $\Xi_2 \equiv 0$, it means the actuator works in a certain fixed missing gain.
Specifically, if we choose $e_i = 0 (i \in \mathcal{I})$, it means that the i-th channel of the control
input is complete failure, and we choose $e_i = 1 (i \in \mathcal{I})$, it denotes that the actuator is
intactness, otherwise, the feature of partial actuator failure can be described by
defining $0 < e_i < 1$.

Remark 2. If Ξ_2 satisfies (6) and$\Xi_2 \neq 0$, then the failure model (4) characterizes the
actuator failure in each channels is time-varying   and   coupled with different levels
of line/nonlinear nature.

Consider the following memoryless output feedback controller

$$u(t) = u(i_k h) = Ky(i_k h) \tag{7}$$

For $t \in [i_k h + \tau_{i_k}, i_{k+1} h + \tau_{i_k+1})$, where K is a controller gain to be designed.
Combining (2), (4) and (7), we can obtain the following closed-loop system with
consideration of the actuator fault for $t \in [i_k h + \tau_{i_k}, i_{k+1} h + \tau_{i_k+1})$ as

$$\dot{x}(t) = Ax(t) + B\Xi_1 KCx(i_k h) + Bg(u(i_k h)) + D\omega(t) \tag{8}$$

Define $\eta(t) = t - i_k h$ in every interval $[i_k h + \tau_{i_k}, i_{k+1} h + \tau_{i_k+1})$, it follows that

$$i_k h = t - (t - i_k h) = t - \eta(t) \tag{9}$$

Then we have

$$\eta_1 \leq \tau_{i_k} \leq \eta(t) \leq (i_{k+1} - i_k)h + \tau_{i_k+1} \leq \eta_2$$

So far, the closed-loop NCS with consideration of the actuator fault can be further
written as

$$\begin{cases} \dot{x}(t) = \mathcal{A}\xi(t) \\ z(t) = C_2 x(t) \end{cases} \tag{10}$$

For $t \in [i_k h + \tau_{i_k}, i_{k+1} h + \tau_{i_k+1})$,   where $\mathcal{A} = \begin{bmatrix} A & 0 & B\Xi_1 KC_1 & 0 & B & D \end{bmatrix}$,
$\xi(t) = [x^T(t) \ x^T(t-\eta_1) \ x^T(t-\eta(t)) \ x^T(t-\eta_2) \ g^T(u(t-\eta(t))) \ \omega^T(t)]^T$.

From the definition of $\eta(t)$, the nonlinear constraint condition in (6) in the time
interval $t \in [i_k h + \tau_{i_k}, i_{k+1} h + \tau_{i_k+1})$ can be described as

$$\xi^T(t)[\Pi_2^T\Pi_2 - \Pi_1^T\Xi_2\Pi_1]\xi(t) \le 0 \qquad (11)$$

where$\Pi_1 = \begin{bmatrix} 0 & 0 & KC_1 & 0 & 0 & 0 \end{bmatrix}, \Pi_2 = \begin{bmatrix} 0 & 0 & 0 & 0 & I & 0 \end{bmatrix}.$.

3   Main Results

In this section, the output feedback reliable controller will be designed. Before we give the Theorems, the following lemma are first introduced, which will be used in the subsequent development.

Lemma 1. [15,16] Let $Y_0(\xi(t)), Y_1(\xi(t)), \cdots, Y_p(\xi(t))$ be quadratic functions of $\xi(t) \in \mathbb{R}^n$

$$Y_i(\xi(t)) = \xi(t)^T T_i \xi(t), i = 0, 1, \cdots, p \qquad (12)$$

with$T_i = T_i^T$. Then, the implication

$$Y_0(\xi(t)) \le 0, \cdots, Y_p(\xi(t)) \le 0 \Longrightarrow Y_0(\xi(t)) \le 0 \qquad (13)$$

holds if there exist $\kappa_1, \cdots, \kappa_p > 0$ such that

$$T_0 - \sum_{i=1}^{p} \kappa_i^{-1} T_i \le 0 \qquad (14)$$

Theorem 1. For some given constants $\eta_1, \eta_2, \gamma, e_i, \alpha_i (i \in \mathcal{I})$ and matrix K, the closed-loop system (10) satisfies H_∞ performance criterion, if there exist real matrices $P > 0, Q_1 > 0, Q_2 > 0$ and $R > 0$ with appropriate dimensions, such that the following inequality hold

$$\begin{bmatrix} \Theta_0 - \kappa^{-1}\Pi_2^T\Pi_2 & * & * & * \\ (\eta_2 - \eta_1)PA & -PR^{-1}P & * & * \\ \mathcal{C} & 0 & -I & * \\ \Pi_1 & 0 & 0 & -\kappa\Xi_2^{-1} \end{bmatrix} \le 0 \qquad (15)$$

where

$$\Theta_0 = \begin{bmatrix} \Gamma_0^{11} & * & * & * & * & * \\ \Gamma_0^{21} & \Gamma_0^{22} & * & * & * & * \\ 0 & R & -2R & * & * & * \\ 0 & 0 & R & -Q_2 - R & * & * \\ B^T P & 0 & 0 & 0 & 0 & 0 \\ D^T P & 0 & 0 & 0 & 0 & -\gamma^2 I \end{bmatrix},$$

$$\Gamma_0^{11} = PA + A^T P + Q_1 + Q_2,$$
$$\Gamma_0^{21} = C_1^T K^T \Xi_1^T B^T P, \Gamma_0^{22} = -Q_1 - R$$
$$\mathcal{C} = [C_2\ 0\ 0\ 0\ 0\ 0]$$

Proof: Choose a Lyapunov functional candidate for the system (10) as

$$V(x_t) = V_1(x_t) + V_2(x_t) + V_3(x_t)$$
$$V_1(x_t) = x^T(t)Px(t)$$
$$V_2(x_t) = \sum_{i=1}^{2} \int_{t-\eta_i}^{t} x^T(s)Q_i x(s)ds$$
$$V_3(x_t) = (\eta_2 - \eta_1)\int_{t-\eta_2}^{t-\eta_1} \int_{s}^{t} \dot{x}^T(v)R\dot{x}(v)dvds$$

Taking the derivative of $V(x_t)$ with respect to t along the trajectory of (10) yields

$$\begin{aligned}
\dot{V}(x_t) &= 2x^T(t)PA\xi(t) \\
&+ \sum_{i=1}^{2} x^T(t)Q_i x(t) + \sum_{i=1}^{2} x^T(t-\eta_i)Q_i x(t-\eta_i) \\
&+ (\eta_2 - \eta_1)^2 \dot{x}^T(t)R\dot{x}(t) - (\eta_2 - \eta_1)\int_{t-\eta_2}^{t-\eta_1} \dot{x}^T(s)R\dot{x}(s)ds
\end{aligned}$$

From Lemma 1 in [15], it yields

$$z^T(t)z(t) - \gamma^2\omega^T(t)\omega(t) + \dot{V}(t) \leq \xi^T(t)[\Theta_0 + (\eta_2 - \eta_1)^2 A^T RA + C^T C]\xi(t)$$

Applying Schur complement and Lemma 1, one can see that Eq.(15) is a sufficient condition to guarantee Eq.(11) and

$$z^T(t)z(t) - \gamma^2\omega^T(t)\omega(t) + \dot{V}(t) \leq 0 \qquad (16)$$

Under zero initial conditions, integrating both side of Eq.(16) yields

$$V(t) \leq \int_0^t [\gamma^2\omega^T(t)\omega(t) - z^T(t)z(t)]dt \qquad (17)$$

Then we can conclude that $\|z(t)\|_2 \leq \|\omega(t)\|_2$ for all nonzero $\omega(t) \in l_2[0,\infty)$. H_∞ performance is established. The proof is completed.

With the result of Theorem 1, the reliable H_∞ control for system(2) with the control law (4) is provided in the following result.

Theorem 2. For some given constants $\eta_1, \eta_2, \gamma, e_i, \alpha_i (i \in \mathcal{I}), \varepsilon$, the closed-loop system (10) satisfies H_∞ performance criterion, if there exist real matrices $P > 0, Q_1 > 0, Q_2 > 0$ and $R > 0$ with appropriate dimensions, such that the following equalities hold

$$\begin{bmatrix}
\Theta_1 - \kappa\Pi_2^T\Pi_2 & * & * & * \\
(\eta_2 - \eta_1)\Pi_3 & -2\varepsilon X + \varepsilon^2\bar{R} & * & * \\
\bar{C} & 0 & -I & * \\
\bar{\Pi}_1 & 0 & 0 & -\kappa\Xi_2^{-1}
\end{bmatrix} \leq 0 \qquad (18)$$

$$C_1 X = W C_1 \qquad (19)$$

Moreover, the controller gain is given by $K = YW^{-1}$, where

$$
\Theta_1 = \begin{bmatrix}
\Gamma_1^{11} & * & * & * & * & * \\
\Gamma_1^{21} & \Gamma_1^{22} & * & * & * & * \\
0 & \bar{R} & -2\bar{R} & * & * & * \\
0 & 0 & \bar{R} & -\bar{Q}_2 - \bar{R} & * & * \\
\kappa B^T & 0 & 0 & 0 & 0 & 0 \\
D^T & 0 & 0 & 0 & 0 & -\gamma^2 I
\end{bmatrix},
$$

$$
\begin{aligned}
\Gamma_1^{11} &= AX + XA^T + \bar{Q}_1 + \bar{Q}_2, \\
\Gamma_1^{21} &= C_1^T Y^T \Xi_1^T B^T, \Gamma_1^{22} = -\bar{Q}_1 - \bar{R} \\
\bar{C} &= [C_2 X\ 0\ 0\ 0\ 0\ 0], \\
\bar{\Pi}_1 &= [0\ \ 0\ \ KWC_1\ \ 0\ \ 0\ \ 0], \\
\Pi_3 &= [AX\ \ 0\ \ B\Xi_1 YC_1\ \ 0\ \ \kappa B\ \ D]
\end{aligned}
$$

Proof: Note that

$$
(\varepsilon R - P)R^{-1}(\varepsilon R - P) \geq 0 \tag{20}
$$

where ε is a positive scalar. Then it is true that

$$
-PR^{-1}P \leq -2\varepsilon P + \varepsilon^2 R \tag{21}
$$

It follows that

$$
\begin{bmatrix}
\Theta_0 - \kappa \Pi_2^T \Pi_2 & * & * & * \\
(\eta_2 - \eta_1)P\mathcal{A} & -2\varepsilon P + \varepsilon^2 R & * & * \\
\mathcal{C} & 0 & -I & * \\
\Pi_1 & 0 & 0 & -\kappa \Xi_2^{-1}
\end{bmatrix} \leq 0 \tag{22}
$$

from Eq. (15).

Defining $\quad X = P^{-1}, \bar{Q}_1 = XQ_1X, \bar{Q}_2 = XQ_2X, \bar{R} = XRX, Y = KW \quad$, $J = diag\{X, X, X, X, \kappa, I, X, I, I\}$, pre- and post-multiplying (22) with J and its transposes, respectively. It can obviously see that (22) is equivalent to (18) under the condition of (19). This completes the proof.

One can see that it is difficult to find a feasible solution by Theorem 2 since Eq.(19) is not a strict inequality. Now we introduce the following algorithm to address this problem.

It is clear that Eq.(19) is equivalent to

$$
\text{trace}\left[(CX - WC)^T(CX - WC)\right] = 0 \tag{23}
$$

which can be converted to the following optimization problem by using Schur complement

$$
\begin{cases}
\begin{bmatrix}
-\sigma I & * \\
WC - CX & -I
\end{bmatrix} < 0 \\
\sigma \to 0
\end{cases} \tag{24}
$$

where the scalar σ is a small enough positive scalar. Then the controller gain can be resolved by (18),(19) and (24).

Remark: To obtain a feasible solution of the SOF controller gain K, in some existed results, the output matrix C_1 is assumed to be invertible or some intelligent optimization algorithms are applied to find a sub-optimum solution. For the sake of technical simplicity, we take the above algorithm, in this paper, to tackle this problem.

4   A Numerical Example

An example of networked control for an unstable batch reactor [18] is used in this section to demonstrate the effectiveness of the proposed approach. The plant matrices are

$$A = \begin{bmatrix} 1.3800 & -0.2077 & 6.7150 & -5.6760 \\ 0.5814 & -4.2900 & 0 & 0.6750 \\ 1.0670 & 4.2730 & -6.6540 & 5.8930 \\ 0.0480 & 4.2730 & 1.3430 & -2.1040 \end{bmatrix},$$

$$B = \begin{bmatrix} 0 & 0 \\ 5.6790 & 0 \\ 1.1360 & -3.1460 \\ 1.1360 & 0 \end{bmatrix}, C_1 = \begin{bmatrix} 0 & 1 & 0 & 1 \\ 1 & 0 & 1 & 0 \end{bmatrix},$$

$$C_2 = \begin{bmatrix} 0 & 1 & 0 & 1 \\ 1 & 0 & 1 & 0 \end{bmatrix}, D = \begin{bmatrix} 1 & 1 & 1 & 1 \end{bmatrix}^T$$

The output $y(t)$ is communicated over the network whose parameters are assumed as $\eta_1 = 5$ ms and $\eta_2 = 50$ ms, respectively. The disturbance $w(t) = 4\exp^{-(2t-2)^2}$. Next, we will study the two different actuator   failure scenarios, which are described as

$$\begin{cases} u_1^F(t) = 0.9u(t) + g_1(u_1(t)) \\ u_2^F(t) = 0.6u(t) + g_2(u_2(t)) \end{cases} \tag{25}$$

where $g(\cdot)$ satisfies $g_1(u_1(t)) = 0.1sat(u_1(t))$ and $g_2(u_2(t)) = 0.02sat(u_2(t))$, respectively.

Under these failure scenarios, the reliable controller K can be obtained as

$$K = \begin{bmatrix} -1.3609 & -0.1032 \\ -0.6563 & 4.4042 \end{bmatrix} \tag{26}$$

by using Theorem 2 together with its algorithm.

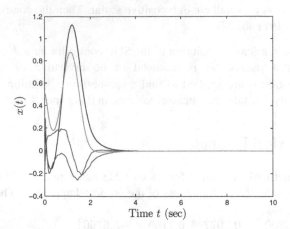

Fig. 1. State response x(t)

Simulation results are provided as Fig. 1 under the initial condition $\varphi(t) = [0.2 \ 0.3 \ -0.2 \ 0.5]^T$. In this study, the control inputs are of 90% and 60% linear deficiency, respectively, meanwhile there are some different levels of saturation adhered on those two actuators. It is obviously observed from Fig. 1 that the reliable controller implemented on the system under the above failure scenarios can maintain the systems performance.

5   Conclusion

In this paper, the reliable output feedback control design for the NCS with actuator failure has been developed. Based on a more general actuator fault model, the reliable H_∞ SOF controller is derived by using Lyapunov method, which guarantees the closed-loop system satisfies a desired H_∞ disturbance attenuation constraint. An illustrative example is given to show the validity of the present control scheme.

Acknowledgement. This work was supported by the National Natural Science Foundation of China (GrantNo. 61273119,61273115,61374038), Research Fund for the Doctoral Program of Higher Education of China (20110092110021) and the Natural Science Foundation of Jiangsu Province of China(Grant No. BK2012469,BK2011253).

References

1. Yang, T.C.: Networked control system: A brief survey. IEE Proceedings-Control Theory and Applications 153(4), 403–412 (2006)
2. Zhang, W., Branicky, M.S., Phillips, S.M.: Stability of networked control systems. IEEE Control Systems 21(1), 84–99 (2001)

3. Yue, D., Han, Q.L., Lam, J.: Network-based robust H_∞ control of systems with uncertainty. Automatica 41(6), 999–1007 (2005)
4. Tian, E., Yue, D., Peng, C.: Quantized output feedback control for networked control systems. Information Sciences 178(12), 2734–2749 (2008)
5. Lien, C.H., Yu, K.W., Chang, H.C., Chung, L.Y., Chen. J.D.: Robust reliable guaranteed cost control for uncertain uncertain T-S fuzzy neutral systems with interval time-varying delay and linear fractional perturbations. Optimal Control Applications and Methods (2014)
6. Li, H., Liu, H., Gao, H., Shi, P.: Reliable fuzzy control for active suspension systems with actuator delay and fault. IEEE Transactions on Fuzzy Systems 20(2), 342–357 (2012)
7. Wu, H.N., Zhang, H.Y.: Reliable mixed l2/h fuzzy static output feedback control for nonlinear systems with sensor faults. Automatica 41(11), 1925–1932 (2005)
8. Aiqing, Z., Huajing, F.: Robust reliable guaranteed cost control for nonlinear singular stochastic systems with time delay. Journal of Systems Engineering and Electronics 19(4), 791–798 (2008)
9. Wang, Z., Huang, B., Burnbam, K.: Stochastic reliable control of a class of uncertain time-delay systems with unknown nonlinearities. IEEE Transactions on Circuits and Systems I: Fundamental Theory and Applications 48(5), 646–650 (2001)
10. Lien, C., Yu, K.: Robust reliable control for uncertain time-delay systems with IQC performance. Journal of Optimization Theory and Applications 138(2), 235–251 (2008)
11. Gu, Z., Tian, E., Liu, J.: Reliable filter design for sampled-data systems with consideration of probabilistic sensor signal distortion. IET Signal Processing 7(5), 420–426 (2013)
12. Wei, G., Wang, Z., Shu, H.: Robust filtering with stochastic nonlinearities and multiple missing measurements. Automatica 45(3), 836–841 (2009)
13. Gu, Z., Wang, D., Yue, D.: Fault detection for continuous-time networked control systems with non-ideal QOS. Int. J. Innov. Comput., Inf. Control 6(8), 3631–3640 (2010)
14. Gu, Z., Tian, E., Liu, J., Huang, L., Zou, H., Zhao, Y.: Network-based precise tracking control of systems subject to stochastic failure and non-zero input. IET Control Theory & Applications 7(10), 1370–1376 (2013)
15. Gu, K., Kharitonov, V., Chen, J.: Stability and robust stability of time-delay systems (2003)
16. Savkin, A., Petersen, I.: Robust state estimation and model validation for discrete-time uncertain systems with a deterministic description of noise and uncertainty. Automatica 34(2), 271–274 (1998)
17. Yang, F., Li, Y.: Set-membership filtering for systems with sensor saturation. Automatica 45(8), 1896–1902 (2009)
18. Walsh, G.C., Ye, H., Bushnell, L.G.: Stability analysis of networked control system. IEEE Transactions on Control Systems Technology 10(3), 438–446 (2002)

A Novel Adaptive Event-Triggered Communication Scheme for Networked Control Systems with Nonlinearities

Jin Zhang[1], Chen Peng[1,2], and Dacheng Peng[2]

[1] School of Mechatronic Engineering and Automation,
Shanghai University, Shanghai, China
`zhangjin1116@126.com`
[2] School of Electrical and Automation Engineering,
Nanjing Normal University, Jiangsu, China
`pchme@163.com`

Abstract. This paper presents a novel adaptive event-triggered communication scheme for networked control systems (NCSs) with nonlinearities. Firstly, a novel adaptive event-triggered communication scheme for NCSs with nonlinearities is proposed, which can adaptively adjust the event-triggered communication threshold with respect to dynamic error to save the limited communication resource while ensuring the desired control performance. Secondly, a model of the considered system is built under consideration of the network-induced delay, adaptive event-triggered communication scheme and nonlinearities in a unified framework. Then, sufficient stability and stabilization criteria are obtained to judge the mean-square sense asymptotically stable for the studied system. Finally, two examples illustrate the effectiveness of the developed method.

Keywords: Networked control systems, adaptive event-triggered communication scheme, nonlinearities.

1 Introduction

With the rapid development of network technologies, networked control systems (NCSs) have received considered attention because of their own advantages such as low cost, increased system flexibility and easy maintenance [1]. However, the insertion of communication network causes some challenging problems on account of limited network bandwidth, which leads to performance degradation or even instability of an NCS. In the past few years, significant consideration has been focused on stability analysis and controller synthesis for NCSs, see, for example, [2], [3], [4] and references therein.

Notice that the network in NCSs is the shared band-limited communication network [5]. Considering the limited network bandwidth, event-triggered communication scheme has emerged as an alternative approach to minimize

M. Fei et al. (Eds.): LSMS/ICSEE 2014, Part II, CCIS 462, pp. 468–477, 2014.

the use of the communication resources while retaining a satisfactory performance of closed-loop system [6], [7], [8]. In recent years, significant consideration has been focused on the event-triggered control for linear system [9], [10], [11], fuzzy system [12], [13] as well as multi-agent system [14], [15]. Note that most of existing ones are based on a "static" event-triggered scheme, that is, $[x(jh) - x(t_kh)]^TV[x(jh) - x(t_kh)] \leq \sigma x^T(jh)Vx(jh)$, where V is a symmetric positive definite matrix, $x(jh)$, $x(t_kh)$ are the newly sampled and last released state, respectively, $jh = t_kh + lh$, $l = 1, 2, \cdots$. Since the trigger parameter σ is a given constant, it cannot dynamically adjust the sampling interval based on the realtime error, i.e., $\|x(jh) - x(t_kh)\|_2$ of the controlled system. How to design an adaptive event-triggered communication scheme to save the limited communication resources while keeping the desired control performance is still open. This is the motivation of this work.

In this paper, we are motivated to develop an adaptive event-triggered communication scheme for a class of uncertain NCSs with nonlinearities. Compared with the traditional communication scheme in the literature, the adaptive communication scheme shows an effective way to keep balance for the system control performance and network bandwidth burden. Then a novel NCS model is build considering the effect of the adaptive communication scheme, transmission delay, nonlinearities. By use of Lyapunov functional method, the criteria for the asymptotical mean-square stability analysis and control synthesis are established in terms of linear matrix inequalities. The main contributions of this paper can be listed as follows:
1) a novel adaptive event-triggered communication scheme for NCSs is proposed to tradeoff the relationship between the control performance and the network bandwidth burden;
2) an integrated model is built under consideration of the adaptive communication scheme, transmission delay, nonlinearities in an unified framework.

2   System Framework

The physical plant is given by the following continuous-time system

$$\dot{x}(t) = Ax(t) + Bu(t) + h(t, x(t)) \tag{1}$$

where $x(t) \in \mathbb{R}^n$ and $u(t) \in \mathbb{R}^m$ are the system state vector and control input vector, respectively; A and B are the parameter matrices with randomly occurring uncertainties satisfying the following condition

$$A = A_0 + \triangle A(t) \ B = B_0 + \triangle B(t) \tag{2}$$

$$[\triangle A(t) \ \triangle B(t)] = GF(t)[E_a \ E_b] \tag{3}$$

where A_0, B_0 G, E_a and E_b are constant matrices with appropriate dimensions and $F(t)$ is an unknown time-varying matrix satisfying $\|F^T(t)F(t)\| \leq I$. The function $h(t, x(t)) : \mathbb{R} \times \mathbb{R}^n \to \mathbb{R}^n$ is assumed to be a piecewise-continuous nonlinear function in both arguments t and x, and satisfies $h(t, 0) = 0$ and the following condition

$$h^T(t, x(t))h(t, x(t)) \leq \kappa^2 x^T(t)H^T Hx(t) \tag{4}$$

where $\kappa > 0$ is the bounding parameter on the uncertain function $h(t, x(t))$ and H is a constant matrix.

For our NCS modeling, the following assumptions can be made, which are common in the NCS research in the literature.

Assumption 1. The sensors are time-driven with a constant sampling period h, while the controllers and actuators are event-driven.

Assumption 2. An adaptive event-triggered communications scheme is embedded in an event generator, which determines the sampled data should be released or not. The released instant sequence is described by the set $\{t_0h, t_1h, \ldots, t_kh\}$.

Assumption 3. The holding time of a zero order hold (ZOH) at the actuator is $[t_kh + \eta_k, t_{k+1}h + \eta_{k+1})$, where η_k is a bounded delay satisfying $\eta_m \leq \eta_k \leq \eta_M$.

2.1 Adaptive Event-Triggered Communication Scheme

If all state variables of the physical plant are measurable, the following event-triggered communication scheme is used to trigger the transmission

$$t_{k+1}h = t_kh + \min\{nh \mid e^T(i_kh)\Phi e(i_kh) > \sigma(i_kh)x^T(i_kh)\Phi x(i_kh)\} \tag{5}$$

where $\Phi > 0$ is a weighting matrix, $e(i_kh)$ is the error between the two states at the current sampling instant i_kh and the latest transmitted sampling instant t_kh, where $i_kh = t_kh + nh$, $n \in \mathbb{R}^+$, h is the sampling period. t_k $(k = 0, 1, 2, \cdots)$ are some integers such that $\{t_0, t_1, t_2, \cdots\} \subset \{0, 1, 2, \cdots\}$. Then it is clear that $t_1h < t_2h < \cdots < t_kh$ with the assumption that packet dropouts and packet disorders do not occur.

Different from some existing ones, we consider an adaptive event-triggered communication scheme herein, in which the trigger parameter $\sigma(t)$ is time-varying and presents a differential function satisfying

$$\dot{\sigma}(t) = d\sigma(t) \tag{6}$$

where $\sigma(0) \in [0, 1]$ is the initial condition and

$$d = \begin{cases} 1, & \text{if } e^T(i_kh)e(i_kh) < \rho \\ 0, & \text{if } e^T(i_kh)e(i_kh) = \rho \\ -1, & \text{if } e^T(i_kh)e(i_kh) > \rho \end{cases} \tag{7}$$

with a non-negative constant ρ. For the theoretical development easier, we assume that the parameter $\sigma(t)$ is bounded and varies within a interval $[\sigma_m, \sigma_M]$, where σ_m an σ_M are the lower and upper bounds of $\sigma(t)$, respectively, and when $\sigma(t) = \sigma_m$ and $d = -1$, the trigger parameter $\sigma(t)$ remains as σ_m until the next time $d = 1$. Moreover, if $\sigma(t) = \sigma_M$ and $d = 1$, $\sigma(t)$ is also assumed to hold until the latest future time $d = -1$.

Remark 1. Notice that if $\sigma(i_k h)$ is a constant, the event-triggered scheme is simplified as the communication scheme in [12]. Therefore, the proposed scheme in [12] is a special case of the proposed scheme in this work. Moreover, one can see from (5) that the next released instant $t_{k+1}h$ depends not only on the error $e(i_k h)$, but also on the latest released state $x(t_k h)$.

Remark 2. Notice that the trigger parameter $\sigma(t)$ changes depending on the comparison between the error function $e^T(i_k h)e(i_k h)$ and ρ. If $e^T(i_k h)e(i_k h) > \rho$, it is clear that the error between the newly sampled vector $x(i_k h)$ and the latest released state $x(t_k h)$ is larger than the threshold $\sqrt{\rho}$. However, the newly sampled vector $x(i_k h)$ cannot be released due to that $\sigma(i_k h)$ is large, which results in deteriorating the control performance of NCSs. From (7), we can obtain $d = -1$ such that the parameter $\sigma(t)$ appears a decreasing trend in $t \in [i_k h, i_k h + h)$. Then the next sampling instant $i_k h + h$ can more easily be the next released instant because of $\sigma(i_k h + h) < \sigma(i_k h)$, which remains the performance to some certain extent. However, if the function (6) and (7) are not considered, the next released instant can be far bigger than $i_k h + h$. In contrast when $e^T(i_k h)e(i_k h) < \rho$, there exits $d = 1$, which means that the parameter $\sigma(t)$ will grow bigger and the bandwidth burden will be saved. Briefly speaking, the proposed adaptive event-triggered scheme could provide a balance relationship between the system performance and the network bandwidth burden effectively.

2.2   Modeling of Closed-Loop NCS

Consider the behavior of ZOH, a state feedback control law is chosen as

$$u(t) = Kx(t_k h), \quad t \in [t_k h + \eta_k, t_{k+1}h + \eta_{k+1}) \tag{8}$$

where K is the controller gain to be determined.

   Inspired by [17], let

$$l_k = \min\{n | t_k h + nh + \eta_k \geq t_{k+1} + \eta_{k+1}\} \tag{9}$$

Then the interval $[t_k h + \eta_k, t_{k+1}h + \eta_{k+1})$ can be expressed as

$$[t_k h + \eta_k, t_{k+1}h + \eta_{k+1}) = \bigcup_{n=0}^{l_k} \Omega_n \tag{10}$$

where $\Omega_n = [t_k h + nh + \eta_k, t_k h + nh + h + \eta_k)$, $n = 0, \cdots, l_k - 2$ and $\Omega_{l_k} = [t_k h + l_k h - h + \eta_k, t_{k+1}h + \eta_{k+1})$.

   Define $\tau(t) = t - i_k h$, for $i_k h = t_k h + nh$, $t \in \Omega_n$. It is clear that $\tau(t)$ is a piecewise-linear function satisfying

$$\eta_m \leq \eta_k \leq \tau(t) \leq h + \max\{\eta_k, \eta_{k+1}\} \leq h + \eta_M \quad t \in \Omega_n \tag{11}$$

Then the control law (8) becomes

$$u(t) = K(x(t - \tau(t)) - e(i_k h)) \ t \in \Omega_n \tag{12}$$

Set $\tau_1 = \eta_m$, $\tau_2 = h + \eta_M$. Combining (1) and (12) leads to the following closed-loop system

$$\dot{x}(t) = Ax(t) + BKx(t - \tau(t)) - BKe(i_k h) + h(t, x(t)) \ t \in \Omega_n \tag{13}$$

where the initial condition of state $x(t)$ is $x(t) = \phi(t)$ with $\phi(t_0) = x_0$, $\phi(t)$ is a continuous function on $t \in [t_0 - \tau_2, t_0 - \tau_1]$. From (5) and the above division (10), one can see that non-triggered data packets satisfy the following inequality

$$e^T(i_k h)\Phi e(i_k h) < \sigma_M x^T(t - \tau(t))\Phi x(t - \tau(t)) \tag{14}$$

3   Main Results

In this section, we develop an approach for stability analysis and controller synthesis of for the closed-loop system (13). The following Lyapunov functional candidate $V(t)$ will be used in deriving our result. Due to page limitation, the detail proof omits herein.

$$V(t) = x^T(t)Px(t) + \int_{t-\tau_1}^t x^T(s)Q_i x(s)ds + \int_{t-\tau_2}^{t-\tau_1} x^T(s)Q_i x(s)ds$$
$$+ \tau_1 \int_{t-\tau_1}^t \int_s^t \dot{x}^T(v)R_1\dot{x}(v)dvds + (\tau_2 - \tau_1)\int_{t-\tau_2}^{t-\tau_1}\int_s^t \dot{x}^T(v)R_2\dot{x}(v)dvds$$
$$\tag{15}$$

Theorem 1. *For given scalars τ_1, τ_2, σ_M and a matrix K, the equilibrium of the system (13) is asymptotically stable in mean-square sense with degree κ, if there exist matrices $P > 0$, $Q_i > 0$, $R_i > 0$ $(i = 1, 2)$, $\Phi > 0$ and U with appropriate dimensions, and scalar $\theta \geq 0$ such that the following matrix inequities hold*

$$\begin{bmatrix} \Gamma_{11} & * & * \\ \Gamma_{21} & \Gamma_{22} & * \\ \Gamma_{31} & \Gamma_{32} & \Gamma_{33} \end{bmatrix} < 0 \tag{16}$$

$$\begin{bmatrix} R_2 & * \\ U & R_2 \end{bmatrix} > 0 \tag{17}$$

where $\Gamma_{11} = [(1,1) = PA_0 + A_0^T P + Q_1 - R_1, (2,1) = R_1, (2,2) = Q_2 - Q_1 - R_1 - R_2, (3,1) = K^T B_0^T P, (3,2) = R_2 - U, (3,3) = U + U^T - 2R_2 + \sigma_M\Phi, (4,2) = U, (4,3) = R_2 - U, (4,4) = -Q_2 - R_2, (5,1) = -K^T B_0^T P, (5,5) = -\Phi, (6,1) = P, (6,6) = -\theta I]$, $\Gamma_{21} = col\{\tau_1 R_1 F_1, (\tau_2 - \tau_1)R_2 F_1, F_2\}$, $\Gamma_{22} = diag\{-R_1, -R_2, -\theta\kappa^{-2}I\}$, $\Gamma_{31} = col\{F_3, F_4\}$, $\Gamma_{32} = [F_5, 0]$, $\Gamma_{33} = diag\{-\epsilon I, -\epsilon I\}$ with $F_1 = [A_0, B_0 K, 0, -B_0 K, 0, I]$, $F_2 = [\theta H, 0, 0, 0, 0, 0]$, $F_3 = [\epsilon G^T P, 0, 0, 0, 0, 0]$, $F_4 = [E_a, 0, E_b K, 0, -E_b K, 0]$, and $F_5 = [\tau_1\epsilon G^T R_1, (\tau_2 - \tau_1)\epsilon G^T R_2, 0]$.

Remark 3. For a special case of a time-triggered scheme, i.e., $\sigma(t) = 0$ in (5), Theorem 1 will lead to less conservative results than those in some existing ones, which will be given in section 4.

Based on Theorem 1, we are in a position to design the controller (8) for the closed-loop system (13).

Theorem 2. *For given scalars τ_1, τ_2, and σ_M, the equilibrium of the system (13) with nonlinear connection function satisfying (4) is asymptotically stable in mean-square sense, and the corresponding controller gain is $K = YX^{-1}$, if there exist matrices $X > 0$, $\tilde{Q}_i > 0$, $\tilde{R}_i > 0$ $(i = 1, 2)$, $\tilde{\Phi} > 0$ and $\bar{U}$ with appropriate dimensions, and scalars $\varepsilon > 0$, $\lambda > 0$ such that the following matrix inequities hold*

$$\begin{bmatrix} \tilde{\Gamma}_{11} & * & * \\ \tilde{\Gamma}_{21} & \tilde{\Gamma}_{22} & * \\ \tilde{\Gamma}_{31} & \tilde{\Gamma}_{32} & \tilde{\Gamma}_{33} \end{bmatrix} < 0 \tag{18}$$

$$\begin{bmatrix} \tilde{R}_2 & * \\ \tilde{U} & \tilde{R}_2 \end{bmatrix} > 0 \tag{19}$$

where $\tilde{\Gamma}_{11} = [(1,1) = A_0 X + X A_0^T + \tilde{Q}_1 - \tilde{R}_1, (2,1) = \tilde{R}_1, (2,2) = \tilde{Q}_2 - \tilde{Q}_1 - \tilde{R}_1 - \tilde{R}_2, (3,1) = Y^T B_0^T, (3,2) = \tilde{R}_2 - \tilde{U}, (3,3) = \tilde{U} + U^T - 2R_2 + \sigma_M \tilde{\Phi}, (4,2) = \tilde{U}, (4,3) = \tilde{R}_2 - \tilde{U}, (4,4) = -\tilde{Q}_2 - \tilde{R}_2, (5,1) = -Y^T B_0^T, (5,5) = -\tilde{\Phi}, (6,1) = I, (6,6) = -I], \tilde{\Gamma}_{21} = col\{\tau_1 \tilde{F}_1, (\tau_2 - \tau_1)\tilde{F}_1, \tilde{F}_2\}, \tilde{\Gamma}_{22} = diag\{-X\tilde{R}_1 X, -X\tilde{R}_2 X, -\lambda I\}, \tilde{\Gamma}_{31} = col\{\tilde{F}_3, \tilde{F}_4\}, \tilde{\Gamma}_{32} = [\tilde{F}_5, 0], \tilde{\Gamma}_{33} = diag\{-\epsilon I, -\epsilon I\},$ *with* $\tilde{F}_1 = [A_0 X, B_0 Y, 0, -B_0 Y, 0, X], \tilde{F}_2 = [HX, 0, 0, 0, 0], \tilde{F}_3 = [\epsilon G^T, 0, 0, 0, 0, 0], \tilde{F}_4 = [E_a X, 0, E_b Y, 0, -E_b Y, 0], \tilde{F}_5 = [\tau_1 \epsilon G^T, (\tau_2 - \tau_1)\epsilon G^T, 0].$

Proof: Similar to the proof in [3], substitute $\theta > 0$ for $\theta \geq 0$. Define $\bar{P} = P/\theta$, $\bar{Q}_i = Q_i/\theta$, $\bar{R}_i = R_i/\theta$ $(i = 1, 2)$, $\bar{\Phi} = \Phi/\theta$, $\bar{U} = U/\theta$, pre-multiplying (16) with $diag\{1/\theta, 1/\theta, 1/\theta, 1/\theta, 1/\theta, 1/\theta, 1/\theta, 1/\theta\}$. Then define $X = \bar{P}^{-1}$, $X\bar{Q}_i X = \tilde{Q}_i$, $X\bar{R}_i X = \tilde{R}_i$ $(i = 1, 2)$, $X\bar{\Phi}X = \tilde{\Phi}$, $X\bar{U}X = \tilde{U}$ and $Y = KX$, pre- and post-multiplying (16), (17) with $diag\{X, X, X, X, X, I, R_1^{-1}, R_2^{-1}, I, I, I\}$, $diag\{X, X\}$ and their transposes, respectively. This completes the proof.

Remark 4. Notice that the derived matrix inequalities (18) can not be solved directly by using Matlab LMI Control Toolbox due to that the non-linear terms $X\tilde{R}_j^{-1}X$ $(j = 1, 2)$. Generally, there are two methods to deal with the above-mentioned non-linear items . The first is simpler linear approach, such as replacing $-X\tilde{R}_j^{-1}X$ $(j = 1, 2)$ with $\rho_j^2 \tilde{R}_j - 2\rho_j X$ $(j = 1, 2)$ [8]; the second is cone complementarity linearisation (CCL) method [18]. In this paper, the latter CCL method is used to solve the original non-convex problems, since the less conservative result can be expected based on the second method than the first method. Since the detailed discusses on CCL algorithm are seen in [3], [18], it is omitted in this paper.

4   Numerical Example

In the following, we will provide two examples to demonstrate the effective of the proposed method.

Example 1. The inverted pendulum introduced in [7] is considered in this paper

$$\dot{x}(t) = \begin{bmatrix} 0 & 1 & 0 & 0 \\ 0 & 0 & -1 & 0 \\ 0 & 0 & 0 & 1 \\ 0 & 0 & 10/3 & 0 \end{bmatrix} x(t) + \begin{bmatrix} 0 \\ 0.1 \\ 0 \\ -1/30 \end{bmatrix} u(t) \tag{20}$$

For the case with adaptive event-triggered communication scheme, i.e. $\sigma(t)$ varies in $[0.2, 0.4]$, applying Theorem 1 with $K = \begin{bmatrix} 2 & 12 & 378 & 210 \end{bmatrix}$, one can obtain that the maximum value of τ_2 is 0.02 and

$$\Phi = \begin{bmatrix} 0.0054 & 0.0270 & 0.8573 & 0.4714 \\ 0.0270 & 0.1947 & 5.0990 & 3.2215 \\ 0.8573 & 5.0990 & 160.2193 & 89.1322 \\ 0.4714 & 3.2215 & 89.1322 & 54.2020 \end{bmatrix} \tag{21}$$

Suppose that $\eta_M = 0$, then the maximum sampling period is 0.02. Taking $h = 0.02s$ with $x_0 = [0.98 \ 0 \ 0.2 \ 0]^T$, Fig.1 shows the state trajectories of x_2 under periodic sampling triggered scheme (PSTS), traditional event-triggered scheme (TETS) with $\sigma = 0.3$ and adaptive event-triggered scheme (AETS) with $\rho = 0.02$. It should be strongly stressed that the trajectories controlled by AETS closely matches to those by PSTS, whereas TETS method yields a poor control performance, in which the resource utilisation by TETS and AETS can be obtained 53.09% and 55.85% improvements, respectively. Based on the simulation results, one can see that the proposed AETS in this paper sends less data packets than those by TETS, and achieves a better control performance than TETS. In a conclusion, the tradeoff between the frequency of data transmission and control performance can be balanced well by AETS.

Example 2. Consider the system (1) with the following matrices

$$A_0 = \begin{bmatrix} 1 & 1 \\ 0 & 0.99 \end{bmatrix}, \quad B_0 \begin{bmatrix} 0 \\ 10 \end{bmatrix}, \quad H = \begin{bmatrix} 1 & 0 \end{bmatrix}, \quad G = \begin{bmatrix} 0.1 \\ 1 \end{bmatrix}, \quad E_a = \begin{bmatrix} 1 & 0 \end{bmatrix}, \quad E_b = -1$$

Case 1: when $\sigma = 0$ (i.e. PSTS)
To compare with the existing work [3], [16], it is assumed that the lower delay bound is $\tau_1 = 0$. The maximum allowable transfer intervals (MATIs) obtained based on the methods in [16], [3] are 0.2509 and 0.2838. It can be shown that applying Theorem 2, one can get a better result of MATIs= 0.3043. Moreover, with given $\tau_2 = 0.2509$, we obtain $\gamma_{min} = 16$ from Theorem 2. Compared with the result of nonlinear bound 0.1636 in [3], the upper nonlinear bound κ is 0.25, which shows the proposed stability conditions less conservative.

Case 2: when σ is time-varying (i.e. AETS)

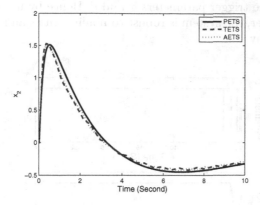

Fig. 1. State trajectories of x_2

For this case, our aim is to design a robust controller over a networked environment under the adaptive scheme in (5). Suppose that the trigger parameter $\sigma(t) \in [0.2, 0.4]$ and the initial trigger parameter is $\sigma(0) = 0.3$. Setting $\tau_1 = 0$, solving the matrix inequities in Theorem 2 with CCL algorithm, we can obtain the upper bound of τ_2 is 0.09, and the corresponding gain matrix K and trigger matrix Φ are given by

$$K = \begin{bmatrix} -0.6779 & -0.6652 \end{bmatrix}, \quad \Phi = \begin{bmatrix} 0.8408 & 0.8249 \\ 0.8249 & 0.8094 \end{bmatrix} \tag{22}$$

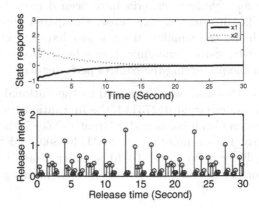

Fig. 2. State response, release instants and release interval

Setting the sampling period $h = 0.06$, thus the random delay η_k belongs a interval $[0, 0.03]$. Based on the obtained parameters and $\rho = 0.007$, considering the nonlinearities and uncertainties with the proposed AETS, the state trajectories, release instants and release interval for $t \in [0, 30]$ are shown in Fig.2. Fig. 3 plots the trajectory of the trigger parameters σ and d. Hence by use of the proposed method in this paper, the system is robust to nonlinearities and uncertainties as well as random delay.

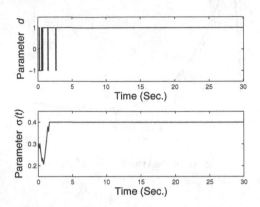

Fig. 3. Trigger parameters σ and d

5   Conclusion

A novel adaptive event-triggered communication scheme has been proposed for a class of uncertain NCSs with nonlinearities. The proposed event-triggered communication scheme specially shows an effective approach to balance the control performance and the network bandwidth burden. Based on the presented model, the stability and synthesis criteria have been derived to guarantee the asymptotically stability in mean-square sense. Compared with some existing event-triggered methods, the simulation examples have shown that the proposed adaptive event-triggered communication scheme has saved the network bandwidth burden while preserving the control performance.

Acknowledgments. This work was supported by the National Natural Science Foundation of China under Grant 61273114, the Innovation Program of Shanghai Municipal Education Commission under Grant 14ZZ087, the Pujiang Talent Plan of Shanghai City, China under Grant 14PJ1403800, and the Natural Science Foundation of Jiangsu Province of China under Grant BK20131403, the open fund project of Jiangsu key Laboratory for 3D printing equipment and manufacturing

References

1. Zhang, W., Branicky, M.S., Phillips, S.M.: Stability of Networked Control Systems. IEEE Control Systems Magazine 21, 84–99 (2001)
2. Antsaklis, P., Baillieul, J.: Special Issue on Technology of Networked Control Systems. Proceedings of the IEEE 95, 5–8 (2007)

3. Peng, C., Tian, Y.C., Tade, M.O.: State Feedback Controller Design of Networked Control Systems with Interval Time-Varying Delay and Nonlinearity. International Journal of Robust and Nonlinear Control 18, 1285–1301 (2008)
4. Yang, R., Shi, P., Liu, G.P., Gao, H.J.: Network-Based Feedback Control for Systems with Mixed Delays Based on Quantization and Dropout Compensation. Automatica 47, 2805–2809 (2011)
5. Liu, Q., Wang, Z., He, X., Zhou, D.: A Survey of Event-Based Strategies on Control and Estimation. Systems Science and Control Engineering: An Open Access Journal 2, 90–97 (2014)
6. Tabuada, P.: Event-Triggered Real-Time Scheduling of Stabilizing Control Tasks. IEEE Transactions on Automatic Control 52, 1680–1685 (2007)
7. Wang, X., Lemmon, M.D.: Self-Triggered Feedback Control Systems with Finite-Gain L_2 Stability. IEEE Transactions on Automatic Control 54, 452–467 (2009)
8. Yue, D., Tian, E., Han, Q.L.: A Delay System Method for Designing Event-Triggered Controllers of Networked Control Systems. IEEE Transactions on Automatic Control 58, 475–481 (2013)
9. Meng, X., Chen, T.: Event Triggered Robust Filter Design for Discrete-Time Systems. IET Control Theory and Applications 8, 104–113 (2014)
10. Hu, S., Yin, X., Zhang, Y., Tian, E.G.: Event-Triggered Guaranteed Cost Control for Uncertain Discrete-Time Networked Control Systems with Time-Varying Transmission delays. IET Control Theory and Applications 6, 2793–2804 (2012)
11. Heemels, W., Donkers, M., Teel, A.R.: Periodic Event-Triggered Control for Linear Systems. IEEE Transactions on Automatic Control 58, 847–861 (2013)
12. Peng, C., Han, Q.L., Yue, D.: To Transmit or not to Transmit: a Discrete Event-Triggered Communication Scheme for Networked Takagi-Sugeno Fuzzy Systems. IEEE Transactions on Fuzzy Systems 21, 164–170 (2013)
13. Jia, X.C., Chi, X.B., Han, Q.L., Zheng, N.N.: Event-Triggered Fuzzy H_∞ Control for a Class of Nonlinear Networked Control Systems using the Deviation Bounds of Asynchronous Normalized Membership Functions. Information Sciences 259, 100–117 (2014)
14. Dimarogonas, D.V., Frazzoli, F., Johansson, K.H.: Distributed Event-Triggered Control for Multi-Agent Systems. IEEE Transactions on Automatic Control 57, 1291–1297 (2012)
15. Yin, X., Yue, D.: Event-Triggered Tracking Control for Heterogeneous Multi-Agent Systems with Markov Communication Delays. Journal of the Franklin Institute 350, 1312–1334 (2013)
16. Yu, M., Wang, L., Chu, T.: Robust Stabilization of Nonlinear Sampled-Data Systems. In: The 2005 American Control Conference, pp. 3421–3426. IEEE Press, Portland (2005)
17. Zhang, X.M., Han, Q.L.: Event-Triggered Dynamic Output Feedback Control for Networked Control Systems. IET Control Theory and Applications 8, 226–234 (2014)
18. Ghaoui, L.E., Oustry, F., AitRami, M.: A Cone Complementarity Linearization Algorithm for Static Output-Feedback and Related Problems. IEEE Transactions on Automatic Control 42, 1171–1176 (1997)

Analytical Model for Epidemic Information Dissemination in Mobile Social Networks with a Novel Selfishness Division

Qichao Xu[1], Zhou Su[1,2,*], Bo Han[3], Dongfeng Fang[1], and Zejun Xu[1]

[1] School of Mechatronics and Automation,
Shanghai University,
NO. 149, Yanchang Road, Zhabei Dist., Shanghai 200072, China
xqc690926910@shu.edu.cn
[2] Faculty of Science and Technology,
Waseda University,
Tokyo 169-8555, Japan
[3] Center of Information and Networks,
Xi'an Jiaotong University,
No.28, Xianning WestRoad,
710049 Xi'an, China
zhousu@ieee.org

Abstract. With the prosperity of smartphone markets, mobile social networks (MSNs) attract much attention. In MSNs, information can be shared among users through their opportunistic contacts. The information forwarding mode needs mobile users to work in a cooperative and altruistic way. However, in the real world, most of users are not willing to forward information because of their selfishness. In this paper, we firstly divide selfishness into two new types: weak selfishness and extreme selfishness. Then we develop an analytical model to evaluate the influence of selfish behavior on the information dissemination process in the MSNs. Numerical results demonstrate weak selfishness and extreme selfishness hinder information dissemination.

Keywords: Mobile social networks (MSNs), Information dissemination, Weak selfishness, Extreme selfishness, Analytical model.

1   Introduction

With the rapid growth of the usage of mobile devices, mobile social networks (MSNs) have been emerged. In such networks, people can share information with others exploiting short range wireless techniques [1]. Actually, an MSN is a special type of delay tolerant network [2], where user can adopt store-carry-and-forward mode to diffuse information. This information dissemination scheme needs all mobile users to work in a cooperative way which is popular to save energy and resources [10][11].

---

* Corresponding author.

M. Fei et al. (Eds.): LSMS/ICSEE 2014, Part II, CCIS 462, pp. 478–484, 2014.
© Springer-Verlag Berlin Heidelberg 2014

However, a large number of people exhibit various degrees of selfishness in MSNs [3]. In previous, the selfishness is mainly divided into two parts: individual selfishness and social selfishness [4]. The first one is from the perspective of an individual, which means this user is not willing to relay information for others [5]. The second one is from the perspective of a community, which means the user is more willing to help others in the same community than forwarding information to the users outside the community [4].

In this paper, we divide mobile users' selfish behavior into new two types: weak selfishness and extreme selfishness. The weak selfishness means an individual who is unwilling to forward information to others sometimes. The extreme selfishness denotes that an individual does not want to help others absolutely. Then, we present a model to mimic the epidemic information dissemination with weak and extreme selfishness through the ordinary differential equations (ODEs), where mobile users share information with their friends upon they encounter. Based on this model, we evaluate the influences of both the weak selfish behavior and extreme selfish behavior on dissemination process. By doing experiments with the real trace, we explain simulation and numerical results. The simulation results prove the accuracy of presented model. Numerical results show that the weak selfishness and extreme selfishness can affect the information dissemination.

The remainder of the work is organized as follows. The system model is introduced in Section 2 and the information dissemination mode is presented in Section 3. Our simulation and numerical results are shown in Section 4. The conclusion of this paper and the future work are given in Section 5.

2   System Model

We assume that there are N users denoted by $n_1, n_2, \cdots n_N$ moving around in a region all the time. The information can be shared between two people who are friends when they meet. This section introduces the social graph with the social ties among mobile users and explains the selfishness division.

2.1   Social Graph

An undirected and unweighted graph $G(V, E)$ is employed to denote the social graph. The symbol V is the set of nodes representing mobile users and E denotes the set of edges referring to social ties among users in social graph. Thus $V = \{n_1, n_2, \cdots n_N\}$. Since we want to study the effect of selfishness, we need to know the distribution of relationship. Let $P(k)$ denote the probability of a node with k friends and k can be called the degree of this node. Existing works have shown that $P(k)$ usually has the power-low distribution in the social graph [6] as follows:

$$P(k) = \begin{cases} 0, & k < m \\ C(m, \gamma)k^{-\gamma}, & m \le k < N \end{cases}, \tag{1}$$

where m is the smallest degree of the network, γ is the skewness of the degree distribution, and $C(m,\gamma)$ is the normalization constant

In MSNs, the mobility patterns of nodes are very sophisticated. We assume that the inter-contact between two nodes follows an exponential distribution with parameter λ, which has been demonstrated by using real trace by some studies [7].

2.2   Selfishness Division

Selfishness is divided into weak selfishness and extreme selfishness. we assume that 1) a mobile node forwards information to one of its friends with the probability p_{in}, which is called weak selflessness level and is opposite to weak selfishness level; 2) a mobile node doesn't accept information from others and doesn't store it absolutely corresponding to a parameter p_1 exponential distribution; 3) a mobile node refuses to forward information to any nodes corresponding to parameter p_2 exponential distribution. In particular, the first one regards to the weak selfishness and the next two assumptions can be seen as extreme selfishness.

3   Information Dissemination Model

In this section, we present a model to mimic information dissemination with selfishness in the MSN through Ordinary differential equations (ODEs).

Let $I_k(t)$ denote the number of k-degree nodes that don't have information at time t. Similarly, let $S_k(t)$ denote the number of k-degree nodes obtaining information at time t. $R_k(t)$ denotes the number of k-degree nodes that refuse to receive and store the information, and $Z_k(t)$ denotes the number of k-degree nodes that have received the information but don't forward it at time t. If we let $N_k(t)$ represent the number of k-degree nodes at time t, the fractions of four types of nodes in $N_k(t)$ at time t are $i_k(t) = I_k(t)/N_k(t)$, $s_k(t) = S_k(t)/N_k(t)$, $r_k(t) = R_k(t)/N_k(t)$, $z_k(t) = Z_k(t)/N_k(t)$, respectively.

Within time interval $[t, t+\Delta t]$, the variation of $s_k(t)$ is as follows.

$$i_k(t+\Delta t) = i_k(t)(1 - P(G,k)), \tag{2}$$

where $P(G,k)$ denotes the probability that a k-degree node n_i without information will receive it or refuse to accept it from any one at within Δt. To derive $P(G,k,t)$, we consider two aspects. One is that node n_i receives information from its friends so that it is classified into $s_k(t)$. The other is that, node n_i also may refuse to help any nodes so that it is classified into $r_k(t)$.

As mentioned above, only when there is an opportunistic link between two nodes, they have a chance to share information. Since the inter-contact between two nodes follows the exponential distribution, the probability that node n_i encounters a node (e.g. n_j) within $[t, t+\Delta t]$ is $1-e^{-\lambda(N-1)\Delta t}$, when the interval is enough small. The probability that node n_j is a friend of node n_i is $k/N-1$. Furthermore, the probability that node n_j has the information is $\sum_{k'=m}^{N-1}(N/N-1)P(k')s_{k'}(t)$, where the k' is the degree of a node. In addition, node n_i can get the information from one of its friends successfully after they encounter with a probability p_{in}. Therefore, the probability $P(A,k)$ that node n_i receives the information time within Δt is

$$P(A,k)= p_i \frac{k}{N-1}(1-e^{-\lambda(N-1)\Delta t})\sum_{k'=m}^{N-1}\frac{N}{N-1}P(k')s_{k'}(t). \qquad (3)$$

Furthermore, we analyze the second aspect. The probability $P(notreceive)$ that node n_i doesn't accept information absolutely within Δt is

$$P(notreceive) = \int_0^{\Delta t} p_1 e^{-p_1 t}dt = 1-e^{-p_1\Delta t}. \qquad (4)$$

Therefore, combining with (3) and (4), we can have

$$P(G,k) = 1-(1-P(A,k))(1-P(nottreceive))$$
$$= 1-(1-p_{in}(k/N-1)(1-e^{-\lambda(N-1)\Delta t})\sum_{k'=m}^{N-1}\frac{N}{N-1}P(k')s_{k'}(t)\times e^{-p_1\Delta t} \qquad (5)$$

From (2), we can obtain

$$\frac{di_k(t)}{dt} = -i_k(t)\lim_{\Delta t\to 0}\frac{P(G,k)}{\Delta t} = -\lambda k p_{in}i_k(t)\sum_{k'=m}^{N-1}\frac{N}{N-1}P(k')s_{k'}(t)-p_1 i_k(t). \qquad (6)$$

Similarly, give by a small time interval Δt, we have

$$\frac{ds_k(t)}{dt} = \lim_{\Delta t\to 0}\frac{i_k(t)P(A,k)-s_k(t)P(notforward)}{\Delta t}$$
$$= \lim_{\Delta t\to 0}\frac{i_k(t)P(A,k)-s_k(t)(1-e^{-p_2\Delta t})}{\Delta t}, \qquad (7)$$
$$= \lambda k p_{in}i_k(t)\sum_{k'=m}^{N-1}\frac{N}{N-1}P(k')s_{k'}(t)-p_2 s_k(t)$$

$$\frac{dr_k(t)}{dt} = \lim_{\Delta t\to 0}\frac{i_k(t)P(notreceive)}{\Delta t}, \qquad (8)$$
$$= p_1 i_k(t)$$

$$\frac{dz_k(t)}{dt} = \lim_{\Delta t \to 0} \frac{s_k(t)P(notforward)}{\Delta t},$$ (9)
$$= p_2 s_k(t)$$

where $P(notforward)$ denotes the probability that a k-degree node with the information refuses to forward it to anyone within Δt.

The numbers of four types of nodes in the whole network are

$$I(t) = N \sum_{k=m}^{N-1} i_k(t)P(k),$$ (10)

$$S(t) = N \sum_{k=m}^{N-1} s_k(t)P(k),$$ (11)

$$R(t) = N \sum_{k=m}^{N-1} r_k(t)P(k),$$ (12)

$$Z(t) = N \sum_{k=m}^{N-1} z_k(t)P(k).$$ (13)

4   Numerical Results

In this section, we employ our model to evaluate the effect of weak selfishness and extreme selfishness.

In the experiment, the number of infected nodes $F(t)$ is considered as a performance metric, which means the average number of nodes that have received information at time t and it can be defined as

$$F(t) = S(t) + Z(t) = N \sum_{k=m}^{N-1} (s_k(t) + z_k(t))P(k).$$ (14)

4.1   Experiment Setup

We assume that the unit time of the system is 0.01 hour. Since existing empirical studies have showed that the average inter-meet time between two individuals is around 5 hours [8], which is also used in some studies [9], we choose the $\lambda = 0.002$ for the experiment. In addition, we do the experiment on the social graph based on a real trace from logs of a MSN system developed by the network center of the university. In the social graph, $\gamma = 2.1$, $N = 802$, $E = 1222$ and $m = 1$.

4.2   Effect of Weak Selfishness

We firstly analyze the effect of weak selfishness. In particular, we let $p_1 = p_2 = 0$. The value of p_{in} is varied from 0 to 1, and $T = 1000, 2000, 5000, 10000$, respectively. It is easy to know that when weak selflessness level p_{in} is larger, the weak selfishness level is lower. Fig. 1 shows the effect of weak selfishness. From the Fig. 1, we observe that the weak selflessness is positively related with the number of nodes having information. In addition, when the dissemination time is larger, the network is much more robust to the weak selflessness. Therefore, since the selfishness is opposite to the selflessness, we can say that the selfishness is negatively related with $F(t)$.

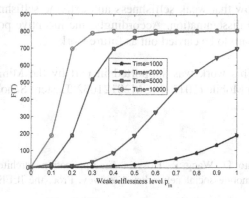

Fig. 1. Effect of weak selfishness

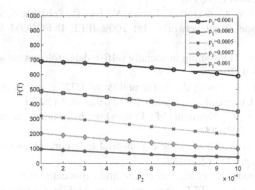

Fig. 2. Effect of extreme selfishness

4.3 Effect of Extreme Selfishness

In order to evaluate the effect of extreme selfishness explicitly, we choose five types of p_1 for comparison. In particular, we let $p_{in} = 1$ and the value of p_2 increases from 0.0001 to 0.001 and $T = 5000$. Based on these settings, we can get Fig. 2. From Fig. 2, we can observe that the extreme selfishness can restrain the information dissemination. For example, the value of $F(T)$ decreases when the value of p_2 is increasing in each curve and the value of $F(T)$ also decreases with the value of p_1 increasing at the same p_2.

5   Conclusion

In this paper, we have presented an analytical model to evaluate effect of selfishness on information dissemination in MSNs. To the best of our knowledge, this is the first analytical model that considers two new types of selfishness: weak selfishness and extreme selfishness. The experiments have been done on the MSN with real traces. Numerical results show that weak selfishness and extreme selfishness have negative effect on information dissemination. Accordingly, the incentive policy, making uses cooperative, is necessary to be carried out as future work.

Acknowledgment. This work was supported in part by the Ministry of Education Research Fund-China Mobile (2012) MCM20121032, Eastern Scholar Program.

References

1. Kayastha, N., Niyato, D., Wang, P., Hossain, E.: Applications, architectures, and protocol design issues for mobile social networks: A survey. Proc. the IEEE 99(12), 2130–2158 (2011)
2. Fall, K.: A delay-tolerant network architecture for challenged Internets. In: 2003 ACM SIGCOMM, pp. 27–34 (2003)
3. Hui, P., Xu, K., Li, V., Crowcroft, J., Latora, V., Lio, P.: Selfishness, altruism and message spreading in mobile social networks. In: 2009 IEEE INFOCOM Workshops, pp. 1–6 (2009)
4. Li, Q., Zhu, S., Cao, G.: Routing in socially selfish delay tolerant networks. In: 2010 IEEE INFOCOM, pp. 1–9 (2010)
5. Karaliopoulos, M.: Assessing the vulnerability of DTN data relaying schemes to node selfishness. IEEE Communication Letter 13(12), 923–925 (2009)
6. Clauset, A., Shalizi, C., Newman, M.: Power-law distributions in empirical data. SIAM Review 51(4), 661–703 (2009)
7. Zhu, H., Xue, G., et al.: Recognizing exponential inter-contact time in VANETs. In: 2010 IEEE INFOCOM, pp. 1–5 (2010)
8. Karagianis, T., Boudec, J., et al.: Power law and exponential decay of inter contact times between mobile devices. IEEE Transactions on Mobile Computing 9(10), 1377–1390 (2010)
9. Sun, H., Wu, C.: Epidemic forwarding in mobile social networks. In: 2012 IEEE ICC, pp. 1421–1425 (2012)
10. Fang, D., Su, Z., Xu, Q.: Analysis of Data Transmission Based on the Priority over Grid Structures. ICIC Express Letters, Part B: Applications, 751–755 (2014)
11. Fang, D., Su, Z., Xu, Q., Xu, Z.: Data scheduling based on pricing strategy and priority over smart grid. In: Fei, M., Peng, C., Su, Z., Song, Y., Han, Q. (eds.) LSMS/ICSEE 2014, Part III. CCIS, vol. 463, pp. 482–488. Springer, Heidelberg (2014)

Distributed Collaborative Control
Based on Adaptive Chaos Mutation PSO with WSAN

Fuqiang Li[1,2], Jingqi Fu[1], Dajun Du[1], and Weihua Bao[3]

[1] School of Mechatronic Engineering and Automation,
Shanghai University. 149 Yanchang Road, Shanghai, China
[2] College of Sciences, Henan Agricultural University. 63 Nongye Road,
Zhengzhou City, Henan Province, China
[3] Shanghai Automation Instrumentation Co., Ltd. 191 Guangzhouxi Road, Shanghai, China
fuqiangli@126.com, {jqfu,ddj}@staff.shu.edu.cn, bwh@saic.sh.cn

Abstract. In large scale wireless sensor and actuator network (WSAN), unreliable wireless and multihop communications make challenges in designing centralized control due to severe packet dropout and latency. To this end, distributed collaborative control (DCC) scheme based on adaptive chaos mutation particle swarm optimization (ACMPSO) is proposed. In this scheme, control task is carried out collaboratively based on hierarchical clustering using only local information. Moreover, DCC is formulated as an optimization problem, and chaos mutation is introduced to improve the optimal performance of PSO. Simulation shows that the improved PSO has better convergence property, and the proposed scheme works well to obtain control objective.

Keywords: wireless sensor and actuator network, distributed collaborative control, adaptive, chaos, particle swarm optimization.

1 Introduction

Wireless Sensor Network (WSN) is formed by lots of low-cost, small-sized, limited-energy, and wireless-communication capable sensors, which provides us with ubiquitous information acquisition [1]. Recently, actuators capable of wireless communication are also introduced into the WSN to form up the WSAN. WSAN is a new type of heterogeneous network consisting of a group of sensors and actuators that are connected with wireless medium [2]. It is an important extension of WSN, allowing actuators to make autonomous decisions and perform actions in response to sensory information. Due to the benefits such as ubiquitous monitoring even in harsh environments, easy installation, low cost, large coverage, self configuration and organization, fast communication and reaction, fault tolerance, etc. [3], WSAN has many promising applications in industrial and agricultural fields, such as industrial monitoring and control [4], inventory management [5], greenhouse automatic control system[6], automated irrigation system[7], etc.

Since centralized control is not competent in large scale WSAN, there are lots of researches concerning distributed control. Sinopoli. et al. addressed some of the issues

M. Fei et al. (Eds.): LSMS/ICSEE 2014, Part II, CCIS 462, pp. 485–495, 2014.

arising from the use of sensor networks in control applications [8]. They took a distributed pursuit-evasion game (PEG) application for example and presented the research challenges within the distributed control system, e.g., no global information. Cao Xianghui et al. considered joint problems of control and communication in WSAN and compared the performances of centralized control scheme and distributed control scheme [9]. Although the distributed control method performed better in many aspects, it suffered from the fact that its control performance was sensitive to two parameters based on global information, rendering it from being a wholly distributed and plug-and-run control method. Chen Jiming et al. proposed and evaluated a new distributed collaborative control scheme for industrial control systems with WSAN, which fully exploited the collaborations between actuators and sensors [10]. Simulation showed that the proposed method effectively achieved control objective. However, authors thought the method discussed was a heuristic method, and it was hard to give an analytical description of the control performance. Mo Lei et al. proposed distributed collaborative processing scheme over WSAN based on distributed parallel genetic algorithm (DPGA) [11]. The Actuator-Actuator coordination problem was formulated as an optimization problem and DPGA was applied as a searching technique to minimize the objective function. In the migration stage of DPGA, each actuator scheduled should generate a subpopulation and receive the optimal individuals migrated from subpopulations in other actuators, so frequent communication was needed.

The main contributions of this paper are as follows. Firstly, in order to improve control performance in large scale WSAN, a distributed collaborative control scheme based on hierarchical clustering is introduced, where sensor cluster and actuator cluster are needed to be formed [12]. Secondly, DCC is formulated as an optimization problem. Due to simplicity and easy implementation, PSO is chosen as the optimization algorithm. Based on the properties of ergodicity, randomicity, and regularity of chaos, ACMPSO is proposed to improve convergence performance of PSO.

The rest of this paper is organized as follows. Section 2 presents system models. ACMPSO is proposed in Section 3. Collaboration mechanism based on hierarchical clustering is described in Section 4. In Section 5, numerical simulation is carried out to evaluate the performance of DCC based on ACMPSO. Finally, the paper is concluded in Section 6.

2   System Formulation

Taking the automated control system in greenhouse for example, we view temperature as state variables. The objective is to control temperatures at different locations to meet set points. Suppose there are n_s sensor nodes and n_a actuator nodes spread throughout the greenhouse to detect plant states and take necessary actions.

Plant dynamics in discrete time domain is given as follows

$$X(k) = AX(k-1) + BU(k) + \varphi(k) \tag{1}$$

Where $A \in \mathbb{R}^{n_s \times n_s}, B \in \mathbb{R}^{n_s \times n_a}, X(k), U(k), \varphi(k)$ are state vector of plants, output vector of actuators and system disturbance respectively. Moreover, we assume states are independent of each other, so $A = diag\left[a_{1,1}, a_{2,2}, \cdots, a_{n_s, n_s}\right]$ is diagonal matrix. $\varphi(k)$ includes measurement noise, actuation disturbances, channel noise, unmodeled dynamics, and other sources of uncertainty [13]. The element $b_{i,j}$ of matrix B indicates the effect actuator A_j has on plant state x_i [11]. So $b_{i,j}$ can be shown as

$$b_{i,j} = \begin{cases} 0 & d_{i,j} > R_j \\ f_B(d_{i,j}) & d_{i,j} \leq R_j \end{cases} \tag{2}$$

Where $d_{i,j}$ describes the Euclidean distance between A_j and x_i, R_j is the maximum action range of A_j, $f_B(d_{i,j})$ is a function which makes $b_{i,j}$ is inversely proportional to $d_{i,j}$.

As a result, the dynamics of plant state x_i can be rewritten as bellow

$$\begin{aligned} x_i(k) &= \sum_{j=1}^{n_s} A_{i,j} x_j(k-1) + \sum_{j=1}^{n_a} b_{i,j} u_j(k) + \varphi_i(k) \\ &= A_{i,i} x_i(k-1) + \sum_{d_{i,j} \leq R_j} b_{i,j} u_j(k) + \varphi_i(k) \end{aligned} \tag{3}$$

As mentioned before, control objective is to meet set points $X^* = \left[x_1^*, x_2^*, \cdots, x_{n_s}^*\right]^T$, thus it is interesting to use certain optimal control strategy to minimize the control objective function defined as

$$\min : \quad J(k) \triangleq \frac{1}{n_s}\left(\sum_{i=1}^{n_s}\left(x_i^* - x_i(k)\right)^2\right)^{\frac{1}{2}}, \quad k = 0, 1, 2, \cdots \tag{4}$$

3   Nodes Coordination

3.1   Sensor-Sensor (S-S) Coordination

At sampling instants, some sensors nearby, whose sensing ranges cover the same plant state, will be activated to form a sensor cluster. For the sake of balancing energy consumption to prolong the network lifetime, the sensor with more energy left will be selected to be cluster head. Data aggregation is then done by cluster head as follow [11]: after reading measurements from cluster members, cluster head computes the mean value. By comparing each measurement with the mean value, some abnormal ones will be distinguished and removed. Then a new mean value is figured out and transmitted to the nearest one of its responsible actuators, which can be shared by all its responsible actuators. In this way, it is not necessary for each sensor to transmit measurement to all its responsible actuators, which brings convenience to quick response and energy saving. Moreover, the responsible actuators of sensor s_i is defined as

$$A^{s_i} = \left\{ A_j \mid b_{i,j} \neq 0, \ j = 1, \cdots, n_a \right\} \tag{5}$$

3.2   Actuator-Actuator (A-A) coordination

Firstly, overlapping degree of s_i is defined as

$$\rho_i \triangleq \left| A^{s_i} \right|, \ i = 1, 2, \cdots, n_s \tag{6}$$

Where $|\cdot|$ is cardinality of a set. Obviously, if $\rho_i > 1$, there are at least two responsible actuators. Without coordination, each responsible actuator computes control law respectively. When responsible actuators act on the plant state simultaneously, the actuation will be doubled or multiplied, which may cause the system unstable. So A-A coordination is imperative to eliminate the actuation overshooting. In order to achieve A-A collaboration, responsible actuators are activated to form actuator cluster. Similarly with sensor cluster, the actuator with more energy left will be selected as cluster head. Then the cluster head works as follow: firstly, it collects information of plant states from cluster members. Then using some optimization algorithm, it figures out optimal control laws for all cluster members. Next, it disseminates control laws to cluster members, which update actuation outputs immediately.

4   Adaptive Chaos Mutation PSO

PSO is a population-based evolutionary computation technique developed by Kennedy and Eberhart in 1995 [14], which imitates foraging behavior of bird flock. In PSO, each solution can be considered as an individual particle in a given search space, which has its own position and velocity. During movement, each particle adjusts its position by changing its velocity based on experiences of its own and companions, until an optimum position is reached [15]. Iteration functions of PSO are given as follows.

$$v_{id} = w v_{id} + c_1 r_1 \left(p_{id}^{Pbest} - p_{id} \right) + c_2 r_2 \left(p_{gd}^{Gbest} - p_{id} \right) \tag{7}$$

$$p_{id} = p_{id} + v_{id} \tag{8}$$

Where v_{id} and p_{id} represent the d_{th} element's speed and position of the i_{th} particle, $d \in \{1, 2, \cdots, n_d\}$, $i \in \{1, 2, \cdots, n_i\}$. n_d is the dimension of searching space and n_i is the number of particles. w is inertia weight. r_1 and r_2 are random numbers with $r_1, r_2 \in [0,1]$. c_1 and c_2 are accelerating factors. p_{id}^{Pbest} and p_{gd}^{Gbest} indicate the optimum positions of the i_{th} particle and all particles respectively.

Although PSO has many benefits, when premature convergence leads to low particle diversity, it is hard to escape from local optimum. In order to improve particle diversity, chaos mutation is introduced. Chaos is the behavior of systems that follow deterministic laws but appear random and unpredictable. Chaotic systems are sensitive to initial conditions, small changes in initial conditions can lead to quite different outcomes. The classic logistic map is employed to produce chaos system as bellow [16].

$$c_{n+1} = \mu c_n (1 - c_n), \ n = 0, 1, 2, \cdots \tag{9}$$

Where $c_n \in [0,1]$ is chaotic variable, $\mu \in [0,4]$ is logistic parameter, which is always set to $\mu = 4$.

Based on the properties of ergodicity, randomicity, and regularity of chaos, chaos variable is used to produce some mutated particles as

$$p_{id} = \gamma \cdot c_n \cdot p_{gd}^{Gbest}, \ i = 1, 2, \cdots, n_i \times P \tag{10}$$

Where γ is a constant coefficient, which is used to specify the search space. Obviously, if γ takes a large value, the search space is large. P is probability, which determines how many particles are replaced by the mutated ones randomly.

By contrast with PSO, due to the introduction of chaos iterations, chaos mutation PSO (CMPSO) needs more computation and energy resources. In order to reduce side effects of CMPSO, PSO is only replaced by CMPSO when PSO is in stagnation behavior, so the ACMPSO scheme is given as

$$ACMPSO = \begin{cases} PSO & f\left(p_{gd}^{Gbest}(k)\right) < f\left(p_{gd}^{Gbest}(k-1)\right) \\ CMPSO & f\left(p_{gd}^{Gbest}(k)\right) = f\left(p_{gd}^{Gbest}(k-1)\right) \end{cases}, \ k = 2, 3, \cdots \tag{11}$$

Where $f(\cdot)$ is the fitness function. If $f\left(p_{gd}^{Gbest}(k)\right)$ is smaller than $f\left(p_{gd}^{Gbest}(k-1)\right)$, PSO is thought to be in formal evolution process. Otherwise, if the two global optimum values are equal, PSO is thought to be in stagnation behavior and would be replaced by CMPSO.

In order to compare convergence performance of the two algorithms, two benchmark functions are employed, where Sphere is unimodal function and Rastrigin is multi-modal function with many local extrema [17]. The information of two benchmark functions is given in Table 1, where p^* denotes global optimum position and $f(p^*)$ denotes global optimum value. In simulation for convergence precision comparison, the number of iterations is set to 150. In simulation for convergence speed comparison, $f_1(p^*) = 4.73 \times 10^{-4}$ and $f_2(p^*) = 2.17 \times 10^{-1}$ are the required global optimum values both algorithms should guarantee for Sphere and Rastrigin respectively. Simulation parameters are set as follows: $w = 0.8$, $c_1 = c_2 = 2$, $n_i = 20$, $n_d = 2$, $\gamma = 0.8$, $P = 0.2$. In order to ensure particles scatter uniformly in solution space, initial population is produced by logistic map. Besides, in order to obtain reliable information, simulation results are given based on the mean value of 50 times independent run.

Table 1. Benchmark functions

name	function	p^*	$f(p^*)$	Initial range
Sphere	$f_1 = \sum_{i=1}^{n_d} p_i^2$	$[0,0]$	0	$[-100,100]$
Rastrigin	$f_2 = \sum_{i=1}^{n_d} \left[p_i^2 - 10\cos(2\pi p_i) + 10 \right]$	$[0,0]$	0	$[-10,10]$

Table 2. Global optimum position and global optimum value

	PSO	ACMPSO
$f_1 : p^*$	$\left[-0.04 \times 10^{-2}, -0.16 \times 10^{-2} \right]$	$\left[-0.01 \times 10^{-5}, 0.23 \times 10^{-5} \right]$
$f_1(p^*)$	4.73×10^{-4}	2.67×10^{-10}
$f_2 : p^*$	$\left[0.20 \times 10^{-1}, -0.81 \times 10^{-1} \right]$	$\left[-0.07 \times 10^{-4}, 0.28 \times 10^{-4} \right]$
$f_2(p^*)$	2.17×10^{-1}	1.29×10^{-5}

Table 3. Iteration number

Required performance	PSO	ACMPSO
$f_1(p^*) = 4.73 \times 10^{-4}$	150	52
$f_2(p^*) = 2.17 \times 10^{-1}$	150	70

After the given number of iterations, Table 2 shows the performance comparison in global optimum position and global optimum value. Comparing with the parameters given in Table 1, it turns out that the convergence precision of ACMPSO is higher. Table 3 presents the number of iterations needed for the two algorithms when required convergence precision is satisfied. Under same requirement of convergence precision, it turns out that less number of iterations is needed for ACMPSO, i.e.,

ACMPSO has faster convergence speed. In sum, by contrast with PSO, ACMPSO has better performance in convergence speed and precision.

Fig.1 presents the entire convergence process of the two algorithms. After certain number of iterations, particle diversity of PSO becomes low and convergence speed slows down, so it is easy to be trapped in local optimum with low convergence precision. Meanwhile, based on adaptive chaos mutation, particle diversity of ACMPSO is improved and the global optimum value keeps converging, so fast convergence speed and high precision can be guaranteed.

5   Numerical Example

In order to present the simulation results concisely, we consider a temperature control system in intelligent greenhouse with two plant states and two actuators, i.e., $n_s = n_a = 2$. DCC based on ACMPSO is employed here to control temperatures at different locations to meet set points. Moreover, the proposed method can be used in more complex WSAN system with more nodes. The initial states are $\begin{bmatrix} 17°C & 18°C \end{bmatrix}$ and the set points are $\begin{bmatrix} 20°C & 20°C \end{bmatrix}$. We use the plant model presented in [10]

$$X(k) = \begin{bmatrix} 0.9 & 0 \\ 0 & 0.9 \end{bmatrix} X(k-1) + \begin{bmatrix} 0.57 & 0 \\ 0.49 & 0.68 \end{bmatrix} U(k) \tag{12}$$

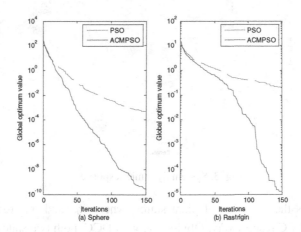

Fig. 1. Evolution of global optimum value: (a) for Sphere, (b) for Rastrigin

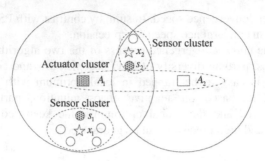

Fig. 2. System topology

The topology of system (12) is shown in Fig.2, where s_1, s_2 denote cluster heads of sensor clusters, A_1 is the cluster head of actuator cluster. Because actuator A_1 influences both states x_1 and x_2, i.e., $s^{A_1} = \{s_1, s_2\}$, while x_2 is affected by both actuators A_1 and A_2, i.e., $A^{s_2} = \{A_1, A_2\}$, it is necessary for A_1 and A_2 to form actuator cluster to accomplish the distributed collaborative control. The iteration number of system (12) is set to 15 and simulation results are given bellow.

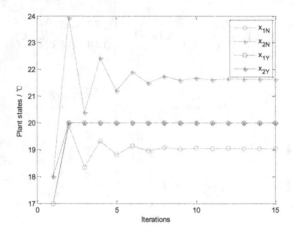

Fig. 3. System dynamic response

Fig.3 describes the evolution of plant states, where x_{iY} and x_{iN} represent states with or without DCC respectively. Obviously, with DCC, both two states converge to set points quickly. However, without DCC, due to overlapping actuations, two states fluctuate intensely at first and can not converge to set points at last.

Convergence performance of control objective function is presented in Fig.4, where J_Y and J_N represent values of control objective function with or without DCC respectively. With DCC, control objective function converges to zero rapidly. By contrast, without DCC, control objective function oscillates at first and can't converge to zero at last.

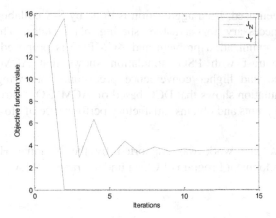

Fig. 4. Evolution of control objective function

Fig.5 shows the actuation outputs of actuators, where u_{iY} and u_{iN} represent outputs of actuators with or without DCC respectively. Without DCC, actuation outputs of A_1 and A_2 vary violently and cause actuation overshooting. Although the outputs of actuators converge at last, but they can not control plant states to meet set points, as shown in Fig.3. Meanwhile, with DCC, we can see that the distributed collaboration among A_1 and A_2 compensates the violent actuation outputs significantly and both actuation outputs converge to their stable states.

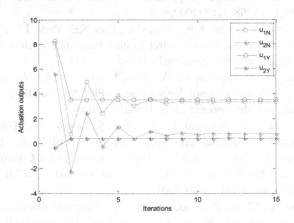

Fig. 5. Evolution of actuation outputs

6   Conclusion

In this paper, in order to control temperature in greenhouse with WSAN, based on hierarchical clustering of sensors and actuators, a distributed collaborative control mechanism has been used. Data aggregation is done by S-S collaboration. And A-A collaboration is necessary for actuators sharing plant states. Then the DCC is formulated as an optimization problem and ACMPSO is proposed as optimization algorithm. By contrast with PSO, simulation shows that ACMPSO has faster convergence speed and higher convergence precision. Comparing with situation without DCC, simulation shows that DCC based on ACMPSO controls plant states to meet well with set points and obtains satisfactory performance of closed loop system.

Acknowledgments. The work was supported by the National High Technology Research and Development Program of China under Grants 2011AA040103-7.

References

[1] Yick, J., Mukherjee, B., Ghosal, D.: Wireless Sensor Network Survey. Computer Networks 52(12), 2292–2330 (2008)
[2] Akyildiz, I.F., Kasimoglu, I.H.: Wireless Sensor and Actor Networks: Research Challenges. Ad Hoc Networks 2(4), 351–367 (2004)
[3] Gungor, V.C., Lambert, F.C.: A Survey on Communication Networks for Electric System Automation. Computer Networks 50(7), 877–897 (2006)
[4] Ramamurthy, H., Prabhu, B.S., Gadh, R., Madni, A.M.: Wireless Industrial Monitoring and Control using a Smart Sensor Platform. IEEE Sensors Journal 7(5), 611–618 (2007)
[5] Gungor, V.C., Hancke, G.P.: Industrial Wireless Sensor Networks: Challenges, Design Principles, and Technical Approaches. IEEE Transactions on Industrial Electronics 56(10), 4258–4265 (2009)
[6] Park, D.H., Kang, B.J., Cho, K.R., Shin, C.S., Cho, S.E., Park, J.W., Yang, W.M.: A Study on Greenhouse Automatic Control System Based on Wireless Sensor Network. Wireless Personal Communications 56(1), 117–130 (2011)
[7] Villa-Medina, J.F., Nieto-Garibay, A., Porta-Gandara, M.A.: Automated Irrigation System Using a Wireless Sensor Network and GPRS Module. IEEE Transactions on Instrumentation and Measurement 63(1), 166–176 (2014)
[8] Sinopoli, B., Sharp, C., Schenato, L., Schaffert, S., Sastry, S.S.: Distributed Control Applications within Sensor Networks. Proceedings of the IEEE 91(8), 1235–1245 (2003)
[9] Cao, X., Chen, J., Xiao, Y., Sun, Y.: Building-Environment Control with Wireless Sensor and Actuator Networks: Centralized versus Distributed. IEEE Transactions on Industrial Electronics 57(11), 3596–3605 (2010)
[10] Chen, J., Cao, X., Cheng, P., Xiao, Y., Sun, Y.: Distributed Collaborative Control for Industrial Automation with Wireless Sensor and Actuator Networks. IEEE Transactions on Industrial Electronics 57(12), 4219–4230 (2010)
[11] Mo, L., Xu, B.: Node Coordination Mechanism Based on Distributed Estimation and Control in Wireless Sensor and Actuator Networks. Journal of Control Theory and Applications 11(4), 570–578 (2013)
[12] Salarian, H., Chin, K.W., Naghdy, F.: Coordination in Wireless Sensor-Actuator Networks: A Survey. Journal of Parallel and Distributed Computing 72(7), 856–867 (2012)

[13] Kellett, C.M., Shim, H., Teel, A.R.: Further Results on Robustness of (Possibly Discontinuous) Sample and Hold Feedback. IEEE Transactions on Automatic Control 49(7), 1081–1089 (2004)

[14] Kennedy, J., Eberhart, R.: Particle Swarm Optimization. In: IEEE International Conference on Neural Networks, pp. 1942–1948. IEEE Press, Perth (1995)

[15] Chuang, L., Yang, C., Li, J.: Chaotic Maps Based on Binary Particle Swarm Optimization for Feature Selection. Applied Soft Computing Journal 11(1), 239–248 (2011)

[16] Xiang, T., Liao, X., Wong, K.: An Improved Particle Swarm Optimization Algorithm Combined with Piecewise Linear Chaotic Map. Applied Mathematics and Computation 190(2), 1637–1645 (2007)

[17] Lu, Y., Li, S., Chen, S., Guo, W., Zhou, C.: Particle Swarm Optimization Based on Adaptive Diffusion and Hybrid Mutation. Journal of Software 18(11), 2740–2751 (2007)

Distributed Fault Detection and Isolation for Discrete Time Multi-agent Systems

Dong Wang and Wei Wang

School of Control Science and Engineering,
Dalian University of Technology,
Dalian, 116023, China
wangdongandy@gmail.com, wangwei@dlut.edu.cn

Abstract. In this paper a distributed fault detection and isolation (FDI) scheme for discrete time multi-agent systems based on a filter is proposed. An FDI filter is designed such that the effects of external input disturbances on the residual signals are minimized and those of fault signals on the residual signals are maximized. These two optimal specifications are represented by H_∞ and H_-. Although sufficient conditions for the existence of such filters are obtained in terms of matrix inequality feasibility conditions, an algorithm to solve them is provided to find a feasible solution such that the filters can be established.

Keywords: Distributed fault detection, Multi-agent systems, H_- performance, H_∞ performance.

1 Introduction

With the increased demand for high safety and reliability in industrial process, fault detection and isolation (FDI) have been one of the hot topics and paid considerable attention [1,2]. One of effective methods for fault detection in the classical model-based FDI scheme is to design an observer or filter to generate a residual signal, which is sensitive to faults and robust to external input disturbances[3,4,5]. By comparing the residual evaluation function with a predefined threshold, we can determine whether the fault occurred in control systems or not. When the residual evaluation function value exceeds the threshold, the fault is detected and an alarm issued. During the past years, a great amount of effort has been devoted in the field. To mention a few, H_∞ techniques have been utilized and the problem of FDI is formulated as H_∞ filtering one [3,4,5].

On the other hand, multi-agent systems have received considerable attention because of their potential applications such as the formation of flight of satellites, unmanned aerial and underwater vehicles, flocking of mobile vehicles, distributed optimization of multiple mobile robotic systems. The development of FDI scheme for multi-agent systems is challenging and interesting due to the distributed nature of such systems and the limitation of shared information with the lower implementation complexity and fewer network resources requirements.

M. Fei et al. (Eds.): LSMS/ICSEE 2014, Part II, CCIS 462, pp. 496–505, 2014.

Despite the imperative need for decentralized and distributed FDI methodologies for multi-agent systems, there exist only few contributions [8]. A distributed Kalman filter is designed to estimate actuator faults for deep space formation flying satellites. The problem of distributed FDI in a network of multiagent systems with double integrator dynamics is considered in [9]. In [7], an observer based scheme is proposed to detect faults affecting both the local dynamics and the subsystems interconnections, where each subsystem exchanges state information with others. In [6], a distributed FDI algorithm is developed for networked robots such that each robot detects faults of others, even if not directly connected to it. The proposed FDI technique relies on overlapping system decompositions into interconnected simpler subsystems, which is monitored by a local FD unit allowed to communicate with each other [10]. A distributed FDI design of heterogeneous network of multi-agent systems is presented for a set of "virtual" systems whose orders are only equal to the sum of the order of each agent and its nearest neighbors agents. In addition, each agent can detect and isolate not only its own faults but also the faults of its nearest neighbors. This is a multi-objective design problem since detection and isolation objectives must be satisfied simultaneously[11]. It should be pointed out that the papers mentioned above are concerned with continuous time multi-agent systems. To the best of the authors knowledge, up to now, little attention has been paid to FDI for discrete time multi-agent systems. In the context of discrete-time nonlinear systems, distributed fault detection scheme is presented to include local and global FDI units, thanks to the introduction of specialized Fault Isolation Estimators and a Global Fault Diagnoser[12].

In this paper, we investigate the problem of distributed FDI for discrete time multi-agent systems based on fault detection filters whose interactions are described by a distributed relative output protocol. There are external input disturbance in the systems detected, which is coupling with the fault signals. A filter is established to generate the residual signal. A multi-objective optimization, namely, two performance indexes H_∞ and H_-, is introduced to represent the effects of disturbance signal on the residual signals and the sensitivity of residual signals to fault signals. It is shown that the problem of distributed FDI under consideration is solvable if certain matrix inequalities are feasible.

The remainder of this paper is organized as follows. In Section 2, the problem statement and definitions are given. Distributed FDI problem for discrete time multi-agent systems based on a filter is analyzed in Section 3. In Section 4, the parameters of filters are constructed. An algorithm for obtaining such parameters are given in Section 5 which is followed by conclusions in Section 6.

Notation. The notation used in this paper is fairly standard. The superscript T stands for the transposition of vectors or matrices. R^n denotes the n dimensional Euclidean space with the norm $\|x\| = (x^T x)^{1/2}$. $l_2[0, \infty)$ is the space of square summable infinite sequences with the norm $\|w_k\|_2 = (\sum_0^\infty \|w_k\|^2)^{1/2}$. In addition, we use $*$ to denote the symmetry entries of symmetry matrices. Matrices, if their dimensions are not explicitly stated, are assumed to be compatible for algebraic operations. $\otimes$ denotes the Kronecker product of matrices. The notation

$P > 0 \ (\geq 0)$ means P is real symmetric positive (semi-positive) definite. I and 0 represent respectively, identity matrix and zero matrix.

2   Problem Formulation

Consider a network of N identical agents with linear or linearized dynamics in the discrete-time setting, where the dynamics of the i-th agent are described by

$$x_i(k+1) = Ax_i(k) + Bw_i(k) + Gf_i(k)$$
$$y_i(k) = Cx_i(k) + Dw_i(k) + Ef_i(k) \tag{1}$$

where $x_i(k) \in R^n$ is the state, $f_i(k) \in R^m$ is the fault input to be detected, $y_i(k) \in R^q$ is the output to be measured, $w_i(k) \in R^p$ is the disturbance input which belongs to $l_2 [0, \infty)$. The matrices A, B, G, C, D and E are of appropriate dimensions.

The communication topology among agents is represented by a directed graph $\mathcal{G} = (\mathcal{V}, \mathcal{E})$, where $\mathcal{V} = 1, ..., N$ is the set of agents and $\mathcal{E} \subset \mathcal{V} \times \mathcal{V}$ is the set of edges between agents. An edge (i, j) in graph $\mathcal{G}$ means that agent j can get information from agent i, but not conversely.

We are interested in, at each time instant, the information available to agent i, which is the relative measurements of other agents with respect to itself:

$$z_i = \sum_{j=1}^{N} a_{ij}(y_i - y_j) \tag{2}$$

where $\mathcal{A} = (a_{ij})_{N \times N}$ is the row-stochastic matrix associated with graph $\mathcal{G}$.

To detect the fault, two tasks require to finish: to construct a residual generator and then to evaluate a residual signal. The first task is to design an observer or a filter to generate the residual signal. The other is a evaluation stage including determining an evaluation function and a threshold. In the present paper, a distributed fault detection filter is constructed for the i-th agent as a residual generator described by

$$\hat{x}_i^+ = \hat{A}\hat{x}_i + \hat{B}z_i, \ r_i = \hat{C}\hat{x}_i + \hat{D}z_i \tag{3}$$

where $\hat{x}_i^+ = \hat{x}_i(k+1)$ is the state at the next time instant, $\hat{x}_i \in R^n$ is the state vector of a fault detection filter, $r_i \in R^t$ is the so-called residual signal, z_i is the relative measurement output error received by an agent over a network, $\hat{A}$, $\hat{B}$, $\hat{C}$ and $\hat{D}$ are positive matrices to be determined. For simplicity, time instants (k) in the variables are neglected.

Let $\bar{x}_i = \begin{bmatrix} x_i^T & \hat{x}_i^T \end{bmatrix}^T$, $\bar{x} = \begin{bmatrix} \bar{x}_1^T, ..., \bar{x}_N^T \end{bmatrix}^T$, $\bar{w} = \begin{bmatrix} w_1^T, ..., w_N^T \end{bmatrix}^T$, $\bar{f} = \begin{bmatrix} f_1^T, ..., f_N^T \end{bmatrix}^T$ and $r = \begin{bmatrix} r_1^T, r_2^T, ..., r_N^T \end{bmatrix}^T$, combining (1), (2) and (3) leads to the augmented systems described by

$$\bar{x}^+ = \tilde{A}\bar{x} + \tilde{B}\bar{w} + \tilde{G}\bar{f}, \ r = \tilde{C}\bar{x} + \tilde{D}\bar{w} + \tilde{E}\bar{f} \tag{4}$$

where

$$\tilde{A} = I_N \otimes \bar{A} + (I_N - \mathcal{A}) \otimes L, \; \bar{A} = \begin{bmatrix} A & 0 \\ 0 & \hat{A} \end{bmatrix}, \; L = \begin{bmatrix} 0 & 0 \\ \hat{B}C & 0 \end{bmatrix}$$

$$\tilde{B} = I_N \otimes \bar{B} + (I_N - \mathcal{A}) \otimes \bar{D}, \; \bar{B} = \begin{bmatrix} B \\ 0 \end{bmatrix}, \; \bar{D} = \begin{bmatrix} 0 \\ \hat{B}D \end{bmatrix}$$

$$\tilde{G} = I_N \otimes \bar{G} + (I_N - \mathcal{A}) \otimes \bar{E}, \; \bar{G} = \begin{bmatrix} G \\ 0 \end{bmatrix}, \; \bar{E} = \begin{bmatrix} 0 \\ \hat{B}E \end{bmatrix}$$

$$\tilde{C} = I_N \otimes \bar{C} + (I_N - \mathcal{A}) \otimes \bar{F}, \; \bar{C} = \begin{bmatrix} 0 & \hat{C} \end{bmatrix}, \; \bar{F} = \begin{bmatrix} \hat{D}C & 0 \end{bmatrix}$$

$$\tilde{D} = (I_N - \mathcal{A}) \otimes \hat{D}D, \; \tilde{E} = (I_N - \mathcal{A}) \otimes \hat{D}E.$$

The following definitions are introduced for the requirements of robust fault detection.

Definition 1 (H_∞ performance). Given a positive scalar γ, system (4) with $\bar{f}(k) = 0$ is said to be asymptotically stable with a prescribed H_∞ norm bound γ under zero-initial condition if the following inequality holds

$$\sum_{k=0}^{\infty} \|r(k)\|^2 \leq \gamma \sum_{k=0}^{\infty} \|\bar{w}(k)\|^2. \tag{5}$$

Definition 2 (H_- performance). Given a positive scalar β, system (4) with $\bar{w}(k) = 0$ is said to satisfy a prescribed H_- performance β under zero-initial condition if the following inequality holds

$$\sum_{k=0}^{\infty} \|r(k)\|^2 \geq \beta \sum_{k=0}^{\infty} \|\bar{f}(k)\|^2. \tag{6}$$

Remark 1. These two definitions are developed for establishing robust fault detection based on the residuals generated by fault detection filters. The first is to employ β to measure the sensitivity of the residuals to fault occurred while the second is to use α to attenuate the disturbance level. The key feature of the proposed FDI scheme is to make the generated residuals sensitive to faulty signals and robust against disturbances in normal case. This strategy will make it possible for the generated residuals to have obvious discrepancies between faulty cases and normal case. It is easily seen that H_- specification measures the sensitivity of r to f. Accordingly, the larger it is, the more sensitive it becomes. These two specifications above are proposed to evaluate the sensitivity and the robustness of fault detection systems in the literature[3,4,5].

The problem of fault detection to be addressed is to design a filter (3) such that system (4) satisfies the following requirements:

1). System (4) is the asymptotical stability and the effects of disturbances on residuals are minimized in a fault-free cases, namely, a minimum γ is found to satisfy H_∞ performance for system (4).

2). System (4) is the asymptotical stability and the effects of fault signal on residuals are maximized in disturbance-free case, namely, a maximum β is found to meet H_- performance for system (4).

After designing the residual generator, a residual evaluation function and a detection logic unit based on the results proposed by [24] are introduced. A square sum value, an average energy of residual signals, over a time length interval L is selected as a residual evaluation function, namely,

$$J_r(L) = \sum_{k=0}^{L} r^T(k)r(k) \tag{7}$$

where L denotes the evaluation time length. If the initial evaluation time is not at zero such as k_1, then, the end time of evaluation function is k_1+L. A threshold is chosen for the decision as follows,

$$J_{th} = \sup_{\bar{w}(k)\in l_2,\ \bar{f}(k)=0} J_r(L). \tag{8}$$

Based on this, the occurrence of faults can be detected by comparing $J_r(L)$ and J_{th} according to the following test:

$$\begin{cases} J_r(L) \leq J_{th}\ No\ alarm \\ J_r(L) \geq J_{th}\ \ Alarm. \end{cases} \tag{9}$$

3   Fault Detection Analysis

In this section, a robust fault detection scheme for system (1) is presented such that system (4) satisfies performance specifications (5) and (6).

3.1 H_∞ Disturbance Attenuation

In this subsection, we will provide sufficient conditions on disturbance attenuation performance of (4).

Theorem 1. For given $\gamma > 0$, system (4) with $\bar{f}(k) = 0$ satisfying the performance specification (5), if there exist a symmetric and positive definite matrix P and a matrix Q such that

$$\begin{bmatrix} -P & 0 & \tilde{A}^T(I_N \otimes Q) & \tilde{C}^T \\ * & -\gamma I & \tilde{B}^T(I_N \otimes Q) & \tilde{D}^T \\ * & * & P - I_N \otimes Q^T - I_N \otimes Q & 0 \\ * & * & * & -I \end{bmatrix} < 0. \tag{10}$$

Proof. We choose the following Lyapunov function candidate

$$V(\bar{x}, k) = \bar{x}^T P \bar{x}. \tag{11}$$

Calculating the forward difference of (11) along the trajectories of system (4) with $\tilde{f}(k) = 0$, we have

$$\Delta V = V(\bar{x}(k+1), k+1) - V(\bar{x}, k)$$
$$= (\tilde{A}\bar{x} + \tilde{B}\bar{w})^T P(\tilde{A}\bar{x} + \tilde{B}\bar{w}) - \bar{x}^T P\bar{x}.$$

When $\bar{w}(k) = 0$, it follows from (12) that

$$\Delta V = \bar{x}^T (\tilde{A}^T P\tilde{A} - P)\bar{x}.$$

It is easy to display that condition (10) implies $\Delta V < 0$, which means that system (4) at the absence of the disturbances is the asymptotical stability.

In order to obtain H_∞ disturbance attenuation performance of system (4), the following performance index is introduced

$$J = \sum_{k=0}^{\infty} \left(r^T(k)r(k) - \gamma \bar{w}^T(k)\bar{w}(k) \right).$$

Based on the zero-initial conditions $\bar{x} = 0$, we can get

$$J = \sum_{k=0}^{\infty} (r^T r - \bar{w}^T \bar{w}) = \sum_{k=0}^{\infty} (r^T r - \bar{w}^T \bar{w} + \Delta V) - V(\infty) + V(0)$$
$$\leq \sum_{k=0}^{\infty} (r^T r - \bar{w}^T \bar{w} + V(\bar{x}(k+1), k+1) - V(\bar{x}, k)) = \zeta^T \Xi \zeta$$

where

$$\zeta = \begin{bmatrix} \bar{x} \\ \bar{w} \end{bmatrix}, \quad \Xi = \begin{bmatrix} -P & 0 \\ 0 & -\gamma I \end{bmatrix} + \begin{bmatrix} \tilde{A}^T \\ \tilde{B}^T \end{bmatrix} P \begin{bmatrix} \tilde{A} & \tilde{B} \end{bmatrix} + \begin{bmatrix} \tilde{C}^T \\ \tilde{D}^T \end{bmatrix} \begin{bmatrix} \tilde{C} & \tilde{D} \end{bmatrix}.$$

It is clear that $J < 0$ if $\Xi < 0$ when $\zeta(k) \neq 0$. By Schur complement lemma, Ξ is negative definite if and only if the following inequality holds

$$\begin{bmatrix} -P & 0 & \tilde{A}^T & \tilde{C}^T \\ * & -\gamma I & \tilde{B}^T & \tilde{D}^T \\ * & * & -P^{-1} & 0 \\ * & * & * & -I \end{bmatrix} < 0. \tag{12}$$

Since P^{-1} is symmetric and positive definite, from the fact that $(P - I_N \otimes Q)^T P^{-1}(P - I_N \otimes Q) \geq 0$, we have

$$-(I_N \otimes Q^T)P^{-1}(I_N \otimes Q) \leq P - (I_N \otimes Q^T + I_N \otimes Q). \tag{13}$$

Performing a congruence transformation to (12) via $diag\{I, I, (I_N \otimes Q), I\}$ yields

$$\begin{bmatrix} -P & 0 & \tilde{A}^T(I_N \otimes Q) & \tilde{C}^T \\ * & -\gamma I & \tilde{B}^T(I_N \otimes Q) & \tilde{D}^T \\ * & * & -(I_N \otimes Q^T)P^{-1}(I_N \otimes Q) & 0 \\ * & * & * & -I \end{bmatrix} < 0. \tag{14}$$

Based on (13), (10) holds if (14) holds, which means the performance index (5) is satisfied.

3.2 H_- Performance

The following theorem presents a gain β as fault sensitivity performance.

Theorem 2. Given a scalar $\beta > 0$, system (4) with $\bar{w}(k) = 0$ satisfies the performance specification (6), if there exist a symmetric and positive definite matrix P and a matrix Q such that

$$\begin{bmatrix} -P - \tilde{C}^T \tilde{C} & -\tilde{C}^T \tilde{E} & \tilde{A}^T (I_N \otimes Q) \\ * & \beta I - \tilde{E}^T \tilde{E} & \tilde{G}^T (I_N \otimes Q) \\ * & * & P - (I_N \otimes Q^T + I_N \otimes Q) \end{bmatrix} < 0. \tag{15}$$

Proof. A Lyapunov function candidate is selected as (11), calculating the forward difference of (11) along the trajectories of system (4) with $\bar{w}(k) = 0$, we have

$$\Delta V = V(\bar{x}(k+1), k+1) - V(\bar{x}, k)$$
$$= (\tilde{A}\bar{x} + \tilde{G}\bar{f})^T P(\tilde{A}\bar{x} + \tilde{G}\bar{f}) - \bar{x}^T P\bar{x}.$$

When $\bar{f}(k) = 0$, it is similar to the proof line of the asymptotical stability in Theorem 1 and straightforward to show that condition (15) implies $\Delta V < 0$, which means that system (4) at the absence of the fault is the asymptotical stability.

In order to obtain H_- performance of system (4), the following performance index is defined

$$J_- = \sum_{k=0}^{\infty} \left(r^{\mathrm{T}}(k)r(k) - \beta \bar{f}^{\mathrm{T}}(k)\bar{f}(k) \right).$$

Based on the zero-initial conditions $\bar{x} = 0$, we can get

$$J_- = \sum_{k=0}^{\infty} \left(r^{\mathrm{T}}(k)r(k) - \beta \bar{f}^{\mathrm{T}}(k)\bar{f}(k) - \Delta V(\bar{x}(k), k) \right) + V(\infty) \geq -\xi^T \Phi \xi$$

where

$$\xi = \begin{bmatrix} \bar{x} \\ \bar{f} \end{bmatrix}, \quad \Phi = \begin{bmatrix} -P & 0 \\ 0 & \beta I \end{bmatrix} + \begin{bmatrix} \tilde{A}^T \\ \tilde{G}^T \end{bmatrix} P \begin{bmatrix} \tilde{A} & \tilde{G} \end{bmatrix} - \begin{bmatrix} \tilde{C}^T \\ \tilde{E}^T \end{bmatrix} \begin{bmatrix} \tilde{C} & \tilde{E} \end{bmatrix}.$$

It is obvious that $J_- > 0$ if $\Phi > 0$ when $\xi(k) \neq 0$. By using Schur complement lemma, Φ is negative definite if and only if the following inequality holds

$$\begin{bmatrix} -P - \tilde{C}^T \tilde{C} & -\tilde{C}^T \tilde{E} & \tilde{A}^T \\ * & \beta I - \tilde{E}^T \tilde{E} & \tilde{G}^T \\ * & * & -P^{-1} \end{bmatrix} < 0. \tag{16}$$

Performing a congruence transformation to (16) via $diag\{I, \ I, \ (I_N \otimes Q)\}$ leads to

$$\begin{bmatrix} -P - \tilde{C}^T \tilde{C} & -\tilde{C}^T \tilde{E} & \tilde{A}^T (I_N \otimes Q) \\ * & \beta I - \tilde{E}^T \tilde{E} & \tilde{G}^T (I_N \otimes Q) \\ * & * & -(I_N \otimes Q)^T P^{-1} (I_N \otimes Q) \end{bmatrix} < 0. \tag{17}$$

Based on (13), the inequality above holds if (15) holds, which means the performance index (6) is satisfied.

4 Design of Fault Detection Filters

Based on the performance analysis in the previous section, sufficient conditions on the existence of fault detection filters are presented, which is solved to obtain the filter parameters via an optimization problem.

Theorem 3. For system (1) and filter (3), given two scalars a and b, the problem of robust fault detection is solvable if there exists symmetric and positive definite matrices P and matrices Q_1, Q_2, M, O, S and H such that system (4) satisfies performance specifications (6) and (5) with the following optimization constraints:

$$\max_{P,\, Q_1,\, Q_2,\, H,\, S,\, M,\, O} a\beta - b\gamma \tag{18}$$

s.t.

$$\begin{bmatrix} -P & 0 & \phi_{11}^T & \phi_{13}^T \\ * & -\gamma I & \phi_{12}^T & (I_N - \mathcal{A})^T \otimes (OD)^T \\ * & * & P - 2I_N \otimes Q & 0 \\ * & * & * & -I \end{bmatrix} < 0 \tag{19}$$

$$\begin{bmatrix} -P - \phi_{21}^T & -\phi_{23}^T & \phi_{11}^T \\ * & \beta I - \phi_{24}^T & \phi_{22}^T \\ * & * & P - 2I_N \otimes Q \end{bmatrix} < 0 \tag{20}$$

where

$$Q = \begin{bmatrix} Q_1 & -Q_2 \\ * & Q_2 \end{bmatrix}, \quad \phi_{11} = I_N \otimes \begin{bmatrix} Q_1 A & -H \\ -Q_2 A & H \end{bmatrix} + (I_N - \mathcal{A}) \otimes \begin{bmatrix} -SC & 0 \\ SC & 0 \end{bmatrix}$$

$$\phi_{12} = I_N \otimes \begin{bmatrix} Q_1 B \\ -Q_2 B \end{bmatrix} + (I_N - \mathcal{A}) \otimes \begin{bmatrix} -SD \\ SD \end{bmatrix}$$

$$\phi_{13} = I_N \otimes \begin{bmatrix} 0 & M \end{bmatrix} + (I_N - \mathcal{A}) \otimes \begin{bmatrix} OC & 0 \end{bmatrix}$$

$$\phi_{21} = I_N \otimes \begin{bmatrix} 0 & 0 \\ 0 & \hat{C}^T \hat{C} \end{bmatrix} + (I_N - \mathcal{A})^T (I_N - \mathcal{A}) \otimes \begin{bmatrix} C^T \hat{D}^T \hat{D} C & 0 \\ 0 & 0 \end{bmatrix}$$

$$\phi_{22} = I_N \otimes \begin{bmatrix} Q_1 G \\ -Q_2 G \end{bmatrix} + (I_N - \mathcal{A}) \otimes \begin{bmatrix} -SE \\ SE \end{bmatrix}, \quad \phi_{23} = 2(I_N - \mathcal{A}) \otimes \begin{bmatrix} 0 & 0 \\ \hat{C}^T \hat{D} C & 0 \end{bmatrix}$$

$$\phi_{24} = (I_N - \mathcal{A})^T (I_N - \mathcal{A}) \otimes E^T \hat{D}^T \hat{D} E.$$

Then, if there is a feasible solution found, the filter gains can be constructed by

$$\hat{A} = Q_2^{-1} H, \quad \hat{B} = Q_2^{-1} S, \quad \hat{C} = M, \quad \hat{D} = O. \tag{21}$$

Proof. According to the property of Kronecker product,

$$(I_N \otimes Q)(I_N \otimes \bar{A}) = I_N \otimes Q\bar{A}, \ (I_N \otimes Q)(I_N \otimes L) = I_N \otimes QL$$
$$(I_N \otimes Q)(A \otimes L) = A \otimes QL, \ (I_N \otimes Q)(I_N \otimes \bar{B}) = I_N \otimes Q\bar{B}$$
$$(I_N \otimes Q)(I_N \otimes \bar{D}) = I_N \otimes Q\bar{D}, \ (I_N \otimes Q)(A \otimes \bar{D}) = A \otimes Q\bar{D}. \qquad (22)$$

Define new matrices as follows

$$Q = \begin{bmatrix} Q_1 & -Q_2 \\ * & Q_2 \end{bmatrix}, \ H = Q_2\hat{A}, \ L = Q_2\hat{B}, \ M = \hat{C}, \ O = \hat{D}. \qquad (23)$$

Based on the definitions above and (22), from (10) we obtain (19).

On the other hand, from (4), we can get

$$(I_N \otimes \bar{C})^T(I_N \otimes \bar{C}) = I_N \otimes I_N \otimes \begin{bmatrix} 0 & 0 \\ 0 & \hat{C}^T\hat{C} \end{bmatrix}$$

$$((I_N - A) \otimes \bar{F})^T((I_N - A) \otimes \bar{F}) = (I_N - A)^T(I_N - A) \otimes \begin{bmatrix} C^T\hat{D}^T\hat{D}C & 0 \\ 0 & 0 \end{bmatrix}$$

$$(I_N \otimes \bar{C})^T((I_N - A) \otimes \bar{F}) = \phi_{23}$$

$$((I_N - A) \otimes \hat{D}E)^T((I_N - A) \otimes \hat{D}E) = \phi_{24}.$$

Letting

$$(I_N \otimes Q)(I_N \otimes \bar{G}) = I_N \otimes Q\bar{G}$$
$$(I_N \otimes Q)(I_N \otimes \bar{E}) = I_N \otimes Q\bar{E}, \ (I_N \otimes Q)(A \otimes \bar{E}) = A \otimes Q\bar{E}$$
$$\phi_{21} = \tilde{C}^T\tilde{C}, \ \phi_{23} = \tilde{C}^T\tilde{E}, \ \phi_{24} = \tilde{E}^T\tilde{E}, \ \phi_{22} = (I_N \otimes Q)\tilde{G}.$$

By means of replacing with the definitions above, we obtain (20) from (15). Combining (19) and (20), the problem of robust fault detection is solvable. In view of (23), the filter parameters are constructed by (21). The proof is completed.

Remark 2. It is seen from (3) that the residual generator is designed for all the neighbour of each agent. Then, each agent has at least a residual generator. This means that it is possible to have more than one residual generator for detecting the fault of an agent. For example, an agent is a neighbour of another two agents. If a fault occurs in the neighbour, these two agents can both detect this fault. Hence, the proposed scheme can isolate the faulty agents[6].

5 Algorithm for Parameter Design

In Thoerem 3, the non-convex matrix inequality conditions are presented to characterize the sensitivity of fault detection by β and the robustness by γ. However, it is difficult to deal with the sensitivity conditions due to the specification (6). In order to construct the filter parameters, the following algorithm is given:

Step 1. In light of Theorem 3, we choose three positive scalars γ, a and b. By solving (19), we can obtain P, Q_1, Q_2, M, O, S and H.

Step 2. Based on this solution in Step 1, β is selected as an unknown parameter. The optimization (18) is combined into solving (20).

Step 3. If there is a feasible solution in (20), then, the filter can be obtained by (21). Or else, we can select smaller scalars γ and b, and a bigger scalar a, and go to Step 1 to solve (19).

6 Conclusion

The fault detection problem of networked control systems with a time delay satisfying Markov chain has been studied in the paper. First, two new criteria have been proposed to evaluate the performance of detecting the faults. Second, sufficient conditions for the existence of the gains of observers, controllers and residual generators of networked control systems have been given. Since such conditions are easy to solve, all these gains can be designed respectively.

References

1. Chen, J., Patton, P.R.: Robust Model-Based Fault Diagnosis for Dynamic Systems. Kluwer, Boston (1999)
2. Frank, P.M., Ding, S.: Survey of robust residual generation and evaluation methods in observer-based fault detection systems. J. Process Control 7, 403–424 (1997)
3. Wang, J., Yang, G., Liu, J.: An LMI approach to H- index and mixed H-/H fault detection observer design. Automatica 43, 1656–1665 (2007)
4. Chadli, M., Abdo, A., Ding, S.: H_-/H_∞ fault detection filter design for discrete-time Takagi Sugeno fuzzy system. Automatica 49, 1996–2005 (2013)
5. Mao, Z.H., Jiang, B., Shi, P.: H_∞ fault detection filter design for networked control systems modelled by discrete Markovian jump systems. IET Control Theory and Applications 1, 1336–1343 (2007)
6. Arrichiello, F., Marino, A., Pierri, F.: A decentralized fault detection and isolation strategy for networked robots. In: 2013 16th International Conference on Advanced Robotics (ICAR), Montevideo, November 25-29, pp. 1–6 (2013)
7. Panagi, P., Polycarpou, M.: Decentralized fault tolerant control of a class of interconnected nonlinear systems. IEEE Transactions on Automatic Control 56, 178–184 (2011)
8. Azizi, S.M., Khorasani, K.: Consensus based overlapping decentralized fault detection and isolation. In: Proc. 3rd Annu. IEEE Int. Syst. Conf., pp. 570–575 (September 2009)
9. Shames, I., Teixeira, A.H., Sandberg, H., Johansson, K.H.: Distributed fault detection for interconnected systems. Automatica 47, 2757–2764 (2011)
10. Boem, F., Ferrari, M., Parisini, T.: Distributed Fault Detection and Isolation of Continuous-Time Non-Linear Systems. European Journal of Control 5-6, 603–620 (2011)
11. Davoodi, M., Khorasani, K., Talebi, H., Momeni, H.:Distributed Fault Detection and Isolation Filter Design for a Network of Heterogeneous Multiagent Systems. IEEE Transactions on Control Systems Technology (in press, 2014)
12. Ferrari, R., Parisini, T., Polycarpou, M.: Distributed Fault Detection and Isolation of Large-Scale Discrete-Time Nonlinear Systems: An Adaptive Approximation Approach. IEEE Transactions on Automatic Control 57, 275–285 (2012)

The Design Of Serial Communication Module between Android and Muti-lower Machines

Yun-xia XI and Xin Sun

School of Mechanical Engineering and Automation, Shanghai University,
Shanghai 200072, China
869862730@qq.com, teachersunxin@163.com

Abstract. Due to the low price of serial communication, it has been widely used among the industrial field. Human machine interface based on Android platform has not been widely used in the industrial field. By the article, Human-machine interface has been designed to achieve the communication between HMI and multiple lower machine.

Keywords: Android, HMI, serial communication, multiple lower machines.

1 Introduction

Serial communication has the low transmission rate while it is relatively inexpensive among communication devices.Therefore serial communication is an instrument or device commonly used on industrial field of communication. If transport the content is not very complicated, serial communication still has a comparative advantage. Upper and lower traditional machine communications industry generally use a PC or PLC to communication with lower machine communication[1].Advantages concerning multiple models,good hardware configuration,openness and so on makes Android the overwhelmed share of the commercial market[2]. However, the application is popular among the industrial field.This paper studies serial communication based on Android platform and makes design of serial communication between Android platform and multi-lower machines.

2 Framework Design

Cortex A8 is used in this paper cor,which uses Samsung S5PV210 as the main processor that runs at up to 1GH so parameter.There are four serial ports in the S5PV210 including UART0, UART1,UART2,UART3[3].All of them,UART0 has been converted level shift, which corresponds to COM0,which makes this development board communicate with PC via the included serial cable [8].The design of lower machines adapts 485 serial port,requiring switch electrical level,thus makes communication between the upper and lower machine.The design as a whole (**Figure 1**). In this paper, it discusses communication module design on the higher machine.

M. Fei et al. (Eds.): LSMS/ICSEE 2014, Part II, CCIS 462, pp. 506–512, 2014.
© Springer-Verlag Berlin Heidelberg 2014

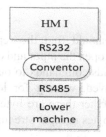

Fig. 1. The whole design framework

3 The Design of Comxmunication between HMI and Singxle Lower Machine

3.1 Module Introduction of HMI

It is mainly used to receive, transmit data by Human-machine interface. We can use buttons to control data communication,.Therefore, in the process of the interface design , we would design some buttons to listen some actions that come from buttons,for example, sending data to the serial port. For sending data ,we need listen whether there is data buffer in serial port constantly.By the way,it's necessary that we explain input stream[6,7].The Java.io.InputStream class is the superclass of all classes representing an input stream of bytes.Applications that need to define a subclass of InputStream must always provide a method that returns the next byte of input[8].

JNI bridges the part of Android or Java and the sections C or C++.So it needs Java code and C or C++ code if we need use completely JNI.While C / C + + code is used to generate the library file, Java code is used to reference the C / C + + libraries and call C / C + + method [3,4].In this paper,JNI connects HMI and hardware.The following shows the flow chart of Android machine to communicate with the next bit by Figure 2. First, man-machine interface which can be accessed via the RS232 serial port hardware by calling JNI.

Fig. 2. Connection Figure between HMI and Lower computer

3.2   Development Process

For program design, firstly we should find the serial port, set the baud rate,and equipment.After Setting up above, open the input and output streams, by which the data is processed(**Fig. 3**). When UI sends commands to output stream, data will be sent to the buffer where data can be saved, waiting for lower machine processing.Then, receiving data needs to detect whether there are bytes in input stream.Lastly, open a thread to judge whether there are bytes in input stream[4].

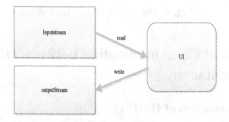

Fig. 3. Working mode of I/O

3.3   Open Serial Port

How to open and close serial port ? Before concerning it,I introduce the class of android.app.Application that can be created by Android system. android.app. Application runs for each program and create a just, it can be said that an Application class singleton mode., And life cycle of application objects ,accompanying the program period ,is the longest for the whole procedure.its life cycle is equivalent to the program's life cycle. Because it is a single example of a global, objects obtained is the same in different Activity, Service object[4] . So we can do data transfer, data sharing, data caching and other operations by Application. Therefore, in the design of open-closed serial write a Application, inheriting android.app.Application. By the design, more   interfaces can open, close the serial port (**Fig. 4**).

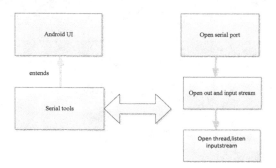

Fig. 4. program flow between HMI and single lower machine

3.4   Handler Mechanism

When you open the serial port, it needs to keep judging there is data in the input buffer. If data is found, start the method sendMessage of Handler object .After Message object receives message , then process buffer data(**Fig. 5**).

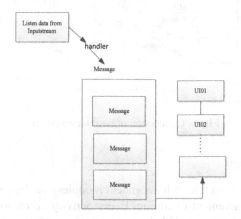

Fig. 5. Start the handler mechanism

4   HMI Design of Communication between Lower Machines

Characteristics of Human machine interface communicating with lower machines requires into independent interface, convenient operation, query. So compared with HMI communicating with a lower computer,we can achieve this effect through multiple label switching different interface. Before doing it,we must know the android.widget.TabHost , which is used to deposit a plurality of Tab tags containers, each tab can correspond to the layout of their own similar in Tab layout telephone book that is a a linear layout interface [4,9]. [5,6]If we use TabHost, the first to obtain TabHost by method of getTabHost from android.widget.TabHost and then by using the method of addTab from android.widget.TabHost to add TabHost Tab, when you switch,every Tab would have an event capturing the event .Set the method of event setOnTabChangedListener from TabActivity object , of course,we can also use a Intent to jump[4,9].

4.1   HMI Communication Design

HMI is divided into five modules, requiring it communicate with the lower machine 5. Star topology uses a centralized communication control strategy [10].

Android tablet is always the initiator of communication, five separate lower machine in a passive state [5]. When HMI makes broadcast-style request to lower machines .Then whether to accept the data, we can judge by address.Then message data is parsed by data format based on the message type .If a message is sent to get a response, then we send the next data(**Fig. 6**).

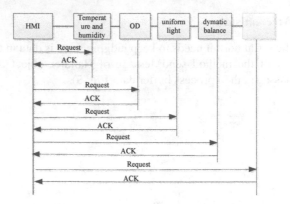

Fig. 6. Connection of hardware system

4.2   Program Design

Because lower machine contains 5 independent modules, we design into 5 independent interfaces,which is convenient operation. Every activity corresponds to an interface. Each interface function can be divided into:monitoring, setting, query function. Every activity add into monitor queue[4,7]. We wrote a receiving method from the listener. We open the serial input and output streams to receive and send data[7]. Start a timer to receive data,and determine whether send data to serial port(**Fig. 7**).

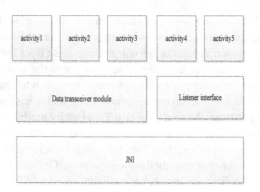

Fig. 7. Muti-machine interface design

Design of sending and receiving data timer is shown in Figure 8. Received data determine whether the data transmission is corrector lost by parsing data. therefore we need to check CRC verification. When the data are not complete or has error ,we need to throw data, the interface need to extract data from queue to sends commands to lower machine .

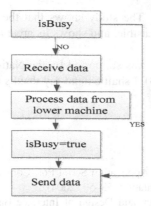

Fig. 8. Design of Data transceiving timer

4.3   Communication Protocol

The communication between host computer and lower computer is used in active response mode, how to let 5 lower machines accurately receive and send from HMI. Therefore, we need to introduce the communication protocol, like in the car on the street to drive "rules, "pedestrians pedestrians" rule [10]. Then we set the rules of communication.

We encapsulate message frame containing the address of the machine at the same time. When we receive data, we judge whether we go on parsing by parsing the first bit .The content length we can judge sent to host computer's command. Verifying CRC is very important index judging whether integrity of data.

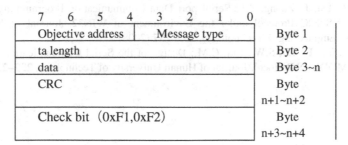

7	6	5	4	3	2	1	0	
Objective address				Message type				Byte 1
ta length								Byte 2
data								Byte 3~n
CRC								Byte n+1~n+2
Check bit （0xF1,0xF2)								Byte n+3~n+4

Fig. 9. Communication protocol

5   Conclusion

This paper designs the communication between single lower machine and HMI. Based on it ,we design module of Human-machine interface, communication mode, communication protocol.The system is used communication between HMI based on Android platform and   lower machines, the system uses the RS-232C communication protocol. The host human computer interface is friendly, portable.maintenance and monitoring is very convenient. The study shows that the industrial human-machine interface based on Android is feasible, and shows its great advantage.

Acknowledgments. This work was supported by the National Instruments for major projects (No.2012YQ15008703). I shall extend my thanks to my teacher Mrs. Sun for all her kindness and help.

References

1. Zhang, B., Wang, B.W.: The Application Serial Port Communication in Process Control. Process Automation Instrumentation, 58–59 (2004)
2. Zhao, L., Zhang, W.: Research and Design of Interface Based on Android Technology. Computer Knowledge and Technology, 8183–8185 (2009)
3. Ren, J.W., Lin, D.D.: Research of Platform Independent Programming Using JNI Technology. Application Research of Computers, 180–184 (2005)
4. Android SDK Document, http://developer.android.com/guide/index.html
5. Peng, X., Tan, Z., Huang, W.J., Wang, X.H.: Design of Industrial Control Monitoring Software Based on Android. Computer Engineering, 86–89 (2013)
6. Liu, A.R., Lu, Y.J., Wang, Z.C.: Fast Algorithm of CRC in Serial Communications. Journal of Henan University (Natural Science), 418–420 (2007)
7. Java API, http://docs.oracle.com/javase/7/docs/api/java/io/InputStream.html
8. Lu, C.: Design of Distributed Temperature-collecting System Based on PC and MCU. Instrument Technique and Sensor, 165–167 (2007)
9. Cheng, T.R., Yang, F., Jia, Z.W.: Discussion of Two Problems in the Serial Communication. Electrical Automation, 123–126 (2006)
10. Chen, C.B., Du, J., Zhang, Z.J.: Serial port Data Communication Programming Method Based on RS232C Protocol under the Environment of WIN32. Journal of Nanchang University (Engineering & Technology), 255–257 (2008)
11. Wen, X.L., Yu, F., Lu, S.W., Liu, C.M.: Design of the Serial Communication Software Based on MODBUS Protocol. Journal of Hunan University of Technology, 222–226 (2008)

Design for MIMO Networked Control Systems Based on Multi-threshold Dead Band Scheduling

Chenyu Zhang, Weihua Fan[*], Ronghua Xie, and Qingwei Chen

School of Automation, Nanjing University of Science and Technology, Nanjing, Jiangsu, 210094
NJBZCY@163.com, fanweihua@njust.edu.cn

Abstract. The control and scheduling co-design of MIMO networked control system are studied in this paper. Considering the MIMO networked control system with multiple sensor nodes, the deadband scheduling strategy with multi-threshold is employed, then the transmit character of the sensor node's data packets is analyzed. To compensate the effect of the deadband and network-induced delay, an observer is employed, and the closed loop system with the observer and dynamic feed back controller is modeled as a discrete time linear system with uncertainty. Then by the Lyapunov function and LMIs method, the sufficient condition for asymptotic stability is presented, together with the design method for the dynamic feedback controller and observer. Finally, a numerical example is given to validate the method.

Keywords: MIMO Networked control systems, multi-threshold deadband scheduling, dynamic feedback control, observer.

1 Introduction

Networked control systems (NCSs) are closed-loop control systems whose sensors, controllers and actuators are connected by network. In recent years, NCSs have attracted wide attention. Commonly, MIMO NCSs refer to those consist of multiple sensor nodes, actuators and controller nodes. Besides those problems in SISO NCSs, more challenging problems are involved in MIMO NCSs. For example, for sending data packets, a sensor node should compete with other sensor nodes and the nodes which do not belong to the control loop, so some sensor nodes cannot transmit data to the controller nodes in some periods because of failure competition. That causes the controller node cannot obtain all recent output data, which will result in bad performance. Also the distributed time-varying network induced delay occurs, which will make the model and analysis more difficult than the SISO one.

Recently, MIMO NCSs have attracted much attention. The modeling, analysis and design of the MIMO NCSs are progressing steadily. In [1], the MIMO NCS was modeled as a discrete-time linear system, and a sufficient condition for stability subject to the bounded packet dropping probability was given. But the controller design approach did not be concerned. In [2], a sufficient condition to asymptotically stability

---

[*] Corresponding author.

M. Fei et al. (Eds.): LSMS/ICSEE 2014, Part II, CCIS 462, pp. 513–522, 2014.

for the MIMO NCS with delay and packet loss was proposed. In [3], the MIMO NCSs where network with limited access channels were modeled as discrete switch system with time delay, and a stability criterion was developed in terms of linear matrix inequalities by constructing a novel piece-wise Lyapunov functional. In [4], the problem of H_∞ performance analysis and controller design was considered for NCSs with multiple communication channels. The idle communication channels were made full use to compensate the negative influences of time delay and packet dropout. In [5], the MIMO NCSs with short time delay was modeled as an asynchronous dynamic system. And the condition for exponential stability of the system was given. But the nonlinear condition is difficult to be solved.

In MIMO NCSs, better control performance usually require more transmission, which will reduce service quality of network. That is a problem to be solved. The control and scheduling strategy co-design is a validate solution. In co-design method, the scheduling strategy is used to improve the network service quality, while the controller is designed to compensate the negative impact of the network. In [7], the scheduling strategy based on time slice was employed to regulate the data transmissions in MIMO NCS. Then the models were given with the fixed short time delay, random short time delay and data packet loss respectively. Also the corresponding conditions for the asymptotic stability were presented respectively, the design methods of controllers were also given.

In NCSs, scheduling policy should guarantee not only the transmissions of control system, but also regulate network traffic that non-control packets, such as alarm message, monitor information, etc. can be transmitted during operation. At present, the common used scheduling policies, such as RM [9], MEF-TOD [10] and MTS [11], focus on important data transmission rather than the reasonable network traffic. Scheduling policy based on dead band [12] meets both requirements, and thus can be applied in MIMO NCSs. However, when using the dead band scheduling, a different threshold value should be set for each nodes, because the effect of data fluctuation to the system performance is different.

Based on the analysis above, we employ a multi-threshold deadband scheduling policy for the MIMO NCSs. Then with the corresponding dynamic feedback controller and observer, the model of the MIMO NCS is given. Based on the model, the stability is analyzed with the Lyapunov functional method and the design methods for the dynamic feedback controller and observer is present in terms of linear matrix inequalities (LMIs).

The paper is organized as follows. In Section 2, formulates the problem, and some preliminary definitions and hypothesis are provided. On the basis, the multi-threshold dead band scheduling is proposed, and the discrete time model is given. In section 3, the sufficient condition to asymptotic stability and the design for the dynamic feedback controller and observer are introduced. In section 4, an example is given to illustrate method. Conclusion and future work is introduced in last section.

2   Problem Formulation

The structure of the MIMO NCSs is illustrated in Fig1. And it is supposed that there are m sensor codes in system. $y_i(k), i = 1, \ldots, m$ denotes the sampled output of plant the kth sampling period, $\hat{y}_i(k), i = 1, \ldots, m$ denotes the data arrived at the controller node through network. $\hat{y}(k), \hat{x}(k)$ denote the input and output of the observer respectively.

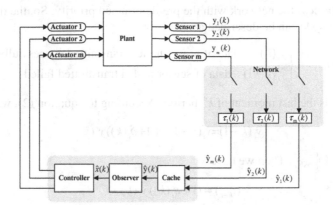

Fig. 1. The structure of MIMO NCS

Suppose the plant is described as in (1)

$$x(k+1) = Ax(k) + Bu(k)$$
$$y(k) = Cx(k)$$

(1)

where, $x(k) \in \mathbf{R}^n$, $y(k) \in \mathbf{R}^m$, $u(k) \in \mathbf{R}^r$ denote the state vector, output vector and input vector respectively. A, B, C are matrices with appropriate dimensions.

Firstly, we give following assumptions.

Assumption 1: The sensor nodes are time-driven, and the sampling period is same, denote as $T, T > 0$;

Assumption 2: The controller node is also time-driven, whose period is same as the sensor nodes;

Assumption 3: The network-induced delay $\tau_i(k) < T, i = 1, \ldots, m$;

Assumption 4: There are no packet dropouts and disorders in the NCSs.

Because multiple sensor nodes in MIMO NCS sample at the same time, the conflict is inevitable. So we employ the deadband schedule strategy to reduce network conflict and regulate network load. Notice that different sensor node measures different output signal, and its range and impact on system performance is different, it is unreasonable to use unique threshold in deadband scheduling. Therefore, multi-threshold deadband scheduling strategy is used. According to the scheduling strategy, the sensor node judges whether the sampled data to be transmitted in this period. If the formula (2) is satisfied, the data packet should be transmitted. Otherwise, it would not be transmitted.

$$|y_i(k) - \tilde{y}_i(k-1)| \geq \delta_i |\tilde{y}_i(k)| \tag{2}$$

where $\tilde{y}_i(k-1)$ is the data transmitted latest time, and δ_i is the threshold of the ith sensor node.

If more than one sensor nodes want to transmit data packet at same period, they will complete to access the network with the pre-determined priority. So, the data received by controller node can be described as:

$$\hat{y}_i(t_{k+1}) = \begin{cases} y_i(k) & \text{data of sensor node i transmitted successfully} \\ \hat{y}_i(k-1) & \text{data of sensor node i transmitted failed} \end{cases} \tag{3}$$

where t_{k+1} is the last moment of kth period. According to equation (2), we have

$$\hat{y}_i(k-1) = \tilde{y}_i(k-1) = (1+\delta_i(k))y_i(k) \tag{4}$$

where $|\delta_i(k)| \leq \delta_i$. Then we have

$$\hat{y}_i(t_{k+1}) = (1+\varepsilon_i(k))y_i(k) \tag{5}$$

where $\varepsilon_i(k) = \begin{cases} 0 & \text{transmitted successfully} \\ \delta_i(k) & \text{transmitted failed} \end{cases}$. Hence,

$$\hat{y}(t_{k+1}) = (I + E(k))y(k) \tag{6}$$

where $\hat{y}(t_{k+1}) = \left[\hat{y}_1^T(t_{k+1}), \cdots, \hat{y}_m^T(t_{k+1}) \right]^T$, $y(k) = \left[y_1^T(k), \cdots, y_m^T(k) \right]^T$, $E(k) = diag\left(\varepsilon_1(k), \cdots, \varepsilon_m(k)\right)$.

At time $t = kT$, the data used to calculate control variable comprise of some components of $\hat{y}(t_k)$ and some components of the historical datum $y(k - d_i)$, $d_i \leq d_{max}$. That is, the controller cannot use the full recently output to calculate. The performance may decline if the controller uses the compound output variable directly. To deal with this drawback, we employ an observer to compensate the effect of the historical datum and network-induced delay, because the observer furnish one-step forecast.

The observer is described in (7).

$$\hat{x}(k+1) = A\hat{x}(k) + Bu(k) + L(\hat{y}(t_{k+1}) - C\hat{x}(k)) \tag{7}$$

where $\hat{x}(k) \in \mathbf{R}^n$ is the state vector of observer, $L \in \mathbf{R}^{n \times m}$ is the gain matrix .

Remark 1: When the observer calculated in the kth period, the available data is $\hat{y}(t_k)$. So actually, the observer calculate as following equation

$$\hat{x}(k) = A\hat{x}(k-1) + Bu(k-1) + L(\hat{y}(t_k) - C\hat{x}(k-1))$$

And the $\hat{x}(k)$ is used to calculate the control variable. Because the above formula is equivalent to (7), we use equation (7) in following process.

Suppose the dynamic feedback controller is in (8):

$$
\begin{aligned}
x_c(k+1) &= A_c x_c(k) + B_c \hat{x}(k) \\
u(k) &= C_c x_c(k) + D_c \hat{x}(k)
\end{aligned}
\tag{8}
$$

where $x_c(k) \in \mathbf{R}^p$ is the state vector of controller, A_c, B_c, C_c, D_c are matrices with appropriate dimensions.

Defined as the observation error $e(k) = x(k) - \hat{x}(k)$. With Eq. (1), (6), (7) and (8), the closed-loop system can be rewritten as

$$
\begin{bmatrix} x(k+1) \\ e(k+1) \\ x_c(k+1) \end{bmatrix} = \begin{bmatrix} A+BD_c & -BD_c & BC_c \\ -LE(k)C & A-LC & 0 \\ B_c & -B_c & A_c \end{bmatrix} \begin{bmatrix} x(k) \\ e(k) \\ x_c(k) \end{bmatrix}
\tag{9}
$$

Note that $E(k)$ is time-varying and can be divided into two parts:

$$
E(k) = \bar{E} + \tilde{E}(k)
$$

where, $\bar{E} = \dfrac{1}{2} diag(\delta_1, \cdots, \delta_m)$, $\tilde{E}(k) = \bar{U}\Delta(k)\bar{V}$, $\bar{U} = \dfrac{1}{2} diag(\sqrt{\delta_1}, \cdots, \sqrt{\delta_m})$,

$\bar{V} = diag(\sqrt{\delta_1}, \cdots, \sqrt{\delta_m})$, $\Delta(k) = diag(\Delta_1, \cdots, \Delta_i)$, $\Delta^T(k)\Delta(k) \le I$. Therefore,

$$
\begin{bmatrix} x(k+1) \\ e(k+1) \\ x_c(k+1) \end{bmatrix} = \begin{bmatrix} A+BD_c & -BD_c & BC_c \\ -L(\bar{E}+\bar{U}\Delta(k)\bar{V})C & A-LC & 0 \\ B_c & -B_c & A_c \end{bmatrix} \begin{bmatrix} x(k) \\ e(k) \\ x_c(k) \end{bmatrix}
\tag{10}
$$

From equation (10), the MIMO NCSs should be modeled as as a uncertain discrete linear system.The objective of the following parts is to design the controller and observer to stabilize the system (10).

3   Main Results

Lemma 1 [13]:For arbitrarily appropriate matrices $W, M, N, F(k)$, where $F^T(k)F(k) \le I$, W is symmetric matrix, then the inequality holds:

$$
W + N^T F^T(k) M^T + M F(k) N < 0
$$

if and only if there exists $\varepsilon > 0$, such that

$$
W + \varepsilon^{-1} N^T N + \varepsilon M M^T < 0
$$

Theorem 1: If there exist positive symmetric matrices P_1, P_2 and P_3, and a scalar $\varepsilon > 0$, such that the following matrix inequality holds,

$$\begin{bmatrix} -P_1+\varepsilon^{-1}C^T\bar{V}^T\bar{V}C & 0 & 0 & (A+BD_c)^T & (-L\bar{E}C)^T & -B_c^T & 0 \\ * & -P_2 & 0 & (-BD_c)^T & (A-LC)^T & B_c^T & 0 \\ * & * & -P_3 & (BC_c)^T & 0 & -A_c^T & 0 \\ * & * & * & -P_1^{-1} & 0 & 0 & 0 \\ * & * & * & * & -P_2^{-1} & 0 & L\bar{U} \\ * & * & * & * & * & -P_3^{-1} & 0 \\ * & * & * & * & * & * & -\varepsilon^{-1}I \end{bmatrix} < 0 \quad (11)$$

where "*" denotes the symmetrical part.
Then the system (10) is asymptotically stable.

Proof: Choose Lyapunov function as following

$$V(k) = x^T(k)P_1x(k) + e^T(k)P_2e(k) + x_c^T(k)P_3x_c(k)$$

The forward difference is calculated as.

$$\begin{aligned} \Delta V(k) &= V(k+1) - V(k) \\ &= x^T(k+1)P_1x(k+1) + e^T(k+1)P_2e(k+1) + x_c^T(k+1)P_3x_c(k+1) \\ &\quad - x^T(k)P_1x(k) - e^T(k)P_2e(k) - x_c^T(k)P_3x_c(k) \end{aligned}$$

Using equation (10), we have

$$\Delta V(k) = \begin{bmatrix} x^T(k) & e^T(k) & x_c^T(k) \end{bmatrix} \prod \begin{bmatrix} x^T(k) & e^T(k) & x_c^T(k) \end{bmatrix}^T$$

where

$$\prod = \begin{bmatrix} \begin{array}{l}(A+BD_c)^TP_1(A+BD_c) \\ +(LE(k)C)^TP_2(LE(k)C) \\ +B_c^TP_3B_c - P_1 \end{array} & \begin{array}{l}-(A+BD_c)^TP_1(BD_c) \\ -(LE(k)C)^TP_2(A-LC) \\ -B_c^TP_3B_c \end{array} & \begin{array}{l}(A+BD_c)^TP_1(BC_c) \\ +B_c^TP_3A_c \end{array} \\ \begin{array}{l}-(BD_c)^TP_1(A+BD_c) \\ -(A-LC)^TP_2(LE(k)C) \\ -B_c^TP_3B_c \end{array} & \begin{array}{l}(BD_c)^TP_1(BD_c) \\ +(A-LC)^TP_2(A-LC) \\ +B_c^TP_3B_c - P_2 \end{array} & \begin{array}{l}-(BD_c)^TP_1(BC_c) \\ -B_c^TP_3A_c \end{array} \\ \begin{array}{l}(BC_c)^TP_1(A+BD_c) \\ +A_c^TP_3B_c \end{array} & \begin{array}{l}-(BC_c)^TP_1(BD_c) \\ -A_c^TP_3B_c \end{array} & \begin{array}{l}(BC_c)^TP_1(BC_c) + \\ A_c^TP_3A_c - P_3 \end{array} \end{bmatrix}$$

If $\prod < 0$, then $\Delta V(k) < 0$, the system (10) is asymptotically stable. By Schur complement and lemma 1, $\prod < 0$ is equivalent to (11). The proof is completed.

By pre-multiplying and post-multiplying (11) with $diag(I,I,I,P_1,P_2,P_3,I)$, we have

$$\begin{bmatrix} -P_1+\varepsilon^{-1}C^T\bar{V}^T\bar{V}C & 0 & 0 & (A+BD_c)^T P_1 & (-L\bar{E}C)^T P_2 & -B_c^T P_3 & 0 \\ * & -P_2 & 0 & (-BD_c)^T P_1 & (A-LC)^T P_2 & B_c^T P_3 & 0 \\ * & * & -P_3 & (BC_c)^T P_1 & 0 & -A_c^T P_3 & 0 \\ * & * & * & -P_1 & 0 & 0 & 0 \\ * & * & * & * & -P_2 & 0 & P_2 L\bar{U} \\ * & * & * & * & * & -P_3 & 0 \\ * & * & * & * & * & * & -\varepsilon^{-1}I \end{bmatrix} < 0 \quad (12)$$

Remark 2: If the matrices A_c, B_c, C_c, D_c, L are known, the inequality (12) is a LMI which can be easily solved by Matlab LMI Toolbox.

When design the controller and observer, the matrices A_c, B_c, C_c, D_c, L are unknown variable to be determined, then (12) is nonlinear. And we cannot transform it to LMI by congruent transformation and variable substitution method.

If there exist a matrix R_1, such that

$$P_1 B = BR_1 \quad (13)$$

and define $Y_1 = R_1 D_c, Y_2 = R_1 C_c, Y_3 = P_2 L$, $P_3 A_c = Q_2, P_3 B_c = Q_1$, Eq. (12) is rewritten as

$$\begin{bmatrix} -P_1+\varepsilon^{-1}C^T\bar{V}^T\bar{V}C & 0 & 0 & (P_1 A+BY_1)^T & (-Y_3\bar{E}C)^T & -Q_1^T & 0 \\ * & -P_2 & 0 & (-BY_1)^T & (P_2 A-Y_3 C)^T & Q_1 & 0 \\ * & * & -P_3 & (BY_2)^T & 0 & -Q_2^T & 0 \\ * & * & * & -P_1 & 0 & 0 & 0 \\ * & * & * & * & -P_2 & 0 & Y_3\bar{U} \\ * & * & * & * & * & -P_3 & 0 \\ * & * & * & * & * & * & -\varepsilon^{-1}I \end{bmatrix} < 0 \quad (14)$$

Let $P_1 = U\begin{bmatrix} P_{11} & P_{21}^T \\ P_{21} & P_{22} \end{bmatrix}U^T$, by singular value decomposition [14], we have

$$\begin{bmatrix} P_{11}\Sigma V \\ P_{21}\Sigma V \end{bmatrix} = \begin{bmatrix} \Sigma VR_1 \\ 0 \end{bmatrix} \quad (15)$$

Obviously, the equation (15) holds if $P_{21}\Sigma V = 0$, $P_{11}\Sigma V = \Sigma VR_1$, which means $P_{21} = 0, R_1 = \Sigma^{-1}V^{-1}P_{11}\Sigma V$. So we have

$$P_1 = U\begin{bmatrix} P_{11} & 0 \\ 0 & P_{22} \end{bmatrix}U^T.$$

Theorem 2: Suppose the matrix B is full-column rank, and $B = U^T \begin{bmatrix} \Sigma \\ 0 \end{bmatrix} V$. If there exist positive symmetric matrices P_2, P_3, P_{11} and P_{22}, a scalar $\varepsilon > 0$ and arbitrarily appropriate matrices Y_1, Y_2 and Y_3, such that the matrix inequality (14) holds, where

$$P_1 B = BR_1$$

$$P_1 = U \begin{bmatrix} P_{11} & 0 \\ 0 & P_{22} \end{bmatrix} U^T \tag{16}$$

Then the system (10) is asymptotically stable. And the evaluated gain matrices are $A_c = P_3^{-1} Q_2$, $B_c = P_3^{-1} Q_1$, $C_c = V^{-1} \Sigma^{-1} P_{11}^{-1} \Sigma V Y_2$, $D_c = V^{-1} \Sigma^{-1} P_{11}^{-1} \Sigma V Y_1$, the evaluated observer gain matrix is $L = P_2^{-1} Y_3$.

4   Simulation

Considering the plant of the MIMO NCS is described as:

$$x(k+1) = \begin{bmatrix} 0.5 & 0.25 & -0.1 \\ 0 & 0.5 & 0.6 \\ 0 & 0.2 & -0.4 \end{bmatrix} x(k) + \begin{bmatrix} 0.1 & 0 \\ 0.5 & 0.2 \\ 0 & 0 \end{bmatrix} u(k)$$

$$y(k) = \begin{bmatrix} 1 & 0 & 0 \\ 0 & 1 & 0 \\ 0 & 0 & 1 \end{bmatrix} x(k)$$

Set the threshold of three sensor nodes are 0.64,0.8,0.9. By calculation, we have

$$\bar{E} = \begin{bmatrix} 0.32 & & \\ & 0.40 & \\ & & 0.45 \end{bmatrix}, \bar{U} = \begin{bmatrix} 0.40 & & \\ & 0.445 & \\ & & 0.475 \end{bmatrix}, \bar{V} = \begin{bmatrix} 0.8 & & \\ & 0.89 & \\ & & 0.95 \end{bmatrix}$$

By theorem 2, we have

$$A_c = \begin{bmatrix} -0.0616 & -0.0148 & 0.0065 \\ -0.0148 & -0.0603 & -0.0480 \\ 0.0065 & -0.0480 & -0.0480 \end{bmatrix}, B_c = \begin{bmatrix} 0.2749 & -0.0047 & -0.0058 \\ -0.0047 & 0.2598 & 0.0097 \\ -0.0058 & 0.0097 & 0.2783 \end{bmatrix},$$

$$C_c = \begin{bmatrix} 0.0687 & 0.0561 & -0.0741 \\ 0.3248 & -0.1876 & 0.0147 \end{bmatrix}, D_c = \begin{bmatrix} 6.1278 & 2.8421 & -8.1654 \\ -5.1789 & -8.2548 & -2.7894 \end{bmatrix},$$

$$L = \begin{bmatrix} -0.8594 & 1.3675 & -0.6104 \\ 0.1245 & -0.9629 & 0.3461 \\ -0.1242 & -0.4648 & 0.1592 \end{bmatrix}.$$

Using above gain matrices, the output trajectories are given in Fig.2. The closed loop system is asymptotically stable.

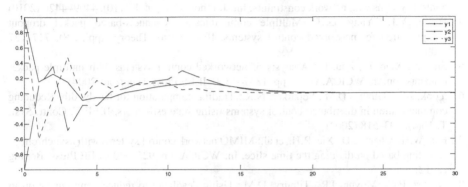

Fig. 2. The output trajectories of closed loop system

5   Conclusion

The modeling and stabilization of the MIMO NCS is studied in this paper. For MIMO NCSs with multiple sensor nodes and synchronized sampling, considering the network-induced delay less than a sampling period, multi-threshold deadband scheduling strategy is employed to improve the service quality of the network. Based on dynamic feedback controller, a discrete-time model of the system is given. Using Lyapunov functional method, a sufficient condition which assure asymptotically stability of the system is presented. And the design approach for the controller and the observer gain are given in terms of LMIs. The simulation demonstrates the efficiency of the proposed methods.

In this paper, only the case of unilateral network is considered, the modeling and analysis of MIMO NCSs with bilateral network, network-induced delay and packet loss will be studied in the future.

Acknowledgments. This work was support by NSFC (grant numbers 61074023), the Jiangsu postdoctoral fund, and the Jiangsu Overseas Research & Training Program for University Prominent Young & Middle-aged Teachers and Presidents.

References

1. Hu, S., Yan, W.Y.: Stability of networked control systems under a multiple-packet transmission policy. IEEE T. Auto. Contr. 53(7), 1706–1711 (2008)
2. Li, J., Zuo, B., Hu, Y., Kang, S.H.: Modeling and stabilization of MIMO networked control system. J. Huazhong Univ. of Sci. & Tech. (Natural Science Edition) 36(10), 15–19 (2008)
3. Liu, X., Xia, Y., Mahmoud, M.S., et al.: Modeling and stabilization of MIMO networked control systems with network constraints. Int. J. Innov. Comput. I. 6(10), 4409–4420 (2010)
4. Wang, Y.L., Yang, G.H.: Multiple communication channels-based packet dropout compensation for networked control systems. IET Control Theory Appl. 2(8), 717–727 (2008)
5. Sun, Z., Xiao, L., Zhu, D.: Analysis of networked control systems with multiple-packet transmission. In: WCICA, vol. 2, pp. 1357–1360. IEEE Press, Hangzou (2004)
6. Yook, J.K., Tilbury, D.M., Soparkar, N.R.: Trading computation for bandwidth: Reducing communication in distributed control systems using state estimators. IEEE T. Contr. Syst. T. 10(4), 503–518 (2002)
7. Fan, W.H., Chen, X.D., Xie, R.H., et al.: MIMO network control systems with asynchronous sampling based on dividing the time slice. In: WCICA, pp. 925–930. IEEE Press, Beijing (2012)
8. Otanez, P.G., Moyne, J.R., Tilbury, D.M.: Using deadband to reduce communication in networked control systems. In: ACC, vol. 4, pp. 3015–3020. IEEE Press, Anchorage (2002)
9. Zhang, W.: Stability analysis of networked control systems. Case Western Reserve University (2001)
10. Branicky, M.S., Pillips, M., Zhang, W.: Stability of networked control systems: explicate analysis of delay. In: AACC, pp. 2352–2357. IEEE Press, Chicago (2000)
11. Zuberi, K.M., Shin, K.G.: Scheduling messages on control area network for real-time CIM applications. IEEE T. Robo. Auto. 13(2), 310–314 (1997)
12. Otanez, P., Moyne, J., Tilbury, D.: Using dead bands to reduce communication in networked control systems. In: ACCC, pp. 615–619. IEEE Press, Anchorage (2002)
13. Yu, L.: Robust Control—approach of LMI. The Tsinghua University Press, Beijing (2002)
14. Zhu, Y.G., Rao, L.: Matrix Analysis and Computation. The Defense Industry Press, Beijing (2010)

Adaptive Distributed Event-Triggered Collaborative Control with WSAN

Fuqiang Li[1,2], Jingqi Fu[1], Dajun Du[1], and Weihua Bao[3]

[1] School of Mechatronic Engineering and Automation, Shanghai University. 149 Yanchang Road, Shanghai, China
[2] College of Sciences, Henan Agricultural University. 63 Nongye Road, Zhengzhou City, Henan Province, China
[3] Shanghai Automation Instrumentation Co., Ltd. 191 Guangzhouxi Road, Shanghai, China
fuqiangli@126.com, jqfu@staff.shu.edu.cn,
ddj@shu.edu.cn, bwh@saic.sh.cn

Abstract. In large scale wireless sensor and actuator network (WSAN), severe packet dropout and latency make challenges in designing centralized control scheme. Moreover, if system runs stably with little fluctuation of states, period sampling always leads to waste of energy and communication resources. To this end, adaptive distributed event-triggered collaborative control scheme is proposed. In this scheme, distributed event-triggered (DET) mechanism is replaced by adaptive DET mechanism with variable triggering threshold in order to reduce updates of controller law further. Based on local information, distributed collaborative control (DCC) is carried out and formulated as optimization problem. And an improved particle swarm optimization (PSO) with linearly decreasing inertia weight is used to make a good tradeoff between global and local search ability. Simulation shows that, comparing with DET scheme, while guaranteeing required closed-loop performance, ADET scheme needs fewer executions of control task.

Keywords: distributed event-triggered, adaptive, collaboration control, particle swarm optimization, linearly decreasing inertia weight.

1   Introduction

WSAN is an important extension of wireless sensor network, allowing actuators to make autonomous decisions and perform actions in response to sensory information [1]. Due to the benefits such as easy installation, self configuration and organization, fast communication and reaction, etc., WSAN has many promising applications in industrial and agricultural fields.

Generally, control task is executed periodically. However, supposing system has been running stably, it is a waste of communication and energy resources if control task is still executed periodically. As a result, it is interesting to consider state based aperiodic event-triggered control where the triggering condition implicitly determines time instances when control law is updated [2-4].

M. Fei et al. (Eds.): LSMS/ICSEE 2014, Part II, CCIS 462, pp. 523–532, 2014.

Since control task is only executed when needed for stability or performance purposes, while guaranteeing desired control performance, event-triggered control is appealing to resource-constrained WSAN. Paulo Tabuada investigated a simple event-triggered scheduler based on feedback paradigm and showed how it leaded to guaranteed performance thus relaxing the more traditional periodic execution requirements [5]. Inspired by [5], Wang and Lemmon proposed a decentralized event-triggering scheme under the assumption that the control system was composed of weakly coupled subsystems [6]. The scheme was completely decentralized, which meant that a subsystem's broadcast decisions were made using its local measurements, and the threshold was chosen only requiring information about an individual subsystem and its immediate neighbors. Manuel Mazo and Paulo Tabuada presented a strategy for the construction of decentralized event-triggered implementations over WSAN of centralized controllers [7]. The proposed method had been shown effective in a distributed event-triggered implementation of a nonlinear dynamic controller for a quadruple water-tank system. Furthermore, since it did not rely on any internal weak coupling assumptions about the system, it could be used to complement the techniques in [6] at the local subsystem level.

The main contributions of this paper are as follows. Firstly, inspired by [7], an adaptive distributed event-triggered scheme (ADET) is proposed, where different triggering threshold can be chosen adaptively according to the system state. Secondly, as for actuators sharing plant states, collaborative control is formulated as an optimization problem and an improved PSO with linearly decreasing inertia weight is used due to the good tradeoff between global and local search ability.

The rest of this paper is organized as follows. System models are presented in Section 2. The ADET scheme is proposed in Section 3. DCC scheme and the improved PSO are described in Section 4. In Section 5, simulation is carried out to evaluate the performance of the proposed scheme. Finally, the paper is concluded in Section 6.

2   System Formulation

Considering the temperature control system in intelligent greenhouse with WSAN, temperatures at different locations are assumed to be state variables. By using event detector, only when triggering condition is violated, current state is transmitted and control law is updated. The infrastructure of event-triggered control system is plotted in Fig.1 [8].

Plant dynamics is given as

$$\dot{x}(t) = Ax(t) + Bu(t) + \varphi(t) \tag{1}$$

Where $A \in \mathbb{R}^{n_s \times n_s}, B \in \mathbb{R}^{n_s \times n_a}$, $x(t), u(t), \varphi(t)$ are state vector of plants, output vector of actuators and system disturbance respectively. Moreover, we assume plant states are independent of each other, so $A = diag\left[a_{1,1}, a_{2,2}, \cdots, a_{n_s,n_s}\right]$ is diagonal matrix. In this setup, we assume that one state is only detected by one sensor and one sensor only measures one state.

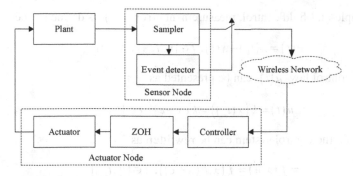

Fig. 1. The infrastructure of event-triggered control system

Control objective is to control temperatures at different locations to meet well with the set points, thus it is interesting to choose some optimal control strategy to minimize the control objective function defined as

$$\min: \quad J(t) \triangleq \frac{1}{n_s} \left| x^* - x(t) \right| \tag{2}$$

Where $|\cdot|$ represents Euclidean norm and $x^* = \left[x_1^*, x_2^*, \cdots, x_{n_s}^* \right]^T$ is set points.

3   Adaptive Distributed Event-triggered Scheme

3.1   Notation

We denote by $\mathbb{N}^+$ the positive natural numbers. A continuous and strictly increasing function $f : [0, \infty[\to \mathbb{R}_0^+$ is of class K_∞ if $f(0) = 0$ and $f(g) \to \infty$ as $g \to \infty$. A function $f : \mathbb{R}^n \to \mathbb{R}^m$ is said to be Lipschitz continuous on compacts if for every compact set $S \subseteq \mathbb{R}^n$ there exists a constant $L > 0$ such that $|f(x) - f(y)| \leq L|x - y|$ for every $x, y \in S$ [5].

3.2   Adaptive Distributed Event-triggered Scheme

Consider a control system

$$\dot{x} = f(x, u), \quad x \in \mathbb{R}^{n_s}, u \in \mathbb{R}^{n_a} \tag{3}$$

Where the feedback control law $u = k(x)$ has been designed rendering the closed loop system $\dot{x} = f(x, k(x + e))$ Input-to-State Stable (ISS) [9] with respect to measure errors $e \in \mathbb{R}^{n_s}$.

In a sample-and-hold control, measurement error $e(t)$ is defined to be

$$e(t) = x(t_k) - x(t), \ t \in [t_k, t_{k+1}[, \ k \in \mathbb{N} \tag{4}$$

And the control law $u(t)$ can be presented as

$$u(t) = k(x(t_k)) = k(x+e), \ t \in [t_k, t_{k+1}[\tag{5}$$

As a result, the control system can be rewritten as

$$\dot{x} = f(x,u) = f(x,k(x+e)), \ t \in [t_k, t_{k+1}[\tag{6}$$

$$\dot{e} = -\dot{x} = -f(x,k(x+e)), \ t \in [t_k, t_{k+1}[\tag{7}$$

If a system is ISS with respect to measurement errors $e \in \mathbb{R}^{n_s}$, there exists an ISS Lyapunov function $V(x)$ satisfying

$$\underline{\alpha}(|x|) \leq V(x) \leq \bar{\alpha}(|x|) \tag{8}$$

$$\frac{\partial V}{\partial x} f(x,k(x+e)) \leq -\alpha(|x|) + \gamma(|e|) \tag{9}$$

Where $\underline{\alpha}, \bar{\alpha}, \alpha$ and γ are K_∞ functions [7].

Consider a condition that

$$\gamma(|e|) \leq \sigma\alpha(|x|), \ \sigma \in]0,1[\tag{10}$$

When the condition (10) is violated at $t \in \{t_k\}_{k \in \mathbb{N}}$, event happens and new information is transmitted to controller. Since $e(t_k) = 0$ leading to $\gamma(|e(t_k)|) = 0$, the inequality (10) is enforced.

Then inequality (9) can be rewritten as

$$\frac{\partial V}{\partial x} f(x,k(x+e)) \leq -\alpha(|x|) + \gamma(|e|) \leq (\sigma-1)\alpha(|x|) < 0 \tag{11}$$

Which ensures Lyapunov function $V(x)$ decreases. That is to say, the event-triggered mechanism defines such a sequence of update times $\{t_k\}_{k \in \mathbb{N}}$ for the controller, rendering the closed loop system asymptotically stable.

Furthermore, if system operates in some compact set $S \subseteq \mathbb{R}^n$ and α^{-1} and γ are Lipschitz continuous on S, for a suitably chosen $\sigma > 0$, the condition (10) can be replaced by

$$|e(t)|^2 \leq \sigma |x(t)|^2 \tag{12}$$

As for distributed control scenario, since no sensor can obtain the information of all plant states needed by condition (12), a set of distributed conditions for sensors to check locally should be proposed. To this end, given a set $\left\{\theta_i \in \mathbb{R}, i = 1, 2, \cdots, n_s\right\}$ satisfying $\sum_{i=0}^{n_s} \theta_i = 0$, inequality (12) can be rewritten as [7]

$$|e(t)|^2 \leq \sigma |x(t)|^2 \Leftrightarrow \sum_{i=1}^{n_s} e_i^2(t) - \sigma \sum_{i=1}^{n_s} x_i^2(t) \leq 0 = \sum_{i=1}^{n_s} \theta_i$$
$$\Leftrightarrow \sum_{i=1}^{n_s} \left(e_i^2(t) - \sigma x_i^2(t)\right) \leq \sum_{i=1}^{n_s} \theta_i \tag{13}$$

So the following implication holds

$$\bigwedge_{i=1}^{n_s} \left(e_i^2(t) - \sigma x_i^2(t) \leq \theta_i\right) \Rightarrow |e(t)|^2 \leq \sigma |x(t)|^2 \tag{14}$$

Which suggests $e_i^2(t) - \sigma x_i^2(t) \leq \theta_i$ as the distributed event-triggered condition.

In order to avoid Zeno behavior [5], the triggerring condition above is only evaluated periodically with sampling interval $h > 0$, which guerantees the existence of a lower bound for interexecution times. Therefore, the update time of controller is denoted by

$$t_{k+1} = t_k + \min\left\{n_h h \middle| e_i^2(t_k + n_h h) - \sigma x_i^2(t_k + n_h h) \geq \theta_i\right\}, \ n_h \in \mathbb{N}^+ \tag{15}$$

If system has not converged to steady state, σ takes a small value to ensure high frequency of control law updates. On the other hand, if system has already been stable without disturbance, in order to reduce executions of control task further, it is interesting for σ to take a larger value on the premise of guaranteeing required closed loop performance. So an adaptive distributed event-triggered condition is proposed as bellow.

$$\begin{cases} e_i^2(t) - \sigma_1 x_i^2(t) \leq \theta_i, \ \dfrac{1}{n_h + 1}\left|X_i'(t_k) - x_i^*\right| \geq \Delta x_i \\ e_i^2(t) - \sigma_2 x_i^2(t) \leq \theta_i, \ \dfrac{1}{n_h + 1}\left|X_i'(t_k) - x_i^*\right| < \Delta x_i \end{cases}, \ t \in [t_k, t_{k+1}[\tag{16}$$

Where $0 < \sigma_1 < \sigma_2$, $x_i^* \in x^*$, Δx_i is steady state error system allowed, $X_i'(t_k) = \left[x_i\left(t_{k-n_h}\right), x_i\left(t_{k-n_h+1}\right), \cdots, x_i(t_k)\right]$ includes measurements of x_i at current and historical triggering instants. n_h is the number of historical data needed, the value of which depends on the size of on-chip memory and the requirement of system performance.

4   Collaborative Control Scheme

4.1   Nodes Collaboration

Firstly, the influenced sensors of actuator A_j is marked by s^{A_j}, and the responsible actuators of sensor s_i is marked by A^{s_i}, which are defined as [10].

$$s^{A_j} = \left\{ s_i \mid b_{i,j} \neq 0, \ b_{i,j} \in B \right\} \tag{17}$$

$$A^{s_i} = \left\{ A_j \mid b_{i,j} \neq 0, \ b_{i,j} \in B \right\} \tag{18}$$

Then, overlapping degree of s_i is defined as

$$\rho_i \triangleq \left| A^{s_i} \right|, \ \forall i \in \{1, \cdots, n_s\} \tag{19}$$

Where $|\cdot|$ is cardinality of a set. Obviously, if $\rho_i > 1$, there are at least two responsible actuators and collaboration is necessary. When an event happens, new measurements of states are allowed to transmit. The responsible actuators form an actuator cluster to share the local information of states and carry out control task collaboratively.

4.2   Linearly Decreasing Inertia Weight PSO

PSO is a stochastic global optimization algorithm, which imitates foraging behavior of bird flock. Iteration functions of PSO are given as follows [11].

$$v_{id} = v_{id} + c_1 r_1 \left(p_{id}^{Pbest} - p_{id} \right) + c_2 r_2 \left(p_{gd}^{Gbest} - p_{id} \right) \tag{20}$$

$$p_{id} = p_{id} + v_{id} \tag{21}$$

Where v_{id} and p_{id} represent the d_{th} element's flying speed and position of the i_{th} particle, $d \in \{1, 2, \cdots, n_d\}$, $i \in \{1, 2, \cdots, n_i\}$. n_i is the number of particles and n_d is the dimension of search space. r_1 and r_2 are random numbers with $r_1, r_2 \in [0,1]$. c_1 and c_2 are accelerating factors. p_{id}^{Pbest} and p_{gd}^{Gbest} indicate the best positions of the i_{th} particle and all particles respectively.

In order to maintain a good tradeoff between global search ability and local search ability, Yuhui Shi and Russell Eberhart proposed a modified PSO by introducing inertia weight, which was a linearly decreased function of time [12]. The iteration functions of the modified PSO are given as

$$v_{id} = w v_{id} + c_1 r_1 \left(p_{id}^{Pbest} - p_{id} \right) + c_2 r_2 \left(p_{gd}^{Gbest} - p_{id} \right) \tag{22}$$

$$p_{id} = p_{id} + v_{id} \tag{23}$$

$$w = \begin{cases} w_{max} & , \ i = 1 \\ \dfrac{(w_{max} - w_{min})}{n_I}(n_I - i) + w_{min} & , \ i = 2, \cdots, n_I \end{cases} \tag{24}$$

Where w_{max} and w_{min} are the maximum and minimum of inertia weight w. n_I is the total number of iterations and i is current iteration count.

5 Numerical Example

Let us consider a temperature control system with two events and two actuators, i.e., $n_s = n_a = 2$. Suppose the initial states are $\begin{bmatrix} 22°C & 24°C \end{bmatrix}$ and the set points are $\begin{bmatrix} 26°C & 26°C \end{bmatrix}$. The plant model is presented as

$$\dot{x}(t) = \begin{bmatrix} -2.11 & 0 \\ 0 & -2.11 \end{bmatrix} x(t) + \begin{bmatrix} 12.01 & 0 \\ 10.33 & 14.33 \end{bmatrix} u(t) \tag{25}$$

Since $s^{A_1} = \{s_1, s_2\}$, $A^{s_2} = \{A_1, A_2\}$, coordination is necessary. In simulation, we set $\theta_i = 0$ and use $|e_i(t)| \le \sigma |x_i(t)|$ as the distributed event-triggered condition.

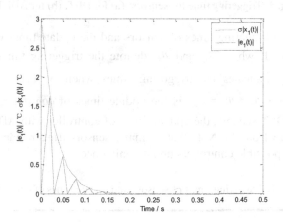

Fig. 2. Evolution of $|e_1(t)|$ and $\sigma |x_1(t)|$ for sensor s_1

In order to describe event triggered scheme vividly, we take situation of s_1 for example, and other cases are similar. As shown in Fig.2, at trigger time $\{t_k\}$, measurement error $e_1(t_k) = 0$. Then it increases until it is reset to zero again at next trigger time t_{k+1}. Therefore, $\left|e_1(t)\right|$ does not exceed the $\sigma\left|x_1(t)\right|$ threshold, i.e., $\left|e_1(t)\right| \le \sigma\left|x_1(t)\right|$. So the asymptotic stability of the control system is promised.

Triggering time instants for DET and ADET scheme are described in Fig.3(a) and Fig.3 (b) respectively, where horizontal axis and vertical axis are used to indicate triggering time and inter-triggering time. Moreover, the inter-triggering time denotes time interval between the current and previous triggering instants for the same sensor. From Fig.3, we can see that the overall triggering times of ADET scheme is less than that of DET scheme.

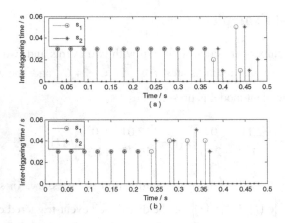

Fig. 3. Triggering time for sensors: (a) for DET, (b) for ADET

Furthermore, the triggering times of sensors and the update times of controller are presented in Table 1, where n_{s_1} and n_{s_2} denote the triggering times of s_1 and s_2 respectively, $n_{s_1 s_2}$ indicates the triggering times when s_1 and s_2 are triggered simultaneously, $n_c = n_{s_1} + n_{s_2} - n_{s_1 s_2}$ is the update times of controller. From Table 1, by contrast with DET scheme, the update times of controller with ADET scheme has been reduced to 15 from 20. Note that, if many sensors trigger independently at the same time, the responsible controllers update only once.

Table 1. Triggering times of sensors and update times of controller

	n_{s_1}	n_{s_2}	$n_{s_1 s_2}$	n_c
DET	15	17	12	20
ADET	11	11	7	15

In Fig.4, $x_i \_DET$ and $x_i \_ADET$ represent state evolutions with DET and ADET scheme respectively. Obviously, the two methods produce almost undistinguishable state trajectories, and both of them can control states to meet well with set points.

In Fig.5, convergence performance of control objective function is presented, where $J \_DET$ and $J \_ADET$ represent values of control objective function with DET and ADET schemes respectively. Apparently, both methods make control objective function converge to zero rapidly. That is to say, comparing with DET scheme, while requiring less executions of control task, ADET scheme still obtains satisfactory closed loop performance.

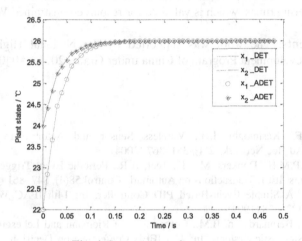

Fig. 4. System dynamic response

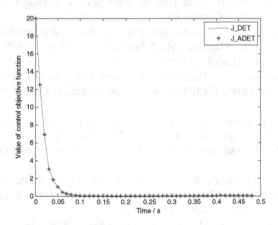

Fig. 5. Evolution of control objective function

6   Conclusion

In this paper, in order to reduce the triggering times further, an adaptive distributed event-triggered scheme with variable triggering thresholds is proposed. In order to achieve control objective, based on local information and actuator cluster, the collaborative control is formulated as an optimization problem. And an improved PSO with linearly decreasing inertia weight is used to balance global and local search ability. Simulation shows that collaborative control based on both DET and ADET schemes makes states meet well with set points and achieves control objective successfully. Comparing with DET scheme, while guaranteeing the required performance of closed loop system, collaborative control based on ADET scheme needs less triggering times, which is valuable for resource-constrained WSAN.

Acknowledgments. The work was supported by the National High Technology Research and Development Program of China under Grants 2011AA040103-7.

References

1. Akyildiz, I.F., Kasimoglu, I.H.: Wireless Sensor and Actor Networks: Research Challenges. Ad Hoc Networks 2(4), 351–367 (2004)
2. Heemels, W.P.M.H., Donkers, M.C.F., Teel, A.R.: Periodic Event-Triggered Control for Linear Systems. IEEE Transactions on Automatic Control 58(4), 847–861 (2013)
3. Årzén, K.E.: A Simple Event-Based PID Controller. In: 14th IFAC World Congress, Beijing, pp. 423–428 (1999)
4. Åström, K.J., Bernhardsson, B.M.: Comparison of Riemann and Lebesgue Sampling for First Order Stochastic Systems. In: 41st IEEE Conference on Decision and Control, pp. 2011–2016. IEEE Press, Las Vegas (2002)
5. Tabuada, P.: Event-Triggered Real-Time Scheduling of Stabilizing Control Tasks. IEEE Transactions on Automatic Control 52(9), 1680–1685 (2007)
6. Wang, X., Lemmon, M.D.: Event-Triggering in Distributed Networked Systems with Data Dropouts and Delays. In: Majumdar, R., Tabuada, P. (eds.) HSCC 2009. LNCS, vol. 5469, pp. 366–380. Springer, Heidelberg (2009)
7. Mazo, M., Tabuada, P.: Decentralized Event-Triggered Control over Wireless Sensor/Actuator Networks. IEEE Transactions on Automatic Control 56(10), 2456–2461 (2011)
8. Peng, C., Yang, T.: Event-Triggered Communication and H_∞ Control Co-design for Networked Control Systems. Automatica 49(5), 1326–1332 (2013)
9. Sontag, E.D.: On the Input-to-State Stability Property. European Journal of Control 1(1), 24–36 (1995)
10. Chen, J., Cao, X., Cheng, P., Xiao, Y., Sun, Y.: Distributed Collaborative Control for Industrial Automation with Wireless Sensor and Actuator Networks. IEEE Transactions on Industrial Electronics 57(12), 4219–4230 (2010)
11. Kennedy, J., Eberhart, R.: Particle Swarm Optimization. In: IEEE International Conference on Neural Networks, pp. 1942–1948. IEEE Press, Perth (1995)
12. Shi, Y., Eberhart, R.: Modified Particle Swarm Optimizer. In: IEEE International Conference on Evolutionary Computation, pp. 69–73. IEEE Press, Anchorage (1998)

Synthesis of Multi-robot Formation Manoeuvre and Collision Avoidance*

Aolei Yang[1],**, Wasif Naeem[2], Minrui Fei[1], Li Liu[1], and Xiaowei Tu[1]

[1] Shanghai Key Laboratory of Power Station Automation Technology,
School of Mechatronic Engineering and Automation, Shanghai University,
200072, Shanghai, China
aolei.yang@gmail.com
[2] School of Electronics, Electrical Engineering and Computer Science,
Queen's University Belfast, BT9 5AH, Belfast, UK

Abstract. This paper presents the synthesis of a multi-robot formation manoeuvre and collision avoidance. Turning-compliant waypoints are first achieved to support the multi-robot formation manoeuvre. The formation-based collision avoidance is then presented to translate the collision avoidance problem into the formation stability problem. The extension-decomposition-aggregation scheme is next employed to solve both the formation control problem and the collision avoidance problem during the multi-robot formation manoeuvre. Simulation study finally shows that the formation control and the collision avoidance can be simultaneously solved if the stability of the expanded formation including unidentified objects can be satisfied.

Keywords: path planning, formation control, manoeuvring waypoints, line-of-sight, collision avoidance.

1 Introduction

Recent advances in technology and in the theory of computing, communication and control now make it feasible to coordinate a number of ground, air and underwater autonomous vehicles to work cooperatively to accomplish a task. Compared to a vehicle performing a solo mission, greater efficiency and operational capability can be achieved through a group of vehicles operating in this way. Such cooperative control of a multi-vehicle system could have enormous impacts in many potential civilian and military applications. These include multi-robot motion, air traffic control, collaborative mapping and exploration, unmanned aerial vehicle (UAV) formation patrol, autonomous underwater vehicle (AUV) formation cruises [7], the monitoring of forest fires, oil fields, pipelines and nuclear

* This work was supported by the Science and Technology Commission of Shanghai Municipality under "Yangfan Program" (14YF1408600), the Shanghai Municipal Commission Of Economy and Informatization under Shanghai Industry-University-Research Collaboration Grant (CXY-2013-71).
** Corresponding author.

M. Fei et al. (Eds.): LSMS/ICSEE 2014, Part II, CCIS 462, pp. 533–542, 2014.
© Springer-Verlag Berlin Heidelberg 2014

power plants, tracking wildlife, border patrol and surveillance, reconnaissance, battle damage assessment and space satellite clustering [4] .

To realise this potential, various cooperative control capabilities need to be developed, one of the most important and fundamental research areas being multi-vehicle formation manoeuvre [6]. The path planning of the formation should be first done to support the formation manoeuvre from the source position to the destination, and the formation controllers should also be designed to stabilise the formation geometry during the formation manoeuvre [2]. Many published paper have addressed the path planning problem, i.e. potential-field algorithm, sampling-based algorithm and A* algorithm. The direction priority sequential selection (DPSS) algorithm was proposed in [11] to overcome some drawbacks of the A* algorithm and to achieve a feasible path with minimal steps. In addition, there are a number of formation control approaches in the literature, i.e. behaviour-based approach [3], leader-following approach [5], virtual structure approach [1], and artificial potential field (APF) approach [8]. Another novel formation control strategy: extension-decomposition-aggregation (EDA) scheme was proposed in [12].

In this paper, the multi-robot formation manoeuvre and its relevant collision avoidance are synthesised to study the feasibility of combining the formation manoeuvre and the proposed formation-based collision avoidance. The turning-compliant waypoints are achieved to support the formation global manoeuvre from the source to the destination. The EDA scheme discussed in [12] is applied to maintain the stability of the formation geometry, and subsequently realise collision avoidance indirectly. Simulation study shows that if the stability of the formation extended by UOs can be satisfied, both the formation manoeuvre problem and the collision avoidance problem can be simultaneously solved. This idea extends the approach of solving the collision avoidance problem.

The remainder of the paper is organised as follows. Section 2 outlines the background material. Section 3 presents the algorithm of achieving turning-complaint waypoints, as well as the collision avoidance approach. Extensive simulation results are presented in Section 4. The paper finally concludes with some discussion and suggestions for future work.

2   Background Material and Foundation

The direction priority sequential selection (DPSS) algorithm is essentially a modified version of the A* path planning algorithm which has been developed by Yang et al. [11] to enhance its performance in terms of improved trajectories and computational cost. The DPSS method is based on a goal direction vector (GDV) and results in the search time reduced by up to 50%. The DPSS method produces waypoints around obstacles that forms a smoother path with less jagged segments than A* algorithm. This scheme consists primarily of two functions: the node sorting search (NSS) and the path fin cutting (PFC). The NSS function is based on a direction priority sorting and can be used to achieve a feasible path with minimum number of *steps*. The route obtained at this step is termed as the raw path. The PFC function is then utilised to optimise the raw path. Since the NSS function focuses on the steps rather than the length/cost, it is necessary to develop some measures to generate smoother paths with less jagged segments and to improve the quality of the raw path.

Complex systems can be decomposed into subsystems, and those individual subsystem solutions are then combined to provide an overall system response. Motivated by this idea, the extension-decomposition-aggregation(EDA) scheme proposed in [10] is described in detail here, and Fig. 1 shows the overall process.

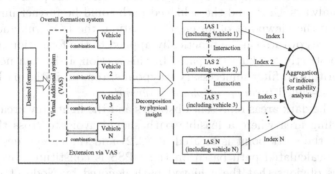

Fig. 1. Process of the extension-decomposition-aggregation

In the **extension** process, a *virtual* additional system (VAS) is employed to build the relationship between isolated vehicles. The VAS is merely an algorithm to act as an "interaction bridge" providing each vehicle with the capability of sensing its local-formation states. For the overall complex formation system, it is natural to **decompose** it into several local-formation subsystems. Using physical insight the overall formation system is then decomposed into N individual subsystems, each being called an individual augmented subsystem (IAS). The initial overall complex formation control problem can be finally redefined in terms of stability and set-point tracking for all the decomposed IASs. A scalar Lyapunov function is next selected as an index for representing the stability of each IAS. These indices are finally **aggregated** to mathematically analyse the subsystem interactions and the stability of the overall formation using Lyapunov stability theory.

3 Manoeuvring Waypoints and Collision Avoidance Approach

During the manoeuvre, there inevitably exists the collision risk between the robots and other robots or obstacles. The formation control approach adopted to maintain the stability of the formation is then innovatively used to solve the collision avoidance problem, and the corresponding method is called as *formation-based collision avoidance*.

3.1   Turning-Compliant Waypoints

The DPSS algorithm has been modified in this work to generate safe and feasible raw paths. The novelty of this work includes the modification of waypoint-by-LOS and the DPSS algorithms to generate turning-compliant waypoints.

Based on the DPSS, a sequence of **nodes** can be calculated in the grid-based map, which actually constitute a feasible **path** from the start node to the destination. However, recalling the assumption that each node in the grid-based map has an eight-connected mapping [11], the turning angle of the path calculated by the DPSS is always $\pm 45°$, and the achieved path could be further improved. In addition, from the viewpoint of an implementation, the LOS guidance method tracking the **waypoints** can be practically applied to the formation manoeuvre compared to strictly tracking the path. In this section, it is then necessary to take an algorithm to filter unnecessary waypoints from an initial path, and the generate the so-called *turning-compliant waypoints*.

In Fig. 2, the dark areas represent obstacles and the light grey area is defined as the guarding area where a feasible path should avoid to cross through. It is given that the start point is located at $(2, 2)$ and the chosen destination is $(18, 18)$. The calculated path based on the DPSS is constituted by the nodes $1 \sim 30$. It is obvious that the achieved path denoted by Node $(1 \sim 30)$ can be segmented to 4 sections of paths which are represented by Node $(1 \sim 3)$, Node $(3 \sim 16)$, Node $(16 \sim 17)$ and Node $(17 \sim 30)$ respectively. It can be observed that the turning angles between two connected path segments are $45°$ or $-45°$, and the corresponding turning points of the overall path are orderly given as: Node 1, Node 3, Node 16, Node 17 and Node 30. The above five turning points can be considered as the **initial turning-compliant waypoints**, but it is not optimal and a few unnecessary turning points can still be filtered from them. After performing a relevant algorithm, the **final turning-compliant waypoints** chosen from the initial turning-compliant waypoints are given as: Node 1, Node 16 and Node 30 which are labelled as "**A**", "**B**" and "**C**" as shown in Fig. 2.

3.2   Formation-Based Collision Avoidance

The idea of formation-based collision avoidance can be described briefly as follows: *when obstacles or UOs invade into the predefined **alert area** of the current robot, the current robot considers these obstacles or UOs as its reaction robot.* In other words, these obstacles are considered as a part of the whole formation topology once there appears the possibility of a collision. Note that the prerequisite that the proposed formation-based collision avoidance approach is feasible and effective is that the formation consisting of the robots and other UOs is stable during the formation manoeuvre. By this way, the collision avoidance problem is then translated into the formation stability problem. If the dynamic formation including the UOs is stable, its collision avoidance problem can be solved by using a formation controller.

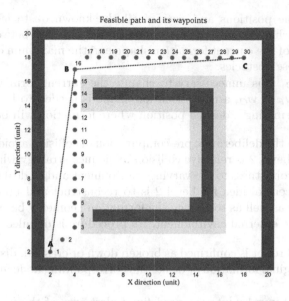

Fig. 2. Turning-compliant waypoints achieved from the path in the map of size 20 units × 20 units

The proposed formation-based collision avoidance strategy contains two levels: deliberative pre-computation of collision avoidance and reflexive collision avoidance strategy, which is shown in Fig. 3. The first level is an offline approach,

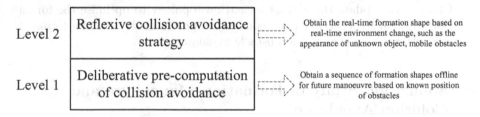

Fig. 3. Formation-based collision avoidance strategy

where a sequence of formation geometries (or *formation shape candidates*: FSCs) can be first pre-calculated under the condition that multi-robot mission map and position of obstacles are known *a priori*. These achieved FSCs are then in turn selected as the real-time *reference formation shape* at specific manoeuvring locations. It is noted that Level 1 is different to the simple path planing, and the procedure of designing the FSC is briefly listed below:

1. Calculate waypoints from start point to destination, and construct a desired motional path as preparation for next formation manoeuvre.

2. Based on the positions and sizes of all the known obstacles, successively search the achieved path and determine the relationship between the lateral width (w_F) of the current formation shape and the maximum clearance (w_C) between these obstacles.

3. If $w_F < w_C$, it is unnecessary to change the current formation geometry, whereas if $w_F \geq w_C$, a new formation shape is needed to be calculated, as well as determining a desired position where formation will be changed.

In contrast to the deliberative pre-computation of collision avoidance manoeuvre in Lever 1, Level 2 is reflexive collision avoidance strategy which is adopted to enhance the robustness to the varying environment and to deal with emergent obstacles. The central idea for Level 2 is to reconstruct and update formation structure online as well as to change the formation geometry based on real-time knowledge of the external environment. Its procedure is detailed as follows:

1. Once a local robot is confirmed as broken down or crashed, discard the robot from the original formation or delete the corresponding node in the formation topology.

2. Once an UO invades into the predefined alert area of the ith robot, the related robot will accept the UO as its reaction target. This means that the original formation system is augmented with the UO and a new formation topology is created. This newly generated formation is termed as an *augmented formation*.

3. Determine a sequence of new formation geometries in real time to make the original formation move away from the newly joined UO as quickly as possible.

4. Change and update the current formation topology to optimise the formation reaction structure, and consequently improve the dynamic performance during formation change and obstacle avoidance.

4   Simulation Study of Formation Manoeuvre and Collision Avoidance

As previously mentioned, the formation controller can be used to maintain the stability of the formation and to avoid collision with UOs during formation manoeuvre. In this section, the decentralised formation controller will be designed, followed by the synthesis of the formation manoeuvre and the collision avoidance.

4.1   Implementation of Decentralised Formation Controller

For simplicity, it is assumed that the dynamic model of the robot is $\ddot{p} = u$, where p denotes the referred position of the robot, and u is the acceleration. The model of the employed IAS is given as $\ddot{\theta} = k_1\theta + k_2u + k_3w$, where θ is the designed variable to reflect the formation change, u is the acceleration of the robot, and

w denotes the dynamics of the local-formation. The complete IAS model can be constructed by combining the above equations, which is given by (1),

$$\begin{cases} \ddot{p} = u \\ \dddot{\theta} = k_1\theta + k_2 u + k_3 w \end{cases} \tag{1}$$

The state vector of each IAS is defined as $\mathbf{x} = [p, \dot{p}, \theta, \dot{\theta}]^T = [\tilde{\mathbf{x}}, \theta, \dot{\theta}]^T$, and (1) can be rewritten as (2).

$$\frac{d}{dt}\begin{bmatrix} p \\ \dot{p} \\ \theta \\ \dot{\theta} \end{bmatrix} = \underbrace{\begin{bmatrix} 0 & 1 & 0 & 0 \\ 0 & 0 & 0 & 0 \\ 0 & 0 & 0 & 1 \\ 0 & 0 & k_1 & 0 \end{bmatrix}}_{A}\begin{bmatrix} p \\ \dot{p} \\ \theta \\ \dot{\theta} \end{bmatrix} + \underbrace{\begin{bmatrix} 0 \\ 1 \\ 0 \\ k_2 \end{bmatrix}}_{B_2} u + \underbrace{\begin{bmatrix} 0 \\ 0 \\ 0 \\ k_3 \end{bmatrix}}_{B_1} w \tag{2}$$

Based on the achieved formation stability result [12], the controllers for maintaining the stability of the IASs should be designed to stabilise the whole formation which may include the UOs. Here, the exogenous input w is derived from the formation change or error, and is considered as a *bounded disturbance* which needs to be rejected by the formation control law for stabilising desired formation. The output feedback H_∞ controller $[A_k, B_k, C_k, D_k]$ in $\begin{aligned} \dot{\hat{x}} &= A_k\hat{x} + B_k y \\ u &= C_k\hat{x} + D_k y \end{aligned}$ is then designed to stabilise the relevant IASs under the bounded disturbance condition. The purpose is to minimise the exogenous impact, i.e. to maximise the robust stability and performance of the IAS. Here, $\hat{x}$ denotes the states of controller, which feeds the measurements y back to the control signal u of the plant so that the closed-loop controlled system is internally stable and satisfies the desired performance specifications. Choosing $k_1 = 24.5$, $k_2 = -2.5$, $k_3 = 8.3$, the matrices of the model in (2) are given: $A = [0, 1, 0, 0; 0, 0, 0, 0; 0, 0, 0, 1; 0, 0, 24.5, 0]$, $B_1 = [0, 0, 0, 8.3]^T$ and $B_2 = [0, 1, 0, -2.5]^T$. The robust Control Toolbox in Matlab can be conveniently used to calculate the relevant output feedback controller matrices. The achieved controller matrices are given in (3). The above controller matrices and parameter values were applied to all the simulations.

$$\begin{aligned}
A_k &= \begin{bmatrix} -4.8716 & -0.0372 & -0.0311 & -0.0000 \\ 0.0634 & -4.8907 & -0.0142 & -0.0000 \\ -0.3507 & 1.1765 & -4.7396 & 0.0001 \\ -0.0020 & 0.0055 & 0.0013 & -5.0000 \end{bmatrix} \\
B_k &= \begin{bmatrix} 9.4332 & 8.2024 & 29.9302 & 6.0602 \\ 4.0572 & 4.6614 & 15.2525 & 3.0928 \\ -55.6087 & -35.5943 & -95.5597 & -19.3688 \\ -0.2921 & -0.1942 & -0.5328 & -0.1080 \end{bmatrix} \\
C_k &= \begin{bmatrix} 0.1730 & -0.1674 & -0.0484 & -0.0000 \end{bmatrix} \\
D_k &= \begin{bmatrix} 19.1812 & 16.6582 & 70.1396 & 14.1985 \end{bmatrix}
\end{aligned} \tag{3}$$

4.2 Formation Manoeuvre and Collision Avoidance

Based on the known obstacles map, the DPSS path planning approach was first employed to calculate a feasible path from start position to goal position.

The turning-compliant waypoints were then obtained based on an algorithm. The multi-robot formation was tasked to navigate all the calculated turning-compliant waypoints: $(6,6)$, $(22,45)$, $(45,80)$, $(81,80)$, $(81,24)$ and $(56,6)$, as well as avoiding collisions with other UOs.

Some obstacles were also assumed to be present along the path and hence formation changes were required to avoid them. Assuming that the obstacle positions were known *a priori*, formation changes were obtained off-line and are shown in Fig. 4, where robot 2 is the reference vehicle (RV) [9] and robots 1

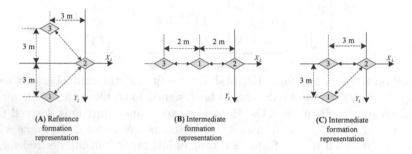

(A) Reference formation representation

(B) Intermediate formation representation

(C) Intermediate formation representation

Fig. 4. Formation representation during multi-robot manoeuvring

and 3 are its designated neighbours (DNs). Specifically, Fig. 4 (A) denotes the reference formation representation (RFR), and Fig. 4 (B) and Fig. 4 (C) for the intermediate formation representations (IFRs). Note that the RFR indicates the geometry when the formation moves normally, whereas the IFR is the geometry designed for avoiding obstacles.

The given map, obstacles and the trajectory of the group of robots are displayed in Fig 5 showing the formation maintenance and the formation change for avoiding collision with obstacles. During the time intervals $t = 24\,s \sim 36\,s$ and $t = 80\,s \sim 92\,s$, the formation geometry shown in Fig. 4 (B) and (C) were respectively adopted to negotiate forward narrow passages. After passing the passages, the group of robots restore to the RFR shown in Fig. 4 (A).

In addition, the formation errors for robot 1 and robot 3 relative to robot 2 are displayed in Fig. 6. As shown in the given figures, the following observations could be made:

1. The multi-robot formation remained stable during the turning manoeuvres and successfully tracked all the given waypoints in addition to switching the formations for avoiding collision.
2. Velocity variation with the formation turning or changing could be observed as it was required to maintain the positions of the robots within the new desired formation as quickly as possible. The speeds of the robots approached the reference speed $2\,m/s$ once the new formation shape was achieved.
3. During the turning manoeuvre and the formation changes, all the transient formation errors asymptotically converged to zero over time.

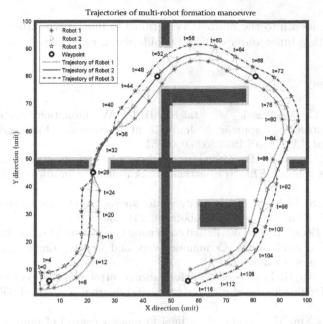

Fig. 5. Trajectories of multi-robot formation manoeuvre

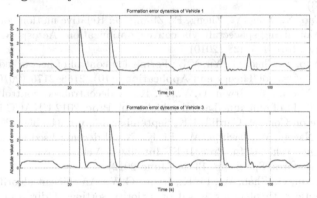

Fig. 6. Formation variation of robot 1 and robot 3

5   Conclusions and Future Work

In this paper, different from the path achieved using the DPSS approach, the turning-compliant waypoints were calculated first, and the so-called formation-based collision avoidance is presented to translate the collision avoidance problem into the stability problem of the formation control. The EDA scheme is applied to stabilise the multi-robot formation geometry when manoeuvring, and to simultaneously realise collision avoidance. This idea provides another approach of solving the collision avoidance problem. Future work includes the application of the proposed idea into the multi-UAV formation manoeuvre, and the study of the multi-vehicle inner/outer collision avoidance in real-time.

References

[1] Askari, A., Mortazavi, M., Talebi, H.: UAV formation control via the virtual structure approach. Journal of Aerospace Engineering (2013), doi:10.1061/(ASCE)AS.1943-5525.0000351

[2] Dörfler, F., Francis, B.: Geometric analysis of the formation problem for autonomous robots. IEEE Transactions on Automatic Control 55(10), 2379–2384 (2010)

[3] Monteiro, S., Bicho, E.: Attractor dynamics approach to formation control: Theory and application. Autonomous Robots 29, 331–355 (2010)

[4] Ren, W., Beard, R.W.: Distributed consensus in multi-vehicle cooperative control: Theory and applications. Communications and Control Engineering. Springer-Verlag London Limited (2008)

[5] Wang, J., Wu, H.: Leader-following formation control of multi-agent systems under fixed and switching topologies. International Journal of Control 85(6), 695–705 (2012)

[6] Wang, J., Xin, M.: Integrated optimal formation control of multiple unmanned aerial vehicles. IEEE Transactions on Control System Technology 21(5), 1731–1744 (2013)

[7] Wu, Y., Cao, X., Xing, Y., Zheng, P., Zhang, S.: Relative motion coupled control for formation flying spacecraft via convex optimization. Aerospace Science and Technology 14(6), 415–428 (2010)

[8] Xue, D., Yao, J., Chen, G., Yu, Y.L.: Formation control of networked multi-agent system. IET Control Theory and Application 4(10), 2168–2176 (2010)

[9] Yang, A., Naeem, W., Irwin, G.W., Li, K.: A decentralised control strategy for formation flight of unmanned aerial vehicles. In: Proc. 2012 UKACC International Conference on Control, Cardiff, UK, September 3-5, pp. 345–350 (2012)

[10] Yang, A., Naeem, W., Irwin, G.W., Li, K.: Novel decentralised formation control for unmanned vehicles. In: Proc. IEEE Intelligent Vehicles Symposium (IV 2012), Alcalá de Henares, Spain, pp. 13–18 (June 2012)

[11] Yang, A., Niu, Q., Zhao, W., Li, K., Irwin, G.W.: An efficient algorithm for grid-based robotic path planning based on priority sorting of direction vectors. In: Li, K., Fei, M., Jia, L., Irwin, G.W. (eds.) LSMS/ICSEE 2010, Part II. LNCS, vol. 6329, pp. 456–466. Springer, Heidelberg (2010)

[12] Yang, A., Naeem, W., Irwin, G.W., Li, K.: Stability analysis and implementation of a decentralised formation control strategy for unmanned vehicles. IEEE Transactions on Control Systems Technology 22(2), 706–720 (2014)

Data Communications for Intelligent Electric Vehicle Charging Stations

David Laverty*, Kang Li, and Jing Deng

School of EEECS, Queen's University Belfast, Ashby Building,
125 Stranmillis Road, Belfast, BT9 5AH, United Kingdom
{david.laverty,k.li,j.deng}@qub.ac.uk

Abstract. This paper describes the communication requirements for an intelligent electric vehicle charge station which can provide inertial support to the electricity grid. The application is described. Telecoms delivery technologies are experimentally assessed, and an open source measurement system is discussed.

Keywords: Telecoms, Latency, Last-Mile, Inertia, Open Source Software.

1 Introduction

Widespread adoption of Electric Vehicles (EVs) presents a variety of challenges to the operation of electrical power systems, notably an increase in system demand, and have the potential to overload residential distribution networks if EV batteries are allowed to charge without co-ordination. Technologies are required which allow EVs to charge in a manner which respects constraints on the electricity network.

The authors propose that the technologies which allow an EV to change within network constraints can be leveraged to provide services of benefit to the grid, such as inertial support during times of system stress and transient activity. This yields an intelligent EV charge station which can use energy stored in the EV battery to simulate system inertia for brief periods of time.

The communication requirements for doing so are feasible but are beyond the level of telecommunication found in the typical electrical utility. This paper describes the concept of the intelligent vehicle charger, outlines the telecoms requirements, assesses delivery technologies, and makes note of cost effective monitoring systems.

2 Intelligent EV Charger with Simulated Inertia

The proposed system uses the EV battery stored energy to provide short term supply of electrical power to the electricity grid in response to changes in the electricity grid frequency. By successful co-ordination of many EV battery charge stations, a

---

* This work was supported by the EPSRC UK-China iGIVE project (EP/L001063/1).

M. Fei et al. (Eds.): LSMS/ICSEE 2014, Part II, CCIS 462, pp. 543–551, 2014.

population of vehicles will have the affect of simulating the inertia of a rotating machine. This is useful in the context of electricity grids with low inertia thanks to the increasing prevalence of electronic machine drives, and renewable generation technologies connected via power electronic interfaces.

The principle of operation is described with reference to Fig. 1. In Fig. 1 we see that there has been an event on the electrical power system which has led to a frequency disturbance. The frequency initially falls, causing governor action at thermal generating plant to increase power output. Over the next number of seconds, the frequency recovers. The intelligent charge station detects that there is a rate-of-change-of-frequency. Based on the difference between current frequency and nominal frequency, and modulated by the rate-of-change-of-frequency, the intelligent charger adjusts its power import from the grid (or exports power to the grid) such as to simulate the response of a synchronous machine experiencing the frequency transient.

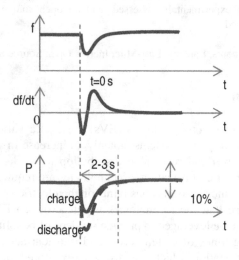

Fig. 1. The charge station responses to a frequency disturbance on the electricity grid by modulating its power import/export in proportion to the rate-of-change-of-frequency of the grid

Many examples of EV battery stored energy being used to supply energy to the electricity grid, Vehicle to Grid (V2G), can be found in literature. Such proposals yield technical solutions to challenges of load balancing in grids featuring large quantities of renewable generation, yet raise concerns with regard to the effect on the EV battery State-of-Charge (SoC) and State-of-Health (SoH). That is to say, firstly it is undesirable that the EV owner should find their vehicle without sufficient charge, and secondly it is undesirable to shorten the life of the battery. The proposed system only makes use of the EV battery for periods of several seconds at a time, not for long periods of time. The effect on SoC is negligible and since these grid transient events happen rarely, the effect on SoH also minimal, being similar to a one-off acceleration of the vehicle in normal driving conditions.

2.1   Example of Transient Event on Low Inertia System

The authors operate an expensive network of phasor measurement equipment which allows for the capture and analysis of grid frequency related phenomena. Details of the system are available in [1, 2].

In April 2013, a disturbance occurred which caused a DC interconnector between the island of Ireland and the island of Great Britain to trip. Each island operates a separate synchronous system. The peak load on the Ireland system is circa 6 GW, whilst in the UK it is circa 60 GW. At the moment that the interconnector disconnected, it was exporting 500 MW of power from Ireland to Great Britain. The effect on frequency on the two systems is shown in Fig. 2. The effect of the disturbance on the Ireland system is much greater than the effect on the Great Britain system. Notably, Ireland was operating with 35% of infeed supplied by wind generation. Wind generators are usually connected by power electronic interface and provide little or no inertia to the system. This is an example of transient effects that can occur on systems with increasing infeeds of renewable generation and thus lessening system inertia.

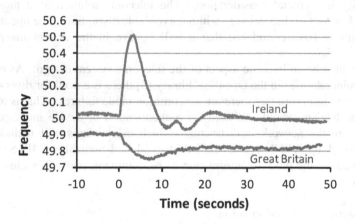

Fig. 2. Frequency disturbances seen on the Ireland and Great Britain electrical systems. This is in response to the disconnection of a DC interconnector exporting 500 MW from Ireland to Great Britain in April 2013.

3   Telecommunication Requirements

Telecommunications are needed with the intelligent vehicle charger for a number of reasons. #There are intrinsic needs for telecoms for the purposes of billing the customer for their electricity consumption, but there are further needs related to grid operations. Firstly, it is necessary to consider the changing of the vehicles within the constraints of the electricity network. These will relate to the available generation capacity, power flow limits in transmission networks and especially local distribution networks, and coordinating the charging to make best use of low-carbon renewable electricity generation.

Secondly, with specific concern for the inertial response of the intelligent charger, it is necessary to know how many vehicles are available to participate in a simulated inertial response, their locations, and their SoC.

The volume of information to be communicated is low. SoH, SoH and identification information will be tens of bytes in size, and does not need to update with great regularity. Several 'slow scan' solutions, including VHF radio, could be employed to communicate directly with the vehicle. However, it is absolutely necessary to make security an intrinsic design consideration in the communication system. Thus, reliable communication that can handle several kilobits/second (kbps) is required.

When inertial response is required, it will be necessary to have low latency communication between multiple vehicles in a swarm so as to quickly determine the optimal response. Thus, it is desirable that the communications should have a latency of less than 1 AC waveform cycle (20 ms in 50 Hz systems, 16.7 ms in 60 Hz systems).

Reliability is a crucial consideration. The telecoms solution used must have a proven track record of data delivery without avoidable disruption. The operator of the system needs to have confidence that it will behave in the correct manner when required.

Security must be an intrinsic aspect of the telecoms system design. As evidenced by recent vulnerabilities in the OpenSSL library [3] there is a need for diversity in the security system to ensure that there are no common mode failures. That is to say that different mechanisms should be used for authentication of device messages to that used to secure the network infrastructure. Such mechanisms will include Virtual Private Networking (VPN) [4] and Multi-Protocol-Label-Switching (MPLS) [5].

To summaries, the types of information to be communicated will include, but not be limited to:

1) Customer billing information
2) Vehicle identity, SoC and SoH
3) Information on distribution network constraints
4) Availability of generation (possibly price sensitive)
5) Forecast of renewable generation
6) Population (or swarm) information on nearby EVs

The authors suggest that the following specification should be considered a minimum set of requirements for the telecoms system:

1) Throughput of the order of kbps or greater
2) Latency of the order of 20 ms or less
3) Reliable
4) Secure

Since it is very difficult to find very low bandwidth telecoms delivery technologies, the next section shall consider latency as a primary design consideration.

4   Delivery Technologies

Several telecoms delivery technologies have been reviewed by the authors regarding their suitability for real-time control in electrical utility applications. These technologies are described in light of data captured by the authors. These technologies are "last mile" delivery technologies, suitable for connection an EV charge station with a backhaul network.

a) ADSL
Asymmetric Digital Subscriber Line, ADSL, works over a voice telephone line (twisted or untwisted copper wire pair) using frequencies beyond those used for voice (3 kHz). Frequencies of 1 MHz are typical on an ADSL enabled telephone line. The service operates between the subscriber's premises (e.g. a home) and the telephone exchange where the termination equipment is hosted (a DSLAM - Digital Subscriber Line Access Multiplexer) [6].

ADSL2+ is the current standard for most newly enabled telephone exchanges. Speeds of up to 24 Mbps downstream and 3.3 Mbps upstream are possible. Speed diminishes as a function of distance due to increased attenuation and decreased signal-to-noise (SNR) ratio.

b) Fibre-to-the-Cabinet (VDSL)
Fibre-to-the-Cabinet (FTTC) is similar to ADSL in that the connection to the consumer premises is achieved using copper pair, however optical fibre runs from the telephone exchange to the local street cabinet. Very-High-bitrate-DSL (VDSL) [7] is used in place of ADSL. VDSL allows speeds of up to 80 Mbps downstream and 20 Mbps upstream at distances of 500 m from the street cabinet.

c) WiMAX
WiMAX, or Wireless Microwave Access (IEEE 802.16 [8]), can operate in either fixed or mobile forms. The fixed flavor of WiMAX is more established and is the subject of this study. A typically WiMax base station can service a radius of approximately 30 km, with the throughput diminishing as a function of distance between the base station and the subscriber. Depending on the frequency band used, WiMAX is usually limited to line-of-sight operation. Whilst LTE has taken over the mobile sector, there is increasing adoption of WiMAX in fixed applications due to the low cost of equipment. A pair of WiMAX transceivers can be purchased for as little as $100, and can operate in license exempt bands.

4.4 3G / LTE

Non-enhanced GSM system allows 9.6 kbps of data at maximum signal quality, and is suitable for voice and short text message only. Enhancements such as EDGE (Enhanced Data Rates for GSM Evolution) and HSDPA (High-Speed Downlink Packet Access) theoretically increase this speed to 14.4 Mbps. Long Term Evolution (LTE) increases speeds further.

Mobile telephone networks do not guarantee service. Although certain subscribers such as emergency services can have their service prioritized, service level agreements for real-time control are prohibitive. This makes them unsuited to carry essential power system signals.

d) Power Line Carrier

Power Line Carrier is a potentially useful technology for delivery of low bandwidth telecoms to EV charge stations in some environments. The technology must be applied on the low voltage side of the distribution transformer, thus only has economies of scales in regions were a large number of customers are connecter to a single distribution transformer (e.g. in urban 230V systems such as in Europe). It is less useful in 110V systems were typically small number of subscribers are connected to the distribution transformer, or in rural areas. In these situations, radio is probably preferred, but unlike slow scan radio systems used in Smart Meter reading the challenge with control systems is providing adequate security in the knowledge that the proposed system can affect grid frequency.

4.1   Latency Accessed by Experimental Trial

The latency of various last-mile telecoms delivery technologies was assessed by the authors in a practical scenario. Each delivery technology was configured to 'ping' a server on a high-speed telecoms link at Queen's University Belfast. The round trip times were recorded over a period of several weeks and the results are presented as a Probability Density Function in Fig. 3. The ping times can be divided by two to yield an approximation for the one-way communication delay. In the previous section, it was stated that a one-way latency of less than 20 ms is desired. Thus round-trip times less than 40 ms is preferred.

Of the technologies surveyed, WiMAX and VDSL performed the best. WiMAX performed with fiber-optic like latency, with 90% of pings returned in under 3 ms, and 99% in less than 10 ms. However, a very small number of pings took as long as 20 ms to be returned. VDSL performed very consistently with 99.7% of pings returned with times between 30 and 32 ms, with a small number requiring up to 40 ms. ADSL returned 97% of pings within 50 ms. 3G requires 250 ms to return 97% of pings, and GPRS requires 1,100 ms. Both 3G and GPRS experienced high packed loss.

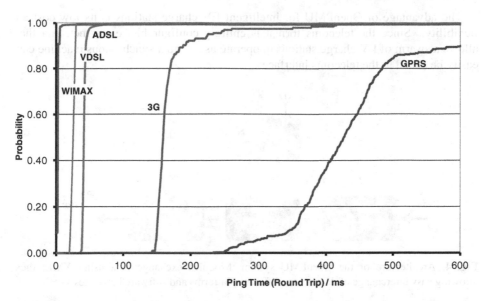

Fig. 3. Latency of various last-mile delivery technologies as assessed experimentally in a practical scenario. Divide times by two for one-way communication.

5   Open Source Measurement System

The authors have designed an Open Source Measurement System appropriate for monitoring power system frequency in the context of intelligent EV charge stations. This is under the umbrella of the open source Phasor Measurement Unit (PMU), OpenPMU [4].

The OpenPMU project is a suite of tools to help researchers create new power system technologies. The project was originally founded at Queen's University Belfast, UK, and subsequently joined by the SmarTS Lab at KTH Stockholm, Sweden.

Phasor Measurement Unit (PMU) technology has now become a mainstream tool of the power systems engineer. PMUs allow all manner of interesting phenomena to be analysed, from transmission system faults to power oscillations induced by wind farms. But often PMUs are more difficult to use than they should be.

The OpenPMU project is concerned with the advancement of Phasor Measurement Unit technology for the purposes of academic research. It does not compete with commercial vendors. Software elements are licensed under a suitable open source license such as BSD, whilst the design drawings are made available under copyleft licenses such as Creative Commons.

The OpenPMU device is shown schematically in Fig. 4. Data is interchanged between the various modules by means of human readable XML datagrams. Since XML is widely supported by toolboxes in many software languages, this allows PMU related data to be read / written on almost any platform. This means that software modules can easily be interchanged and interoperability quickly tested.

The advantage of OpenPMU for intelligent EV charge stations is its low cost an flexibility. Since the telecoms format is entirely configurable, extra metadata that allows a swarm of EV charge stations to operate as a virtual synchronous machine can easily be added to the telecoms interface.

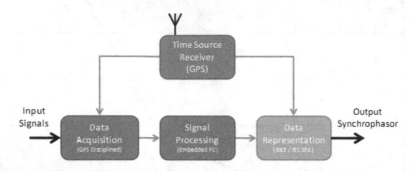

Fig. 4. Architecture of the OpenPMU system. Modules exchange data using XML, thus allowing easy interchange of data across different platforms and software languages.

6   Conclusions

The authors have outlined the telecoms bandwidth and latency requirements for an intelligent electric vehicle charging station. The charging station is to participate in a novel technique which simulates the inertial response of a synchronous machine when there is a transient on the electricity grid. Commonly available telecoms delivery technologies are shown to provide connectivity with the desired latency, although mobile telecoms technology performance has not been demonstrated feasible for real-time control. A measurement system based on open source software is recommended for this application due to the flexibility to modify the telecoms module to support commands necessary to have multiple EV change stations participate as a swarm. The components to achieve the system are available and the challenges of interoperability will be tested experimentally in future work.

References

1. Guo, Y., Li, K., Laverty, D.M.: A statistical process control approach for automatic anti-islanding detection using synchrophasors. In: 2013 IEEE Power and Energy Society General Meeting (PES), July 21-25, pp. 1–5 (2013)
2. Laverty, D.M., Vanfretti, L., Al Khatib, I., Applegreen, V.K., Best, R.J., Morrow, D.J.: The OpenPMU Project: Challenges and perspectives. In: 2013 IEEE Power and Energy Society General Meeting (PES), July 21-25, pp. 1–5 (2013)
3. Heartbleed Vulnerability in OpenSSL, http://heartbleed.com/ (accessed April 2014)
4. Friedl, S.: An Illustrated Guide to IPsec, http://www.unixwiz.net/techtips/iguide-ipsec.html

5. Armitage, G.: MPLS: The magic behind the myths. IEEE Communications Magazine 38(1), 124–131 (2000)
6. ITU Rec. G.992.5, G.992.5: Asymmetric Digital Subscriber Line (ADSL) transceivers - Extended bandwidth ADSL2 (ADSL2plus), http://www.itu.int/rec/T-REC-G.992.5/en
7. ITU Focus Group on Full Service VDSL (2006), Internet: http://www.itu.int/ITU-T/studygroups/com16/fs-vdsl/index.html
8. IEEE Std. 802.16-2001 IEEE Standard for Local and Metropolitan area networks Part 16: Air Interface for Fixed Broadband Wireless Access Systems. IEEE Std 802.16-2001, pp.1–322 (2002)
9. Laverty, D.M., Best, R.J., Brogan, P., Al Khatib, I., Vanfretti, L., Morrow, D.J.: The OpenPMU Platform for Open-Source Phasor Measurements. IEEE Transactions on Instrumentation and Measurement 62(4), 701–709 (2013)

Human-UAV Coordinated Flight Path Planning
of UAV Low-Altitude Penetration on Pop-Up Threats

Peng Ren, Xiao-guang Gao, and Jun Chen

Institute of Electronic Engineering, Northwestern Polytechnical University,
710129 Xi´an, China
2728428@qq.com, r122p@163.com

Abstract. Human-UAV coordinated flight path planning of UAV low-altitude penetration on pop-up threats is a key technology achieving manned and unmanned aerial vehicles cooperative combat and is proposed in this paper. In the most dangerous environment, human's wisdom, experience and synthetic judgments can make up for the lack of intelligence algorithm. By using variable length gene encoding based on angle for the flight paths planning, and combining artificial auxiliary decision with novel intelligence algorithm, it makes the best possible use of the human brain to guide solution procedures of the flight path planning on pop-up threats. A lot of simulation studies show that the on-line three-dimensional flight paths by this technology can meet the requirements of UAV low-altitude penetration, efficient implementation of threat avoidance, terrain avoidance and terrain following. This method has a certain practicality.

Keywords: human-UAV coordinated, flight path planning, low-altitude penetration, pop-up threats.

1 Introduction

X-47B UAV is for non-operational use, its precision navigation algorithms and more intelligent automatic control system have been used to create the first operational carrier-based unmanned aircraft without ground operator's control. X-47B still executes the task by computer programs, so that the single self tactical decision and aerial battle is not able to be performed completely by itself. As the increasingly perfectness of integrated air defense system (IADS), it is impossible that carrying out the mission of low penetration and precision strike only relying on UAV. In the modern air battles, manned and unmanned aerial vehicles each have advantages and are interdependent, they complement each other and divide of labor and cooperation for getting the more operational effectiveness. The manned and unmanned aerial vehicles cooperative combat is a revolutionary new combat method to achieve effective breaking through IADS and would be the main air combat pattern in the future.[1-5] Human-UAV coordinated operation would fully utilize human's wisdom, experience and synthetic judgments, combining the special ability of UAV to carry out the mission of surveillance and precision strike in the most dangerous environment.[6-8]

M. Fei et al. (Eds.): LSMS/ICSEE 2014, Part II, CCIS 462, pp. 552–561, 2014.

Advanced technologies have enabled some UAVs to execute simple mission tasks without human interaction, many tasks are pre-planned using reconnaissance or environment information. In the most dangerous environment, UAV must make full use of the terrain masking and the IADS shadow to penetrate. With the battlefield situation dynamic development, the pop-up threats are inevitable. Because of huge three-dimension flight path points, vast searching space, strong real-timeliness and various constraint missions, the method of solution is extremely complex and difficult. The final solution is difficult to be guaranteed by using conventional intelligent algorithms, such as genetic, particle swarm optimization, ant colony and A* algorithm, et al.[9-12]

Human-UAV coordinated flight path planning is a critical technology of manned and unmanned aerial vehicles cooperative combat. It is proposed in this paper, combining artificial auxiliary decision with novel intelligence algorithm. Human's wisdom, experience and synthetic judgments can make up for the lack of intelligence algorithm. In the most dangerous environment, it makes the best possible use of the human brain to guiding solution procedures of the flight path re-planning on pop-up threats range. By using variable length gene encoding based on angle for the flight paths planning, the flight performance constraints of UAV are exactly integrated into the novel intelligence algorithm, the result of flight paths can meet the requirements of UAV mobility. It can make the UAV automatic driving more easily and reliably. The novel intelligence algorithm comes from evolutionary algorithm and is a kind of niche adaptive pseudo parallel genetic algorithm. The method can re-plan the on-line three-dimensional flight paths when the starting and ending positions are set by human. A lot of simulation studies show that the on-line three-dimensional flight paths by this technology can meet the requirements of UAV low-altitude penetration, efficient implementation of threat avoidance, terrain avoidance and terrain following. This method has a certain practicality and contributes to realize the manned and unmanned aerial vehicles cooperative combat.

2   Mathematical Descriptions of Flight Path

2.1   Represent Flight Path Based on Angle

Only the proper mathematical descriptions of flight path reflect the essence of solution to the problems, the accuracy and convergence of algorithm would be improved. In the presented algorithm, a chromosome integer code with changeable length based on angle is used to represent flight path. (l_i, φ_i, μ_i) represents the sequence point (x_i, y_i, z_i) of flight path taking place at discrete times $t_i (i = 1, 2, \cdots, N)$ in the three-dimensional space, respectively. Where l_i denotes length of the ith section path. μ_i and φ_i are yaw and pitch angle of the ith section path, respectively. It is shown in Fig.1.

This method representing flight path based on angles is able to satisfy the demands of UAV's flight performance, such as constraints of navigation and control. The pitch and yaw angle are also flight control variables and enable the UAV to track a trajectory as long as changes. So automatic pilot is achieved conveniently.

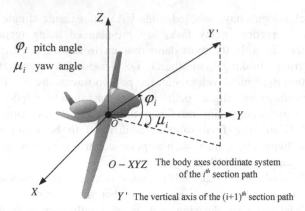

$O - XYZ$ The body axes coordinate system
of the ith section path

Y' The vertical axis of the $(i+1)^{th}$ section path

Fig. 1. Pitch and yaw angle

2.2   Genetic Representation

(l_i, φ_i, μ_i) is converted into chromosome genes in a certain arrangement. $g_1, g_2, \cdots, g_{i-1}, g_i, \cdots, g_{n-1}$ is a chromosome with changeable length representing flight path from starting point to end point. Δl ($l_i = \Delta l$) is equal distant-interval, g_i denotes combination of φ_i and μ_i about Δl, the value range of g_i is integer from 0 to 127. The whole flight path is separated into n equal-distant sections. Every flight path has variable-length chromosomes and their genes have different numbers. The value for n is confirmed by the nodes of flight path. According to UAV's flight performance, the values for yaw and pitch angle is confirmed by itself flight overload, $\varphi_i \leq \pm 70°$, $\mu_i \leq \pm 30°$. The direction interval is 10°, the value for μ_i' is {0, ±1, ±2, ±3} and the value for φ_i' is {0, ±1, ±2, ···, ±7}, then it corresponds to μ_i' and φ_i' by 3-bit and 4-bit binary codes, respectively. The leftmost bit (most significant bit) is sign bit, "0" for positive, "1" for negative. A 7-bit binary codes can be used to represent the integer from 0 to 127, the combination of μ_i' and φ_i' is ability to represent g_i. So g_i can denote combination of φ_i and μ_i about Δl, it is shown in Fig.2.

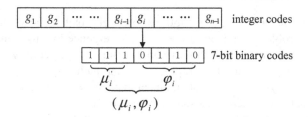

Fig. 2. Chromosome gene codes

For instance, when g_i=118, the binary codes are '1110110'. After decomposing, μ_i'=111, φ'=0110, then μ_i=−30°, φ_i=60°. The section i path downward sloping at 30° and counter-clockwise rotating at 60° is represented.

3   The Algorithm of Human-UAV Coordinated Intelligent Flight Path Planning

Taking low-altitude penetration especially with unknown pop-up threats, the accurate mathematical model couldn't be established. Generally, only relying on conventional intelligence algorithms, satisfied optimal solutions are not effectively and quickly obtained. Therefore, in the conditions of current technological level, the algorithm of human-UAV coordinated intelligent flight path planning is proposed in this paper, combining artificial auxiliary decision with novel intelligence algorithm.

3.1   Artificial Auxiliary Decision

Most often UAV is controlled by itself onboard equipments and its tasks are managed by ground control station. As the separation of management and control function, there are communication time-delay including the delay of remote control commands and telemetry signals. On-line path re-planning may not be achieved so that UAV could not avoid the new threat. During the course of combat, the target information obtained by UAV is transmitted to manned aerial vehicles via data link. As for human, manned aerial vehicles have better battlefield situational awareness. Human has been endowed by nature with special abilities about evaluation of air combat threat and the decision-making behavior. Human-UAV coordinated would fully utilize human's wisdom, experience and synthetic judgments to help UAV path re-planning and avoiding threat.

There are two situations:

UAV is outside the pop-up threats coverage
As the pre-planned flight path that UAV has not been reached yet is inside the pop-up threats coverage, utilize human's experience and synthetic judgments to accurately determine starting and ending point of re-planning path through comprehensive assessing the situation and threats, then apply the novel intelligence algorithm to re-plan path for avoiding threats effectively and quickly. If the flight path points are so numerous, which need more computing time only relying on novel intelligence algorithms, it is difficult to guarantee accessibility and timeliness. In order to avoid threats effectively, set few navigation points manually between the starting and ending points according to the actual situation and flight characteristics, then the path is departed to calculate respectively. If the pre-planned flight path is outside the pop-up threats coverage, it is to continue on its way as pre-planned.

UAV is inside the pop-up threats coverage
Immediately launching emergency response, escape routes by computer-generated as the scheduled angle interval in all directions, then human selects the most suitable escape route by comprehensive assessing the situation and threats. The next step is set the intersection of the selected escape route and the threats coverage region boundary as starting point, and set ending point on the pre-planned path manually. Then re-plan the route between the new starting and ending point. Through these steps, utilize the human's wisdom, experience, synthetic judgments and the novel intelligence algorithm to guide UAV safely and reasonably through the pop-up threats coverage area.

3.2   The Novel Intelligence Algorithm

The novel intelligence algorithm based on evolutionary algorithm principles is introduced, which meet the need of low-altitude penetration. It divides the whole group into several sub-groups to search multi-orientatedly in the solution space so that both the depth and the extent of the search can be assured. Every subgroup evolves in different processors independently and synchronously, each search own solution space respectively. Then exchange and update the best information periodically according to proper migration strategy, and assembled to find the overall optimum solution. By this way, the diversity of population is maintained and the operation speed of the algorithm is enhanced. To further restrain the premature convergence, a kind of niche adaptive genetic mechanism is applied in the subgroup. The individual in each generation can be divided into several categories, and the best individuals are selected from each category to form a new population. Then the adaptive crossover and mutation operators are used as genetic operators to generate the newer population among the different populations. The niche technique based on sharing function can effectively prevent premature, maintain population diversity and improve the search efficiency so that as more optimal solutions would be achieved.[11, 13] The novel intelligence algorithm flow chart is shown in Fig. 3.

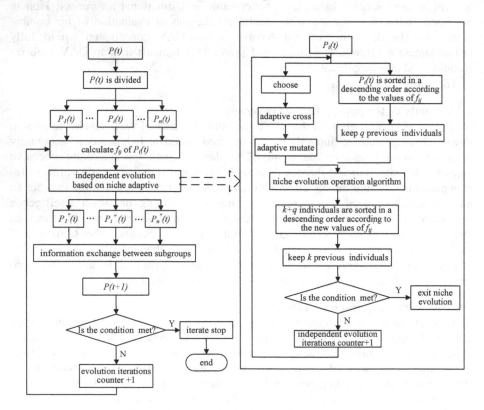

Fig. 3. Novel intelligence algorithm flow chart

4   Mathematical Descriptions of Threat

In the most dangerous environment, UAV must quickly and effectively to avoid various static and pop-up threats coming from IADS. It would be difficult to calculate the amount of threat about IADS, the analytical solutions to it are not obtained generally. There are some simplified hypotheses for the mathematical descriptions of threat without influence on the effects of the path planning algorithm. f_{TAi} is the threat index of UAV facing threat in the ith section path and can be expressed as follows:

$$f_{TAi} = \sum_{k=1}^{m+n} f_{TAi\_k} \tag{1}$$

where f_{TAi_k} is the threat index of the facing the NO. k threat in the ith section path, m is the number of static threats and n is the number of pop-up threats. f_{TAi_k} is given by:

$$f_{TAi\_k} = \begin{cases} 0 & (d_{i\_k}^4 > d_{Rmax\_k}) \\ \frac{\alpha_k * d_{Rmax_k}^4}{d_{i\_k}^4 + d_{Rmax}^4} & (d_{i\_k}^4 \leq d_{Rmax\_k}) \end{cases} \tag{2}$$

Here, d_{Rmax_k} is the maximum effective distance of the NO. k threat, d_{i_k} is the distance between the NO. k threat and UAV in the ith section path, α_k is the multiplier.

Such simplified hypotheses don't influence the research of path planning algorithms. It is easy for computers and has a certain practicality.

5   Fitness Function

The fitness function is a chromosome in terms of physical representation, and evaluates its fitness based on desired traits of the solution. And, the fitness function must accurately measure the quality of the chromosomes in the population. The value of fitness function represents the cost of a given path, the fitness function of the jth chromosome is defined as follows:

$$C_j = \sum_{i=1}^{n} (w_L * l_i + w_H * h_i + w_{TA} * f_{TAi}) + w_d * d(n) \tag{3}$$

where l_i is length, h_i is altitude and f_{TAi} is the threat index of the ith section path, n denotes length of the jth chromosome, w_L, w_H, w_{TA} and w_d are all weight. $d(n)$ is heuristic function that is used instead of direct distance between $T(x_n, y_n, z_n)$ and $G(x_{goal}, y_{goal}, z_{goal})$, $T(x_n, y_n, z_n)$ denotes the end of the jth chromosome, $G(x_{goal}, y_{goal}, z_{goal})$ is end point, then $d(n)$ is defined as follows:

$$d(n) = \sqrt{(x_n - x_{goal})^2 + (y_n - y_{goal})^2 + (z_n - z_{goal})^2} \tag{4}$$

6   Experimental and Simulation Results

To test and evaluate the performance of the proposed algorithm, a number of design and aircraft simulation tests are conducted in the current work. The details of the experimental and simulation results are described in the sections below.

DEM was used, which generates 80×80km three dimension terrain including gradient, slope aspect, relative elevation, etc. It can meet with the request of track-planning. In the simulate airspace, starting point is $S(x_0, y_0, z_0)= (0,0,250)$ and end point is $G(x_{goal}, y_{goal}, z_{goal})=(76000,76000,250)$ (unit: m). Flight speed $v=50$m/s (constant), the minimum path length $l_{min}=400$m, the maximum pitch $\varphi_{max}=\pm70°$and yaw angle $\mu_{max}=\pm30°$, the maximum path length $L_{max}=200$km, the minimal safe flight height $H_0=20$m, the maximum safe flight height $H_{max}=100$m. The population size is 80, the maximum number of pre-plan evolution iterations is 300, the maximum number of re-plan evolution iterations is 50.

6.1   Re-planning Paths as UAV is Outside the Pop-Up Threats Coverage

Fig.4 shows re-planning paths elevation chart by the novel intelligence algorithm and human intervention as UAV is outside the pop-up threats coverage. 6 known threats and effective distance $d_{Rmax}=12000$m, representing by 6 same shape and size hollow circle. One pop-up threat and effective distance $d_{tRmax}=6000$m, representing by a red circle. From this figure, we can see that UAV could avoid threats and terrain along pre-planned path I and II. When the pop-up threat is found and covers pre-planned path I that UAV should fly along, it is necessary to re-plan the path for UAV immediately. Now utilize human's wisdom and experience to set starting and end points of the re-planning path. Starting point is on the pre-planned path I that UAV has not been reached, end point is on the pre-planned path I or II that is outside the pop-up threats coverage. Then re-plan the new section paths by the novel intelligence algorithm. As we can see that UAV could avoid threats and terrain along the new section paths. The next step is artificial select one new suitable section path according to the mission requirement of UAV and comprehensive situation assessment. If the pop-up threats are detected until later and the computing time for re-planning path is not enough, set few navigation points manually between the starting and ending point, then the re-planning path is departed to calculate respectively for avoiding threats effectively.

Fig.5 shows re-planning paths profile, the horizontal axis d_{xoy} represents the paths shadow on horizontal plane, the vertical axis Z represents flight height. From this figure, we can see that UAV could avoid threats and terrain along re-planning paths even further. Fig.6 shows re-planning paths three-dimensional map, it can be more complete and clear to show the result for path planning of UAV penetration.

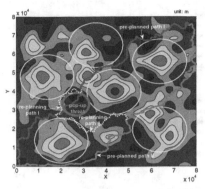

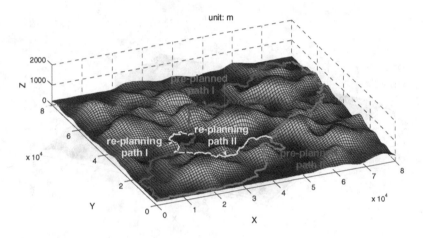

Fig.4. Re-planning paths elevation chart   **Fig.5.** Re-planning paths elevation profile

Fig. 6. Re-planning paths three-dimensional map

6.2   Re-planning Paths as UAV is Inside the Pop-Up Threats Coverage

Fig.7 shows escape paths elevation chart as UAV is inside the pop-up threats coverage. The known and pop-up threats, pre-planned paths are depicted as the same with Fig.4. When the pop-up threat is detected and UAV is flying along pre-planned path inside coverage areas of it, promptly launch emergency response program, five scheduled escape paths on the different directions are implied by the dashed lines in Fig.8, utilize human's wisdom and experience to select a most suitable path for escaping according to comprehensive situation assessment. Then put the intersection of threat coverage border and escape path as starting point, artificial set end point on the pre-planned path I or II that is outside the threats coverage for the next re-planning path, it is implied by the solid line. From this figure, we can see that along two escape paths UAV can avoid threats and terrain. Finally UAV is guided on the pre-planned paths. Fig.8 and Fig.9 are escape paths profile and three-dimensional map, respectively. They show that the re-planning escape paths can meet the requirements of UAV low-altitude penetration, efficient implementation of threat avoidance, terrain avoidance and terrain following.

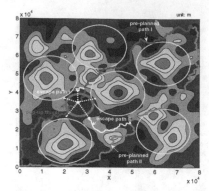 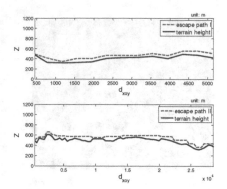

Fig.7. Escape paths elevation chart   **Fig.8.** Escape paths profile

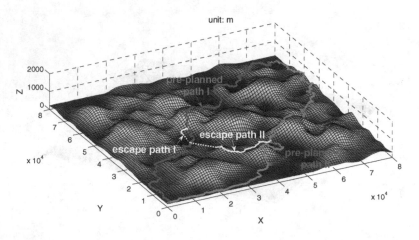

Fig. 9. Escape paths three-dimensional map

7   Conclusions

In this paper, the algorithm of human-machine coordinated intelligent flight path planning is proposed. By using variable length gene encoding based on angle for the flight paths planning, and combining artificial auxiliary decision with novel intelligence algorithm, it makes the best possible use of the human brain to guide solution procedures of the flight path planning on pop-up threats. On the basis of lots of simulation studies, we can reach the following conclusion.

1) The method representing flight path based on angles is able to satisfy the demands of UAV's flight performance and enable the UAV to track a trajectory with changes.

2) Human's wisdom, experience and synthetic judgments can make up for the lack of intelligence algorithm. In the most dangerous environment, it makes the best possible use of the human brain to guide solution procedures of re-planning flight path on pop-up threats.

3) When the threats and terrain information are known, artificial set starting and end points, then applying the novel intelligence algorithm to re-plan the flight paths that can meet the requirements of UAV low-altitude penetration.

Finally, manned and unmanned aerial vehicles cooperative combat is a revolutionary new combat method and would be the main air combat pattern in the future. Human-UAV coordinated operation would fully utilize human's wisdom, experience and synthetic judgments to guide solution procedures of the flight path planning on pop-up threats. The simulation result shows that this method is feasible and effective. It is necessary to make the research more deeply and more widely.

References

[1] Murray, C.C., Park, W.: Incorporating Human Factor Considerations in Unmanned Aerial Vehicle Routing. IEEE Systems, Man, and Cybernetics Society 43(4), 860–874 (2013)

[2] Savla, K., Temple, T., Frazzoli, E.: Human-in-the-loop vehicle routing policies for dynamic environments. In: Proc. 47th IEEE Conf. Decision Control, Cancun, Mexico, pp. 1145–1150 (2008)

[3] Donald, W.: Manned/unmanned common architecture program (MCAP): A review. In: Proceedings of the 22nd Digital Avionics Systems Conference, Indianapolis, United States, pp. 6.B.4/1–6.B.4/7 (2003)

[4] Valenti, M., Schouwenaars, T., Kuwata, Y., et al.: Implementation of a manned vehicle-UAV mission system. AIAA-2004–5142 (2004)

[5] Schouwenaars, T., Valenti, M., Feron, E., et al.: Linear programming and language processing for human/unmanned aerial vehicle team missions. Journal of Guidance, Control, and Dynamics 29(2), 303–313 (2006)

[6] Myung, H., James, K., Takeo, K.: Efficient Two-Phase 3D Motion Planning for Small Fixed Wing UAVs. In: IEEE International Conference on Robotics and Automation, Roma, Italy, pp. 1035–1041 (2007)

[7] Chingtham, T.S., Sahoo, G., Ghose, M.K.: An unmanned aerial vehicle as human-assistant robotics system. In: IEEE International Conference on Computational Intelligence and Computing Research, Coimbatore, India, pp. 1–6 (2010)

[8] Ren, P., Gao, X.G.: Operational Effectiveness analyses of AH/UAV Cooperative Surveillance against Mission. Flight Dynamics 29(3), 92–96 (2011)

[9] Fu, X.W., Gao, X.G.: 3D Flight Path Planning Based on Bayesian Optimization Algorithm. Acta Armamentarii 28(11), 1340–1345 (2007)

[10] Peng, J.L., Sun, X.X., Zhu, F., et al.: 3D Path Planning with Multi-constrains Based on Genetic algorithm. In: The 27th Chinese Control Conference, Kunming, China, pp. 94–97 (2008)

[11] Shen, Z.H., Zhao, Y.K., Wang, X.R.: Niche Pseudo-parallel Genetic Algorithms for Path Optimization of Autonomous Robot. Modern Electronics Technique 206(15), 85–87,90 (2005)

[12] Li, X., Wei, R.X., Zhou, J., et al.: A Three Dimensional Path Planning for Unmanned Air Vehicle Based on Improved Genetic Algorithm. Journal of Northwestern Polytechnical University 28(3), 343–348 (2010)

[13] Zhang, C., Wu, J., Liang, Z.A.: Adaptive Pseudo-Parallel Genetic Algorithms for Solving Double Standard Solid Transportation Problem. Mathematics in Practice and Theory 37(11), 21–28 (2007)

Switched Topology Control and Negotiation of Distributed Self-healing for Mobile Robot Formation

Jianjun Ju[1,2], Zhe Liu[1,2], Weidong Chen[1,2], and Jingchuan Wang[1,2]

[1] Department of Automation, Shanghai Jiao Tong University
[2] Key Laboratory of System Control and Information Processing,
Ministry of Education of China, Shanghai 200240, China
{JianjunJu,liuzhesjtu,wdchen,jchwang}@sjtu.edu.cn

Abstract. In this paper, we investigate robot failure problem in mobile robot formation. A recursive and distributed self-healing algorithm is proposed to restore network topology when one or more robots fail. Firstly, a switched topology control method is introduced to restore the synchronization and connectivity of mobile robot network recursively. Then, a negotiation mechanism is further presented which achieves individual control in switched topology process. This mechanism only needs local information interactions between neighbors. Finally, the effectiveness of the proposed algorithm is validated by results of both simulations and real experiments.

Keywords: mobile robot formation, switched topology control, distributed negotiation, self-healing.

1 Introduction

With the development of sensor network and multiple mobile robotics, reliable formation performance is becoming one of fundamental requirements for many typical tasks such as environment exploration and cooperative surveillance [1, 2]. However, robots may fail due to mechanical breakdown or force majeure when executing tasks in remote or hostile environment. Therefore, in order to improve the robustness of mobile robot formation, the system needs the ability to detect failures and to restore its topology when robots fail, which is also called self-healing capacity.

In recent years, there are three typical categories of self-healing algorithms. The first category is direct self-healing algorithm [3, 4], which makes decisions based on global information and chooses a single robot to repair network topology. The algorithm is not distributed and communication traffic will increase rapidly with the growth of the number of robots. The second one is density self-healing algorithm which is widely used in multi-robot coverage tasks. In [5], the network connectivity is repaired by sharing maps of interested region and specified gateway node when covering interested region. However, the algorithm doesn't consider dynamic changes of the network topology in self-healing process. The third category is recursive self-healing algorithm [6-9] which achieves the self-healing behavior by specified switched topology rules in

M. Fei et al. (Eds.): LSMS/ICSEE 2014, Part II, CCIS 462, pp. 562–574, 2014.

a local manner. In [6, 7], short route table is used to recursively restore the connectivity of wireless sensor and actor network (WSAN). However, in real applications, it's very difficult to maintain route information in real time for dynamic networks such as mobile robot formation. In [8], a distributed self-healing method is proposed based on the 1-hop neighbors information. However, this method requires all the nodes in the network to execute an additional movement, thus leading to large energy consumption.

Our basic idea is to repair the synchronization and connectivity of mobile robot network recursively by switched topology control and negotiation based individual control. Some prior works were reported in [9]. However, [9] requires a temporary expansion of communication range in order to establish additional communication connections to select repair robot, thus increasing the requirement of individual communication ability. What's more, additional communication traffic also increases energy consumption. The main contribution of this paper is to introduce a distributed self-healing algorithm by negotiation based individual control. The convergence analysis is proposed and the validity of proposed method is demonstrated by simulations and real experiments.

2   Switched Topology Control

2.1   Network Topology

Consider a network of $n(n \geq 3)$ robots, whose topology can be defined as a graph $G = (V, E)$, where $v_i \in V, i = 1, 2, ..., n$ represents mobile robots in the network and $(v_i, v_j) \in E, i, j = 1, 2, ..., n$ represents a communication connection between robot i and robot j. The coupling matrix $A = (a_{ij}) \in \Re^{n \times n}$ represents the coupling configuration of the network topology, where

$$a_{ij} = a_{ji} = \begin{cases} 1, (v_i, v_j) \in E \\ 0, (v_i, v_j) \notin E \end{cases}, i \neq j, i, j = 1, 2, ..., n \ . \tag{1}$$

Define neighbor set of robot i as follows:

$$N_e(i) \triangleq \{ j \mid a_{ij} = 1, j \neq i, j = 1, 2, ..., n \} \ , \tag{2}$$

and degree of robot i as follows:

$$d_i = \sum_{j \in N_e(i)} a_{ij} \ , \tag{3}$$

then diagonal elements of the coupling matrix A is set as $a_{ii} = -d_i, i = 1, 2, ..., n$.

In this paper, K-neighbors topology is applied to describe mobile robot network, its definition is given as follows:

$$\forall v_i \in V, \exists K \in \mathbb{Z}^+, d_i \leq K, i = 1, 2, ..., n \ . \tag{3}$$

Remark 1. Some basic assumptions of mobile robot network in this paper are given as follows:
1.   Robot can only obtain information from its neighbors.
2.   Robot failures will not cause disconnection of the whole network.

Remark 2. The dynamics of each mobile robot in this paper is the same with our previous work, more details can be found in [9].

2.2   Preparations

Before introducing the main results, some lemmas are given as follows:

Lemma 1[10]. The second largest eigenvalue λ_2 of coupling matrix A demonstrates the stability and robustness of network. Smaller λ_2 indicates better stability and robustness.

Lemma 2[11]. In K-neighbors topology, if a small proportion of robots whose degrees are sufficiently small fail, the second largest eigenvalue λ_2 of coupling matrix A will almost be fixed.

Lemma 3[9]. If a robot with smaller degree fills the blank location of failed robot, the stability and robustness of the network will be improved.

Then some evaluation indices are given as follows:
(1) Recovery Time t_r

Recovery time can be divided into two parts:

$$t_r = t_d + t_m \ . \tag{4}$$

t_d is the communication time which depends on the property of self-healing algorithm, t_m is the motion time which is related to the distance between the failed robot and the repair robot.
(2) Power Consumption P

Power consumption P can be calculated by the following formula[9]:

$$P = \sum_{i=1}^{n} \sum_{j \in N_e(i)} \beta d_{ij}^2 \ , \tag{5}$$

where n is the number of robots, d_{ij} represents the communication distance between robot i and robot j, and β is a positive parameter.

2.3   Direct Self-healing Algorithm versus Recursive Self-healing Algorithm

In this section, a brief description will be given to introduce the difference between direct and recursive self-healing algorithm. Fig. 1 shows both self-healing processes of a network with 8-neighbors topology. In direct self-healing algorithm(Fig. 1(a)), the blank location of failed robot is filled by robot with the lowest degree in the network. While in recursive algorithm(Fig. 1(b)), the blank location is filled by one of its neighbors(the repair robot), then the new blank location that repair robot left is filled by one of repair robot's neighbors(the next repair robot). This process is repeated recursively until robot with the lowest degree in the network becomes repair robot. Compared with direct algorithm, the recursive one has shorter recovery time and lower power consumption and can be implemented in distributed framework.

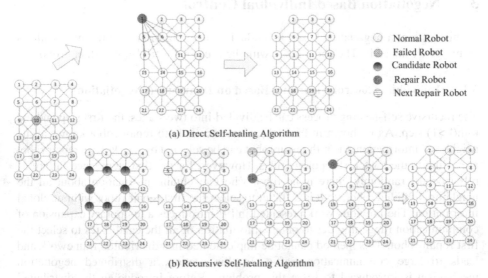

(a) Direct Self-healing Algorithm

(b) Recursive Self-healing Algorithm

Fig. 1. Direct and recursive self-healing algorithms

2.4 Switched Topology Rules

The proposed topology control for self-healing process can be described through the following rules:

Rule 1 Neighbors of failed robot become candidates of repair robot.

Rule 2 The robot with the lowest degree among candidates becomes repair robot if its degree is no larger than failed robot.

Rule 3 If there exist more than one robots with the same smallest degree, then only one robot is randomly chosen to be repair robot.

Theorem 1. Under the rules mentioned above, the topology can be recovered from robots failure, and the stability and robustness of the network will remain almost unchanged.

Proof:

Denote the k-th repair robot as $R_{repair}(k), k = 1, 2, ...$, then degree of $R_{repair}(k+1)$ is no higher than degree of $R_{repair}(k)$ according to Rule 2. Therefore, in each step of switched topology control, failed robot is replaced by a robot with the lower degree recursively. According to Lemma 3, it can be concluded that the stability and robustness of the network are improved at each self-healing step. At the end of switched topology control, the robot with the lowest degree becomes failed robot. And according to Lemma 2, the second largest eigenvalue λ_2 of the coupling matrix A remains almost unchanged. And according to Lemma 1, the stability and robustness of the network remain almost unchanged.

3   Negotiation Based Individual Control

In this section, a negotiation based individual control algorithm is presented with its convergence analysis. The comparison with the previous work [9] is further proposed.

3.1   Individual Control Algorithm Based on Distributed Negotiation

The recursive self-healing process can be divided into two stages, the first step and the k-th(k>1) step. As is shown in Fig. 1(b), since the (k-1)-th repair robot knows all its neighbors' information, the k-th repair robot can be selected by the (k-1)-th repair robot according to the switched topology rules. However, for the first step, since failure of robot occurs randomly, there is no robot who can obtain information about all the neighbors of failed robot, so it is difficult to select the first repair robot without global information. The previous work presented in [9] introduces a temporary expansion of communication range in order to obtain information of all the candidates to select the first repair robot. This method cannot be implemented in a distributed framework and leads to large communication traffic. In this section, a distributed negotiation mechanism is introduced to solve this problem. Before introducing the distributed negotiation algorithm, some notations are given as follows:

(A) Election Information Packet(P_{elect}):

```
P_elect{
   current_lowest_degree;//lowest degree that robot knows in
current step
   }
```

In the first step, all candidate robots receive, update and send this packet recursively in local interactions with their neighbors. At the beginning, *current _lowest _degree* is equal to the degree of robot itself.

(B) Maximum Time of Election(T_{elect}): The upper bound of negotiation time consumed during the first step.

(C) Repair Information Packet(P_{repair}):

```
P_repair{
   current_repair_robot;
   next_repair_robot;
   }
```

In k-th(k>1) step, current repair robot sends this packet to inform next repair robot, next repair robot receives this packet and then fill the blank location of current repair robot.

Now the first step repair robot election algorithm based on distributed negotiation mechanism is described in view of robot R as follows:

Step1 If robot R detects that one of its neighbor R_f fails, go to Step2; otherwise, return to Step1;

Step2 If robot R receives a packet P_{repair}, go to Step3; otherwise, go to Step4;

Step3 If robot R is $P_{repair}.next\_repair\_robot$, go to Step7; otherwise, ignore robot failure and return to Step1;

Step4 Robot R becomes a candidate of the first step repair robot, sends and receives P_{elect}, go to Step 5;

Step5 Define P_{elect_send} and P_{elect_recv} as packets P_{elect} that robot R sends and receives respectively. In each step, if $P_{elect\_send}.current\_lowest\_degree > P_{elect\_recv}.current\_lowest\_degree$, sets $P_{elect\_send} = P_{elect\_recv}$ and sends P_{elect_send} to its neighbors again. If $T = T_{elect}$, go to Step 6; otherwise, return to Step5;

Step6 If $P_{elect\_recv}.current\_lowest\_degree \geq P_{elect\_send}.current\_lowest\_degree$ for all P_{elect_recv}, then robot R becomes repair robot, go to Step7; otherwise, ignore robot failure and go to Step1;

Step7 Robot R decides the next repair robot $P_{repair}.next\_repair\_robot$. If $P_{repair}.next\_repair\_robot$ exists, robot R sends P_{repair} to its neighbors, and fills the blank location, return to Step1; otherwise, Robot R fills the blank location and the self-healing process is accomplished.

Remark 3. In Step5, to avoid the situation that $P_{elect\_send}.current\_lowest\_degree = P_{elect\_recv}.current\_lowest\_degree$, two random numbers are generated to decide who has a "higher" degree.

Remark 4. T_{elect} depends on the degree of failed robot d_{R_f} and the communication time T between candidates, and it satisfies that $T_{elect} \leq Td_{R_f} / 2$, the derivation will be given in the rest of this paper.

Fig. 2 shows an example of the first step repair robot election algorithm. When robot 10 fails (as shown in Fig. 1), its neighbors (robot 5, 6, 7, 9, 11, 13, 14 and 15) become candidates of the first step repair robot. Through local interaction and distributed negotiation mentioned above, robot 9 is selected as the first repair robot.

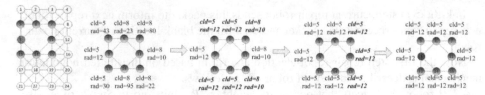

Fig. 2. Example of the first step in recursive self-healing process. *cld* represents the current_lowest_degree, *rad* represents the random number. Bold and italic format means an update of the election information packet.

Theorem 2. Considering a mobile robot network with K-neighbors topology, the proposed negotiation mechanism can elect the first step repair robot within KT/2, where T is the communication time between two candidate robots.

Proof: Note that the degree of the failed robot is no more than K. As shown in Fig. 3, when a robot fails, the worst situation occurs when its neighbors locate in a ring. Without loss of generality, robot R is assumed as the selected repair robot which has the lowest degree among all candidates. Then in the worst situation, within KT/2, all candidates can receive P_{elect} of robot R and then lose the eligibility. So within KT/2, the first step repair robot can be selected even in the worst situation.

3.2   Performance Analysis

In this section, the proposed algorithm is compared with our previous work in [9].

● Recovery Time

According to formula(4), since the motion time t_m of two algorithms are equal, we only compare the communication time t_d. For simplicity and without loss of generality, as shown in Fig.3, a 6-neighbors model is chosen as the network topology for comparison. Firstly, we suppose that in the previous algorithm (as is shown in Fig.3 (b)) two candidate robots need T_b seconds to communicate with each other while in the proposed method (as is shown in Fig.3 (c)) two candidate robots need T_c seconds.

In [9], the repair robot selection process includes two steps, the first step is to select a decision-maker with a temporary expansion of communication range and the second step is to choose the repair robot by the selected decision-maker. Therefore, the total selection time t_b is equal to $2T_b$. And for the proposed method in this paper, the total negotiation time t_c is equal to $KT_c/2$. Considering that $4 \leq K \leq 8$ in practical application and $T_b > T_c$ due to the communication distances and bandwidth, the communication time of the proposed method t_c is almost equal to that of the previous work t_b.

● Power Consumption

Denote D as the distance between two candidate robots. In our previous work, the power consumption includes two parts: P_{b1} for decision-maker selection and P_{b2} for repair robot selection. In decision-maker selection step, all candidate robots need to communicate with each other, while in repair robot selection step only the decision-maker need to communicate with other candidates. According to formula(5) and Fig.3(b), we have:

$$P_{b1} = 6[2\beta D^2 + 2\beta(\sqrt{3}D)^2 + \beta(2D)^2] , \qquad (6)$$

$$P_{b2} = 2\beta D^2 + 2\beta(\sqrt{3}D)^2 + \beta(2D)^2 . \qquad (7)$$

Therefore, the total power consumption P_b is

$$P_b = P_{b1} + P_{b2} = 84\beta D^2 . \qquad (8)$$

In the proposed negotiation based method, as is shown in Fig.3(c), each of the candidate robots communicates with two other candidates in a ring every T_c seconds, and the negotiation will finish in $KT_c / 2$, thus the total power consumption P_c is

$$P_c \le \frac{K}{2} \bullet K \bullet (2\beta D^2) = 36\beta D^2 . \qquad (9)$$

It is obvious that $P_b \gg P_c$, so the proposed method has lower power consumption than the previous work in [9].

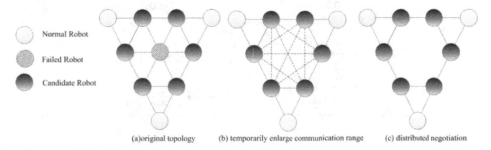

Normal Robot

Failed Robot

Candidate Robot

(a)original topology   (b) temporarily enlarge communication range   (c) distributed negotiation

Fig. 3. Topology of two self-healing algorithms

4   Simulation and Experiment

4.1   Simulation

Without loss of generality, a mobile robot formation of 30 robots with 6-neighbors topology is used in simulations. The sampling time period is 0.05s, the communication time $T_b = 0.3s$ and $T_c = 0.25s$, and the distance D = 10. Fig.4 and Fig.5 show switched topology processes of the previous work in [9] and the proposed method respectively, where red dots represent mobile robots and blue lines represent communication connections. Fig. 6 shows the total power consumption curves of both algorithms, the blue solid curve represents the power consumption of the algorithm in [9] and the red dashed curve represents the power consumption of the algorithm proposed in this paper. At $t_1 = 1.0s$, robot 14 fails; at t_2 and t_3, the previous algorithm in [9] and the proposed algorithm in this paper select the first step repair robot respectively. As shown in Fig. 6, both algorithms recover robot formation from the failure at almost the same time. However, the proposed distributed negotiation based algorithm has much lower power consumption between t_1 and t_3, which suggests that the proposed algorithm need less energy to select the first repair robot due to its distributed negotiation framework.

Remark 5. Since recursive self-healing algorithm outperforms direct self-healing algorithm in [9], comparisons in this paper are restricted to recursive algorithms.

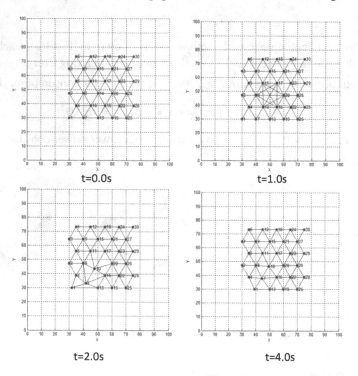

Fig. 4. Switched topology of self-healing algorithm based on communication range expansion

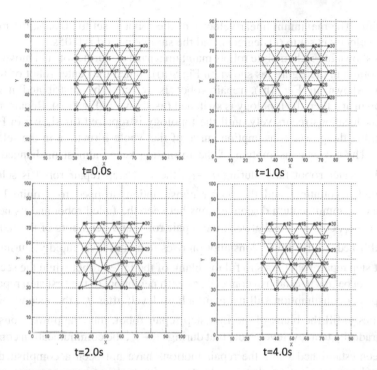

Fig. 5. Switched topology of self-healing algorithm based on distributed negotiation

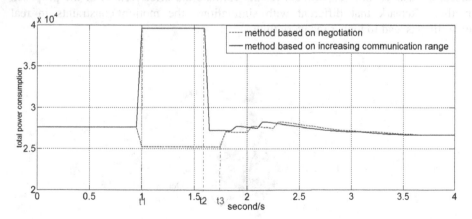

Fig. 6. Total power consumption of two algorithms

4.2   Experiment

We further conduct a real experiment with mobile robot "Frontier III" [12] shown in Fig.7. Since it is very difficult to obtain actual power consumption of robots, we obtain positions of robots through two omni-directional cameras mounted on the ceil, then the power consumption can be calculated by formula(5). ZigBee modules are used for local interaction. The maximum linear velocity of the robot is set to 30cm/s, the maximum angular velocity is set to 60 degree/s, and the sample period is 0.05s.

As is shown in Fig.8, mobile robots maintain a triangle formation and move from the top to bottom in the figure (as shown in Fig.8(a)). When the center robot of formation fails (as shown in Fig.8(b)), repair robot is selected by distributed negotiation mechanism and then the topology is switched (as shown in Fig.8(c)). Repair robots fill the blank locations recursively until the topology is repaired (as shown in Fig.8(d)). Fig.9 shows the power consumption curve of the whole formation in the self-healing process. The power consumption of the formation keeps almost unchanged until t_1 when the center robot fails; during $t_1 - t_2$, the first step repair robot is selected by distributed negotiation, and the power consumption dropped to a quite low level because of communication disconnections between the failed robot and its neighbors; during $t_2 - t_3$, connections between the repair robot and neighbors of the failed robot are established, so the power consumption rises to a high level rapidly; during $t_3 - t_4$, the first step repair robot goes to fill the blank of the failed robot while the second step repair robot begins to establish connections with neighbors of the first step repair robot, so the power consumption still maintain a high level; after t_4, the second step repair robot goes to fill the blank of the first step repair robot, thus the power consumption drops gradually to a low level. Note that during $t_2 - t_4$, the communication connections have been established while the repair motions have not been accomplished. So the additional communication distance leads to the high power consumption. As is discussed above, these experiment results validate the effectiveness of the proposed method. Remark that different with simulations, the motion constraints in real experiments lead to longer recovery time.

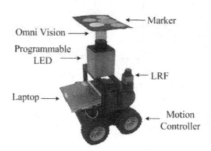

Fig. 7. "Frontier III" mobile robot

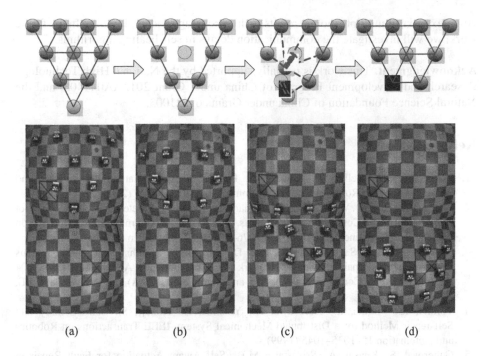

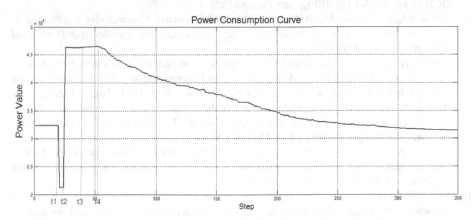

Fig. 8. Results of self-healing algorithm based on distributed negotiation

Fig. 9. Total power consumption curve of self-healing algorithm in real experiment

5   Conclusion

In this paper, a switched topology control and distributed negotiation based self-healing algorithm for mobile robot formation is proposed. Compared with existing algorithms, the proposed method has lower power consumption and can be implemented in a distributed framework. The proposed method can also be implemented in other complex tasks of multiple mobile robots such as coverage and exploration. In the future work, we will investigate the communication delays in self-healing algorithms.

Acknowledgment. This work is partially supported by the National High Technology Research and Development Program of China under Grant 2012AA041403, and the Natural Science Foundation of China under Grants 61221003.

References

1. Grocholsky, B., Keller, J., Kumar, V., Pappas, G.: Cooperative Air and Ground Surveillance. IEEE Robotics and Automation Magazine 13, 16–25 (2006)
2. Zhang, H., Chen, W., Wang, J., Li, K.: Human Interest Oriented Heterogeneous Multi-robot Exploration under Connectivity Constraints. In: 2013 IEEE International Conference on Robotics and Biomimetics (ROBIO), pp. 976–981. IEEE, Shenzhen (2013)
3. Corke, P., Hrabar, S., Peterson, R., Rus, D., Saripalli, S., Sukhatme, G.: Autonomous Deployment and Repair of a Sensor Network Using an Unmanned Aerial Vehicle. In: 2004 IEEE International Conference on Robotics and Automation (ICRA), pp. 3602–3608. IEEE, New Orleans (2004)
4. Tomita, K., Murata, S., Kurokawa, H., Yoshida, E., Kokaji, S.: Self-assembly and Self-repair Method for a Distributed Mechanical System. IEEE Transactions on Robotics and Automation 15, 1035–1045 (1999)
5. Ganeriwal, S., Kansal, A., Srivastava, M.B.: Self Aware Actuation for Fault Repair in Sensor Networks. In: 2004 IEEE International Conference on Robotics and Automation (ICRA), pp. 5244–5249. IEEE, New Orleans (2004)
6. Abbasi, A.A., Younis, M., Akkaya, K.: Movement-assisted Connectivity Restoration in Wireless Sensor and Actor Networks. IEEE Transactions on Parallel and Distributed Systems 20, 1366–1379 (2009)
7. Abbasi, A.A., Younis, M., Baroudi, U.: Restoring Connectivity in Wireless Sensor-Actor Networks with Minimal Node Movement. In: 7th International Wireless Communications and Mobile Computing Conference (IWCMC), pp. 2046–2051. IEEE, Istanbul (2011)
8. Younis, M., Lee, S., Gupta, S., Fisher, K.: A Localized Self-healing Algorithm for Networks of Moveable Sensor Nodes. In: 2008 Global Telecommunications Conference (GLOBECOM), pp. 1–5. IEEE, New Orleans (2008)
9. Zhang, F., Chen, W.: Self-healing for Mobile Robot Networks with Motion Synchronization. In: IEEE/RSJ International Conference on Intelligent Robots and Systems (IROS), pp. 3107–3112. IEEE, San Diego (2007)
10. Li, X., Chen, G.: Synchronization and Desynchronization of Complex Dynamical Networks: An Engineering Viewpoint. IEEE Transactions on Circuits and Systems I: Fundamental Theory and Applications 50, 1381–1390 (2003)
11. Wang, X.F., Chen, G.: Synchronization in Scale-free Dynamical Networks: Robustness and Fragility. IEEE Transactions on Circuits and Systems I: Fundamental Theory and Applications 49, 54–62 (2002)
12. Liu, Z., Chen, W., Wang, J.: Action Selection for Active and Cooperative Global Action Selection based on Localizability Estimation for Multi-robot Global Localization, http://www.paper.edu.cn/en_releasepaper/content/4563143

Design of Embedded Redundant Gateway
Based on Improved TCP Congestion Algorithm

Qin Lai[1], Jingqi Fu[1], and Weihua Bao[2]

[1] School of Mechatronics Engineering and Automation, Shanghai University,
Shanghai, 200072, China
[2] Shanghai Automation Instrumentation Co., Ltd.
191 Guangzhouxi Road, Shanghai, China
lairuqin@163.com, bwh@saic.sh.cn

Abstract. In wireless sensor networks, gateway is a hub in data transmission between wireless nodes and Personal Computer (PC). Once the gateway fails, the entire network will paralyze. To ensure the stability of the network, a methodology for the design of redundant gateway based on improved Transmission Control Protocol (TCP) congestion algorithm is proposed in this paper. In this methodology, two gateways physically connected by Ethernet are logically divided into a primary one and a secondary one. Secondary gateway is the backup of the primary gateway. The paper gives an improve method of congestion algorithm, which is applied in this scheme. It can better satisfy the high demands of real-time communication in the embedded system when compares with the traditional TCP congestion algorithm. Experiment shows the effectiveness of improved TCP congestion algorithm, when the primary gateway fails. Secondary gateway can rapidly replace the primary gateway and it ensures a better reliability of the network.

Keywords: TCP congestion algorithm, redundant gateway, embedded system.

1 Introduction

Industrial wireless communication technology is another hotspot following the Industrial Ethernet field bus. In fact, both the field bus and industrial Ethernet have formed a multi-standard coexist situation. The Gateway, which converts data using different protocols for communication, plays a vital role in the heterogeneous network access protocol conversion process. Industrial wireless network is a wireless network with a central. As the center of the network, once the network hub fails, the entire network will paralyze. Therefore, the gateway is the bottleneck of network reliability. To improve the reliability and stability of the network, increasing its redundant fault-tolerant ability is required. In order to improve the reliability of the entire network, this paper presents a design methodology of redundant gateway, which gives specific ideas and implementation. The two gateways are physically connected by Ethernet interfaces and logically parted into primary and secondary. As the backup of the primary gateway, the secondary gateway will replace the primary gateway when it

M. Fei et al. (Eds.): LSMS/ICSEE 2014, Part II, CCIS 462, pp. 575–584, 2014.

fails. Then an improved TCP congestion algorithm, which not only can automatically adjust to meet the network congestion, but also achieve optimal use of the buffer and make real-time increased, has been proposed for dealing with the problem of real-time communication in embedded gateway.

In the design and algorithm application of Gateway, papers [1-4] gave a general method of gateway design. Xu et al. [1] proposed a design of Zigbee/Industrial Wireless Gateway for industrial wireless communications and industrial Ethernet. However, the multi-master station and multi-slave station are not implemented, and there is no essential change in network communication efficiency. Aiming at the requirement of easy networking, low power consumption and low cost, Zhao et al. [2] put forward a design method of wireless gateway based on Zigbee protocol. The hardware and software designs of the wireless gateway had been given and a reliable operation test had been conducted to corroborate the feasibility of the method. Nonetheless, redundancy gateway has not been considered. Zhang et al. [3] designed the Wireless Networks for Industrial Automation-Process Automation (WIA-PA) redundant gateway, which was a big improvement for the reliability of the network. But algorithm optimizations for wired communication between gateways were not mentioned. Wei et al. [4] analyzed three typical TCP congestion control algorithm performances, and pointed out the need to improve the TCP congestion control. In paper[5], a new adaptive control algorithm based on fuzzy theory was proposed to solve unnecessary performance degradation , it was based on feedback theory method, finally built an adaptive control model. Zhou et al. [6] introduced a practical real time data transmission system based on embedded TCP/IP technology. Both hardware and software system were analyzed in this article. Elaboration of embedded TCP/IP technology in the operation system, the embedded TCP/IP technology, the reliability and security etc. essential technology are major presented in the combination of two characteristics, including systems control object dispersive and the network management. Du et al. [7] analyzed the embedded communications protocol stack based on TCP/IP protocol, improves the TCP congestion control problem in communication, and effectively improves the real-time nature of the TCP/IP protocol. In paper [8], embedded real-time systems were analyzed and the TCP congestion control algorithm embedded real-time communication system was under discussion. The algorithm was improved in the TCP buffer dynamically adjusting module and two-dimensional table based packet scheduling module.

Therefore, based on the summary of above papers, an improved TCP congestion algorithm that can fulfill the characteristics of the embedded system including limited resources and high real-time requirements is proposed in this paper to improve the real-time communication system based on embedded. The algorithm combines the initial priority packet, the size of the package and the wait time. Packet will be discarded to ease congestion after queue length reaches to a certain length. Finally, the effectiveness of the addressed methods is shown by a simulation.

2   Design of Redundant Gateway

For a single gateway based on the Z-Stack network, a gateway is added to backup real-time network information on gateway. Once primary gateway fails, the other gateway will replace the original gateway to maintain the network and the uninterrupted operation of the network system.

For management requires, the two gateways are defined as a primary gateway and secondary gateway to distinguish from each other. The primary gateway is responsible for the process of wireless network setting up and maintenance while the secondary gateway is responsible for backup network information and monitoring operation of the primary gateway. Once the primary gateway fails, the secondary one will replace the main gateway to manage and maintain the network operation immediately.

Primary gateway linked to the PC via an Ethernet connection. The role of the main gateway for all network configuration parameters is configured in the PC. Then, the primary gateway and secondary gateway linked to the same router through the connected system shows in Fig.1.

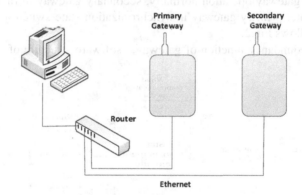

Fig. 1. The system connection diagram

To distinguish each gateway behaviors in different situations, it defines four kinds of state:

Idle state: initialization of system electrification, the idle state, expecting the next state.

Synchronization State: Monitor broadcasters from other running gateway to get network time and synchronize to the network.

Backup State: Secondary gateway particular state, in this state, the gateway is in charge of receiving the backup data of primary gateway and save it, monitoring the primary gateway operation at the same time.

Running State: responsible for the formation and maintenance of the network, updates the backup gateway backup information, in response to Modbus request command of the Distributed Control System (DCS) system at the same time.

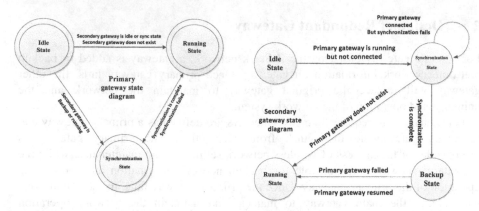

Fig. 2. Primary and secondary gateway state diagram

In normal running condition, primary gateway should be in the running state, secondary gateway should be in backup state. When primary gateway fails, secondary gateway change from backup state to running state and replace the primary gateway. While primary gateway operation normally, secondary gateway in running state turn into backup state, primary gateway in synchronization state switch to running state. State switch follows Fig.2.

To realize redundancy function of gateway, software structure of design is given in Fig.3.

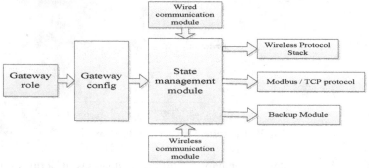

Fig. 3. Software structure of redundant gateway

Wired communication module mainly completes the communication between primary gateway and secondary gateway. The communication content including queries each other's role and state, periodic monitoring of abnormal operation. After above, it transmit the results to the state management module.

Based on the opposite side information and its own role, state management module gives a judgment whether to use wireless monitoring, operate wireless protocol stack or response to Modbus command of terminal equipment. At the same time start the backup module to judge whether to request an active backup or to dynamic send backup data.

Wireless communication module will complete synchronize time with existing gateway in synchronization status and send the results to the state management module. In backup state, the wireless communication module will monitor primary gateway in wireless work and send abnormal information to state management module. The wireless protocol starts to accomplish wireless network management and the real-time data transmission when the gateway is in running state.

3   Improved TCP Congestion Algorithm

3.1   Traditional TCP Congestion Algorithm

TCP/IP congestion control is mainly consisted of four core algorithms [9]: slow start, congestion avoidance [8, 10], fast retransmit and rapid recovery [11]. TCP congestion control algorithm is suitable in the Internet, while traditional TCP congestion algorithm is not suitable for high-speed network and embedded real-time network. The three main reasons are listed below:

(1) Traditional TCP congestion control mechanism reacts relatively poor in the high-speed network and embedded real-time network.
(2) Traditional TCP always put the packet loss as congestion, and it assumes that packet link error caused by the loss is negligible;
(3) Traditional TCP cannot use all the bandwidth of the network. This is mainly due to the long time TCP needed from a packet loss to recovery of bandwidth in the Additive Increase Multiplicative Decrease (AIMD) algorithm.

The main reason for congestion is the resources that the network provided are not sufficient to meet the needs of users. When load beyond the network can bear, the congestion will appear. Although the network layer cannot control the occurrence of congestion, this task is mostly done by TCP. TCP congestion control paid more attention to the reliability of the connection. The bandwidth will halve under the TCP congestion control after any packet loss is detected. This will dramatically reduce the network utilization.

The most effective solution to the congestion is to reduce the rate of data transmission for solving the congestion problem. It requires focusing on two potential problems: network capacity and capacity of the receiver. And they need to be separately treated. To ensure a good real-time performance of embedded system, solving the congestion control problem in the communications of embedded systems is required.

3.2   Improvement of Congestion Control Algorithm

This scheme is an improvement of RED [12](random early detection) algorithm. The core idea of the algorithm is to detect congestion by monitoring the average queue length of the router. Once congestion approached, it'll randomly select packets to discard in order to ease the congestion of the network. In RED algorithm, packets are equally selected when randomly discard all packets. It's not suitable for the measurement and control system in this paper. RED algorithm has only a captain threshold. The improved scheme set three different priorities of packets according to different priorities and set up three thresholds of queue length.

Algorithm design idea is as follows:

In embedded real-time monitoring system, different monitor nodes collect data in different types. Important data require to process in high priority, while collected data from other monitoring node can be delayed sometime to process. The data in the same priority cannot simply rely on rules of short data first or first come first process. They need to take the size of the packet and the waiting time for processing into account.

In this article, the initial priority of packets is set in three categories. The first category is the alarm message. When the message is sent back to control node, it should have the highest priority to treat. The second category is the message that changing constantly. Some message from monitor node constantly changes or fluctuates overtime. For such message, system requires to know real-time changing information. Such packets have a relatively high priority. The third category is messages not always change, or change little. This information is relatively stable, so the priority of packets can be lower.

After a packet sent back to the management platform from monitor points, it is required to wait in packet distribution processor first. The processor would assign an initial priority. It depends on where it comes (from which monitor node). And then set final priority according to the size of the packet and its waiting time.

The priority formula is:

$$P = I + r \cdot \sqrt{(L + \lambda T - 1)(L + \lambda T - 2)} / 2 \tag{1}$$

Where, P is for packet priority, I is for the initial priority packets, L is for the size of the packet, and T is for packet waiting time. As the initial priority packet I was the first consideration of the packet scheduling scheme, the higher the initial priority I was, the higher the final priority P of the packet should be.

In formula (1), λ is the impact factor of time, its value is

$$\lambda = \frac{\alpha t_{RTTSimple} + (1 - \alpha) t_{RTTused}}{t_{RTT}} \tag{2}$$

Where, $t_{RTTSimple}$ is the sampling period of round-trip time, $t_{RTTused}$ is last round-trip time, t_{RTT} is this round-trip time. All the time above is getting through timestamp. α is the decay factor which generally set values in $0.1 \sim 0.125$.

In formula (1), r is discard probability in packet discard device, its value is

$$r = \begin{cases} 0 & q_{av} \leq \min_{th} \\ \max_{av} (\dfrac{q_{av} - \min_{th}}{\max_{th} - \min_{th}}) & \min_{th} < q_{av} < \max_{th} \\ 1 & q_{av} \geq 3 \end{cases} \tag{3}$$

Where, $\min_{th}$ and $\max_{th}$ represent for average queue length of low threshold and high threshold respectively. q_{av} is the average length of queue, which calculated in the method of exponentially weighted moving average :

$$q_{av}^n = (1-\omega_q)q_{av}^{n-1} + \omega_q \cdot q \qquad (4)$$

q is the queue length, q_{av}^n and q_{av}^{n-1} are current and previous average queue length, ω_q is weight constant, it determines influence degree current average queue length affect in settlement of average queue length.

Under the condition of the same initial priority, packet size L and waiting time T needs to be considered comprehensively. If in the same sum, the smaller packet has higher priority. For packet waiting time T, system uses a counter as a time stamp. the Counter plus 1 when each packet arrivals. At the same time, the current value of the counter as the timestamp is assigned to the packet.

Packet joins the buffer queue after setting the ultimate priority. When the queue length reaches first threshold of queue length, low priority packets are to be discarded. If the speed of the discarded packets is lower than the speed of queued packet, the queue will lengthen. When the queue length reaches the second threshold of queue length, medium priority packets are to be discarded. If the queue continues to lengthen till the queue reaches the third threshold of queue length, high priority packets are to be discarded.

The description of the algorithm:

1) Returned packet from collection point set initial priority by packet distributed processor. Then calculate the ultimate priority according to the size of the initial priority packets and its waiting time.

2) The packet whose priority has already calculated is added to the buffer queue by distributing processor and waiting to be processed.

3) When the queue length reaches 200000, low priority packets will be discarded.

4) If the queue length continues to increase to 250000, medium priority packets will be discarded.

5) If the queue length continues to increase to 300000, discard high priority packets will be discarded.

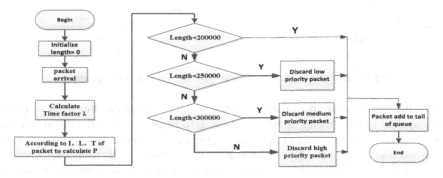

Fig. 4. Flow chart of improved congestion control algorithm

Algorithm process as showed in Fig.4, the length is the queue length. I was the initial priority of the packet. L is the size of packet, T for packet waiting time.

3.3   Experimental Testing and Performance Analysis

Simulation experiment was performed based on NS2. The simulation structure model as showed in Fig.5, bottleneck between nodes 1 and 2 is 1 m, delay is 1-5ms.

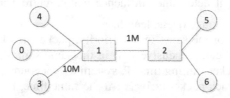

Fig. 5. Experimental network topology

Fig.6 (a) shows the algorithm in this paper. Three kinds of different priority packet discard rate curve with three different thresholds of queue length. From Fig.6 (a), the packet loss rate decreased when the threshold of queue length increased. When queue length of low priority is set to 200000, packet loss rate of a low priority is 9%. It also can be perceived the rate of medium priority and high priority when the queue length is set to 250000 and 300000 respectively. Packet loss rate of the corresponding priority packet satisfied the requirement of important data less discarding in the retransmission.

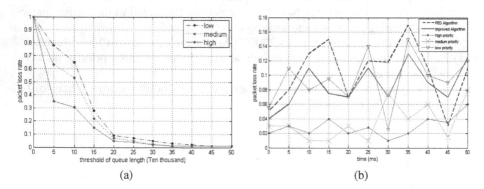

(a)   (b)

Fig. 6. Packet loss rate. (a) Different priority. (b)Different algorithm.

According to the analysis of the simulation results, the suitable threshold of queue length applied in this article can be determined. In Fig.6 (b), from the comparison of packet loss rate in the process of operation between this algorithm and classical RED congestion algorithm, it is easy to see that this algorithm has lower packet loss rate because of the classified priority in packets and the corresponding threshold instead of single priority First In First Out(FIFO) in RED algorithm. This enhanced algorithm can alleviate network congestion situation, so as to improve the real-time performance of the whole network. In addition, Fig.6 (b) shows each packet loss rate of the three kinds of different priority packets. It's clear to see that, compared with packet of low priority. Packet of high priority has a lower rate and less quantity in packet loss. It can ensure important data arrive in time, reduce errors of important data caused by network congestion due to time consuming in retransmission. However, the selection of threshold may have certain influence on the performance of the algorithm. The three queue threshold in this article may not be the best, so the performance of the improved algorithm may be affected.

Users observe data uploaded by node via PC software. Data receiving, storage, analysis and display of Modbus/TCP communication are operated by PC as showed in Fig.7 (b). Users could view information, online time, packet loss rate of nodes and manage the nodes. Compared with gateway without redundant function, gateway designed in this article made monitor system more stable. Test system consists of two gateways and ten temperature and humidity nodes, within the range of 50m, continuous 1200h. Fig.7 (a) shows the daily loss rate is less than 5%. Transient packet loss rate is 1% in Fig.7 (b). This system is easy to form networks. Stability for data transmission and the work time of the node can be maintained for a longer time. It meets the demands of the industrial control field.

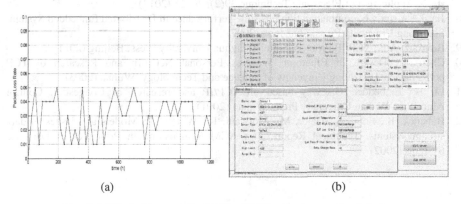

(a)   (b)

Fig. 7. (a) Packet loss rate in 1200h. (b)PC software monitor interface.

4   Conclusion

This article introduces a wireless network system. It has successfully verified the work of design gateway and realized the stability of network communication through redundant gateway. The improved TCP congestion control algorithm performs better than traditional TCP congestion control algorithm in high-speed embedded real-time communication network. It has certain reference value in embedded system application which required large transmission and high transmission speed. Through PC monitoring users can observe and process data like temperature data collected by each node and control wireless network through the wired network. The gateway has the characteristics of low cost, low power consumption, high reliability and high anti-interference ability. It can be easily applied in modern industrial and agricultural monitoring.

Acknowledgments. The work was supported by the National High Technology Research and Development Program of China under Grants 2011AA040103-7.

References

1. Xu, H., Liu, K.: Design of Gateway Between Zigbee and Industrial Ethernet on Modbus Protocol. Microcomputer Information 25(6-2), 281–283 (2009)
2. Zhao, C., Wang, Y., Wang, K.: Design of the Industrial Wireless Gateway based on ZigBee Protocol. Process Automation Instrumentation 34(2), 89–91 (2013)
3. Zhang, Q., Xiao, J.: Chapter 29: Design and Implementation of Redundant Gateway for WIA. Instrument Standardization & Metrology 5, 24–28 (2011)
4. Wei, J., Guo, X.: Research on TCP Congestion Control Technical. Modern Electronics Technique 15, 67–70 (2009)
5. Wu, X., Zhang, Z.: Adaptive TCP congestion algorithm based on fuzzy loss discrimination in heterogeneous networks. Journal of Computer Applications 33(7), 1809–1812 (2013)
6. Zhou, B., Wu, W., Tu, B.: The TCP/IP Protocols of embedded system. Control & Automation 23(3-2), 52–54 (2007)
7. Du, W., Wang, B.: Research and design of embedded real-time communication protocol stack. Application of Electronic Technique 39(2), 26–28 (2013)
8. Jia, Z.: A study of communication real time performance in Distributed Embedded System. Shandong University Press, Shandong (2007)
9. Behrouz, A.F.: TCP/IP Protocol Suite. Tsinghua University Press, Beijing (2011)
10. Chen, T.: Design of Bluetooth Wireless Personal Network and Research of TCP performance. Huazhong University of Science and Technology, Wuhan (2003)
11. Li, B.: Embedded Gateway based on TCP/IP and research of Congestion Control Algorithm. Wuhan University Press, Wuhan (2001)
12. Babcock, B., Babu, S., Dater, M., Motwani, R., Widow, J.: Models and issues in data stream systems. Technical Report, 21st ACM Symposium on Principles of Database Systems (2002)

Author Index